D0880903

FROM HYDROCARBONS TO PETROCHEMICALS

FROM HYDROCARBONS TO PETROCHEMICALS

Lewis F. Hatch
Sami Matar

Gulf Publishing Company
Book Division
Houston, London, Paris, Tokyo

FROM HYDROCARBONS TO PETROCHEMICALS

Library of Congress Cataloging in Publication Data

Hatch, Lewis Frederic, 1912-
From hydrocarbons to petrochemicals.

Includes bibliographical references and indexes.
1. Petroleum chemicals. I. Matar, Sami, 1922- joint author. II. Title.
TP692.3.H36 661'.804 80-24679
ISBN 0-87201-374-X

Contents

To Dr. Abdul Aziz Al-Gwaiz,
a friend, scholar, administrator,
and an excellent chemist.

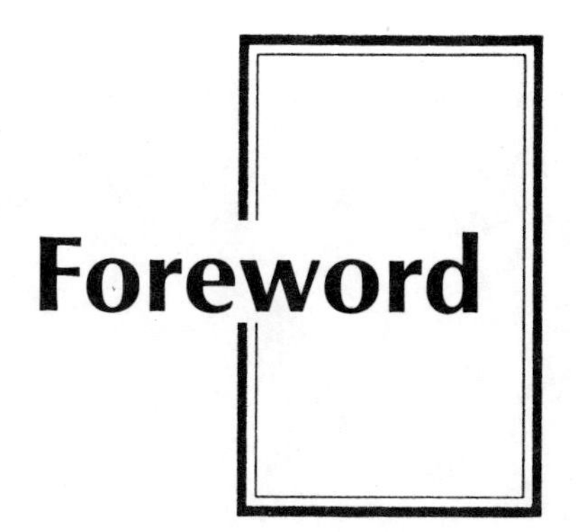

Foreword

The natural sciences are the root of all technology—they are responsible for its creation, understanding, and application. Basic to the technology of petrochemicals production is the natural science of chemistry. Chemistry in itself is not complex, but *using* chemistry to perform specific tasks is often impossible without a good knowledge of that science. Professor Hatch has contributed much to the science of chemistry with teachings, writings, and his intense desire that students learn as much as possible about this fundamental natural science.

In the early 1950s, when many students were coming to the University of Texas at Austin from the oil-producing states, Professor Hatch began a series of lectures under the theme of "understanding the chemistry of petrochemical reactions." These appeared in 1953 and 1954 as a series of articles in *Hydrocarbon Processing* (known then as *Petroleum Refiner*), and as a book in 1955 entitled *The Chemistry of Petrochemical Reactions*.

When Professor Hatch became associated with the University of Petroleum and Minerals at Dhahran, Saudi Arabia, I had the good fortune to meet with him in Baghdad, Iraq in 1975. Much had changed in the petroleum sector and our understanding of chemistry had greatly advanced; for these reasons I prevailed upon him to write the series of articles which comprise this book. As Professor Hatch is a teacher, much of the material in this book is derived from his lectures at UPM. Thus, this book performs two things: It informs by discussing commercial processes and, at the same time, teaches the underlying chemistry so essential to the realization of commercial petrochemicals.

This book is much more than a "guide" or "handbook," but it is not intended to be a complete treatise on organic chemistry. The book gives one an awareness of the importance of petrochemicals and the role they will play in the developing world, yet it will serve as only an introduction to the constantly changing world of chemicals from petroleum.

Professor Hatch has been my friend and teacher since the early 1950s. As a student of the petrochemical sector, I have learned much from his books and articles. It is my privilege to share this book with you. Please put it to immediate use. Life itself is a learning process, and as Justice Oliver Wendell Holmes said in 1884: "I think that as life is action and passion, it is required of a man that he should share the passion and action of his time, at the peril of being judged not to have lived."

Thomas C. Ponder
Petrochemicals Editor,
Hydrocarbon Processing
Houston, Texas
January, 1981

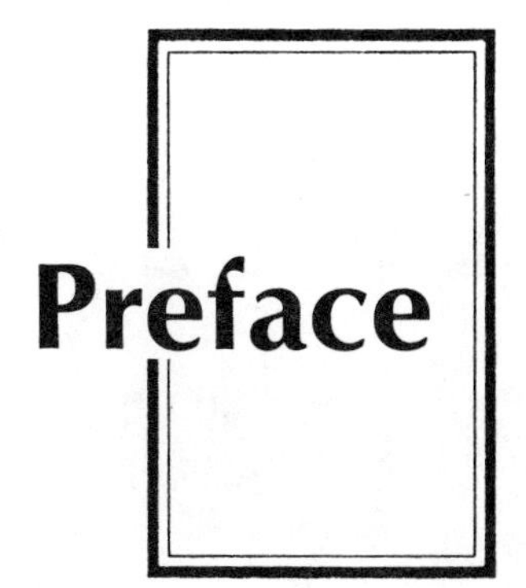

Preface

The petrochemical industry had its inception some 60 years ago with the production of isopropyl alcohol (isopropanol) from propylene. Over 30 years ago the first course in chemicals from petroleum was offered at the University of Texas by one of the authors of this book. Shortly thereafter, the term "petrochemical" was coined to describe chemicals obtained directly or indirectly from natural gas and petroleum hydrocarbons and utilized in chemical markets. The hard-rock chemists objected to the term because "petro" is derived from the Greek *petros*, stone. The word "petrochemical" has now become legitimate; it has even become a household word—and well it should—as carpets, linoleum, draperies, kitchen utensils, pills, and other everyday goods come from petrochemicals.

From these small beginnings, there has developed a $25-billion industry with billion-pound-per-year petrochemical plants scattered throughout the world. The production of petrochemicals is one of the dominant industries of developed countries and a driving force and future goal for developing countries, especially for oil- and gas-producing countries of the Middle East. This book is directed toward the chemistry of petrochemical reactions rather than the political and economic aspects of the world-wide petrochemical industry.

The first several chapters relate to the raw materials—natural and associated gas and crude oil—and to their composition and processing. Special emphasis is given to the chemical reactions taking place in the various refinery processes. These reactions produce the reactive olefins and aromatics needed for further processing to petrochemicals. The non-hydrocarbon petrochemicals obtained directly or indirectly from refinery processes are discussed. The more or less direct conversion of paraffin hydrocarbons to petrochemicals is the topic of two chapters—one on the utilization of methane and the other on the utilization of the higher paraffin hydrocarbons.

The chemical utilization of olefins and diolefins is at the heart of the petrochemical industry. There is always a problem as to how far one should go downstream from an olefin to a derivative of a derivative which was a derivative of the olefin. How many consecutive reactions and how many uses should be noted? In general, no more than second or third generation petrochemicals are discussed, and special emphasis is placed on the chemistry involved in their formation and their various uses.

With the advent of catalytic reforming, aromatic hydrocarbons have become available in large quantities from a stable source, petroleum. This has led to a renewed interest in both benzene and the methylbenzenes as petrochemical raw materials. Benzene and the simple methylbenzenes—toluene and the xylenes—are where the action is.

Only six percent of a barrel of oil is utilized for the production of petrochemicals. The material in this book shows how natural and associated gas as well as crude oil are transformed into consumer goods. Petrochemicals are growing rapidly; this book is designed to show where the substantial growth is and, to some extent, where it likely will be in the future.

All data in the book apply to the United States unless otherwise noted. Various units are used interchangeably, as they are in the industry. In general, temperatures are in degrees Celsius (°C). Pressures are given in pounds per square inch (psi), or pounds per square inch gauge (psig), although atmospheres and kilograms per square centimeters are sometimes used. Tons refer to 2000 pounds and metric tons to 2200 pounds. Percent data are given in mole percent unless otherwise noted. It is hoped that this variety of units will not be too disruptive.

The industry is full of abbreviated names such as NGL for natural gas liquids. These will be identified the first time they are used, and then should become part of the vocabulary of all persons working in the field.

LEWIS F. HATCH
SAMI MATAR

1

Petroleum Gases

A petrochemical is any chemical compound obtained from petroleum or natural gas or derived from petroleum or natural gas hydrocarbons and utilized in chemical markets. The petrochemical industry is a $2-billion enterprise employing over 30,000 workers. This dominant position in the chemical industry is based on an abundant supply of low cost raw materials of high purity. Most petrochemical reactions result in products with a higher molecular weight than the starting organic compound—for example, the addition of relatively inexpensive HOCl, H_2O, and O_2 to olefins. Compounds react by number, *moles*, but are purchased by weight. Also important is the large number of similar chemical reactions which result in wide adaptability and interchange of equipment and technology.

The three main routes to petrochemicals start with either synthesis gas (Syn Gas), olefins, or **BTX** (benzene, toluene and the xylenes). These primary feedstocks provide raw material for a great many petrochemical intermediates and consumer products.

Synthesis gas is a mixture of carbon monoxide and hydrogen obtained by either steam reforming of natural gas or by partial oxidation of natural gas, naphtha or heavy fuel oil. In fact, it can be obtained by similar treatment of almost any organic material. Synthesis gas is, directly or indirectly, the precursor to a host of petrochemicals from fertilizers to methanol to single cell proteins. (**SCP**). *Olefins* are obtained by steam cracking (pyrolysis) various feedstocks including ethane, naphtha, gas oil and even crude oil. These olefins are the reactive link between the relatively unreactive paraffin hydrocarbons of petroleum and consumer petrochemicals.

Benzene, toluene and the xylenes (**BTX**) are the aromatic hydrocarbons of most value as precursors for aromatic petrochemicals. They are also precursors for several non-aromatic petrochemicals such as caprolactam for nylon 6 and adipic acid for nylon 6/6. Aromatics are present in petroleum but the primary source is catalytic reforming streams. Aromatic hydrocarbons are rarely synthesized separately except from other aromatics; for example, the dealkylation of toluene to benzene.

The evaluation of natural gas, associated gas and crude oil as a base upon which to build a petrochemical complex requires a knowledge of their composition. As with any endeavor, it is necessary to know the territory and it is best to start with the smallest territory—natural gas and associated gas.

NATURAL GAS

Natural gas, as the term is used by the petroleum industry, is a mixture of gaseous hydrocarbons with methane as the major constituent. It usually contains small amounts of non-hydrocarbon gases such as nitrogen, carbon dioxide and hydrogen sulfide. Natural gas is found in porous reservoirs either associated with crude oil, *associated gas,* or in gas reservoirs with no oil present, *non-associated gas.* Natural gas and especially associated gas are of great importance not only as a source of energy but also as a basic raw material for the petrochemical industry.

Dry natural gas is a gas that does not contain readily condensable hydrocarbons. If the proportion of hydrocarbon liquids is more than 0.3 gal./Mcf of gas, the gas is considered *wet natural gas* and it is economically feasible to remove the condensable hydrocarbons. A wet gas has also been designated as one containing 2 or more gallons of condensables per 1,000 cu ft. A typical dry natural gas, after non-hydrocarbon gases have been removed, has a methane content in the range of 85 to 95 percent. A wet gas will have a lower methane content. *Sour natural gas* contains an appreciable amount of hydrogen sulfide and carbon dioxide. Table 1-1 shows the composition of a wide variety of natural gases.

TABLE 1-1—Composition of Various Natural Gases*

Component	Salt Lake U.S.	Webb Texas U.S.	Kliffside U.S.	Sussex England	Lacq France
Methane..........	95.0	89.4	65.8	93.2	70.0
Ethane...........	0.8	6.0	3.8	2.9	3.0
Propane..........	0.2	2.2	1.7	....	1.4
Butanes..........	....	1.0	0.8	....	0.6
Pentanes & heavier	....	0.7**	0.5	....	....
Hydrogen sulfide..	....	....	....	1.0***	15.0
Carbon dioxide....	3.6	0.6	....	2.9	10.0
Nitrogen..........	0.4	0.1	25.6	....	....
Helium...........	....	....	1.8	....	....

* Vol. percent.
** Isopentane 0.22%, *n*-pentane 0.14%, hexanes 0.22%, heptanes 0.13%.
*** Carbon monoxide.

TABLE 1-2—Typical Analysis of Feed Gas and Product Gas to Pipeline[6]

Component Mole %	Feed	Pipeline Gas
N_2	0.45	0.62
CO_2	27.85	3.50
H_2S	0.0013	—
C_1	70.35	94.85
C_2	0.83	0.99
C_3	0.22	0.003
C_4	0.13	0.004
C_5	0.06	0.004
C_{6+}	0.11	0.014

The methods used to treat natural gas depend upon the types of compounds present. A dry gas, low in hydrogen sulfide, needs little or no treatment except to adjust the water content (humidity). If the gas is dry but sour, it is necessary to remove the hydrogen sulfide and the carbon dioxide. Treatment of wet sour gas is usually carried out in the following sequence.

Acid Gas Removal. The acid gases in natural gas are hydrogen sulfide and carbon dioxide. Their removal is effected by the treatment of the natural gas with an easily regenerated base. Mono and diethanolamines (MEA and DEA) are frequently used for this purpose. The sour gas passes through the amine where sulfides, carbonates and bicarbonates are formed. The ethanolamines are regenerated by steam treatment. Diethanolamine is currently in favor because of its lower corrosion rate, smaller amine loss potential, less utility requirements and minimal reclaiming needs.[1] However, **DEA** reacts reversibly with 75 percent of the carbonyl sulfide, COS, while **MEA** reacts irreversibly with 95 percent of the COS to form a degradation product which must be disposed of.

Sulfolane (tetramethylene sulfone) and potassium carbonate are used in other processes. The **Sulfinol Process** uses a mixture of sulfolane and alkanolamines.[2] The **Merox Process** utilizes a caustic solvent-containing catalyst to convert mercaptans to caustic-insoluble disulfides, RSSR. The **Perco Process** converts mercaptans to disulfides by use of cupric chloride impregnated with Fullers earth. A maximum mercaptan content of 0.05 percent can be treated by this process.

Diglycolamine, DGA, $HOCH_2CH_2OCH_2CH_2NH_2$ is used in the Econamine Process and is reported to improve acid gas removal.[3] It is claimed to have more flexibility than other ethanolamines. The low freezing point of its aqueous solutions makes it suitable for cold climates. Its treating capability at elevated temperatures is advantageous in warm climates.[4]

Bulk quantities of carbon dioxide can be removed from natural gas containing as much as 44 percent carbon dioxide and 66 ppm hydrogen sulfide by the **Selexol** (dimethyl ether of polyethylene glycol) **Process.**[5] This process is based on the physical absorption of the carbon dioxide rather than on a chemical reaction. It works best when the acid gas concentration and operating pressures are high and the temperatures are low. No chemical bond is formed with the acid gases. Table 1-2 shows the analysis of feed and produce gas to pipeline;[6] Fig. 1-1 shows a Selexol plant for removal of carbon dioxide. This plant is designed to process 100 MM cfd of inlet gas at a pressure of 900 to 1,100 psig and a temperature of 65-120°F. Removal of CO_2 and H_2S takes place in two countercurrent absorption columns. Feed gas at 40°F and 1000 psig is contacted with Selexol. Design solvent loading is a function of partial pressure of acid gas, contact temperature, number of absorption stages and leanness of Selexol entering the absorber.

Sepasolv MPE is a special mixture of oligoethylene glycol methyl isopropyl ethers with a mean molecular weight of 316. As a physical solvent, its solubility characteristics differ fundamentally from those of chemical solvents. Particularly when hydrogen sulfide partial pressure is greater than one bar, physical solvents are reported to offer substantial advantages over chemical methods.[21] The amount of dissolved gas increases virtually linearly with pressure.

Physical solvents are utilized increasingly to remove acid components from natural gas and coal gasification products, and for the purification of synthesis gas.

Mercaptans and low concentrations of hydrogen sulfide and carbon dioxide can be removed by a *caustic wash* process. The caustic concentrations are between 5 and 15 wt. percent. Chemical costs make the process uneconomical for high acid-gas concentrations. Fig. 1-2 gives a generalized representation of a gas treating plant used in refinery operations.[2]

Molecular sieves (zeolites) are used to absorb water, hydrogen sulfide, mercaptans, RSH and carbonyl sulfide. They are effective in extracting carbon dioxide only from vapor streams. In general, a two bed system is used to provide continuous operation, one in use while the other is being regenerated. The zeolite is regenerated by passing sweet, dry fuel gas heated to 400-600°F over it and then cooling the zeolite with the treated product. Zeolites are competitive when the quantity of hydrogen sulfide and carbon disulfide is low.

Humidity Adjustment. It is necessary to adjust the humidity of natural gas to prevent hydrate formation. **Hydrates** are solid, white compounds formed from a physical-chemical reaction between the hydrocarbon gases and water under the pressures and temperatures normally used in gas pipe lines. Water is removed by treatment of the gas with various alcohols or glycols such as *ethylene glycol,* $HOCH_2CH_2OH$, *diethylene glycol,* **DEG**, $HOCH_2CH_2OCH_2CH_2OH$ and *triethylene glycol,* **TEG**, $HOCH_2CH_2OCH_2CH_2OCH_2CH_2OH$, in which water is soluble and hydrocarbons are not. Glycol is used to dehydrate vapor phase ethane by countercurrent contact. Ethylene glycol is used for injection processes because of its lower cost and good physical properties. Triethylene glycol is used in vapor phase processes because of its low vapor pressure which results in less glycol loss. Humidity can also be adjusted by use of solid absorbents such as silica gel or molecular sieves. A mixture of ethylene glycol and ethanolamine is used for removing acid gases and dehydrating natural gas.

Glycol solutions for dehydration are used because of reliability of operation, simplicity of equipment, and low cost for both chemicals and utilities. Maintenance problems of the glycol solution have been reported in depth by Grosso and Pearce.[20]

Condensable Hydrocarbon Removal. Condensable hydrocarbons are removed from natural gas by cooling the gas to a low temperature by heat exchange with liquid propane. The cold gas is then washed with cold hydrocarbon liquid which dissolves the condensable hydrocarbons. The uncondensed

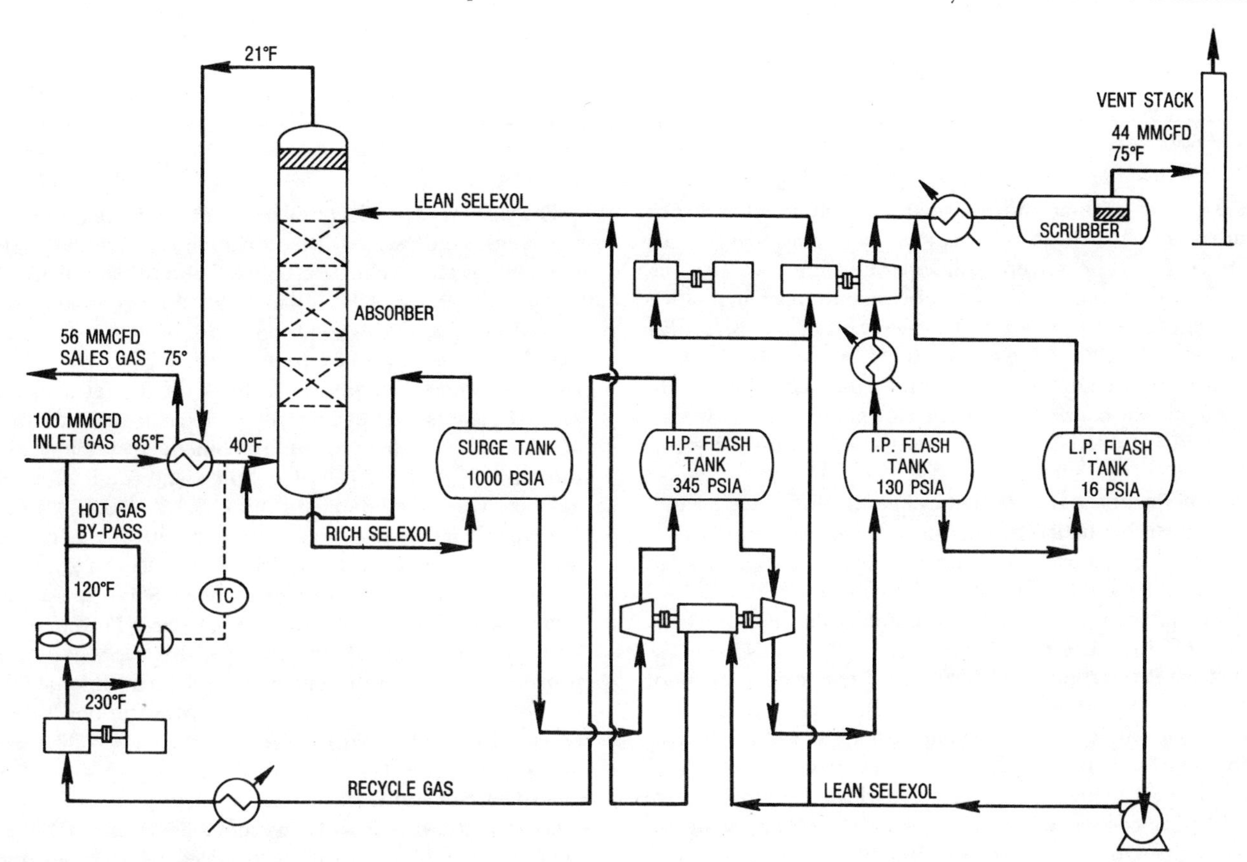

Fig. 1-1—Pikes Peak carbon dioxide removal plant.[5]

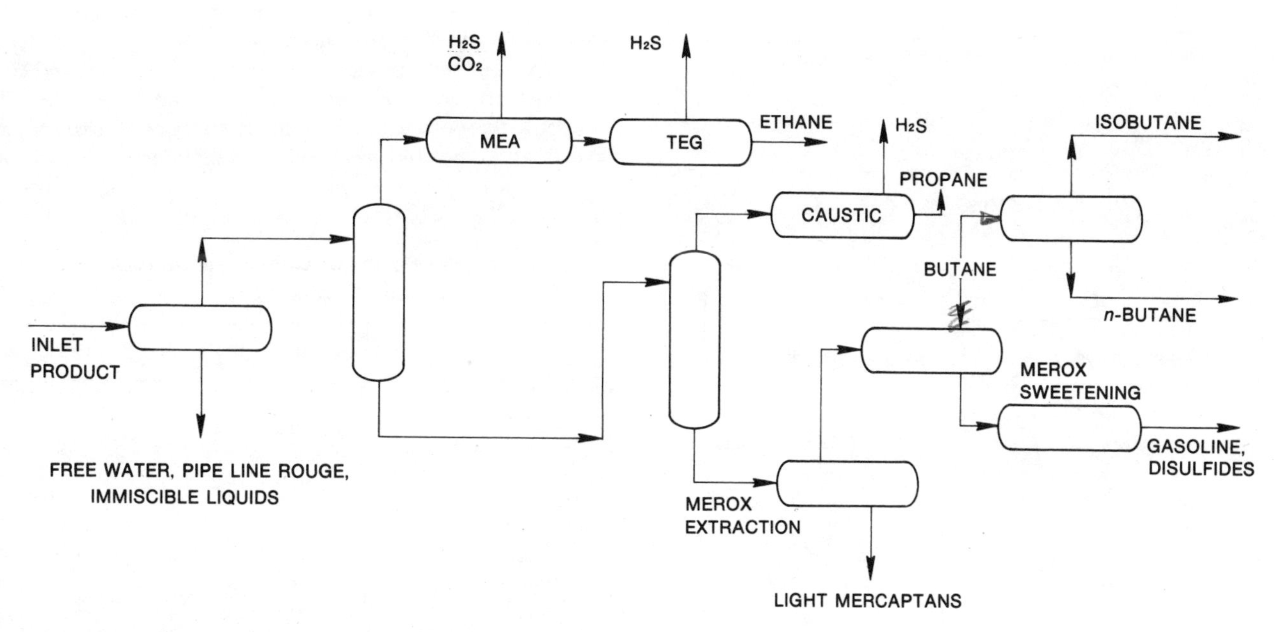

Fig. 1-2—A generalized schematic of a typical gas treating plant.[2]

TABLE 1-3—Physical Properties of Natural Gas Components

Hydrocarbon Name	Hydrocarbon Formula	Specific Gravity	Boiling Point, °C	Calorific Value Btu/cf	Carbon/Hydrogen (by weight)	Critical Temperature, °C	Critical Pressure, Atm.
Methane	(CH_4)	0.554*	−161.5	1100	3.0	−82.3	46.5
Ethane	(C_2H_6)	1.049*	−88.6	1800	4.0	32.2	48.3
Propane	(C_3H_8)	1.562*	−42.1	2300	4.5	96.8	42.1
Isobutane	(C_4H_{10})	0.557	−11.1	2940	4.8	13.5	36.5
n-Butane	(C_4H_{10})	0.579	− 0.5			152.3	37.0
Isopentane	(C_5H_{12})	0.6211	27.9		5.0	187.8	32.9
n-Pentane	(C_5H_{12})	0.626	36.1			197	33.1
n-Hexane	(C_6H_{14})	0.659	68.7		5.1	234.5	29.9
n-Heptane	(C_7H_{16})	0.684	98.4		5.25	267	26.9

* Air = 1.000.

gas is mainly methane with a small amount of ethane. The condensable hydrocarbons, *natural gas liquids* (**NGL**), are stripped from the solvent and separated into two streams. One of the streams, which contains mainly propane with a lesser amount of butanes, is liquefied and becomes *liquefied petroleum gas* (**LPG, LP**-gas). About 80 percent of the **LPG** produced in the United States comes from natural gas.[7] The second stream consists mainly of C_5 and heavier hydrocarbons and is added to reformer gasoline to raise its vapor pressure.

Physical Properties. After natural gas is treated for humidity, condensable hydrocarbons and, if necessary, for hydrogen sulfide and carbon dioxide, its physical properties are very similar to methane, its main constituent. Table 1-3 contains physical properties of the individual hydrocarbons commonly found in natural gas.

Chemical Properties and Utilization. Because 95 percent of natural gas is used as a fuel (Fig. 1-3), its calorific value is very important. The value depends upon the ratio of hydrocarbons present in the gas and ranges between 1,000 and 1,100 Btu/scf. The term "1 million British Thermal Units" (MMBtu) is customarily used to establish the sale value of natural gas. One thousand cubic feet of gas (Mscf) is nearly equivalent to 1 million British Thermal Units (MMBtu).

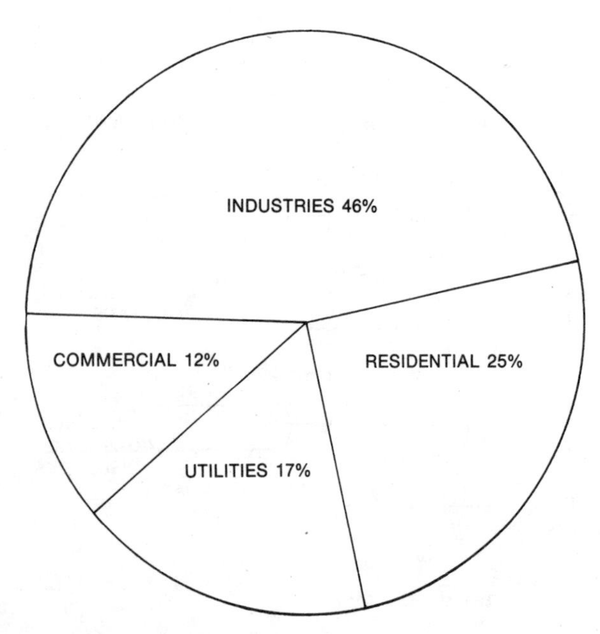

Fig. 1-3—The utilization of natural gas as a fuel.

The primary chemical utilization of methane (natural gas) is through synthesis gas. Direct chemical utilization of methane is limited by its chemical stability and the simplicity of the methane molecule, CH_4. It includes the production of hydrogen cyanide (HCN), carbon disulfide (CS_2), chlorinated methanes, and single cell proteins (**SCP**).

Reserves and Resources. The United States estimated natural gas reserves as of January 1, 1979, were 200,302 × 10^9 cu ft.[8] Gas production remained essentially constant during 1977 and 1979 at 19.4 - 19.3 trillion cu ft.[9] The world gas reserves are centered in the U.S.S.R. (36.4%) and Iran (20.0%).[8] The United States is third with 8.0 percent, followed by Algeria (4.2%) and Saudi Arabia (3.2%).

Natural gas resources of the United States are based on various assumptions and educated guesses. The current best figure is 1.019 quadrillion cu ft.[10] If this gas is converted to proven reserves, it would give the U.S. an additional 50 year supply of natural gas at the present consumption rate. The potential United States gas supplies through 2000 are summarized in Table 1-4.

Liquefied Natural Gas

Natural gas that is to be liquefied receives further treatment so that water vapor is reduced to 10 ppm, carbon dioxide to 100 ppm and hydrogen sulfide to 50 ppm. Solid absorbents are used for this purpose.[11]

The two methods used to liquefy natural gas are the expander cycle and the mechanical refrigeration cycle. In the expander cycle process, part of the gas is expanded from the high transmission pressure to a lower pressure. This causes a temperature lowering of the gas. Through heat exchangers, the cold gas cools the incoming gas which in a

TABLE 1-4—Potential United States Gas Supplies, 1980-2000

Source	1978 Actual	1980	1985	1990	1995	2000
	Trillion cu ft					
Conventional (Lower-48)	19.1	18-19	16-18	15-17	13.5-15.5	12-14
Potential supplements						
Canadian	0.9	1.4	1.4	1.1	1.0	0.8
SNG	0.3	0.5	0.5	0.5	0.5	0.5
LNG imports	0.1	0.4	1.6	2.0	2.5	3.0
Mexican gas	...	0.2	0.5	1.0	1.0	1.0
Alaskan gas						
Southern	...	...	0.1	0.2	0.3	0.6
North Slope	...	...	0.7	1.4	2.2	3.0
Coal gasification	...	...	0.1	0.6	1.8	3.3
New technologies	...	...	0.9	1.8	3.2	5.0
Subtotal supplements	**1.3**	**2.5**	**5.8**	**8.6**	**12.5**	**17.2**
TOTAL	**20.4**	**20.5-21.5**	**21.8-23.8**	**23.6-25.6**	**26.0-28.0**	**29.2-31.2**

Source: American Gas Association

TABLE 1-5—Properties of a Representative Mixture of Liquefied Natural Gas (LNG)

Density, lb./cf.	27.00
Boiling point, °C.	−158
Calorific value, Btu/lb.	21200
Specific volume, cf/lb.	0.037
Critical temperature, °C*	−82.3
Critical pressure, psi*	−673

* Critical temperature and pressure for pure liquid methane.

TABLE 1-6—Composition of Associated Gases from Different Parts of the Middle East and the North Sea*

Component	Lybia	Saudi Arabia Abqaiq	Iran	North Sea	"Typical"[13]
Methane	66.8	62.24	74.9	85.9	51.06
Ethane	19.4	15.07	13.0	8.1	18.52
Propane	9.1	6.64	7.2	2.7	11.53
Butanes	3.5	2.40	3.1	0.9	4.37
Pentanes & heavier	1.2	1.12	1.5	0.3	2.14
Hydrogen sulfide	....	2.80	....	....	2.20
Carbon dioxide	....	9.20	0.3	1.6	9.68
Nitrogen	....	....	....	0.5	0.50

* Vol. percent.

similar manner cools more incoming gas until the liquefaction temperature of methane is reached and liquid natural gas is produced.

The mechanical refrigeration cycle is the more widely used process for liquefying natural gas. In this process three separate liquid refrigerants, propane, ethane and methane, are used in a cascade cycle. When these liquids evaporate, the heat required is obtained from the natural gas being liquefied. The refrigerant gases are recompressed, cooled and recycled as liquid refrigerants.

Table 1-5 lists important properties of liquefied natural gas. Fig. 1-4 is a schematic diagram of a plan for the liquefaction and transportation of Alaskan natural gas.[12]

ASSOCIATED GAS

Most oil producing reservoirs have varying amounts of paraffin hydrocarbon gases dissolved in the oil. *Associated gas* is the gas in excess of that which can be carried in the crude oil at atmospheric pressure. Most of the oil produced in the Middle East is accompanied by associated gas. Table 1-6 contains the analysis of several associated gases along with a "typical" gas.[13] The most important feature of associated gas is the relatively high concentration of ethane, the prime feedstock for ethylene production. As with non-associated gas, an absorption process is used for recovering condensable hydrocarbons including most of the ethane.

Associated gas is a coproduct of crude oil production and something must be done with it. It can be either flared, reinjected into the formation or utilized as a raw material for chemical processes. This situation does not exist with natural gas production which is independent of crude oil production. Associated gas removed from crude oil during the stabilization phase has long been considered a waste product in many of the large oil fields in the Middle East and North Africa.

Within the OPEC countries, over 60 percent of the associated gas is flared. Saudi Arabia utilizes about 17 percent of its daily 4×10^9 cubic feet production of associated gas. With today's world energy situation and a growing shortage of convenient chemical plant feedstocks in the highly industrialized areas, these associated gases represent a feedstock and an energy source of great potential value.

Saudi Arabia is currently building an extensive gathering system to make their associated gas available for both energy and petrochemicals. Associated gas from Umm Shaif, Zakum and El Bundug fields will be processed at the Das Island LNG and LPG plant located 100 miles off the coast of Abu Dhabi and Dubai.[14] This plant is designed to produce **LNG** as the primary product from the associated gases with **NGL** and **LP**-gas as valuable coproducts. Fig. 1-5 indicates the product distribution of the expected 12,136 ton daily production. The relative quantities will vary depending upon the particular associated gas being processed. The sulfur comes from the hydrogen sulfide in the gas.

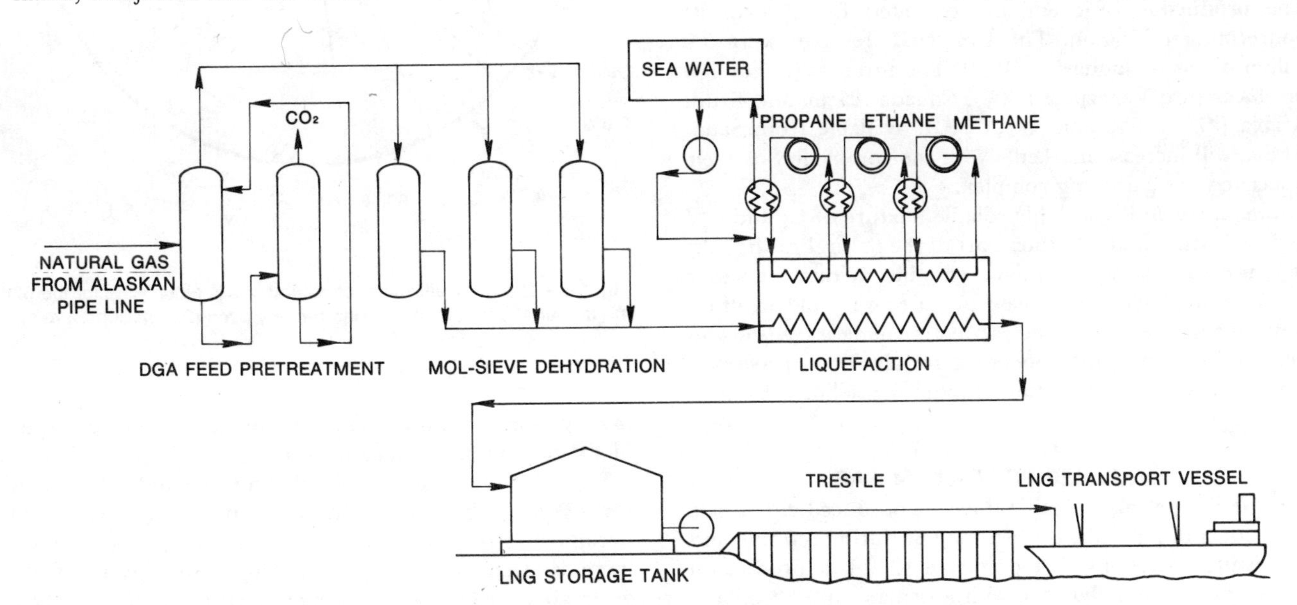

Fig. 1-4—A typical plan for the liquefaction and transportation of Alaskan natural gas.[12]

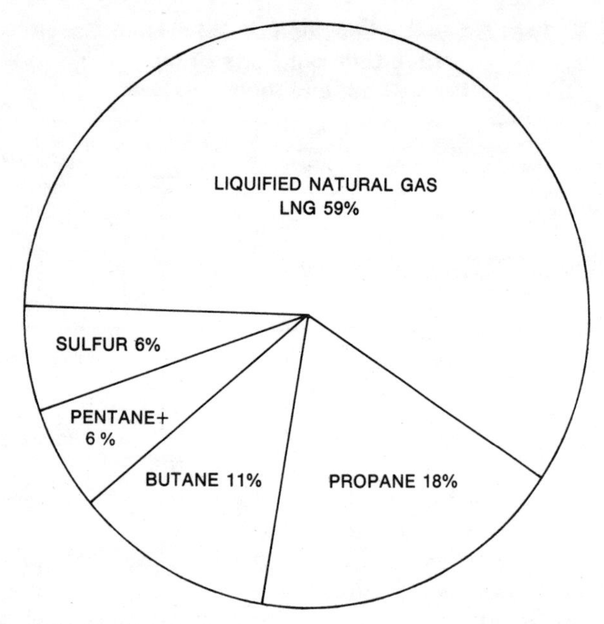

Fig. 1-5—Product distribution from the Das Island associated gas liquefaction plant.[14]

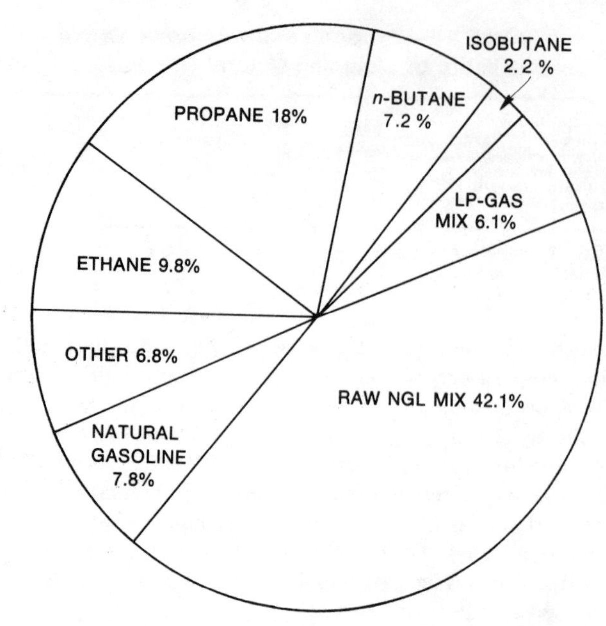

Fig. 1-6—Typical product distribution for natural gas liquids, NGL.

The direct utilization of associated gas as an energy source or petrochemical feedstock is best carried out in the area of production. However, its transportation to industrial areas in the form of a gas or a liquid is being investigated.

NATURAL GAS LIQUIDS

Natural gas liquids, **NGL**, consist of the hydrocarbons condensed from natural gas and associated gases—the ethanes through the pentanes (Fig. 1-6). Propane and the butanes are frequently referred to as *liquefied petroleum gas,* **LPG** or **LP**-gas, and their data are usually reported separately.

The use distribution of the 526 million barrels of **NGL** utilized in 1977 is shown in Fig. 1-7.[15] These data exclude the 234 million barrels of **NGL** utilized in refinery operations. The production of chemicals accounted for 45% of its nonrefining utilization. The U.S. **NGL** reserves were 5.9 billion bbl as of January 1, 1979. The three major foreign producers are Venezuela (32%), Canada (27%), and Saudi Arabia (27%). The amount of **NGL** available from Saudi Arabia will increase markedly with the completion of their associated gas gathering complex.

Natural gasoline is a highly volatile mixture of C_4 and C_{5+} hydrocarbons and constitutes part of the natural gas liquids. It is normally added to motor gasoline to raise its vapor pressure and increase the ease of starting in cold weather. Natural gasoline is also used as a petrochemical feedstock to provide isobutane and isopentane for alkylation processes. The average daily production of natural gasoline is 817,000 barrels.

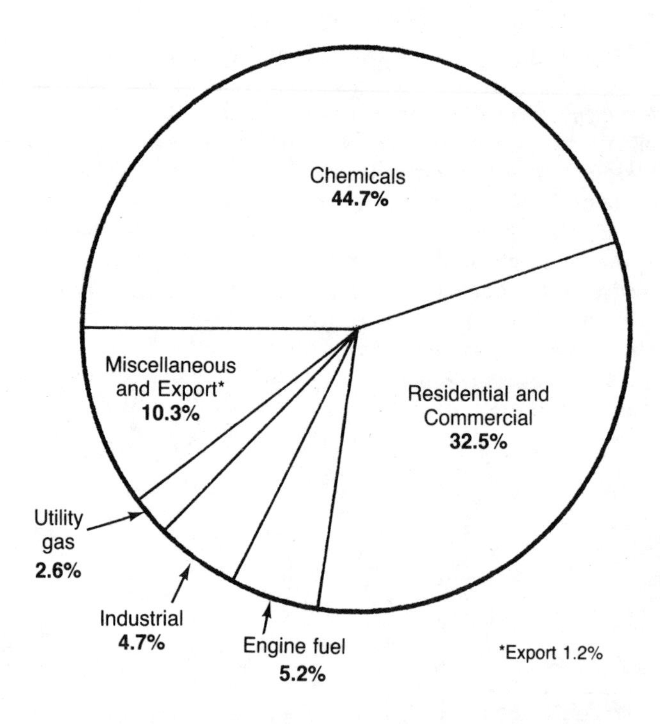

Fig. 1-7—The 1977 utilization of natural gas liquids, NGL, for the 526 million barrel demand, other than refinery uses (234 million barrels).[15]

LIQUEFIED PETROLEUM GAS

Liquefied petroleum gas, **LP**-gas, is produced either as a by-product of natural gas processing or during refining and processing operations. The predicted 1980 U.S. production of **LP**-gas is 477 million bbl with a demand of 533 million bbl.[16] The difference between supply and demand will come mainly from the Middle East and Canada. The world surplus of **LP**-gas will be 16 million tons.

Fig. 1-8 shows the U.S. supply and demand relationship for **LP**-gas.[17] Production from gas processing plants will decline by 3-4% per year while refinery production will increase. Demand for retail **LP**-gas, mostly propane, will grow about 2.5% per year. Before 1985, there will be a surplus of *n*-butane.

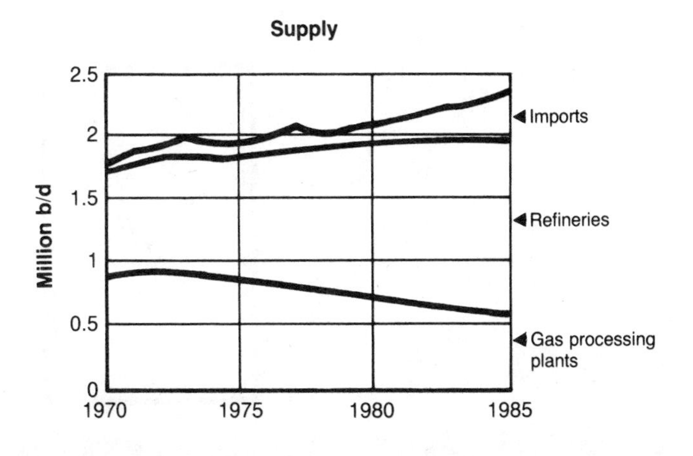

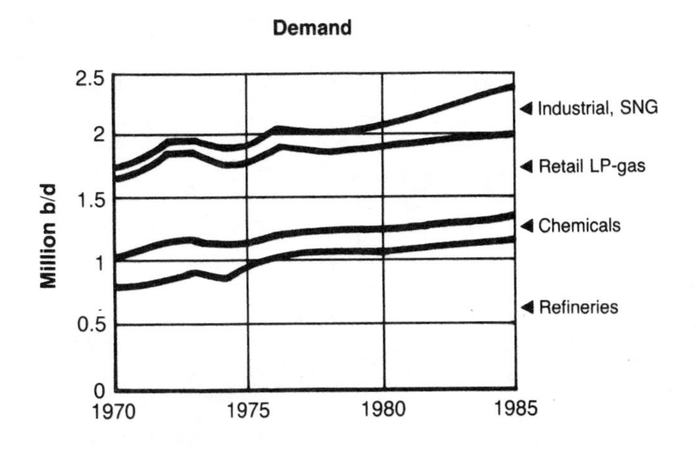

Fig. 1-8—The United States supply and demand picture for LP-gas, 1970-1985.[17]

HEATING VALUE

The final chemical composition and the heating value are dependent on the extent to which the natural gas or associated gas has been treated for condensable hydrocarbons. Table 1-7 shows the composition of two **LNG**, one from Algeria and the other from Badak, Indonesia, in comparison with a representative U.S. gas.[18,19] The heating value of the Algerian **LNG** is slightly higher because it has a greater percentage of heavier gas constituents. Since some countries supplement their gas needs by importing, vaporizing and using **LNG** as natural gas, the combustion interchangeability is important.

TABLE 1-7—Composition of Two Liquefied Natural Gases and a Typical U.S. Natural Gas

Constituents, %	Algerian LNG	Indonesian LNG	Rochester Natural Gas
Methane	87.55	90.75	95.93
Ethane	8.14	4.94	2.46
Propane	2.30	2.92	0.31
Butane	1.14	1.31	0.11
Pentanes	0.02	0.07	0.03
Nitrogen	0.85	0.01	0.42
BTU/ft^3-gross	1127.	1122.	1032.
Specific gravity	0.631	—	0.583

The *Wobbe index* is a parameter used to classify gases into similar heating values and operating performances. The index is calculated by dividing the heating value of the gas by the square root of its specific gravity. A Wobbe index performed on two gas mixtures with results that fall within 10% of each other indicates that the two gases are interchangeable. The revaporized Algerian gas had a heating value of 1,156 BTU/ft^3 and a sp. gr. of 0.648. The Wobbe index was 1,436. The Rochester domestic gas had a Wobbe index of 1,352.

LITERATURE CITED

1. Tuttle, R. and K. Allen. *The Oil and Gas Journal*, Aug. 9, 1976, pp. 78-82.
2. *Hydrocarbon Processing and Petroleum Refiner*, Vol. 43, No. 3, 1954, p. 150.
3. Freireich, E. and R.N. Tennyson. *The Oil and Gas Journal*, Aug. 23, 1976, pp. 131-132.
4. *The Oil and Gas Journal*, July 17, 1978, p. 73.
5. Raney, D.R., *Hydrocarbon Processing*, Vol. 55, No. 4, 1976, pp. 73-75.
6. *Hydrocarbon Processing*, Vol. 57, No. 4, 1978, p. 122.
7. *Hydrocarbon Processing*, Vol. 53, No. 5, 1974, p. 85.
8. Seaton, Earl. *The Oil and Gas Journal*, June 25, 1979, pp. 65-69.
9. *The Oil and Gas Journal*, June 11, 1979, p. 38.
10. *The Oil and Gas Journal*, Apr. 9, 1979, pp. 82-83.
11. *Gas Making and Natural Gas*, New York: Ben Johnson & Co. Ltd., 1972.
12. Wall, J.D. *Hydrocarbon Processing*, Vol. 54, No. 4, 1975, pp. 141-144.
13. Aalund, L.R. *The Oil and Gas Journal*, July 19, 1976, pp. 98-100.
14. Maguire, D.R. Second Arab Conference on Petrochemicals. United Arab Emirates, Abu Dhabi, March 15-22, 1976.
15. Cannon, R.E. *The Oil and Gas Journal*, July 19, 1976, pp. 98-100.
16. *The Oil and Gas Journal*, Mar. 22, 1977, pp. 74-75.
17. *The Oil and Gas Journal*, Mar. 12, 1979, pp. 32-33.
18. *The Oil and Gas Journal*, April 11, 1977, pp. 62-64.
19. Kennedy, T.L. *The Oil and Gas Journal*, May 29, 1978, pp. 51-56.
20. Grosso, Silvano and R.L. Pearce. *The Oil and Gas Journal*, Sept. 24, 1979, pp. 176, 178, 180, 183, 186, 188; *The Oil and Gas Journal*, Oct. 1, 1979, pp. 86, 89-90, 92, 94.
21. Wolfer, W., E. Schwartz, W. Vodrazka and K. Volkamer, *The Oil and Gas Journal*, Jan. 21, 1980, pp. 66-70.

2

Petroleum Composition and Classification

"Petroleum, n. A natural, yellow-to-black, thick, flammable liquid hydrocarbon mixture found principally beneath the earth's surface and processed for fractions including natural gas, gasoline, naphtha, kerosine, fuel and lubricating oils, paraffin wax and a wide variety of derived products. Also called 'crude oil'."[1]

Another definition for *petroleum* is a non-homogeneous mixture of substances of which the main constituents are hydrocarbons along with various quantities of sulfur, oxygen and nitrogen derivatives of hydrocarbons. Petroleum may also contain dissolved gases in varying amounts, and small amounts of metallic compounds. Petroleums differ widely among themselves in their physical properties. This is due to the variations in the ratios of hydrocarbon types present. Non-dissolved water is commonly associated with petroleum.

The ratios of the different types of hydrocarbons in a specific crude oil are important in determining the refining processes to be used. For example, the naphtha cut from atmospheric distillation of a crude oil with a high cycloparaffin (naphthene) content would be better used for the production of aromatics by a catalytic reforming process than for the production of ethylene by pyrolysis. The petrochemical industry has grown to such an extent that petrochemical feedstock and fuel products are competing for the same petroleum fractions. Competition will be more severe in the future.

HYDROCARBONS IN PETROLEUM

Table 2-1 lists the major hydrocarbon types isolated by API Project 6 from its reference petroleum (Ponca City).[2] The data from this work led to the following conclusions:[3]

1. All petroleums contain substantially the same hydrocarbon compounds.

2. The principal compounds in the gasoline fraction of each petroleum may be placed into five main classes: nor-

TABLE 2-1—Major Types of Hydrocarbons Isolated by API Project 6[2]

Type Name	Type Formula*	%
Paraffin Hydrocarbons (Alkanes)		
Normal (*n*)	CH_3-CH_2-R	14
Iso-	$CH_3-CH(CH_3)-R$	18 (Iso- and Branched)
Branched	$CH_3-CH_2-CH(CH_3)-R$	
Cycloparaffins (Naphthenes, Cycloalkanes)		
Alkylcyclopentanes	CH_2 / CH_2 $CH-R$ / CH_2-CH_2	10
Alkylcyclohexanes	CH_2 / CH_2 $CH-R$ / CH_2 CH_2 / CH_2	6
Bicycloparaffins	CH_2 H CH_2 / CH_2 C CH_2 / CH_2 C CH_2 / CH_2 H CH_2	5
Aromatic Hydrocarbons		
Alkylbenzenes	–R	18
Aromatic-cycloparaffin	CH_2 / $CH-R$ / CH_2 / CH_2	5
Fluorenes	CH_2 –R CH_2	3
Binuclear aromatics	–R	17
Tri- and tetranuclear aromatics		4

* R is usually CH_3-.

mal (straight chain) paraffins, branched paraffins, alkylcyclopentanes, alkylcyclohexanes and alkylbenzenes.

3. Within each of these five classes, the individual compounds occur in proportions which are usually of the same order of magnitude for different petroleums.

These conclusions concern primarily the gasoline fractions. It should be noted, however, that the ratios among hydrocarbon classes vary widely from one crude oil to another and that crude oil also contains polynuclear hydrocarbons in addition to the types noted for the gasoline fraction.

Within the last few years marked advances have been made in the isolation and identification of individual hydrocarbons in petroleum. Isolation methods include the formation of adducts and the use of silica gel, molecular sieves (zeolites) and gas-liquid chromatography (Table 2-2). Identification methods include the use of gas-liquid chromatography, ultraviolet spectroscopy, mass spectroscopy and nuclear magnetic resonance (Table 2-3). Even with all the help one gets from these and other analytical methods, the task of isolation and identification is formidable.

n-Paraffin Hydrocarbons. All normal paraffin hydrocarbons from C_1 to C_{33} have been isolated from petroleum. The presence of a high ratio of C_{10}-C_{14} *n*-paraffin hydrocarbons in kerosine (Table 2-4)[10] and C_{14}-C_{18} *n*-paraffins in gas oil is important since these hydrocarbons are used to produce biodegradable detergents.[11,12] They are also gaining importance as carbon sources for single cell protein (SCP) production.[13]

n-Butane is isomerized to isobutane which is used in various alkylation processes. *n*-Pentane and *n*-hexane, usually as mixtures, are isomerized to the iso-structure because of the higher octane number of branched isomers. Platinum is the catalyst of choice to effect this type of isomerism.

Isoparaffins and Branched Paraffins. Isoparaffins and branched paraffins from C_4 to C_{33} have also been isolated from petroleum. *Isoparaffins* are those with a methyl group on the number 2 carbon atom, $CH_3—CH(CH_3)—R$. *Branched* (or branched chain) *paraffins* have an alkyl group(s) (usually methyl) farther down the carbon chain. Isoparaffins are the most abundant followed in order by the 3-methyl and the 4-methyl monosubstituted compounds. Di- and tri-substituted paraffins are less abundant and are mainly in the higher boiling fractions of petroleum. In the kerosine range the *n*-paraffins are more important and also more abundant than the isoparaffins (Table 2-4).[10]

Cycloparaffins-Naphthenes. *Cycloparaffins* (*naphthenes*) of various types have been isolated from crude oil. They are saturated cyclic hydrocarbons, many of which have methyl groups. The lower boiling petroleum fractions contain appreciable quantities of cyclopentanes and cyclohexanes. In contrast to branched chain paraffins, the most abundant cycloparaffins are those with two or more substituents.

The presence of a high percentage of cyclopentanes and cyclohexanes in a gasoline fraction is important because they are the precursors of aromatic hydrocarbons. The dehydrogenation of cyclohexane and methylcyclohexane to benzene and toluene, respectively, and the isomerization followed by dehydrogenation of methylcyclopentane and dimethylcyclopentane to benzene and

TABLE 2-2—Techniques Used for Separating Hydrocarbons from Petroleum

Method	Result
Fractional distillation	Separation of a liquid hydrocarbon mixture into fractions (cuts) with narrow boiling ranges.
Low temperature distillation	Separation of a mixture of volatile hydrocarbons (gaseous hydrocarbons) into narrow boiling fractions (cuts) and individual compounds.
Vacuum distillation	Separation of mixtures of high boiling hydrocarbons into narrow boiling fractions (cuts) and residue.
Silica gel adsorption*	Separation of hydrocarbon types (paraffins, olefins, naphthenes and aromatics) from their mixture in light petroleum fractions and refinery streams.
Sulfonation	Removal of olefins and aromatics by compound formation from petroleum light distillates (gasoline and kerosine).
Urea[4] and thiourea[5] adducts	Separation of *n*-paraffins and isoparaffins from a mixture of light and medium boiling hydrocarbon streams.
Molecular sieve[6,7,8]	Separation of *n*-paraffins from kerosine and gas oil fractions.
Preparative gas chromatography	Separation of individual hydrocarbons from hydrocarbon mixtures obtained from any one of the previous methods. This is also used to identify individual compounds.

* By use of fluorescent indicator or nitrogen tetroxide[9].

TABLE 2-3—Techniques Used for Identifying Petroleum Hydrocarbons

Method	Application
Infrared spectroscopy	Limited to indicating the presence (or absence) of aromatic compounds in a hydrocarbon mixture. Absorption at 1600-1610 cm^{-1} indicates the presence of aromatics. Could be used to determine methyl and methylene group content.
Ultraviolet spectroscopy	Limited to indicating the presence of olefinic and aromatic compounds in a hydrocarbon mixture. Large absorption around 210-220nm is indicative of aromatic compounds.
Mass spectroscopy	Used to determine the molecular weight of a hydrocarbon by determining the mass of the parent ion. Useful in differentiating between isomeric hydrocarbons through the various fragments produced.
Nuclear magnetic resonance	Useful for determining the structure of hydrocarbons through the chemical shift, intensities of proton signals and splitting.
Gas chromatography	Used to determine the components in hydrocarbon mixtures by comparing peak retention times with those of pure compounds.
Physical properties	For characterization of individual compounds. Refractive index, density, boiling point, melting point, etc.

TABLE 2-4—Normal Paraffin and Isoparaffin Hydrocarbons Isolated from Kuwait Kerosine (149-260°C)[10]

Carbon Number	Percent*	
	n-Paraffins	Isoparaffins
C_9	3.11	—
C_{10}	5.6	1.5
C_{11}	5.6	1.2
C_{12}	5.6	1.2
C_{13}	5.6	0.9
C_{14}	2.49	0.6

* Based on 100 ml. of kerosine.

toluene are the major reactions in the catalytic reforming of straight run gasoline.

Bicycloparaffins are those compounds having two or more saturated rings that are fused with two vicinal carbons in common (Table 2-1). They also may have separated rings with or without a paraffin chain linkage. Most of the bicycloparaffins are in kerosine and gas oil fractions.

Aromatic Hydrocarbons. *Aromatic hydrocarbons* are compounds with at least one benzene ring in the molecule. Mononuclear aromatics are present mainly in the naphtha fraction of petroleum and those boiling up to 190°C have been isolated. 1,2-Dialkylbenzenes are less abundant than the more thermodynamically stable 1,3-dialkyl and 1,3,4-trialkylbenzenes.

The octane number of aromatic hydrocarbons is much higher than normal paraffin hydrocarbons, branched paraffins and cycloparaffins of the same carbon number.

Binuclear aromatic hydrocarbons contain two benzene rings. The naphthalenes belong to this group with 1-methyl and 2-methylnaphthalene being the most abundant. Binuclear aromatics are present in the middle distillates and the polynuclear aromatics are in the higher boiling fractions. That condensed ring aromatic hydrocarbons and heterocyclic compounds are the major components in asphaltenes has been confirmed by mass spectroscopic techniques.[14] There are indications that metal petroporphyrins are intimately incorporated into the structure of asphaltenes.

Recent interest in the *polynuclear aromatic hydrocarbons* has stemmed more from their possible carcinogenic properties than from their industrial utilization. The wide industrial use of petroleum pitch, however, has added emphasis to their importance. Various polynuclear aromatics separated and identified from commercial petroleum pitch volatiles are listed in Table 2-5.[15]

Only a few aromatic-cycloparaffin compounds have been isolated and identified. Among these are tetralin and indole from kerosine and light oil.

TABLE 2-5—Polynuclear Hydrocarbons Isolated from Petroleum Pitch Volatiles[15]

Type	Formula*	Wt., %
Methylphenanthrenes	—CH3	0.22 ± 0.06
Pyrene		0.14 ± 0.04
Methylpyrenes	—CH3	0.46 ± 0.14
Chrysene		0.22 ± 0.05
Methylchrysenes	—CH3	0.59 ± 0.04
Dimethylchrysenes	—CH3 —CH3	0.38 ± 0.04
1,2-Benzanthracene		0.13 ± 0.06
1,2-Benzopyrene		0.10 ± 0.03
3,4-Benzopyrene		0.19 ± 0.05
	Total	2.47 ± 0.28

NON-HYDROCARBON COMPOUNDS

Various types of non-hydrocarbon compounds occur in crude oil and refinery streams. The most important are the organic sulfur, nitrogen and oxygen compounds, in that order. Traces of metallic compounds are also present and can cause problems in certain catalytic processes. For catalytic reforming it is important to control the sulfur and vanadium content of the feed to avoid poisoning the catalyst.

Sulfur Compounds. The concentration of sulfur compounds varies from one crude to another. So called "sour" crude oils contain hydrogen sulfide, but many technologists carelessly refer to "high sulfur" oils as sour oils. Crude oils are classified as "sour" if they contain as much as 0.05 cu. ft. of dissolved hydrogen sulfide per 100 gallons. For "high sulfur" crude oils, the element analysis of sulfur may be low but the percentage of compounds containing sulfur is high. For example, it is estimated that crude oil with only 5 percent by wt. sulfur may have half its compounds containing sulfur. It has been observed that the higher the density of a crude oil, the greater the sulfur content.

Sulfur compounds in crude oil are complex and usually thermally unstable. They break down during refining processes to form hydrogen sulfide and simple organic sulfur compounds. Extensive research has been carried out to identify the sulfur compounds in light refinery streams.[16] Table 2-6 lists the types of sulfur compounds in crude oil and refinery streams.

Certain types of sulfur compounds are corrosive because they are acidic. Compounds of this type, such as hydrogen sulfide and mercaptans, can be removed by chemical treatment. The sulfur in non-acidic sulfur compounds are usually removed as hydrogen sulfide during hydrotreating.

Nitrogen Compounds. The nitrogen content of most crude oils is low, usually less than 0.1 percent by wt. The nitrogen content of the higher boiling fractions, however, is appreciably higher. The nitrogen compounds are quite thermally stable and for this reason only trace amounts of nitrogen are in light refinery streams.

There are several major types of hydrocarbon-nitrogen compounds and they are considerably more complex than the hydrocarbon-sulfur compounds. Separation of the

TABLE 2-6—Types of Sulfur Compounds in Petroleum and its Distillates

Type	Formula	Occurrence
Hydrogen sulfide	H_2S	*, **, ***
Mercaptans		
Aliphatic	R-SH	*, **, ***
Aromatic	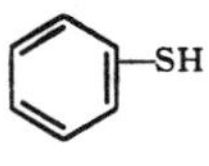 (benzene ring)—SH	***
Sulfides		
Aliphatic	R—S—R	** ***
Cyclic	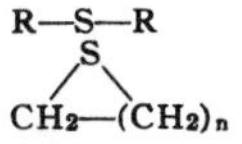 S bridging CH_2—$(CH_2)_n$	**; ***(?)
Disulfides		
Aliphatic	R—S—S—R	**
Aromatic	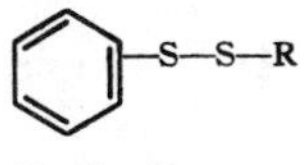 (benzene ring)—S—S—R	***
Polysulfides	R—S_n—R	**, ***
Thiophene and homologs	 (thiophene ring, S)	***

Occurrence: * In crude oil
** In straight run products
*** In cracked products.

TABLE 2-7—Types of Nitrogen Compounds in Petroleum

Type	Representative Compound*
Basic Nitrogen Compounds	
Pyridines	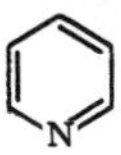 (pyridine ring, N)
Quinolines	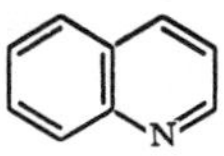 (quinoline, N)
Isoquinolines	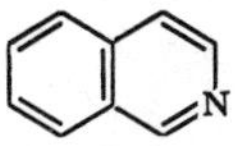 (isoquinoline, N)
Acridines	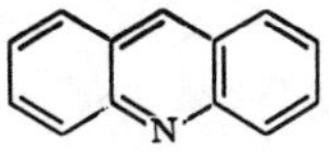 (acridine, N)
Non-basic Nitrogen Compounds	
Pyroles	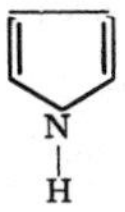 (pyrrole, N—H)
Indoles	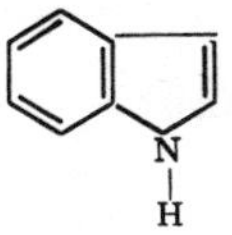 (indole, N—H)
Carbazoles	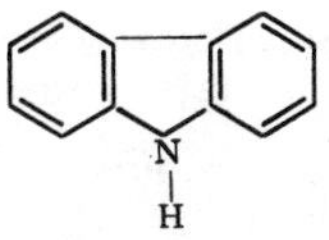 (carbazole, N—H)
Porphyrin	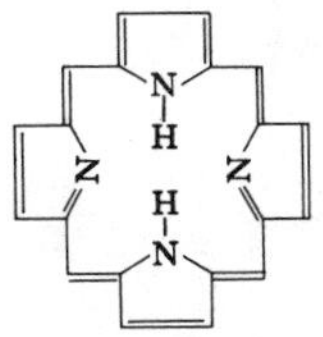 (porphyrin, N—H, N, N, H—N)

* Many of these compounds contain methyl groups.

types is difficult and the compounds are susceptible to alteration and loss during handling. An extensive study of both nitrogen and oxygen compound types in the 200-700° F boiling range has been made on distillates from Welmington, Calif., petroleum at the Union Research Center.[17]

The nitrogen compounds in petroleum may be classified according to whether they are basic or non-basic. Table 2-7 lists some of the important types of nitrogen compounds that have been isolated. These compounds, however, are not utilized as petrochemicals. Hydrotreating is used to reduce the nitrogen content of catalytic feeds since some nitrogen compounds poison catalysts.

Oxygen Compounds. As with the nitrogen compounds in petroleum the oxygen compounds are, in general, more complex than the sulfur compounds. They are usually carboxylic acids, phenol and cresols (cresylic acids) with lesser amounts of non-acidic types such as esters, amides, ketones and benzofurans (Table 2-8). Asphalts contain many highly-oxygenated compounds. Because of the acid nature of most oxygen compounds, they are readily separated from crude oil and its fractions. The total acid content in petroleum varies from 0.03 percent (Iraqi and Egyptian petroleum) to 3 percent in some California petroleums.[18] *Naphthenic acids,* which represent the primary acidic portion of crude oil, are of considerable importance as petrochemicals. The naphthenic acids in the gas oil fractions are straight-chain alkyl cycloparaffin carboxylic acids. The acids are obtained as sodium naphthenates by caustic extraction.

While sulfur and nitrogen content must be controlled in most feedstocks used in catalytic processes, oxygen compounds do not present a serious problem.

Metallic Compounds. Metals in crude oils (Table 2-9) are either in the form of salts dissolved in water suspended in the oil or in the form of organometallic compounds and metallic soaps. Metallic soaps of calcium and magnesium are surface active and act as emulsion stabilizers.

The presence of vanadium in catalytic process feeds is undesirable because the vanadium poisons the catalyst. The presence of vanadium is monitored by emission or atomic absorption techniques.

CLASSIFICATION OF CRUDE OILS

Knowledge of the constituents of a crude oil to be processed is important to a refiner if the purpose is to produce chemicals or to modify process variables. In most refineries where atmospheric distillation is one of the processes used to produce conventional fuels, detailed knowledge of the constituents is not necessary. For this reason, plus the economy of a simple process to determine the characteristics of a crude, a broad classification of crudes has been developed based on some simple physical and chemical properties.

TABLE 2-8—Types of Oxygen Compounds in Petroleum

Type	Formula*
Acidic	
Aliphatic carboxylic acids	$CH_3(CH_2)_n-C(=O)-OH$
Branched aliphatic carboxylic acids	$CH_3(CH_2)_n-CH(R)-C(=O)-OH$
Monocyclic naphthenic acids	R-cyclopentyl (CH_2, CH_2, CH_2, CH_2, CH) $-CH-CH_2-C(=O)-OH$
	R-cyclohexyl (CH_2, CH_2, CH_2, CH_2, CH_2, CH) $-CH-CH_2-C(=O)-OH$
Bicyclic naphthenic acids	R-bicyclic (CH_2 H CH_2 / C / C / CH_2 H CH_2) $-CH-CH_2-C(=O)-OH$
Polynuclear naphthenic acids	
Aromatic acids	R-phenyl$-C(=O)-OH$
Binuclear aromatic acids	R-naphthyl$-C(=O)-OH$
Polynuclear aromatic acids	
Phenol	phenyl-OH
Cresols	phenyl(OH)$-CH_3$
Non-acidic	
Esters	$CH_3(CH_2)_n-C(=O)-O(CH_2)_nCH_3$
Amides	$CH_3(CH_2)_n-C(=O)-NH(CH_2)_nCH_3$
Ketones	$CH_3(CH_2)_n-C(=O)-(CH_2)_nCH_3$
Benzofurans	R-benzofuran
Dibenzofurans	R-dibenzofuran

* R is usually $-CH_3$

Crude oils are generally characterized as belonging to one of four types depending upon the relative amounts of waxes and asphalts present. The wax content shows the degree to which the crude is paraffinic. The presence of asphalts indicates an aromatic crude. The following classification of crude oil is in general use.

TABLE 2-9—Trace Elements Present in Crude Oils

Source	Elements Present
Canada	Fe, Al, Ca, Mg
Ohio	Fe, Al, Ca, Mg
California	Fe, Ni, V
Egypt	Fe, Ca, Ni, V
Iraq	Fe, Ni, V
Venezuela	Fe, Ni, N

1. *Light paraffinic*—Crudes with a relatively low wax content
2. *Paraffinic* —Crudes with a high wax content and relatively low asphalt content in the residue
3. *Naphthenic or asphaltic* —Crudes with trace amounts of waxes and high asphalt content in the residue
4. *Aromatic* —Crudes with high aromatic content.

The term "mixed crude" is used when the crude has average properties between paraffinic and naphthenic types.

The Correlation Index, BMCI. The *correlation index*, BMCI(CI), was developed by the U. S. Bureau of Mines and is equal to zero for paraffins and 100 for benzene. **BMCI** values relate the average boiling point of a distillation fraction to its density. Using the boiling points and densities of pure hydrocarbons, the following simple relation was developed:

$$\text{BMCI} = 48640/K + 473.7/d - 456.8$$

K = the mid-boiling point of a fraction in Kelvin degrees
d = the specific gravity of the fraction at 60/60°F.

It is possible to classify crudes as paraffinic, mixed or aromatic using the **BMCI** values which are calculated from the physical properties of the crude. Lower values indicate predominance of paraffinic and higher values indicate predominance of aromatic. The **BMCI** is considered an excellent index of the various weighted average of the myriad of components of the various types present. For example, it is a simple but powerful designation of the relative quality of gas oils as pyrolysis feedstocks.[19]

CHEMICAL AND PHYSICAL DETERMINATIONS OF CRUDE OIL QUALITY

Simple physical and chemical data indicate the quality of crude oil. Some of the general determinations in use are discussed in the following sections. Table 2-10 lists the analyses of four crudes from the Middle East.[20]

Specific Gravity—Density. The *density* of a liquid, the weight per unit volume, is useful to determine the quality of crude oil. However, a more usual unit used by the petroleum industry is *specific gravity*. Specific gravity is the ratio of the weight of a given volume of a material to the weight of the same volume of water measured at the same temperature. Usually crudes and products are measured on a volume basis

TABLE 2-10—Analyses of Crude Oils from Different Parts of the Middle East[20]

Tests	(1) Jalo (Lybia)	(2) Alamein (Egypt)	(3) Safaniya (Saudi Arabia)	(4) Amer (Egypt)
Specific Gravity @ 60/60°F.	....	0.8576	......	0.9255
Gravity API @ 60°F.......	33.8	33.41	27.3	21.39
Water Content, vol. %......	0.05	0.2	Trace	0.1
Salt Content, wt. %........	0.003	0.007	......	0.005
Total Sulfur, wt. %.........	0.68	0.86	2.87	4.5
Hydrogen Sulfide, ppm......	....	....	....	6311
Mercaptan Sulfur, ppm.....	....	....	....	86
Pour Point, °F.............	9	35	Below −15	30
Conradson Carbon Residue, wt. %..................	1.4	4.9	8.4	4.5
Ash Content, wt. %........	.0012	0.004	0.013	0.05
Asphaltene Content, wt. %..	0.25	3.37	....	8.06
Paraffin Wax Content, wt. %	4.0	3.3	24.0	1.35
Vanadium Content, ppm....	9	15	41	67
Phenol Content, wt. %......	....	....	....	0.0082
Heating Value, gross Btu/lb.	....	....	18660.0	....

and then changed to a weight basis by multiplying the volume by the specific gravity.

Specific gravity is sometimes a rough measure used in differentiating crudes, since crudes with low density are usually paraffinic. The specific gravity of a liquid can be expressed as the API gravity which gives numbers rather than fractions. The specific gravity of a liquid can be changed to API gravity using the following relationship:

$$\text{API (Degrees)} = \frac{141.5}{\text{Sp gr } 60/60} - 131.5,$$

Pour Point. The pour point of a crude oil or a product is the lowest temperature at which an oil is observed to flow under the conditions of the test. The importance of pour point data is that it is an indication of the amount of waxes present in crude oils. The test is also important for diesel fuels and lubricating oils used in cold areas. Specified pour points for these products should be below the lowest temperature reached in these areas.

Sulfur. Determination of sulfur in crudes is important since it gives the refiner an indication of the amount of sulfur compounds in the products and whether the crude oil needs further treatment. The method used is to oxidize the sulfur compounds to sulfur dioxide by burning the crude, or the product, in an atmosphere of air or oxygen. Sulfur dioxide is then further oxidized to produce sulfuric acid and the acid is titrated with a standard alkali. Trace sulfur compounds in light fractions can be determined by a calorimetric method.[21]

Sulfur compounds in the gasoline range are harmful since they reduce the effectiveness of lead alkyls added to increase the octane number. Some sulfur compounds also change under engine conditions to corrosive sulfur compounds which reduce the lifetime of the engine.

Carbon Residue. Carbon residue is an indication of the tendency of a crude oil to form carbonaceous and metal residues which do not burn or evaporate easily at the conditions of the test. Carbon residue determination is an important test for diesel fuels, lubricants and fuel oil. During engine operation, carbon residue in lubricating oils increases because of oxidation and polymerization that takes place. This produces asphaltenes and gums which tend to form coke on heating.

Ash Content. Tests for ash content show the amount of ash left after burning all the liquid and volatile material in a crude oil or a product. The ash is usually metallic salts, metal oxides or silicon oxides and salts (see Table 2-9 for common metals found in petroleum).

Crude Oil Reserves

Total world oil reserves as of January 1, 1978, were 646 billion barrels. Fig. 2-1 shows the geographical distribution of these reserves;[22] Table 2-11 shows where these oil reserves are located.[23] The United States proven reserves as of December 31, 1978, were 27.8 billion barrels.[24] Table 2-12 shows the United States petroleum supply/demand situation

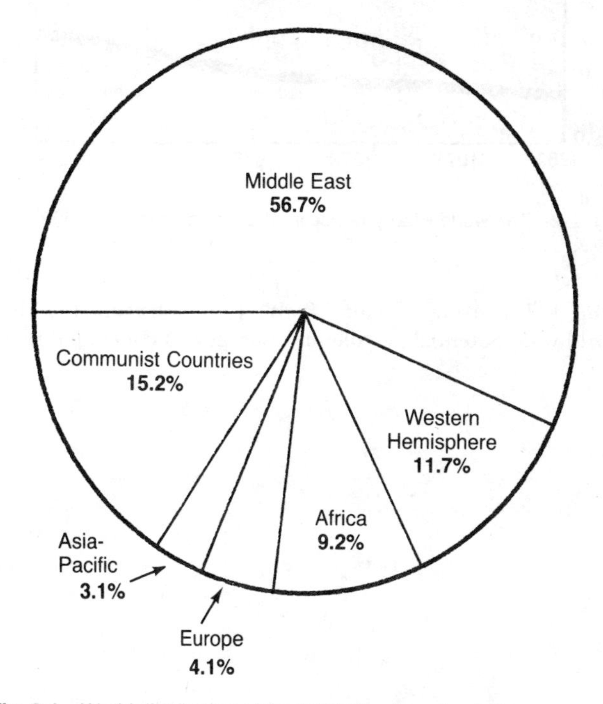

Fig. 2-1—World distribution of the 646 billion barrels of oil reserves as of January 1, 1978.[22]

TABLE 2-11—The Major Oil Reserves by Country: 10^6 Barrels as of January 1, 1978[23]

Country	Reserves
Saudi Arabia	150.0
U.S.S.R.	75.0*
Kuwait	67.0
Iran	62.0
Iraq	34.5
United Eimirates	32.4
United States	29.5
Lybia	25.0
China	20.0
United Kingdom	19.0
Nigeria	18.7
Venezuela	18.2
Mexico	14.0
Indonesia	10.0

*This figure includes proven, probable, and some possible oil reserves.

TABLE 2-12—United States Petroleum Supply/Demand Situation—1975-2000[25]

	(Millions of barrels daily) 1975	1980	1985	1990	1995	2000
Total liquids demand	17.0	17.0	17.0	17.0	17.0	17.0
Production from known fields (excluding Prudhoe Bay)	10.5	7.5	4.0	2.0	1.0	0.5
Production from anticipated discoveries (including Prudhoe Bay)	...	4.0	7.0	7.5	8.0	8.0
Imports	6.5	4.5	2.0	1.0	0.5	—
Syncrude and/or alternate energy	...	1.0	4.0	6.5	7.5	8.5

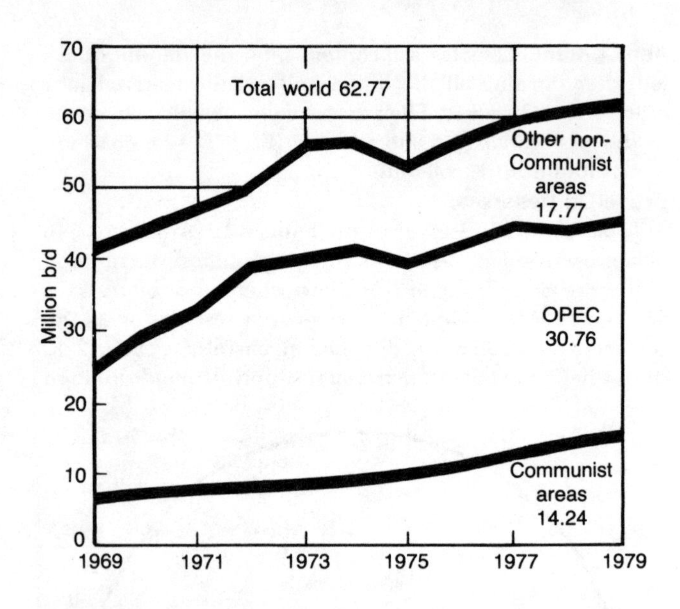

Fig. 2-2—The world-wide production of crude oil between 1969 and 1979.[26]

from 1975 through 2000.[25] P. W. J. Wood estimated the world-wide potential petroleum reserves and concluded that "there's a trillion barrels of oil awaiting discovery." The world-wide production of crude oil from 1969 through 1979 is shown in Fig. 2-2.[26]

LITERATURE CITED

1. *The American Heritage Dictionary of the English Language*, ed. William Morris. Boston: American Heritage Publishing Co., Inc., and Houghton Mifflin Co., 1969.
2. Hatch, Lewis F. *Hydrocarbon Processing*, Vol. 48, No. 2, 1969, pp. 77-88.
3. Rossini, F.D. *J. Chem. Ed.*, Vol. 37, 1960, p. 554.
4. Sinha, A.K., and K.A. Venkatachalam. *Indian J. Technology*, No. 6, 1968, p. 26.
5. Nippon Mining Co. *The Oil and Gas Journal*, Oct. 16, 1972, p. 142.
6. Fetterly, L.C., and El Cerrito. U. S. Patent 2,670,344 (1954).
7. O'Connor, J.G., and M.S. Morris. *Anal. Chem.*, Vol. 32, 1960, p. 701.
8. Brunnock, J.V. *Anal. Chem.*, Vol. 38, 1966, p. 1648.
9. Chen, N.Y., and S.J. Luki. *Anal. Chem.*, Vol. 42, 1970, p. 508.
10. Al-Zaid, K.A.H. Master's Thesis, 1976; Kuwait University; Ijam, M.J., K.A.H. Al-Zaid, and C. Parkanyi. *Hydrocarbon Processing*, Vol. 58, No. 9, 1979, pp. 145-147.
11. Kerfoot, O.C., and H.R. Flammer. *Hydrocarbon Processing*, Vol. 54, No. 3, 1975, pp. 74-78.
12. Hatch, Lewis F. *Hydrocarbon Processing*, Vol. 54, No. 3, 1975, pp. 79-82.
13. Dimming, W., and R. Seipenbusch. *Hydrocarbon Processing*, Vol. 54, No. 9, 1975, pp. 169-172.
14. Gallegos, E.J. *7th World Petroleum Congress Proceedings*, Vol. 4, (1967), p. 256.
15. Grienke, R.A., and I.C. Lewis, Preprints, Division of Petroleum Chemistry, Inc. Am. Chem. Soc., Vol. 20, No. 4, 1975, p. 787.
16. Rall, H.C., C.J. Thomson, H.J. Coleman, and R.L. Hopkins. *Proc. Am. Petrol. Inst.*, Vol. 42, Sec. VIII, 1962, p. 19.
17. Snyder, L.R., *Accounts Chem. Res.*, Vol. 3, No. 9, 1970, pp. 290-299.
18. Osman, A.M., A.I. Khodair, A.A. Swelim, and A.A. Abdel-Wahab. Second Arab Conference on Petrochemicals, Paper No. 14 (P-4), Abu Dhabi, March 15-22, 1976.
19. Green, E.J., S.B. Zdonik, and L.P. Hallee. *Hydrocarbon Processing*, Vol. 54, No. 9, 1975, pp. 164-168.
20. Tiratsoo, E.N. *Oilfields of the World*. Beaconsfield, England: Scientific Press Ltd., 1973, p. 13.
21. Smith, H.M. *Bureau of Mines Technical Paper, 610*, (1940).
22. *International Petroleum Encyclopedia*. Tulsa: The Petroleum Publishing Co., 1978, p. 270.
23. *International Petroleum Encyclopedia*. Tulsa: The Petroleum Publishing Co., 1978, p. 4.
24. *The Oil and Gas Journal*, May 7, 1979, pp. 56-58.
25. Moody, John. *The Oil and Gas Journal*, Aug. 28, 1978, 185-186, 188, 190.
26. *The Oil and Gas Journal*, Feb. 25, 1980, pp. 27-31.

3

Refinery Processes

Fuels are the most important product of petroleum refineries, but the increasing demand for petrochemicals presents a viable possibility of a refinery designed specifically to produce petrochemicals and petrochemical feedstocks. It has been stated repeatedly that it is possible to convert completely and economically a barrel of crude oil into chemicals, making no fuel whatsoever.

Stork, Abraham and Rhoe made an extensive investigation of well-established petroleum and petrochemical processes for moderate to high conversion of crude oil to petrochemicals.[1] They state that nearly complete conversion of crude oil to petrochemicals by these processes is technically feasible, but under normal circumstances hardly economic. The most economic facilities will produce a combination of petrochemicals and some fuel products. Figs. 3-1 and 3-2 illustrate two different approaches to the production of petrochemicals from crude oil.[1] In 1979, 6% of a barrel of crude oil went into petrochemicals. It is predicted that in 1990 the figure will be 11%.

A highly efficient and self-supporting energy-balancing process for the manufacture of petrochemicals from crude oil has been reported by Sinkar.[2] The process is based on the integration of the following two commercially proven processes: heavy oil cracking (HOC) and thermal pyrolysis of the resulting hydrocarbons. The major products are olefins (ethylene, propylene and mixed C_4's) and aromatic hydrocarbons (benzene, toluene and xylenes, BTX). The yield of aromatics can be increased substantially by catalytically reforming the virgin naphtha and the naphtha produced by the HOC unit instead of cracking them to produce olefins.

Veba Oel AG has perfected the integration of refining and petrochemical production to a high degree.[3] This was effected by the integration of a petrochemical complex (Scholven) with a refinery (Horst) as shown in Fig. 3-3.[3] Table 3-1 gives the various products and the respective production capacities.

The relationship between the petroleum industry and the petrochemical industry has been succinctly put by F. Perry Wilson (Union Carbide): "As our fuel supplies diminish we may have to face the fact that oil is not just for burning. At the moment, the petrochemical industry may be one of the smallest customers in comparison with other markets, but we may yet become the largest." Fig. 3-4 shows the predicted refinery product distribution for 1985 and 1990.[4]

REFINERY PROCESSES

Refinery processes are either simple, such as those used to separate crude oil into fractions, or more complex processes where chemical reactions take place and the structure of the constituents changes. After completion of any chemical process, physical processes are commonly used to separate the products.

Physical Processes

The most important physical processes are distillation (atmospheric and vacuum), absorption, adsorption and extraction.

Distillation. *Distillation* is a physical separation process based on differences in boiling points of the components in a mixture. Crude oil is primarily a complex mixture of hydrocarbons, many of which have the same or nearly the same boiling points. Consequently, except for the lowest boiling hydrocarbons, it is not possible to separate crude oil into pure compounds by distillation. Crude oil is, therefore, separated into mixtures having a rather narrow boiling range. This distillation, called *topping* or *skimming*, may be accomplished in a single distillation column (tower) or a series of columns.

Side streams are drawn from the columns and are partially cooled by heat exchange with fresh feed. The feed is further heated in a crude oil heater. The efficiency of the fractionation depends upon the effective number of plates in the

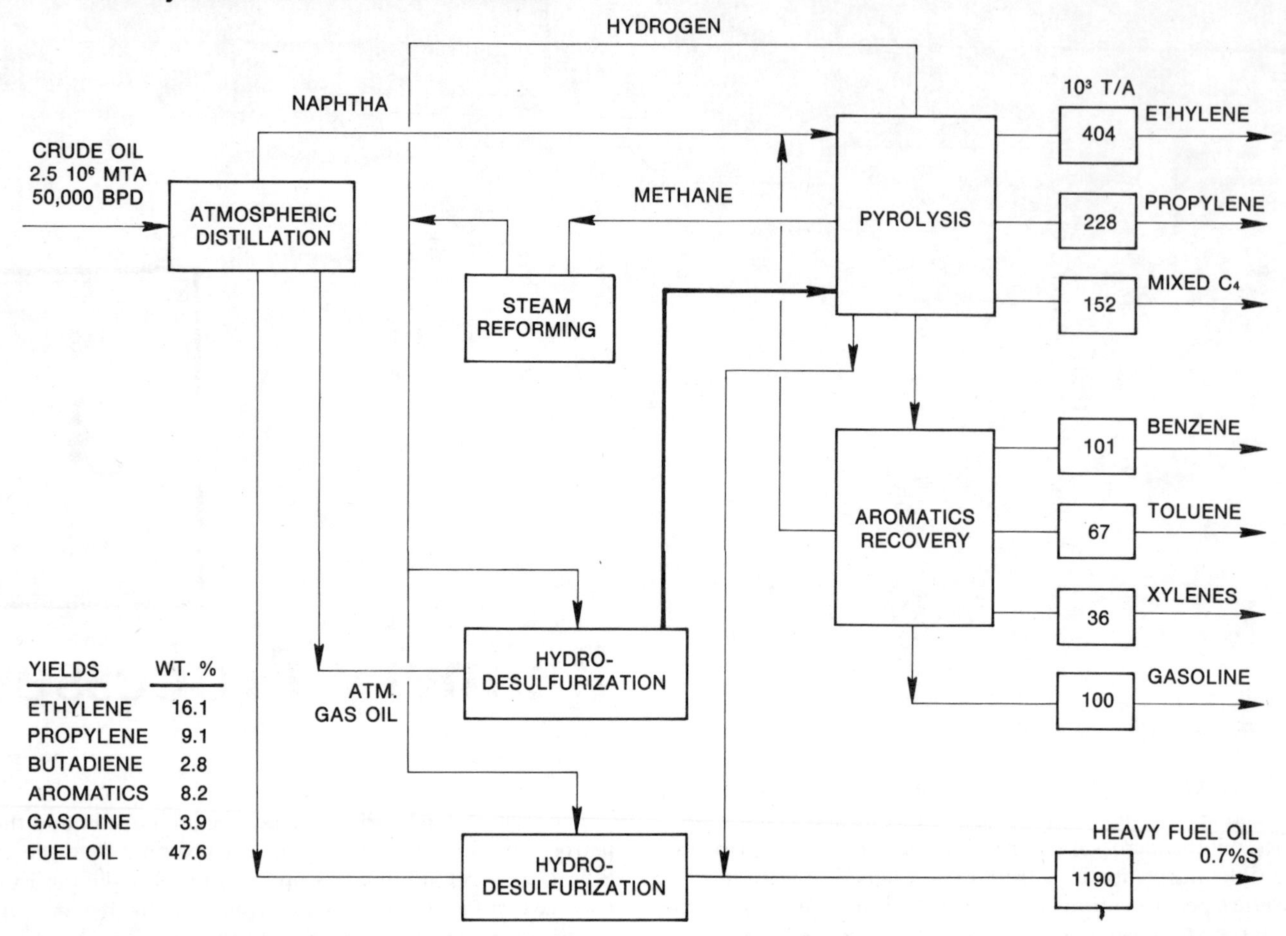

Fig. 3-1—Atmospheric distillation and pyrolysis of all distillates for fuels and petrochemicals.[1]

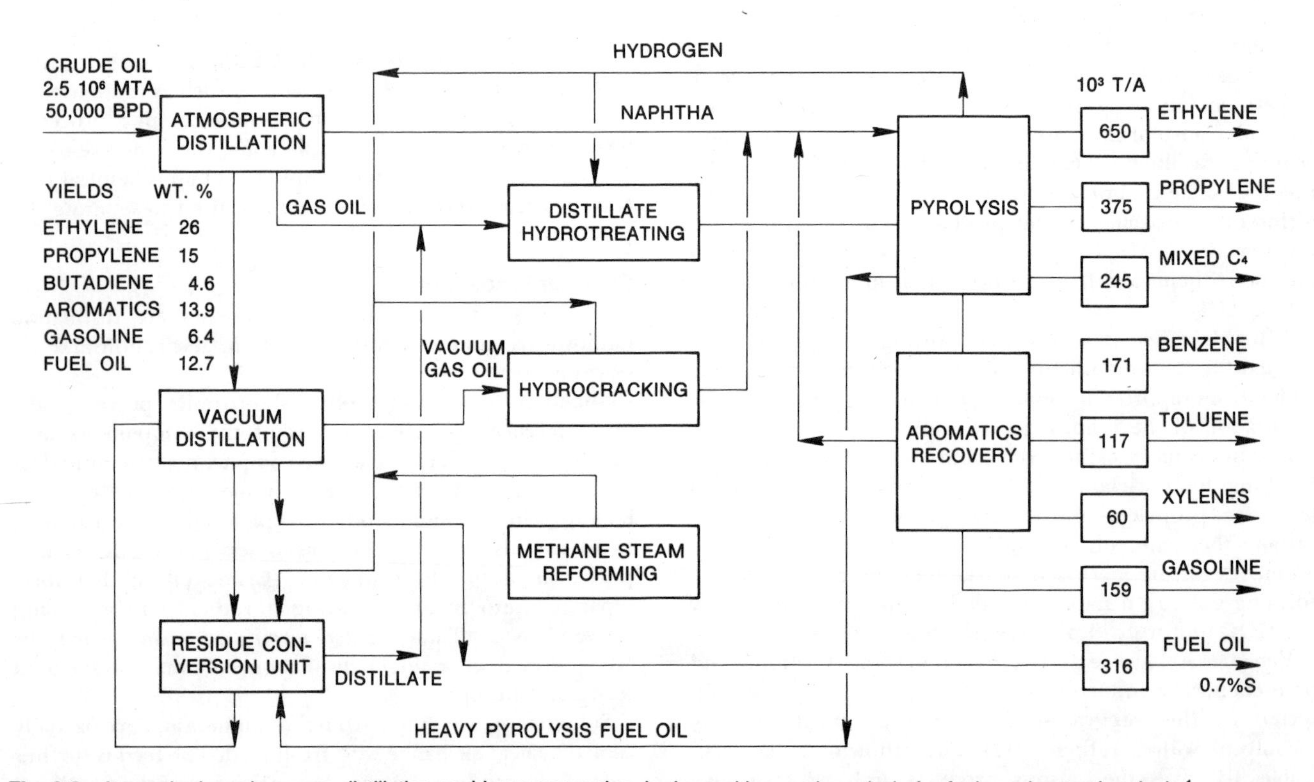

Fig. 3-2—Atmospheric and vacuum distillation, residuum conversion, hydrocracking, and pyrolysis for fuels and petrochemicals.[1]

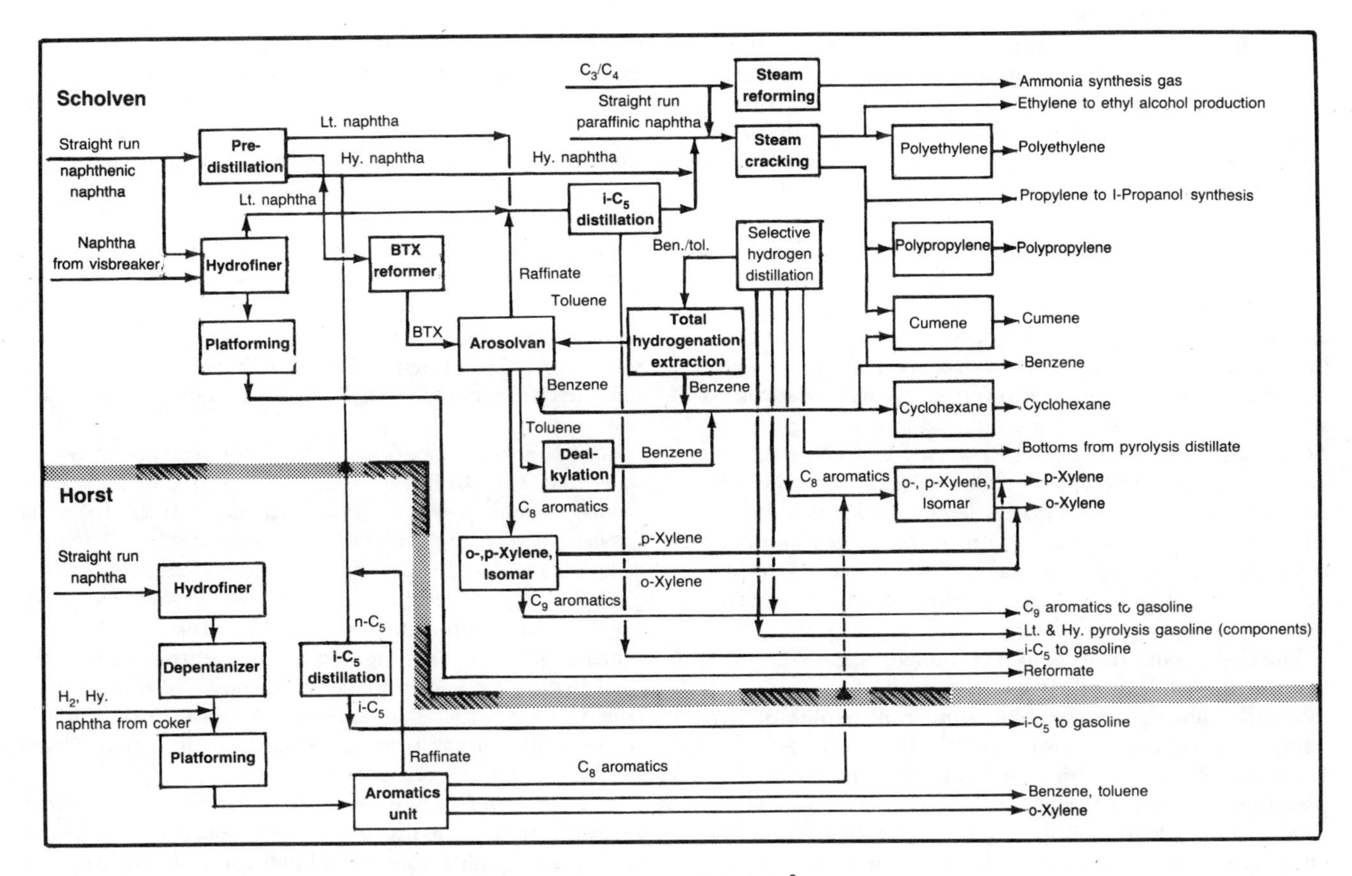

Fig. 3-3—The integrated refinery-petrochemical complex of Horst and Scholven.[3]

column and the reflux ratio. In practice, the reflux ratio is varied over a wide range to achieve the specific separations desired.

Distillation of crude oil is initially carried out at *atmospheric pressure*. The side stream products are gasoline (which is stabilized by the removal of low boiling hydrocarbons), naphtha and gas oil and diesel oil. The residue is drawn from the bottom of the distillation tower and may be used as a fuel, charged to a vacuum distillation unit, or to a catalyst cracking

TABLE 3-1—The Products from the Horst and Scholven Complex[3]

	Scholven	Horst
	1,000's metric tons/year	
Crude distillation	7,500	5,500
Catalytic reforming	810	1,260
Catalytic cracking	220 (FCC)	620 (TCC)
Thermal cracking	850	...
Coking	...	1,300
Asphalt	210	...
Benzene	270	30
Sulfur	40	12
Sulfuric acid	330	...
Ethylene	550	...
Polyethylene	120	...
Polypropylene	70	...
Cyclohexane	100	...
N paraffins (charge)	...	360
Cumene	330	...
Toluene	...	60
Orthoxylene	80	29
Xylenes	120 (para)	120 (meta & para to Scholven)
Phthalic acid anhydride	...	100
Ammonia	330 (nitrogen)	...
Methanol	150 (200 max.)	...
High pressure hydrorefining of coke works benzene	350	...
Nitric acid	85	...
Fumaric acid	...	4

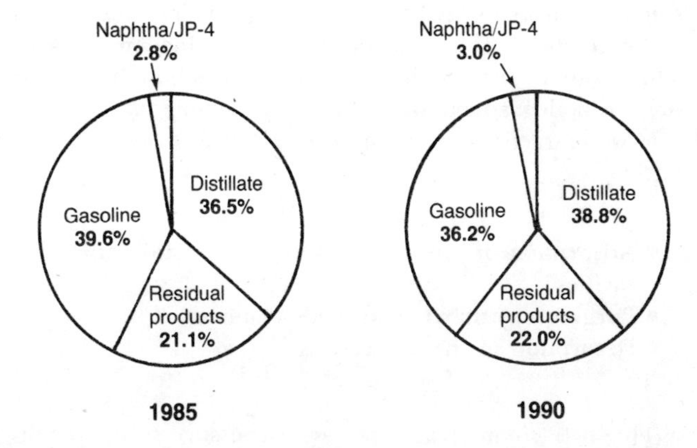

Fig. 3-4—The predicted composite refinery product distribution for the United States in 1985 and 1990.[4]

or hydrocracking unit. The residue is usually hydrotreated to remove metals and sulfur before being cracked.

Vacuum distillation is used to increase the amount of middle distillates and to produce lube oil base stock and asphalt. The use of a vacuum avoids cracking the long-chain hydrocarbons present in the feedstock.

The feed is charged to the distillation tower in an atmosphere of super-heated steam which reduces the partial pressure of the hydrocarbons. *Gas oil* is the top product and the temperature is controlled by refluxing part of this gas oil. The other distillation products are *lube oil base stocks*. The bottom product is *asphalt*.

Absorption. *Absorption* is a process used to collect gases in a liquid absorbent. *Physical absorption* is used to obtain condensable hydrocarbons from natural gas. A hydrocarbon liquid is the absorbent which preferentially absorbs gases with higher molecular weights than ethane. A recent absorption solvent, Selexol, absorbs acid gases without forming chemical bonds. The best absorption is achieved at high operating pressures and when the concentration of the acid gases is high.[5]

Chemical absorption is a term used when a chemical reaction takes place between the gas being absorbed and the absorbent. Refinery gas streams are usually freed from acid gases by absorbing them in basic absorbents such as ethanolamines. The reaction can be reversed by heat or steam to recover the ethanolamine and the acid gas.

Adsorption. *Adsorption* is a process used to free petroleum gases from trace amounts of undesired gases or vapors by adsorbing them on a solid material. The solid should have a large surface area and have the property of preferentially concentrating the adsorbed gas on its surface. Molecular sieves, silica gel and alumina are common solid adsorbents used in the petroleum industry.

The need to improve the octane ratings of gasoline without the use of lead alkyls has increased the use of 5-A molecular sieves to extract *n-paraffins* from naphthas.[6] Extraction of *n*-paraffins also provides improved reformer feedstocks. An added advantage is the use of the extracted *n*-paraffins as steam-cracking feed for ethylene production. The *nonlinear* (denormal) *isoparaffins* and *naphthenes*, on the other hand, are good steam-cracker feed for propylene and butadiene production, respectively.

The use of long-chain *n*-paraffins for the production of single cell proteins (**SCP**) has recently provided incentive for the extraction of *n*-paraffins from kerosine and gas oil. These sources also provide *n*-paraffins for use in the production of biodegradable detergents and flexible plastics.

Most extraction processes use a three stage operating cycle:

- Adsorption of *n*-paraffins into the 5-A molecular sieve cavities
- Purging to remove void-space material
- Desorption of the *n*-paraffins.

The two commercial processes, pressure-swing or displacement, differ basically in the choice of phase (vapor or liquid) and in the method of desorption. The *n*-paraffins can be desorbed by an increase in temperature, decrease in pressure or displacement by another material.

The IsoSiv Process for gas oil, for example, is an *isobaric, isothermal process* operating at *ca* 370° C and at less than 100 psig.[7] Desorption is achieved by stripping with a lower molecular weight *n*-paraffin (*n*-pentane, *n*-hexane) which is displaced when feed is readmitted to the adsorbent and is easily distilled from the product. The *n*-paraffins are suitable for **SCP** production by fermentation. The purity of the *n*-paraffins obtained by this process is not less than 99.2%.

The *pressure-swing process* (**BP**) is reported to offer significant advantages of flexibility and lower costs over displacement in the molecular sieve separation of *n*-paraffins.[8] "Pressure-swing process (**MS2**) produces the exceptionally high *n*-paraffin purities required for protein manufacture. High denormal (isoparaffin) purity provides gasoline-octane improvement and improved reformer and steam-cracking feedstocks. These give the process tremendous potential . . ."[8]

Solvent Extraction. Liquid solvents are used to extract desirable, or undesirable, compounds or mixtures from petroleum streams. For example, lube oil stocks are extracted by use of *liquid propane*. The liquid propane dissolves paraffinic hydrocarbons and leaves aromatic and asphaltic material. The process is called *propane deasphalting*. Other extractive processes in the petroleum industry use various solvents such as phenol, furfural, liquid sulfur dioxide, dimethyl sulfoxide, dimethylformamide, ethylene glycol and sulfolane.

Ethylene glycol has a greater affinity for aromatic hydrocarbons than for paraffinic hydrocarbons and extracts them from reformer gasoline. The raffinate is freed from the ethylene glycol by distillation. Other solvents used for this purpose are liquid sulfur dioxide and sulfolane.

Sulfolane (tetramethylene sulfone) is a versatile extractant for the production of high purity **BTX** aromatics (benzene, toluene, xylenes) and high-octane concentrates. It is also used to produce low aromatic smoke-point kerosines. Sulfolane is used with alkanolamines (*Sulfinol Process*) in the removal of acid gas from gas streams such as natural gas, synthesis gas and hydrogen.

The *Edeleanu Process* is also used to extract aromatics from kerosine. It is an extraction process based on the higher solubility that sulfur dioxide exhibits for aromatic hydrocarbons in comparison with its solubility for paraffins and naphthenes. This process has been reviewed by Chopra and Mukhopadhyay.[9]

A variety of solvents are used to produce asphaltenes, resins and oils from feedstocks as varied as partially topped crude to vacuum reduced and blown asphalt. The solvent is recovered under supercritical conditions which is said to be more economical than recovery by conventional evaporation. This process is called *Residuum Oil Supercritical Extraction* (**ROSE**). **ROSE** can be used for the preparation of heavy oils for catalytic cracking and hydrocracking.

Urea dewaxing is used to produce lube oils with low pour points (transformer oils, refrigerator oils) as well as a number of other special products. The process involves combining urea with straight chain paraffin hydrocarbons to form solid inclusion compounds, *adducts*. The feedstocks require no pretreatment for the removal of sulfur and nitrogen compounds because these compounds do not interfere with the process. The adduct formation takes place at 25-40°C in the presence of methylene chloride; the urea is in aqueous solution. After filtration, the urea adduct is decomposed into an aqueous urea solution and a wax phase. The urea solution is filtered, purified, and recycled to the reactor. Fig. 3-5 is a flow diagram of a urea dewaxing process.[10] Typical product quality is shown in Table 3-2.[10]

Conversion Processes

Conversion processes in a petroleum refinery are varied in nature but all have two things in common—to produce more gasoline and better gasoline. An increase in the amount of gasoline from a barrel of crude oil is obtained by various cracking processes such as thermal cracking, catalytic cracking and hydrocracking. Cracking is the production of

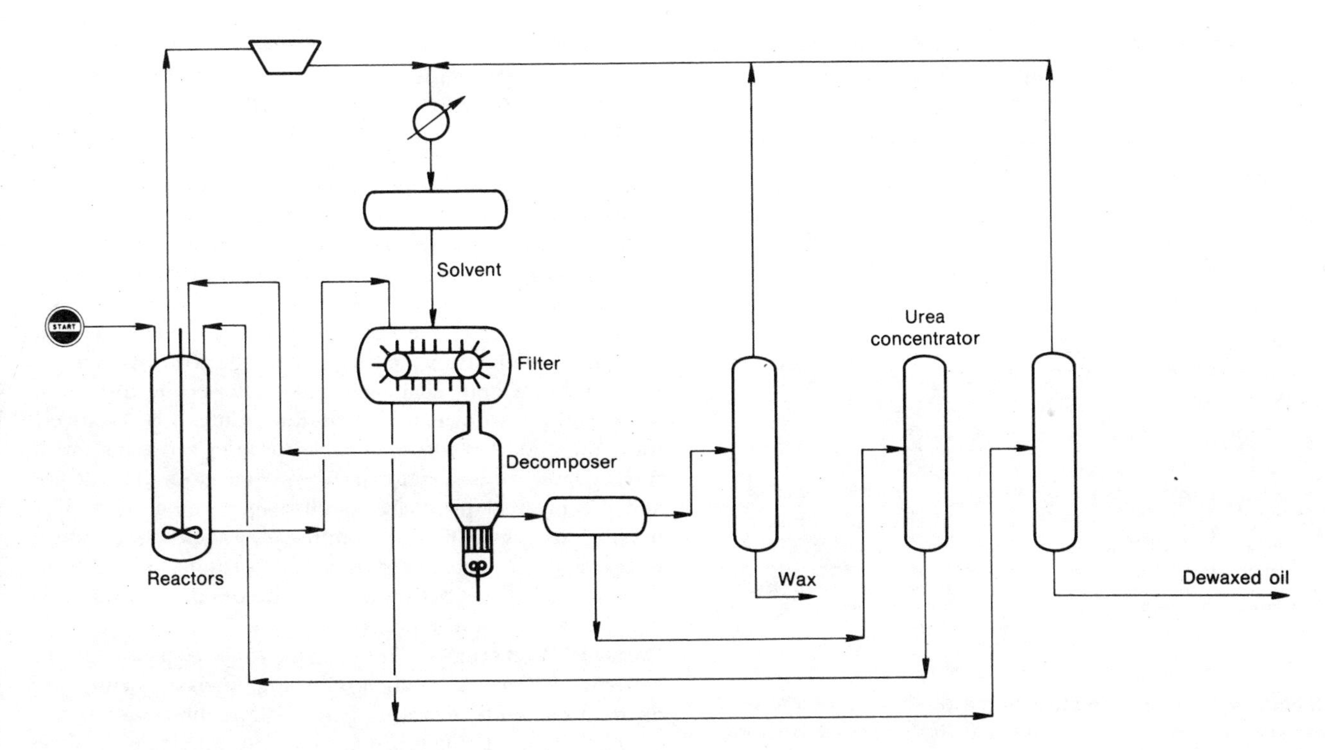

Fig. 3-5—Flow diagram for urea dewaxing.[10]

smaller molecules from larger ones. The four major gasoline making and octane boosting processes are catalytic cracking, catalytic reforming, hydrocracking, and alkylation.

Isomerization, alkylation and dimerization produce gasoline compounds of high octane number. Catalytic reforming improves the quality of gasoline without substantially changing the quantity of gasoline. Still other conversion processes such as hydroprocessing (hydrotreating and hydrorefining) remove undesirable components to provide feedstock for other refinery processes.

A petroleum refinery is a complex of many interrelated processes. These processes are generally classified as either thermal conversion processes or catalytic conversion processes.

Thermal Conversion. *Thermal cracking* was the first process used to increase gasoline production. After the introduction of catalytic cracking, which gives higher yields and better quality, thermal cracking was given other roles in refinery operations. These include viscosity breaking (*visbreaking*) and coking. An important role not directly related to gasoline production is steam cracking of various feedstocks to produce olefins.

Visbreaking of heavy residues is a mild, flexible thermal process which may be used for several different purposes. These are:

- Reduction of the fuel oil production.
- Reduction of fuel oil production with a simultaneous improvement of its pour point by including a mild thermal-cracking step.
- Production of maximum 350°C and lighter distillates.
- Production of maximum distillates including feedstocks for other conversion processes—for example, vacuum gas oil for catalytic cracking.

An analysis of these objectives for visbreaking is essential, since they have an impact on the processing scheme selected and upon the overall economics of the project. The various parameters have been reviewed and evaluated by Rhoe and deBlignieres.[11] Table 3-3 gives data for a typical visbreaking process to produce a minimum of gasoline and a maximum of furnace oil. The trends in product markets and some applications of visbreaking have been reviewed by Notarbartolo, Menegazzo, and Kuhn, with special emphasis on increasing the yield of light products.[12] They conclude that the installation of visbreaking units improves operational flexibility of any refinery and allows product state adjustment to better suit existing market needs.

Coking is a severe thermal cracking process used to obtain light products and coke from topped crude and heavy residues which cannot be directly fed into a catalytic cracking

TABLE 3-2—Typical Product Data Obtained by Urea Dewaxing[10]

Crude source	Mid East*		Nigerian	
Distillate feedstock				
Density @ 15°C, g/ml	0.847	0.890	0.908	0.923
Viscosity @ 100°C, cS	3.33	1.9	3.48	7.55
Pour point, °C	24	–12	12	32
Adduct, paraffins, wt %	22.6	11.7	14	13.6
Boiling range, °C	340–430	290–330	330–400	400–450
Dewaxed oil				
Yield on feed, wt %	75	88	85	85
Density @ 15°C, g/ml	0.864	0.902	0.927	0.944
Viscosity @ 100°C, cS	5.8	1.9	3.57	9.0
Pour point, °C	–24	–60	–48	–25
Slack wax				
Density @ 70°C, g/ml	0.772	0.766	0.780	0.790
Congealing point, °C	44	21	39	53
Oil content, ASTM, wt %	15	16	18	15
n-Paraffins, wt %	77	76	75	79
Carbon number	C_{20}–C_{28}	C_{15}–C_{20}	C_{19}–C_{25}	C_{24}–C_{30}

* Distillate feedstock from Mid East crude was solvent treated.

TABLE 3-3—Analysis of Feed and Products from Viscosity Breaking*

Charge Inspections	Libyan Residue
Gravity, °API	24.4
Vacuum Engler, corrected °F	
IBP	510
5%	583
10%	608
20%	650
Pour point (max), °F	75
Visc. SUS @ 122° F	175.8
Product yield, vol. %	
Gasoline, 100% C_4, 330 EP	10.8
Furnace oil, 805° F EP	42.7
Fuel oil	46.3
Gas, C_3 & Lighter (wt. %)	2.1
Properties of products	
Furnace oil	
Pour point (max.), °F	+5
Flash (PMCO), °F	150
Fuel oil	
Pour point (max.), °F	+40
Flash (PMCC), °F	150
Visc., SFS @ 122° F	67.5
Stability (ASTM D-1661)	No. 1

* To produce a minimum of gasoline and a maximum of furnace oil (**Hydrocarbon Processing,** 1974 Refining Processes Handbook p. 123).

unit because of their high metal and asphaltene content. The products from coking units are usually hydrodesulfurized (hydrotreated) to reduce sulfur content and to hydrogenate olefins. Manzanillia S., *et al*, have presented a new process to desulfurize petroleum coke.[13] This process is reported to be able to effect an almost 90% desulfurization.

Both *delayed coking* and *fluid coking* are used to upgrade residuals but delayed coking is less flexible. The main advantage of fluid coking over delayed coking is the production of more distillate with lower metal content. Treatment of these distillates is less costly than those produced by delayed coking. The coke from fluid coking contains more ash (highly metallic). The addition of a gasifier increases distillates. *Flexicoking* integrates fluid coking with coke gasification and converts vacuum resids (residuums) and tar materials into liquid and gaseous products and coke.

Coking of resids has been used in the past almost exclusively to increase light and middle distillates with the co-product coke considered a low value byproduct. Extensive research has improved the quality of petroleum coke to where it is now a source of industrial carbon. Petroleum cokes with low sulfur and metal content are used as electrolytic anodes, and for the manufacture of synthetic graphites as well as many other uses. The 1978 world green petroleum coke capacity is given in Table 3-4.[14] Chapter 4 contains a more detailed description of petroleum coke, its types, and uses.

TABLE 3-4—The 1978 World Green Petroleum Coke Capacity*[14]

	(Thousands of short tons)				
	Delayed	Needle	Fluid	Total	% of total
North America (U.S. & Canada)	12,910	460	1,595	14,965	81
Europe & Middle East	1,820	160	....	1,980	11
Latin America	885	....	120	1,005	5
Asia	265	225	...	490	3
Total	**15,880**	**845**	**1,715**	**18,440**	
%	86	5	9	...	100

*Excludes Communist bloc

Steam-cracking is a thermal-cracking process in the presence of steam and is used primarily for the production of olefins. The feedstock is versatile and includes ethane, propane, gas oil, and even crude oil. Steam-cracking for olefin production is discussed in Chapter 7.

Steam-cracking of vacuum residue is a new process similar to delayed coking in that it can accept heavy residues with high sulfur and metal content. It can be considered as an upgrading process for profitable disposable of the accumulation of high sulfur, heavy vacuum resids. The process is a thermal-cracking one using superheated steam continuously to strip the lighter products. A typical reaction cycle consists of a two hour and one hour soaking period followed by the quenching and blow-down of the product pitch. The relatively short time cycle makes it possible to build a reactor much smaller than the conventional coker drum. Both the reactant and the product pitch are handled in a liquid state. The molten pitch is rolled and continuously cooled to produce flakes. Fig. 3-6 is a flow diagram of a resid-cracking unit. Tables 3-5, 3-6, and 3-7 show the properties of the feedstock, liquid products, and pitch, respectively.[15]

Catalytic Conversion. The primary catalytic processes are catalytic reforming, catalytic cracking, hydrocracking, hydrotreating and hydrorefining. Alkylation, dimerization, polymerization and isomerization are also important catalytic processes but not of the same magnitude as the cracking processes. They involve individual molecules.

Catalytic cracking processes are concerned with producing more gasoline from a barrel of oil. Hydrotreating and hydrorefining are used primarily to desulfurize feedstock for catalytic reforming and catalytic cracking. Hydrotreating (hydroskimming) is also used to desulfurize middle distillate stocks. Catalytic reforming, alkylation, dimerization, polymerization and isomerization are processes used to produce a higher quality gasoline. These processes are being expanded more rapidly than cracking processes because of the shift toward low lead and unleaded gasoline.

Fig. 3-7 shows the United States refinery process distribution at the start of 1979 with a total capacity of 19 million bpsd.[16] A crude capacity profile of the top eight United States refining company's process capacity distributions is shown in Table 3-8.[16] In 1978 the petroleum industry spent 360 million dollars on catalysts.[17]

Catalytic Reforming. The feed to a catalytic reforming unit is naphtha, either virgin (straight run) or produced from a thermal or catalytic hydrocracking unit. The boiling range of the feed varies according to the process and the products needed. In a fuel refinery where the criteria is to increase the octane number of the product (gasoline) by the production of aromatics (Table 3-9), the boiling range of the feed is not as sensitive a variable as in the case of a petrochemical refinery. If the petrochemical market requires benzene and toluene rather than xylenes, a lower boiling naphtha cut is selected. A relatively higher boiling naphtha is needed to produce a high ratio of xylenes.

Benzene, toluene and the xylenes (**BTX**) are produced from naphtha during catalytic reforming by two distinct types of reactions. The dehydrogenation of naphthenes to benzene from cyclohexane, to toluene from C_7 naphthenes and to the xylenes from C_8 naphthenes reaches thermodynamic equilibrium quickly. A low pressure and, to a lesser extent, a high temperature increases the total yield of **BTX**

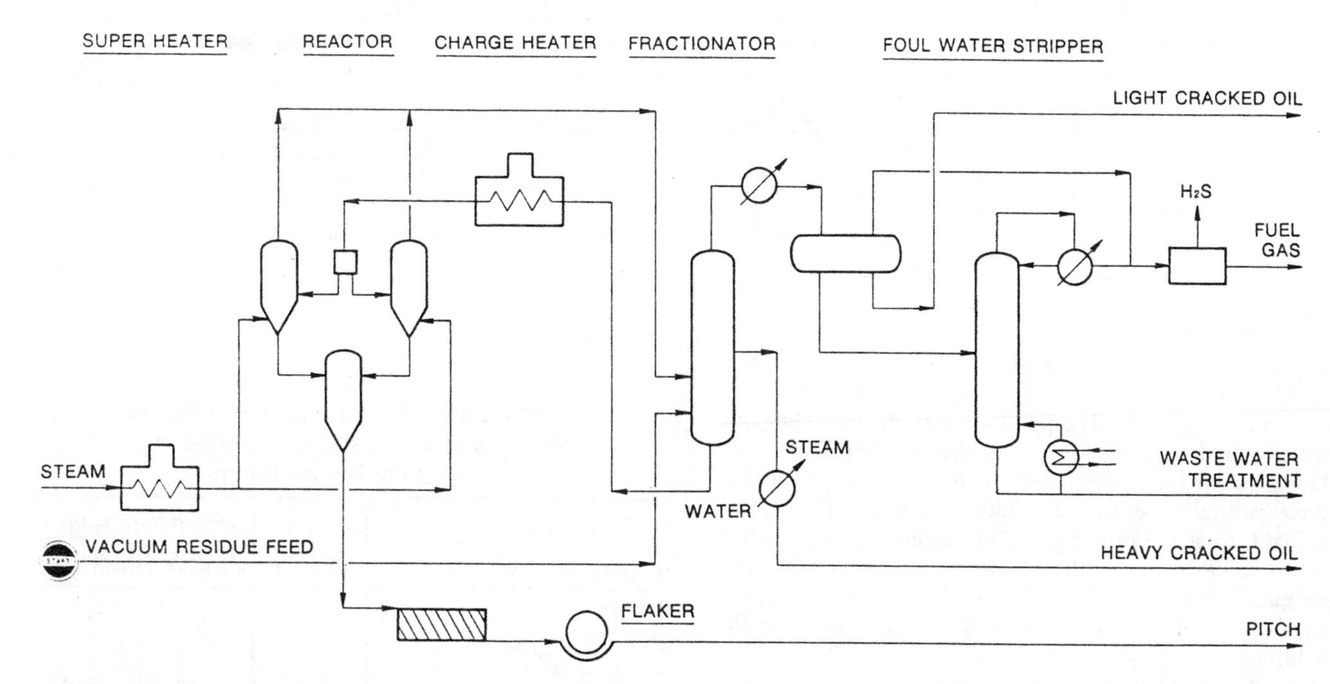

Fig. 3-6—Schematic diagram of steam cracking process for vacuum residue.[15]

TABLE 3-5—Properties of Typical Feedstocks[15]

	Vacuum Residue of Khafji crude	Vacuum Residue of Iranian Heavy
Gravity at 15°C (API)	6.46	6.73
Specific gravity 15/4°C	1.025	1.023
Penetration at 25°C (mm/10)	100	60
Asphaltenes (wt. %)	12.0	11.8
Conradson Carbon Residue (wt. %)	21.4	20.2
Element analysis:		
C (wt. %)	81.14	85.28
H (wt. %)	10.18	10.70
N (wt. %)	0.38	0.58
S (wt. %)	5.30	3.43
V (ppm)	178	345
Ni (ppm)	53	115

TABLE 3-6—Typical Properties of Product Oils[15]

	Light Fraction	Heavy Fraction
Specific gravity 15/4°C	0.7714	0.9241
ASTM Distillation (°C)		
I.B.P.	40	233
95%	238	528
E.P.	249	546
Br. Co.	89.4	....
Diene value	5.6	....
S (wt. %)	1.22	2.82
Fe (ppm)	0.2 less	0.2 less
Ni (ppm)	0.2 less	0.2 less
V (ppm)	0.1 less	0.1 less
Asphaltene (ppm)	10	100
Total Insolubles (ppm)	0.5	6.5
Pour point (°C)	....	30
% in total product oil	22	78

TABLE 3-7—Typical Properties of Product Pitch[15]

Feedstock	High Flow Point Pitch	Low Flow Point Pitch
Flow point (°C)	220	160
Fixed carbon (wt. %)	60	45
Solvent insolubles (wt. %)		
n-Heptane Ins.	80	65
Benzene Ins.	55	32
Quinoline Ins.	18	4
Specific gravity (15/4°C)	1.25	1.17
H/C Atomic ratio	0.85	1.05

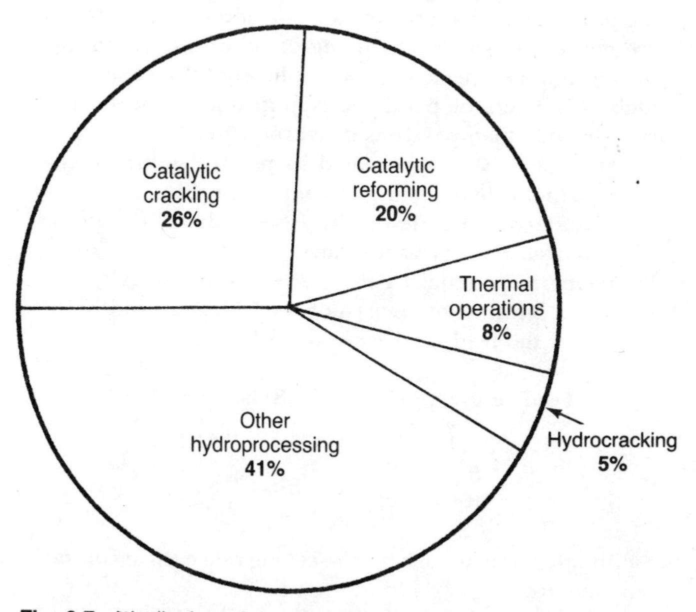

Fig. 3-7—Distribution of the 19 million b/d refining capacity as of January 1, 1979.[16]

by this type of reaction. Kinetics and thermodynamics influence both paraffin dehydrocyclization, the other type of reaction for **BTX** formation, and the hydrodealkylation of aromatics. These are rather slow reactions and thermodynamic equilibria are usually not reached.

Catalysts. The catalysts generally used for catalytic reforming are dual functional with a *hydrogenation-dehydrogenation site* and an *acid site*. Most of the gasoline reforming processes use an alumina for the acid site. The initial hydrogenation-dehydrogenation catalyst was platinum (*Platforming*). By 1972 a bimetallic platinum/rhenium (Pt/Re) catalyst (*Rheniforming*) was used in 27% of the United States' reforming capacity. Currently 68% utilize bimetallic catalysts.

TABLE 3-8—Refining Capacities of the Top Eight United States Petroleum Companies* [16]

Company	Cat cracking**		Cat reforming		Hydrocracking		Other hydro processes		Alkylation		TOTAL
	b/sd	%***	b/sd	%	b/sd	%	b/sd	%	b/sd	%	b/sd
Exxon	502.2	24.8	314.5	15.6	73.9	3.7	1,051.6	52.0	79.2	3.9	2,021.4
Shell	398.7	25.1	290.8	18.3	77.5	4.9	744.5	46.9	76.4	4.8	1,587.9
Chevron	258.0	17.6	309.5	21.1	192.0	13.1	663.2	45.4	41.1	2.8	1,463.8
Amoco	487.0	36.9	267.8	20.3	59.0	4.5	436.8	33.1	68.1	5.2	1,318.7
Texaco	452.2	34.8	259.4	20.0	38.9	3.0	476.1	36.6	69.0	5.4	1,295.6
Mobil	357.0	28.0	264.5	20.8	50.7	4.0	536.1	42.1	65.3	5.1	1,273.6
Gulf	333.9	32.0	212.5	20.4	37.5	3.6	382.9	36.7	76.4	7.3	1,043.2
Arco	132.0	13.4	228.0	23.2	88.0	8.9	520.6	52.9	16.2	1.6	984.8

*1,000 b/d
**fresh feed
***percent of combined capacities

The advantages of this type of catalyst are its abilities to operate at high capacity, lower pressure (*ca* 150 psig), higher severity and/or for longer runs[18] (Table 3-10). Type E Pt/Re bimetallic catalyst reforms a predominantly paraffinic Arabian naphtha at 125 psig to give a 14 percent increase in total benzene/toluene yield (31 vol. percent) over a 200 psig operation.[19] A naphthenic California feed showed a 7 percent increase in benzene/toluene yield under comparable conditions.

Both the composition of the reformer feed and its pretreatment are important variables. The composition of the feed influences the process conditions which in turn govern the product distribution. For example, less severe conditions are needed to obtain a high degree of aromaticity using a feed with a high percentage of naphthenes than one with a high percentage of paraffins. Naphthenes dehydrogenate more readily than paraffins dehydrocyclize.

The feed must be pretreated to remove sulfur because both platinum and platinum/rhenium catalysts are sulfur sensitive. High sulfur levels decrease catalyst cycle life because the sulfur decreases catalyst activity, thus increasing the temperature required for a given octane severity. The following information regarding cycle length and sulfur content of the feed has been reported.[20]

Feed sulfur, ppm	Relative cycle life
1.0	1.0
5	0.8
10	0.5

Sulfur also increases the catalyst-coking rate even at constant temperature.

Organic chlorides of concentrations in the 1-10 ppm range may present a costly problem because of corrosion. "It just hurts."[21]

Reforming reactions occur at different rates and degrees of conversion depending on the temperature, hydrogen partial pressure, and catalyst used. Aromatics in the feedstock normally do not undergo significant conversion although dealkylation and transalkylation occur to a minor extent. Reactivity is a strong function of carbon number—it increases as the carbon number increases. The rate of dehydrocyclization of heptane is about four times faster than for hexane, and the rate of hydrocracking is about one third greater. Selectivity to aromatics increases with an increase in carbon number.[22]

TABLE 3-9—Boiling Points and Octane Numbers of Various Hydrocarbons in the Gasoline Boiling Range

Hydrocarbon	Boiling point, °F	Octane Number Clear: Research method F-1	Octane Number Clear: Motor method F-2
n-Butane	0.5	....	....
n-Pentane	97	61.7	61.9
2-Methylbutane	82	92.3	90.3
2,2-Dimethylbutane	122	91.8	93.4
2,3 Dimethylbutane	137	103.5	94.3
n-Hexane	156	24.8	26.0
2-Methylpentane	146	73.4	73.5
3-Methylpentane	140	74.5	74.3
n-Heptane	208	0.0	0.0
2-Methylhexane	194	42.4	46.4
n-Octane	258	−19.0*	−15.0*
2,2,4-Trimethyl pentane (isooctane)	211	100.0	100.0
Benzene	176	....	114.8
Toluene	231	120.1	103.5
Ethylbenzene	278	107.4	97.9
Isopropylbenzene	306	....	....
o-Xylene	292	120.0*	103.0*
m-Xylene	283	145.0	124.0*
p-Xylene	281	146.0*	127.0*

* Blending value of 20% in 60 octane number reference fuel.

TABLE 3-10—Operating Conditions and Analyses Of Feed and Products Using Pt/Re Catalyst (Rheinforming)[18]

Feedstock	Hydrofined Arabian Naphtha	
Boiling Range, °F	180—310	
Composition, LV %		
Paraffins	68.6	
Naphthenes	23.4	
Aromatics	8.0	
Sulfur, ppm	<1.0	
Nitrogen, ppm	<0.5	
Operating conditions		
Reactor outlet pressure, psig	100	200
Reactor temperature, °F	900—1000	
Space velocity, LV %	1—4	
Product yield		
Hydrogen, scf/bbl. feed	1480	1210
C_1-C_3, scf/bbl. feed	230	380
C_5^+ Reformate, LV %	77.8	73.0
Reformate properties		
C_5^+ Research octane clear	98	99
Composition, LV %		
Paraffins	32.2	31.7
Naphthenes	1.7	0.5
Aromatics	66.1	67.8
Aromatics Breakdown		
C_6	1.9	2.6
C_7	15.7	18.8
C_8	28.3	28.9
C_9	19.0	15.9
C_{10}	1.2	1.6

The following are the most important reactions that take place in catalytic reforming.

Aromatization. The dehydrogenation of *cyclohexanes* to benzene and methylbenzenes is comparatively fast and endothermic ($\triangle H = 49.8$ kcal/mole).

$$\text{cyclohexane} \rightleftarrows \text{benzene} + 3H_2$$

The production of benzene is favored by high temperature and low pressure.

The aromatization of *methylcyclopentanes* is a two-step process. First the methylcyclopentanes are isomerized to cyclohexanes and then the cyclohexanes are dehydrogenated. For example, methylcyclopentane is isomerized to cyclohexane and then the cyclohexane is dehydrogenated to benzene.

$$\text{methylcyclopentane } (CH_3) \rightleftarrows \text{cyclohexane} \rightleftarrows \text{benzene} + 3H_2$$

The isomerization reaction is rate controlling.

Table 3-11 compares benzene formation from cyclohexane and from methylcyclopentane, both of which are important components of naphtha.[23] The conversion of methylcyclopentane to benzene is appreciably less selective than the conversion of cyclohexane to benzene. This lack of selectivity is common with two-step, catalyzed reactions.

Normally the reaction conditions in catalytic reforming are not severe enough to cause appreciable cyclization and aromatization of *paraffin hydrocarbons*. This is especially true when bimetallic catalysts are used. Table 3-12 indicates the influence of temperature on the reforming of *n*-hexane over platinum catalyst.[23] With C_6 and C_7 paraffin hydrocarbons the amount of aromatics at equilibrium is low, even at temperatures as high as 480°C. This is more pronounced at high pressures.

Isomerization. Carbonium ion intermediates formed by the acid components in the catalyst can rearrange through hydride, H^-, and/or methide, CH_3^-, shifts to form an isomer of the original compound. The following equations represent the isomerization of *n*-hexane to isohexane. Dehydrogenation:

$$CH_3{-}CH_2{-}CH_2{-}CH_2{-}CH_2{-}CH_3 \rightleftarrows CH_3{-}CH{=}CH{-}CH_2{-}CH_2{-}CH_3 + H_2$$

Protonation:

$$CH_3{-}CH{=}CH{-}CH_2{-}CH_2{-}CH_3 + H^+ \rightleftarrows CH_3{-}CH_2{-}\overset{+}{C}H{-}CH_2{-}CH_2{-}CH_3$$

Methide-hydride shift:

$$CH_3{-}CH_2{-}\overset{+}{C}H{-}CH_2{-}CH_2{-}CH_3 \rightleftarrows CH_3{-}\underset{+}{\overset{CH_3}{\overset{|}{C}}}{-}CH_2{-}CH_2{-}CH_3$$

Proton elimination:

$$CH_3{-}\underset{+}{\overset{CH_3}{\overset{|}{C}}}{-}CH_2{-}CH_2{-}CH_3 \rightleftarrows CH_3{-}\overset{CH_3}{\overset{|}{C}}{=}CH{-}CH_2{-}CH_3 + H^+$$

Hydrogenation:

$$CH_3{-}\overset{CH_3}{\overset{|}{C}}{=}CH{-}CH_2{-}CH_3 + H_2 \rightleftarrows CH_3{-}\overset{CH_3}{\overset{|}{C}H}{-}CH_2{-}CH_2{-}CH_3$$

An increase in temperature decreases the amount of doubly-branched isomers. At all temperatures and carbon numbers, the equilibrium favors isoparaffins over *n*-paraffins.[24]

Hydrocracking. Hydrocracking is essentially a cracking reaction whereby higher molecular weight hydrocarbons pyrolyze to lower molecular weight paraffins and olefins in the presence of hydrogen. The hydrogen saturates the olefins formed during the cracking process. Since the hydrocarbon molecules in naphtha are relatively low molecular weight, hydrocracking forms molecules boiling in the gas range. Thus hydrocracking competes with the aromatization reactions by eliminating hydrocarbons that can be aromatized.

Hydrocracking is a hydrogen-consuming reaction that leads to higher gas production and lower yields than thermal cracking. The following is a representative hydrocracking reaction.

$$R{-}\overset{CH_3}{\overset{|}{C}H}{-}CH_2{-}CH_3 + H_2 \longrightarrow RH + CH_3{-}CH_2{-}CH_2{-}CH_3$$

TABLE 3-11—Reforming of Cyclohexane and Methylcyclohexane to Benzene Over a Platinum Catalyst[23]

Process conditions	Methyl-cyclopentane		Cyclohexane		
Pressure, psig	100	100	100	100	100
LHSV, $hr.^{-1}$	4	2	4	2	4
Temperature, °F	925	900	800	750	750
Percent conversion	60.5	85.4	85	68	65
Selectivity to benzene	61.7	74	97	95	92

TABLE 3-12—Reforming of *n*-Hexane to Benzene Over a Platinum Catalyst[23]

LHSV	Temp., °F	% Conversion	Selectivity to benzene	Selectivity to isohexane
2	885	80.2	16.6	58
2	932	86.8	24.1	36.9
2	977	90.4	27.4	23.4

Dealkylation, like hydrocracking, is a hydrogen-consuming reaction. This reaction does not take place during reforming to an appreciable extent. At higher hydrogen partial pressure, dealkylation is more favorable. The following represents a typical dealkylation reaction.

$$C_6H_5R + H_2 \rightarrow C_6H_6 + RH$$

Process Variables. Product distribution in a catalytic reforming process not only depends upon the constituents of the feed and the type of catalyst used but also on process variables such as reactor temperature, pressure and space velocity of the feed.

All the reaction rates increase with increase in *temperature*. The dehydrogenation of naphthenes to aromatics, however, is an endothermic reaction; thus, an increase in reactor temperature preferentially increases the quantity of aromatics at equilibrium. A side effect of the increase in temperature is the increase in hydrocracking and carbon disposition on the catalyst. More regeneration cycles are then needed because carbon deposits deactivate the catalyst by decreasing its porosity.

The yield of aromatics is markedly affected by *pressure*. The lower the pressure, the higher the yield, when all other variables remain unchanged.[26] The dehydrogenation reaction produces both aromatic hydrocarbons and hydrogen with a fourfold increase in gas volume.

$$C_6H_{12} \rightleftarrows C_6H_6 + 3H_2$$

The LeChatelier principle, which can be used to predict the direction of shift in an equilibrium system, states "if an outside stress is applied to a system at equilibrium, the system will shift in the direction which will tend to counteract the outside stress." However, low pressure operation shortens the catalyst's life. The published data show that the effect of lowering pressure is greatest when operating under the most severe conditions when the yields are lowest. The effect of pressure tends to be linear at high octane levels where it has the greatest commercial significance.[27] Hydrocracking generally increases when the hydrogen partial pressure increases. The same effect is noted for dealkylation.

The LeChatelier principle can be used to explain why an endothermic reaction involved in an equilibrium system such as the dehydrogenation of naphthenes is favored by an increase in temperature.

Reactions taking place in a reactor not only depend upon the temperature and pressure but also on the time the hydrocarbons are in contact with the catalyst. The *space velocity*, expressed as the volume of hydrocarbon feed per hour per volume of the catalyst (**LHSV**), is an important variable. Because naphthenes are aromatized more easily than paraffins, feeds which are high in naphthenes can be used with a higher space velocity than feeds high in paraffins. Space velocities range between 2 and 4.

Table 3-13 shows the effect of changing process variables on the product distribution.[28]

Hydrodealkylation. This process may be used with catalytic reforming to increase benzene production. Although toluene, the xylenes, and the C_9+ aromatics are valuable octane boosters for the gasoline pool, they are not as valuable as benzene for chemical use. These aromatics may be charged to a hydrodealkylation unit to produce more benzene. The main reaction is hydrodealkylation. With toluene, one mole of hydrogen is consumed with the production of one mole of benzene and one mole of methane.

$$C_6H_5CH_3 + H_2 \rightarrow C_6H_6 + CH_4$$

With higher aromatics, more hydrogen is consumed per mole of benzene produced and methane production is increased. The overall hydrogen consumption depends upon the mole fraction of the different aromatics in the feed. The unconverted toluene and higher aromatics are recycled through the catalyst with the fresh feed. Fig. 3-8 shows a flow diagram for the hydrodealkylation process (Detol) of Air Products and Chemicals, Inc.[25]

Catalytic Cracking. Fluid catalytic cracking (**FCC**) is characterized by a remarkable versatility and flexibility. Various product yields are maximized through manipulation of numerous process variables. This accounts for its dominant position to gasoline production from a barrel of crude oil.

The main advantage of catalytic cracking over thermal cracking is the increased gasoline production and the formation of hydrocarbon components having high antiknock properties. Also less undesirable olefinic hydrocarbons are formed. These olefins have a tendency to form gum by polymerization. Another advantage of catalytic cracking processes is the production of C_3 and C_4 hydrocarbons for LPG uses. Catalytic cracking also produces less methane and C_2 hydrocarbons than thermal cracking. These differences are due mainly to the presence of an acidic catalyst which promotes carbonium ion intermediates that are more selective than the free radicals formed in the thermal cracking reaction. A *carbonium ion* (carbon cation) is a group of atoms that contains a carbon atom bearing only six electrons and a positive charge. A *free radical* is an atom or group of atoms possessing an odd (unpaired) electron. Catalytic cracking also involves *hydride ions* and *methide ions*.

Carbonium ion	Methide ion	Hydride ion	Free radical
$R:\overset{R}{\underset{R}{\ddot{\underset{..}{C}}}}{}^{+}$	$H:\overset{H}{\underset{H}{\ddot{\underset{..}{C}}}}:^{-}$	$H:^{-}$	$R:\overset{R}{\underset{R}{\ddot{\underset{..}{C}}}}$

Initially, the feedstocks used for catalytic cracking were light and heavy gas oils. Technological developments in catalysts, catalyst regeneration and treatment have made it possible for catalytic cracking units to accept a wide range of feedstocks varying from naphthas to heavy crude residues. Table 3-14[6] lists the analysis of feeds and products.

TABLE 3-13—Effect of Platformer Process Variables on Product Distribution [28]

Feedstock Hydrotreated Kuwait naphtha 180-350° F ASTM 0.731 sp gr 60/60 Charge 5-8 H_2 mol ratio Early catalyst life					
Process conditions	**I**	**II**	**III**	**IV**	**V**
Reactor pressure	High	Intermediate	Intermediate	Low	Intermediate
Catalyst type	UOP type PT-5	UOP type PT-5	UOP type PT-5	UOP type PT-5	UOP type PT-6
LHSV, $hr.^{-1}$	1.0	2.0	2.0	1.5	2.0
Average temperature, °F	975	962	937	966	939
Product yields					
Hydrogen, cu. ft./bbl. of feed	230	655	614	1354	187
$C_1 + C_2$, wt. %	13.1	9.3	6.28	4.71	8.32
$C_3 + C_4$, wt. %	23.1	16.2	14.01	8.28	27.29
C_5^+ wt. %	63.31	73.11	78.45	84.23	63.79
Debutanized product properties					
Sp gr 60/60	0.792	0.788	0.7742	0.8123	0.766
RON clear	99.6	98.0	92.5	100.1	94.9
RON + 3ml TEL/US gal.	104.2	102.7	98.9	....	102.1
Reid vapor pressure, psi	6.1	4.9	3.3	....	5.7
C_6 plus component breakdown**					
C_6 paraffins, LV %	9	25.7	22.9	13.7	25.4
C_7 paraffins, LV %	1.2	9.5	13.3	5.2	4.10
C_8—C_{10} paraffins, LV %	1.8	4.5	10.3	2.2	0.0
C_6 naphthenes, LV %	0.8	1.6	2.1	2.3	2.8
C_7 naphthenes, LV %	1.5	1.9	2.0	2.5	2.4
C_8—C_9 naphthenes, LV %	0.5	0.3	0.6	0.5	0.0
C_6 aromatics, LV %	6.3	4.7	2.8	6.6	5.5
C_7 aromatics, LV %	20.6	13.6	10.0	17.5	18.4
C_8 aromatics, LV %	30.6	18.4	15.2	23.7	22.9
C_9 aromatics, LV %	21.2	14.3	13.6	18.1	14.6
C_{10}^+ aromatics, LV %	6.7	5.5	7.2	7.7	3.9

* Results obtained from pilot plant operation, adapted from two tables
** By UOP Low Voltage Spectrographic Method

These feedstocks are sometimes pretreated to decrease the *metallic* and *asphaltene* content. Hydrotreating, solvent extraction and propane deasphalting are the important treatment processes used. Excessive asphaltene and aromatic content in the feed are precursors for carbon formation on the catalyst surface which substantially reduces its activity and produces gasoline of reduced quality.

It has been reported, however, that distillates with low metal and asphaltene content can be cracked directly without prior treatment.[29] Current cracking catalysts are tolerant to metal poisoning. However, the increasing pressure to process crudes and resids with high metal content has revived interest in the effect of metals on cracking catalysts.

Several reactions take place in a catalytic cracking unit with varying degrees of importance. These include carbon-carbon bond breaking, disproportionation, olefin formation and cyclization. Most of the scientific information about these reactions was obtained from the cracking of pure hydrocarbons. In general, the cracking of aromatics is very slow compared to other hydrocarbons in the feed. Normal paraffins are difficult to crack with the rate of cracking increasing with chain length.[30] Naphthenes crack more read-

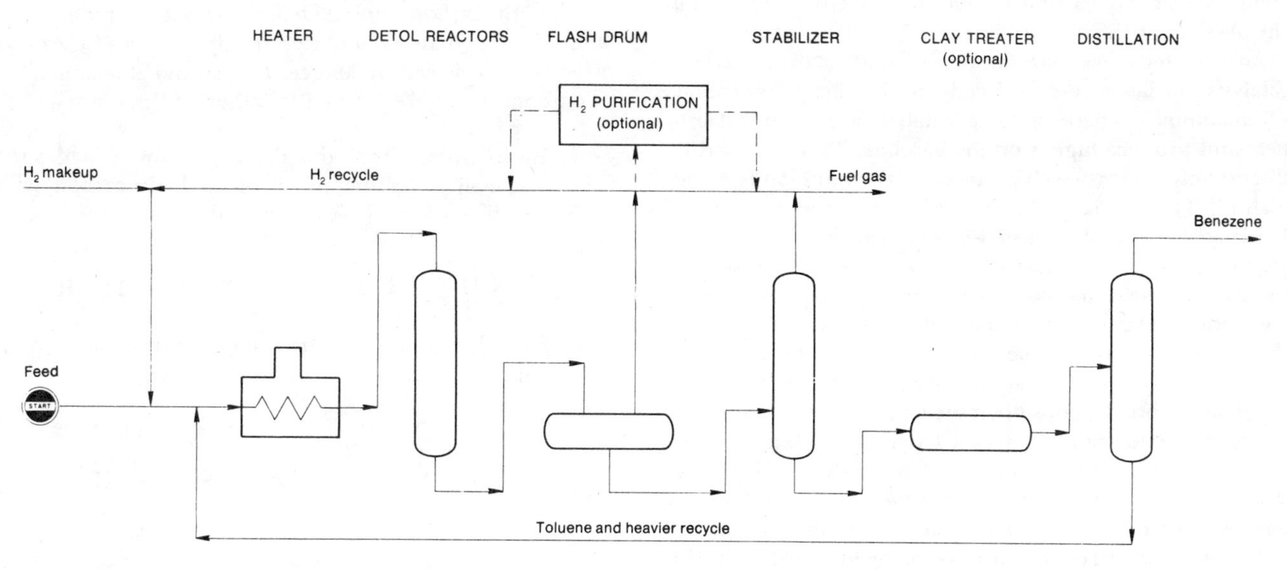

Fig. 3-8—The Detol hydrodealkylation process.[25]

TABLE 3-14—Feed and Product Analyses From a Fluid Catalytic Cracking Operation Using Two Types of Feed[6]

	Kuwait (virgin gas oil)		Kuwait (375° F + HDS resid)	
Feed inspections				
Gravity	22.9		26.0	
Sulfur, wt. %	2.90		0.10	
Rams, carbon wt. %	0.3		2.2	
Boiling range, ASTM °F				
10%	640		......	
50%	815		832	
70%	880		963	
90%	975		......	
Vol. % crude	30		52.3	
Product yields				
Conversion, vol. %	78		82.4	
Gasoline: C_5 — 430° F TBP	56.8		62.5	
Total C_4	18.5		20.4	
Butenes	9.0		10.4	
iC_4/nC_4 ratio	3.8		3.3	
Total C_3	12.8		13.0	
Propene	8.8		8.7	
Light gas oil	16.5		13.8	
Decanted oil	5.5		2.0	
C_1—C_2, wt. %	4.5		3.1	
Ethylene	0.9		0.8	
Coke, wt. %	6.4		6.8	
Product inspections				
Gasolines	Full range		Full range	
Aromatics, vol. %	29.0		31.0	
Olefins, vol. %	30.0		28.0	
Sulfur, wt. %	0.29		......	
Research Octane				
Clear	92.5		94.0	
+ 3g TEL	98.0		100.5	
Motor Octane				
Clear	81.0		81.5	
+ 3g TEL	85.5		88.0	
Gas oils	Light	Decant	Light	Decant
°API	15.0	−2.0	17.0	−1.0
Sulfur, wt. %	4.4	7.0	0.25	0.8

TABLE 3-15—Commercial Fluid Cracking of 25.6° API Mixed Canadian-Northwest U.S. Gas Oil[34]

Equilibrium catlyst in plant	Synthetic high alumina	50% synthetic high alumina 50% zeolite
Conversion, vol. %	65.6	73.0
Coke, wt. % FF	7.3	7.4
H_2 + C_1 + C_2, wt. % FF	3.0	2.9
C_3, vol. % FF**	6.2	5.5
C_3, vol. % FF	2.2	2.2
i-C_4, vol. % FF	4.8	6.6
n-C_4, vol. % FF	1.1	1.5
C_4, vol. % FF**	6.7	7.8
C_5 — 430° F gasoline, vol. % FF	50.5	55.4
Light cycle oil, vol. % FF	24.0	16.5
Slurry oil, vol. % FF	10.5	10.5
C_5 — 430° F gasoline		
Octanes, F-1	92.4	91.2
V-1 + 3cc TEL	97.3	97.3
F-2	77.7	77.8
F-2 ± 3cc TEL	82.2	83.4

* 11.6—11.7 K factor, 1.5—2.0 wt. % S, 893° F cracking temperature, 66% recycle on fresh feed, constant fresh catalyst addition rate of 0.09 lbs./bbl. FF
** Olefins.

ily than paraffins, and olefinic compounds more readily than naphthenes. These reactions are catalyzed by an acid catalyst and take place in the reactor section of a fluidized or moving bed system.

Catalysts, in general, reduce the activation energy needed for the chemical reactants to become a part of an activated complex—a complex which would not have been formed in the absence of the catalyst.

Acid-treated clays were the first used as catalytic cracking catalysts but later were replaced with synthetic amorphous silica-aluminas. These synthetic catalysts are more stable to heat and produce higher octane gasoline. This is caused by their strong acidity which promotes carbonium ion intermediates.[31] The strong acidity is attributed to the presence of Lewis acid sites and Brönsted-Lowry acid sites.[32,33]

Current cracking catalysts incorporate zeolites (molecular sieves) with the alumina-silica matrix. This type of catalyst has better selectivity than the older catalysts. Table 3-15 shows the properties of the products obtained by the use of alumina catalysts and an alumina-zeolite catalyst.[34]

Zeolite acidity depends upon the type of cations present in the zeolite. The acidity may vary from 0 to 10 times that of silica-alumina catalysts.[35] Zeolites have activities similar to strong acid catalysts such as phosphoric acid and boron halides, both of which promote carbonium ion formation.

An important structural feature of zeolite catalysts is the presence of small and large windows or holes in the crystal lattice. The walls of these interstices are composed of silica and alumina tetrahedra. Each aluminum atom and each silicon atom is connected by a single valence bond to each of four oxygen atoms.[35,36] Because aluminum has a positive valence of 3, there is a negative charge in the copolymer for every aluminum atom. These negative sites are balanced by metal cations which results in a strong dipole in the catalyst. The Lewis acid arises from the presence of AlO_4 tetrahedra.

Regeneration of cracking catalysts yields carbon monoxide (about 10 percent) in the flue gas which represents both a loss of energy and atmospheric pollution. Two new **FCC** catalysts, **PCZ** (partial-combustion zeolite) and **CCZ** (complete-combustion zeolite), have been proposed to help solve this problem.[37] These catalysts promote the oxidization of the carbon monoxide to carbon dioxide in the catalyst regenerator. The result is improved catalyst regeneration which increases cracking activity and the yields of useful products. This is accomplished by reduction of the carbon deposit on the regenerated catalyst. The added heat to oxidize these carbon deposits is generated through the combustion of carbon monoxide. All this and reduction of atmospheric pollution, too! Recent advances in FCC catalysts have been reviewed by Magee, Ritten, and Rheaume.[38]

Carbonium ions are formed in either of three ways.

1. By a proton from the Brönsted-Lowry acid site in the zeolites adding to an olefin hydrocarbon which was formed by a cracking process.

$$\text{R}-\text{CH}=\text{CH}-\text{R}+\text{H}^{+}\rightleftarrows \text{R}-\overset{+}{\text{C}}\text{H}-\text{CH}_2-\text{R}$$

2. By a Lewis acid abstracting a hydride ion from a paraffin, isoparaffin or naphthene hydrocarbon.

$$-\underset{\underset{\text{O}}{|}}{\overset{\overset{\text{O}}{|}}{\text{Si}}}-\text{O}-\underset{\underset{\text{O}}{|}}{\overset{\overset{\text{O}}{|}}{\text{Al}}}+\text{R}\;\Big|\;:\text{H}\rightleftarrows-\underset{\underset{\text{O}}{|}}{\overset{\overset{\text{O}}{|}}{\text{Si}}}-\text{O}-\underset{\underset{\text{O}}{|}}{\overset{\overset{\text{O}}{|}}{\text{Al}}}:\text{H}^{-}+\text{R}^{+}$$

Lewis acid

3. By a carbonium ion from 1 or 2 abstracting a hydride ion from another hydrocarbon to form another carbonium ion.

$$R_3C{:}H + R'^+ \rightleftharpoons R_3C^+ + R'{:}H$$

Although reactions taking place on the catalyst sites are not fully understood, it is believed that the first reaction is carbon-carbon bond breaking. This takes place by a free radical pathway. The free radical can then produce an olefinic hydrocarbon. The olefinic hydrocarbon can be adsorbed on the zeolite where it adds a proton to form a carbonium ion.

Carbonium ions from normal paraffins or olefins can *isomerize* through either a hydride shift, a methide shift or both.

1, 2-Hydride shift

$$R-CH_2-\overset{+}{C}H-\ddot{C}H_2 \rightleftharpoons R-CH_2-\overset{+}{C}H-CH_3$$

The carbonium ion stability trend is toward the center of the molecule which accounts for the isomerization of *alpha* olefins to internal olefins under carbonium ion conditions. This also accounts for the preponderance of C_3, C_4 and C_5 olefins over ethylene from catalytic cracking.

Methide shift-hydride shift:

$$R-\overset{+}{C}(H)-CH_2-CH_3 \rightarrow R-C(CH_3)(\ddot{H})-\overset{+}{C}H_2 \rightarrow R-\overset{+}{C}(CH_3)-CH_3$$

The initial shift is followed by a hydride shift to give the more stable tertiary carbonium ion. This type of reaction accounts for the conversion of normal paraffins to branched paraffins.

Many reactions take place starting from a carbonium ion in addition to hydride and methide shifts. Among these are the important *beta fission* and *cyclization*.

Carbon bond fission (beta fission) is the cracking reaction of catalytic cracking.

$$R-CH_2-CH_2-\overset{+}{C}H-CH_2-CH_2-R' \nearrow R-\overset{+}{C}H_2 + R'-CH_2-CH_2-CH=CH_2$$

$$R-CH_2-CH_2-\overset{+}{C}H-CH_2-CH_2-R' \searrow R-CH_2-CH_2-CH=CH_2 + R'-\overset{+}{C}H_2$$

The carbon-carbon beta fission takes place on either side of the carbonium ion with the smallest fragment usually containing at least three carbon atoms. The new primary carbonium ion rearranges before it can undergo beta fission, which is another reason for the low yield of ethylene from catalytic cracking. The butyl carbonium ion gives propylene and a methylcarbonium ion by beta fission.

$$CH_3-CH_2-\overset{+}{C}H-CH_3 \rightarrow CH_2=CH-CH_3 + \overset{+}{C}H_3$$

An isopropyl carbonium ion, $CH_3-\overset{+}{C}H-CH_3$, has nowhere to go by beta fission.

Cyclization can take place by carbonium ion addition to a carbon-carbon double bond in the same molecule.

$$R-\overset{+}{C}H-CH_2-CH_2-CH_2-CH=CH_2 \rightarrow \text{cyclo}[R-CH-CH_2-CH_2-CH_2-\overset{+}{C}H-CH_2]$$

Cyclic olefins and aromatics can be produced by a two-step hydrogen transfer sequence. The first is a proton transfer and the second is a hydride transfer.

$$\text{cyclo}[R'-\overset{+}{C}(CH_2)_5] + R-CH=CH-R \xrightarrow{H^+} \text{cyclo}[R-C=CH-(CH_2)_4] + R-CH_2-\overset{+}{C}H-R$$

$$\text{cyclo}[R-C=CH-(CH_2)_4] + R-CH_2-\overset{+}{C}H-R \xrightarrow{H^-} \text{cyclo}[R-C=CH-\overset{+}{C}H-(CH_2)_3] + R-CH_2-CH_2-R$$

A continuation of this sequence converts the cyclic carbonium ion to an aromatic hydrocarbon.

Process. Either fluidized bed or moving bed reactors are used for catalytic cracking. The catalyst in both processes are similar in their chemical composition but differ in their shape.

In the *fluidized bed process*, **FCC**, (riser cracking), the catalyst is in the form of a powder with an average particle size of 60 microns. The catalyst size is important since it acts as a liquid with the reacting hydrocarbon mixture. The preheated feed enters the reactor section with hot regenerated catalyst through one or more risers where cracking takes place. Products from the reactor section pass to a fractionator for separation into product streams. Fig. 3-9 is a flow diagram for a Kellogg fluid catalyst cracking process.[39]

Fluid catalytic cracking produces unsaturates, especially in the light hydrocarbon range, C_3-C_5, which are used as petrochemical feedstocks and for alkylate production. In general, at constant conversion, low catalyst activity and high

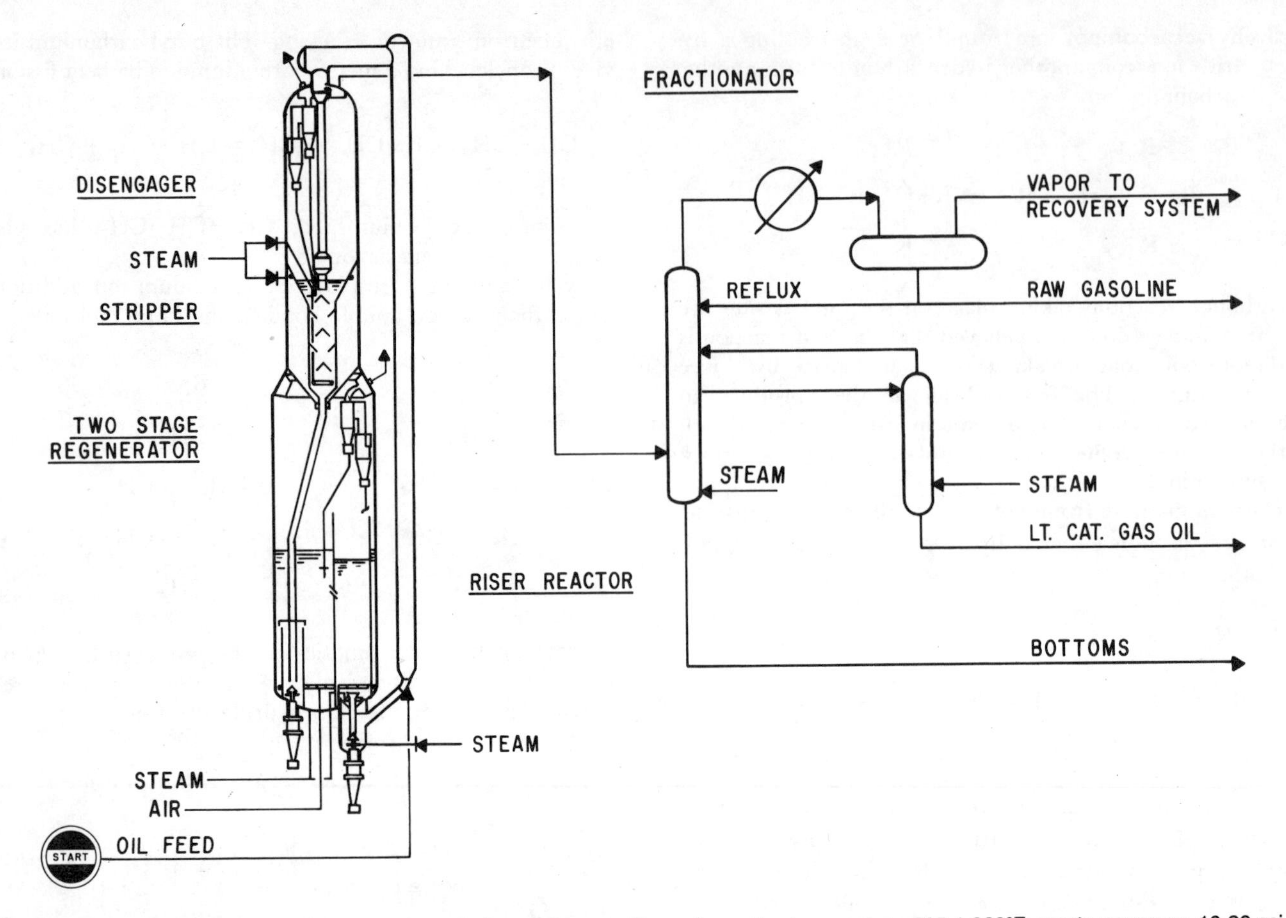

Fig. 3-9—Fluid catalytic cracking, Orthoflow with typical operating conditions of: reaction temperature, 885-1,020°F; reactor pressure, 10-30 psig; regeneration temperature, 1,250-1,400°F; regeneration pressure 15-35 psig.[39]

reactor temperatures maximize gasoline octane and light olefins. Gasoline and light-cycle oil selectivity is maximized with high activity catalysts and lower temperatures.

In the *moving bed process* the preheated feed meets the hot catalyst which is in the form of beds that descend by gravity to the regeneration zone. Various process additives such as barytes (barite, $BaSO_4$), or a mixture of magnesium oxide, calcium carbonate and calcium phosphate are introduced with the catalyst in the moving bed. They form a coating on the catalyst surface that reduces metal erosion and catalyst attrition.[40] As in fluidized bed cracking, conversion of aromatics is low and hydrogenation may be required to produce jet fuels.

Some typical yields for hydrocracking and catalytic cracking are given in Table 3-16.[41]

Resid Cat-Cracking. The need to get more light products out of a barrel of crude coupled with the expected change in patterns of gasoline demand prompted refiners to look for new processes. Resid cat-cracking is one, but until recently it has been limited because of the high cost of controlling metal contamination of the catalyst. Metal contamination of catalysts creates problems such as an increase of coke, hydrogen, and methane production.[42] This reduces the liquid yield used for high conversion gasoline production. Hydrodesulfurization of resids with high metal and asphaltene content does not reduce the catalyst contamination, as the heavy metals and asphaltenes concentrate and thus reduce the desulfurization catalyst life. Different approaches by refiners have been used for demetallizing the feed and improving the cat-cracking catalyst life. Gas oils containing 1-2 parts per million of nickel plus vanadium are currently processed in fluid cat-cracking units.

TABLE 3-16—Cracking Processes Compared[41]

Yield, %	Catalytic cracking	Hydrocracking
C_2 & Lt (FOE)	4.0	2.0
C_3	13.4	2.0
C_4	20.2	3.3
Naphtha	61.3	58.2
Distillate	12.7	52.7
Residue	4.6	....

Basis: Feedstock of a typical vacuum gas oil having 600-1,000°F ASTM distillation. Hydrocracker operated to yield naphtha and distillate.

Demetallization. Fig. 3-10 shows the Demet III flow scheme used in connection with cat-cracking.[43] In this process, part of the catalyst stream is continuously withdrawn from the FCC regenerator, processed in the demetallization unit, and then returned to the FCC unit. No catalyst addition is required. Both chemical and physical treatment are used to free the catalyst of metals. An ion exchange resin is used for removing the metals. The results of a catalyst replacement rate using feeds with different metal content in the Demet III process are shown in Fig. 3-11.[43]

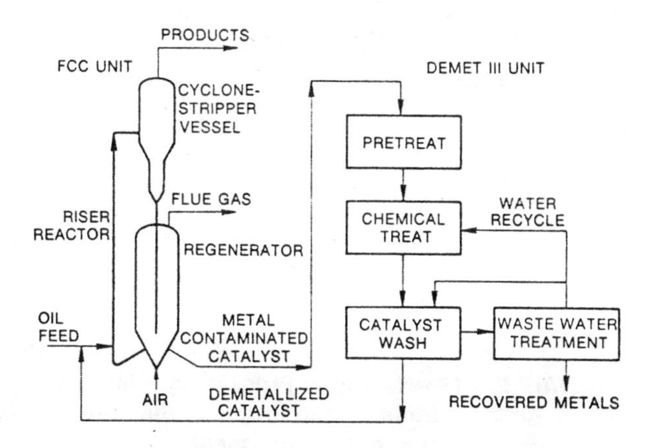

Fig. 3-10—Demet III flow scheme for catalytic cracking.[43]

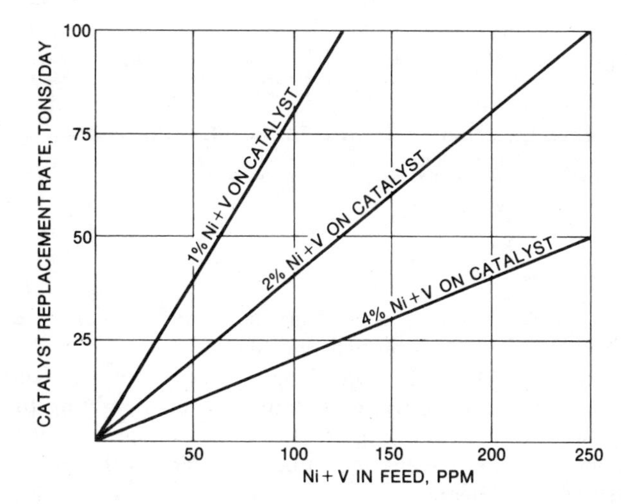

Fig. 3-11—Catalyst replaced to reduce metals content.[43]

Vacuum resids can be demetallized by using a selective organic solvent which separates heavy resids into demetallized oil with low metal and asphaltene content and asphalt with high metal content. Depending upon the metal level, this oil could be hydrodesulfurized and then blended with gas oil for catalytic cracking. Fig. 3-12 shows the Demex residuum treatment.[44] Demetallized oils which contain low to moderate amounts of metal (up to 200 ppm nickel plus vanadium) can be processed by direct hydrocatalysis.[45] The solvent used in this process is less selective than pentane. It has been reported that an Iranian vacuum resid containing 500 ppm metal can be reduced to 10 ppm by this process with yields of 30%, and to 60 ppm with yields of 65%.[46] Other solvent extraction processes use propane and pentane. All solvent extraction processes suffer the same limitation—the higher the demetallization the lower the yield.[45] The results of bench scale solvent extraction of different resids are shown in Table 3-17.[47]

Phillips Petroleum Co. developed a new method, **Metal Passination**, to deactivate harmful metals that are deposited on cat-cracking catalysts and that are responsible for decreased liquid yields and increased hydrogen and coke yields. The treating agent, which is an oil soluble compound containing antimony, deposits contaminant metals on the catalyst and deactivates them without deactivating the catalyst. The most important metals that catalyze coke and hydrogen formations are nickel, vanadium, and iron. It appears that the percent antimony is important. For example, when 0.5 wt.% antimony was deposited, gasoline yield increased 14%, coke yield decreased 22%, and hydrogen yield decreased 57%, as shown in Fig. 3-13.[48]

Filtrol F-87 is a product of Davidson and is claimed to withstand up to 5000 ppm nickel and vanadium without sacrificing performance. Conventional gas oil fluid cat-cracking is generally conducted with a 1000 ppm nickel plus vanadium content, although it has been reported that Phillips is running a heavy-oil cracking unit with a reduced crude

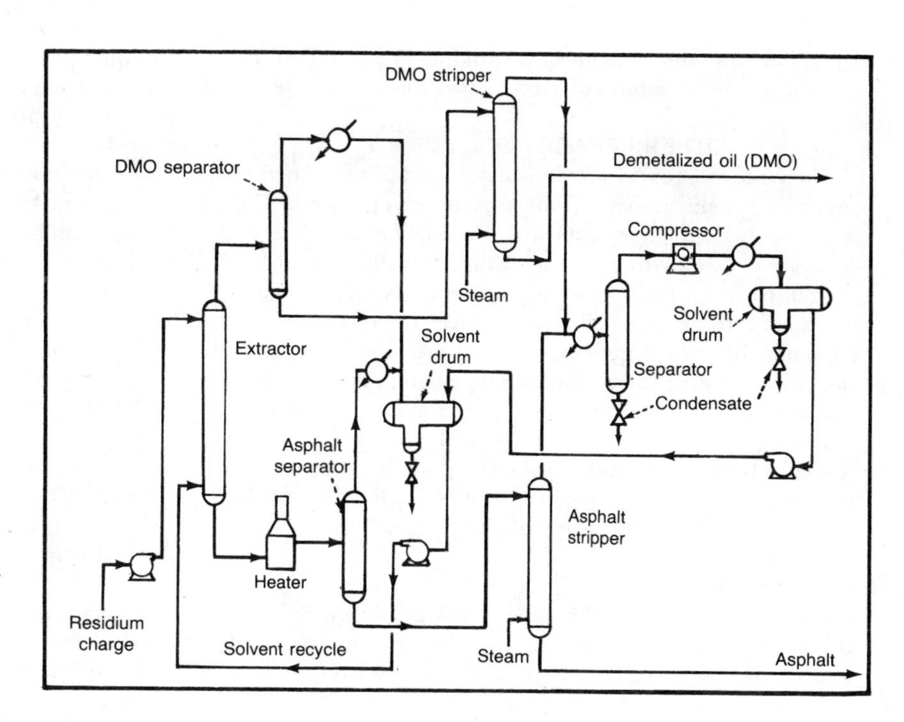

Fig. 3-12—Demex residuum-treatment route typically cuts metals content by 85%.[44]

TABLE 3-17—Summary of Results of Bench Scale ROSE Pilot Operations[47]

	East Texas Residuum				Empire Residuum				Gibson Residuum				Arabian Light Residuum			
Yields, wt. %																
Asphaltenes		15.4				7.5				1.7				22.8		
Resins		7.0				13.1				11.7				8.7		
Oils		77.6				79.4				86.6				68.5		
Distribution of metals, nitrogen, and sulfur.	**Feed**	**Asphal-tenes**	**Resins**	**Oils**	**Feed**	**Asphal-tenes**	**Resins**	**Oils**	**Feed**	**Asphal-tenes**	**Resins**	**Oils**	**Feed**	**Asphal-tenes**	**Resins**	**Oils**
Vanadium, ppm	11	63	17	2.3	4	31	15	1.7	9	68	40	4.5	68	230	80	7.5
Nickel, ppm	14	57	22	3.5	15	38	30	5.4	14	78	48	4.5	13	47	16	1.6
Nitrogen wt%	0.5	0.8	0.7	0.4	0.5	0.8	0.7	0.5	0.3	0.6	0.6	0.2	0.5	0.8	0.5	0.4
Sulfur wt%	1.1	2.2	1.3	0.6	0.8	1.3	1.1	0.7	1.2	1.9	1.7	0.9	3.9	5.3	4.6	3.1

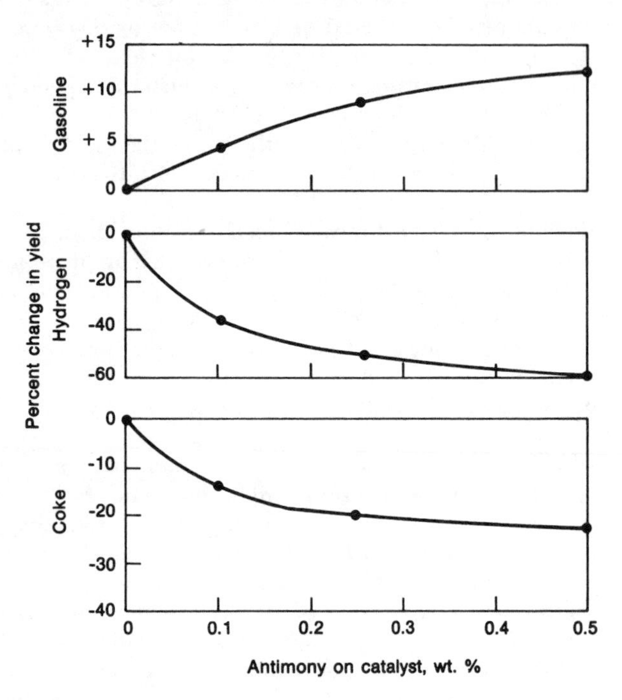

Fig. 3-13—Antimony passivates metals on FCC catalyst. These data for 75 vol % conversion of West Texas topped crude on equilibrium catalyst from heavy oil FCC unit at 950°F.[48]

containing 23 ppm nickel plus vanadium. The metal on the catalyst maintains a level of 1.1% before being discarded.

OTHER REFINERY PROCESSES

There are various refinery processes not directly related to producing more gasoline from a barrel of crude oil. The most widely used one is catalytic reforming. Other processes of this type are hydrocracking, hydrogen refining and hydrogen treating. Hydrocracking both increases the amount of gasoline from crude oil and improves the quality of the gasoline. Other hydrogen processes improve refinery feedstocks by reducing their sulfur, nitrogen and metal content.

Isomerization, alkylation and dimerization are gasoline improving processes. They usually involve specific chemical reactions between essentially pure compounds. They do not substantially increase the quantity of gasoline but do contribute substantially to the quality of the gasoline.

Hydrogen Processing

Hydrogen processing is a general term covering refinery processes with one thing in common—the use of hydrogen to improve refinery streams and products. The separation into three broad headings is arbitrary and observed more in the breach than in the observance (Table 3-18).

TABLE 3-18—Hydrogen Processing—Various Refinery Streams and Products that are Hydrogen Processed*

Hydrocracking	Hydrogen Refining (Hydrorefining) (Hydrofining)	Hydrogen Treating
Distillate	Resid desulfurization	Reformer pretreat
Resid upgrading	Resid visbreaking	Naphtha desulfurization
Lube manufacture	Heavy gas-oil desulfurization	Naphtha saturation
Others	Cat cracking (and cycle pretreat)	Others
	Middle distillate	
	Others	

*There is no sharp line of demarcation between hydrocracking and hydrogen refining or between hydrogen refining and hydrotreating.

Hydrocracking. "Hydrocracking processes are inherently among the most versatile modern petroleum refining processes . . . They have gained widespread acceptance—attaining a permanent and important place in the modern petroleum scheme."

This is not surprising because hydrocracking processes have great flexibility and produce a wide range of products in high yields. Commercially attractive conditions of pressure, temperature and run length are made possible by the dual function catalysts used. As a plus factor, five barrels of product are produced from four barrels of feedstock.

Hydrocracking is specially adapted to the processing of low value stocks such as those that are not suitable for a catalytic cracking or reforming unit because of high metal, nitrogen and/or sulfur content. The process is also suitable for high aromatic feeds which cannot be processed easily by conventional catalytic cracking processes. From these low-value feedstocks come a wide variety of products which include gasoline, kerosine, middle distillate fuels, lubricating oils, catalytic cracker feedstocks, petrochemical feedstocks and **LPG**. These products contain low concentrations of photochemical-active hydrocarbons such as olefins and they are very low in sulfur and nitrogen compounds.

Table 3-19 contains feed and product analyses from hydrocracking for maximum of either gasoline or jet fuel.[49] Typical conditions for hydrocracking are:

Temperature:	500-900°F
Pressure:	500-3,000 psig
Space velocity:	0.5-1.0 v/h/v

TABLE 3-19—Feedstocks and Product Analyses from Hydrotreating Unit (one to maximize gasoline production and the other to maximize jet fuel)*[43]

Inspection	Maximum gasoline	Maximum Jet fuel
Feedstock		
Distillation range, °F	660-1020	570-900
Sulfur content, wt. %	1.35	3.7
Pour point, °F	90	86
Product	Gasoline	Jet fuel
Yield, wt. % on feed	29.3	64.3
Specific gravity	0.760	0.801
Sulfur content, ppm	10	1000
Research Octane number clear	57	....
Freezing point, °F	....	76
Stage number	One	One
Liquid recycle	With	With
Hydrogen consumption, scf/bbl.	1400	2100

*Charge: Straight run, thermally or catalytically cracked gas oil deasphalted vacuum resid.

Fig. 3-14 represents a single-stage Unicracking-JHC process in which feedstocks are passed over a hydrotreating catalyst in the first reactor where organic nitrogen and sulfur compounds are converted to ammonia and hydrogen sulfide. The effluent from this reactor is hydrocracked in the second reactor. The conversion is between 40 and 70 percent to products boiling below a given temperature. The unconverted feedstock, after appropriate separation, is recycled to the hydrocracking reactor.[50]

The cracking of vacuum gas oil (660°-1,020°F) to middle distillates (300°-700°F) has been investigated extensively.[51] Different catalysts and processing conditions are required compared to those used for hydrocracking to gasoline. "Precious metal-zeolite catalysts, while intrinsically more active, did not come out well when product-yield distribution was taken into account." The composition of the catalyst that *did* come out well was not reported.

Most commercial operations use a single stage for maximum middle-distillate optimization despite the possibility of having more than one reactor vessel in series for capacity considerations. Two modes of operation are possible—a once-through mode of operation and a total conversion of the fractionator bottom through recycling. In the once-through mode, low sulfur fuel oil is produced and the fractionator bottoms are not recycled. Table 3-20 shows the results obtained from a once-through operation with 100% conversion.[52] In a two-stage operation, feedstock is first desulfurized and denitrified. Partial conversion takes place simultaneously. After fractionation, the residue gives the refiner a greater flexibility to work in maximum diesel or maximum jet fuel production with minimum variation in feedstock capacity and molecular weight range.[53] Table 3-21 shows the results of a two-stage hydrocracking process of Arabian light-vacuum distillate (350-550°C).[53]

Hydrocracking (hydrorefining) is also used to produce lubricant base stocks. The French Petroleum Institute hydrorefining process is reported to have an "enormous" advantage compared with solvent extraction processes.[54] The hydroprocessing route is able to produce new types of oils having entirely modified structural properties and considerably improved viscosimetric properties. A typical **IFP** hydrorefining process is given in Fig. 3-15.

A wide range of catalysts are available for the various hydrocracking processes. These dual function catalysts provide a high surface area cracking site and a hydrogen-

TABLE 3-20—Yield and Quality Comparison—100% Conversion vs. Once-Through Operation[52]

Product yields	100% conversion bbl/100 bbl feed	Once-through bbl/150 bbl feed
C_7-149°C (300°F)	32.8	31.0
149-288°C (300-550°F)	66.9	69.4
C_7-288°C (550°F)	99.7	100.4
288°C + (550°F +)	0.0	48.5
C_7-149°C product		
Gravity, °API	60.0	58.3
Aromatics, LV-%	4.1	5.8
149-288°C product		
Gravity, °API	44.7	42.8
Aromatics, LV-%	14.2	18.2
Smoke point, mm	24.8	20.8

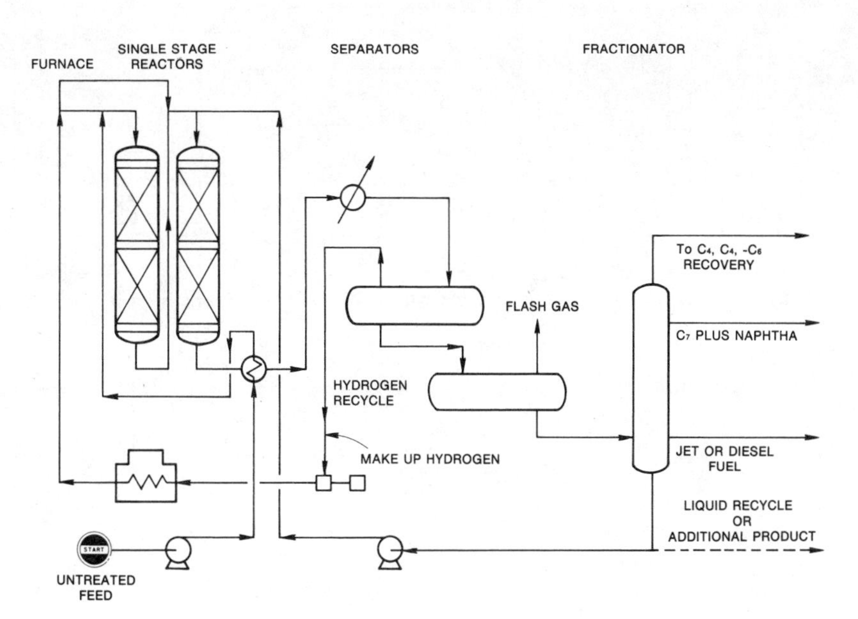

Fig. 3-14—Single-stage Unicracking-JHC unit.[50]

TABLE 3-21—Two-Stage Hydrocracking of the Arabian Light Vacuum Distillate 350-550[53]

Characteristics		First stage	Maximum jet fuel	Maximum diesel oil
YIELD, wt % feed				
$H_2S + NH_3$		2.20	2.20	2.20
$C_1 + C_2$		0.50	0.60	0.58
$C_3 + C_4$		1.40	8.77	3.40
Light gasoline		2.40	14.09	7.48
Heavy gasoline		3.20	16.92	13.50
Jet fuel		16.50	60.52	....
Diesel oil		25.60	....	75.36
Residue		50.00	....	....
Total		101.80	103.10	102.52
H_2 consumption, norm cu. m./ton		201	347	282
PRODUCTS				
Jet fuel				
Specific gravity		....	0.795	....
Freezing point, °C		....	–60°C	....
Smoke point, mm		....	25	....
Sulfur, ppm		....	5	....
Diesel oil				
Specific gravity		....	....	0.820
Cetane number		....	....	58
Pour point, °C		....	....	–30
Sulfur, ppm		....	....	10
Residue	Feedstock			
Specific gravity	0.907	0.840	....	....
Viscosity, cS at 210°F	5.50	4.25	....	....
Sulfur, wt %	2.00	<0.001	....	....
Nitrogen, ppm	800	1	....	....

ation-dehydrogenation site. The hydrogenation-dehyrogenation component is usualy cobalt, nickel, molybdenum, tungsten, vanadium, palladium, platinum or their oxide. Combinations of two or more of these components can be used. The base metals are usually converted from the oxide to the sulfide by pretreatment with hydrogen sulfide or by treatment with sulfur-containing feedstock. Noble-metal-containing catalysts are used, after reduction, either in the metallic form or sulfided form.

The high surface area of molecular sieves, zeolites, incorporated into the catalysts compared to amorphous components results in improved catalyst performance. This includes higher activity for a given process operation, greater tolerance of organic nitrogen and ammonia, lower deactivation rates resulting in longer cycle length and the ability to convert higher boiling point feedstocks.

Probably the most important difference between amorphous (silica-alumina, silica-magnesia, etc.) and molecular sieve catalysts is the greater acidity of the latter.[55] Both types of catalyst are believed to have the same type and strength of acid sites but the molecular sieve catalysts contain about 10 to 20 times as many sites as the amorphous catalysts. It is probable that this greater population of acid sites gives rise to the greater ammonia resistance of molecular sieve catalysts. Another possibility is that the concentrations of reactants within the zeolite pores near the active sites are much greater than for amorphous catalysts.

The choice of process and catalyst depends upon the required products and on feed specifications. Catalysts with strong acidic activity promote isomerization and cracking reactions leading to high iso/normal ratios.[56] This is especially needed when maximizing gasoline and **LPG** production. Catalysts with more hydrogenation activity and less acid activity will reduce the isomerization reactions. This is desirable when jet fuels and lubricating oils are produced. Jet fuels produced by hydrocracking have less naphthenes than those produced from catalytic cracking plus severe hydrogenation.

An important factor in hydrocracking is the need for large amounts of hydrogen. Normally, the hydrogen is obtained from the catalytic reformer. In areas where associated gas is abundant, such as Saudi Arabia, hydrogen can be obtained by the cracking of this gas.

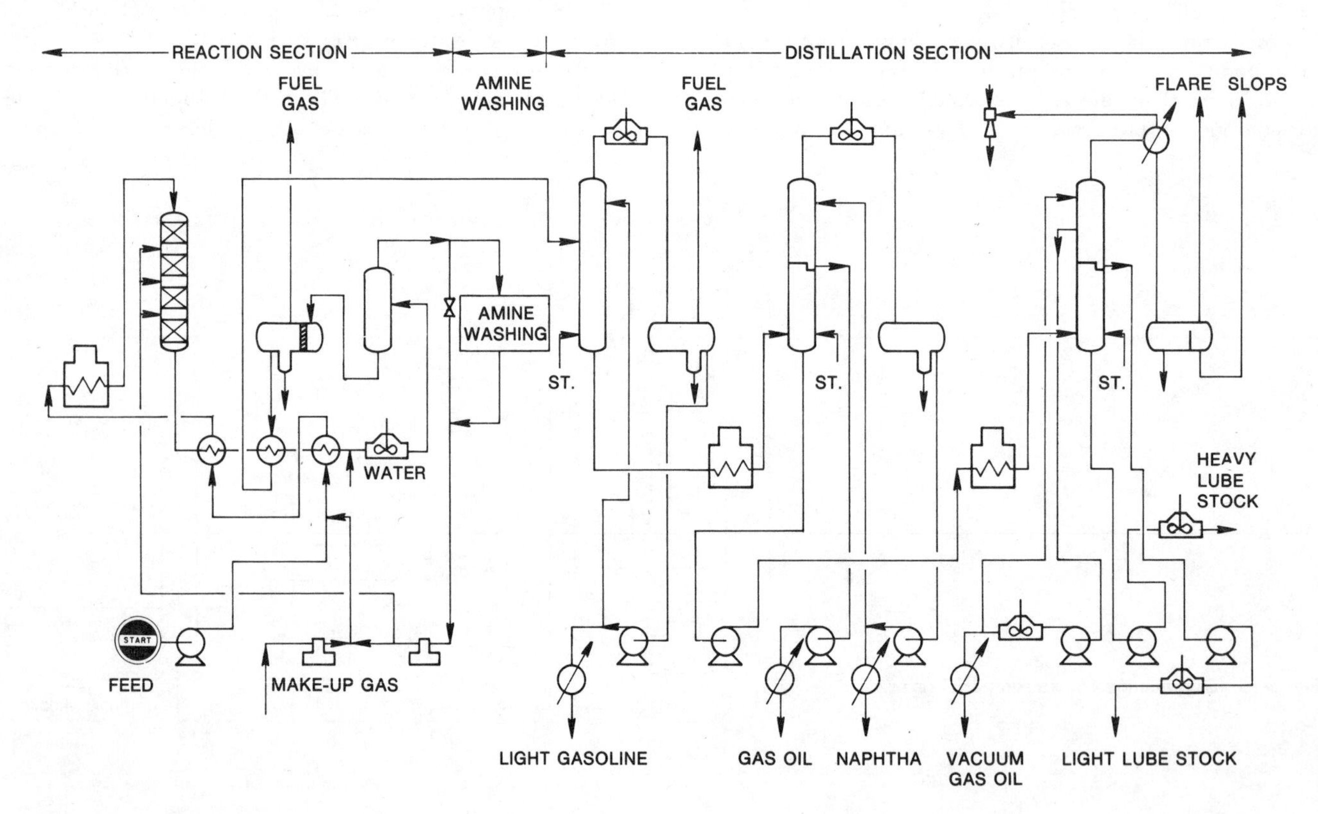

Fig. 3-15—Typical IFP lube hydrorefining process.[51]

Hydrotreating. The refinery of the future will convert crude oil exclusively to gasoline and middle distillates plus petrochemicals, lubricants and special products. Alternate energy sources will replace other refinery products such as the heavier, less valuable, petroleum-based fuels. The petroleum products will be converted to gasoline and middle distillates by *hydrotreating* (hydrofining, hydrodesulfurization) followed by catalytic cracking.

Hydrotreating is a process designed to reduce the sulfur content of atmospheric resid (residuum), vacuum gas oils and vacuum resids where a large fraction of the sulfur is in high-molecular-weight asphaltenic compounds. Two basic hydrotreating routes are used for removal of sulfur from heavy oil—vacuum gas oil desulfurization and direct resid desulfurization.

The process is relatively simple with the choice of desulfurization process depending largely upon the character of the feed and the desired sulfur level of the product (Table 3-22).[57] Feed characteristics of particular importance are sulfur, metals (Ni and V) and asphaltene content. For straight-run diesel feed a temperature of 650°F and a pressure of 700 psig will give a reduction of sulfur from 1 wt. percent to 0.01 wt. percent. Other processes use lower pressures and higher temperatures for desulfurization of straight-run gas oil and similar feeds. Fig. 3-16 represents the Unicracking/HDS process for the hydrodesulfurization of resid.[58] The hydrogen sulfide is subsequently removed by di-isopropylamine, **DIPA**.

Heavy atmospheric gas-oil with a boiling range of 450° to 1,000°F is reported to be hydrotreated at 525 psi and 610°F to effect a sulfur reduction from 1.2 wt. percent sulfur to 0.4 wt. percent sulfur, a 65 percent desulfurization.[59] The catalyst was cobalt-molybdenum. This type of catalyst is favored over older nickel-molybdenum catalysts.

Hydrotreating makes it possible to go deeper into a barrel of crude oil which is an extremely important process in view of the ever increasing cost of crude oil and the inevitable depletion of petroleum reserves and resources.

TABLE 3-22—Low Sulfur Fuel Oil (LSFO) Production from Arabian Light Residuum[57]

Process	VGO*	VRDS**	VGO+ VRDS	RDS***
Feed sulfur, wt. %	2.3	4.1	2.9	2.9
Product sulfur, wt. %	0.1	1.28	0.5	0.5
Product yields				
C_1–C_4 wt. %	0.59	0.56	0.58	0.58
H_2S, NH_3, wt. %	2.44	3.00	2.55	2.55
C_5^+, wt. %	97.51	97.34	97.46	97.67
C_5^+, LV %	100.6	102.0	101.0	101.5
Hydrogen consumption				
scf/bbl.	330	720	450	550
scf/lb. sulfur	47	71	56	69

*Vacuum gas oil hydrotreater
**Vacuum residuum hydrotreater
***Atmospheric residuum desulfurization hydrotreating

Isomerization

Isomerization is a small volume but important refinery process. Normal butane is isomerized to isobutane to be used for the alkylation of isobutylene and other olefins for the production of high octane hydrocarbons such as isooctane (2,2,4-trimethylpentane). The five- and six-carbon fraction (C_5/C_6) of natural gasoline and other refinery streams is isomerized to give an octane-enriched product to blend with low octane gasolines. These isomerates also increase the volatility of the gasoline. Fig. 3-17 shows the effect of temperature on the isomerization of hexanes.[60]

The butane isomerization process was developed early in World War II to produce isobutane as a feedstock for aviation alkylates. A Friedel-Crafts catalyst, $AlCl_3$, was used at temperatures in the range of 100-200° F. At these temperatures a favorable equilibrium is rapidly obtained. Friedel-Crafts catalysts, because of their great reactivity, tend to react indiscriminately with reactants, products and impurities. Reaction conditions must be very carefully controlled.

Fixed bed, dual function catalysts activated by either inorganic or organic chlorides are now the preferred

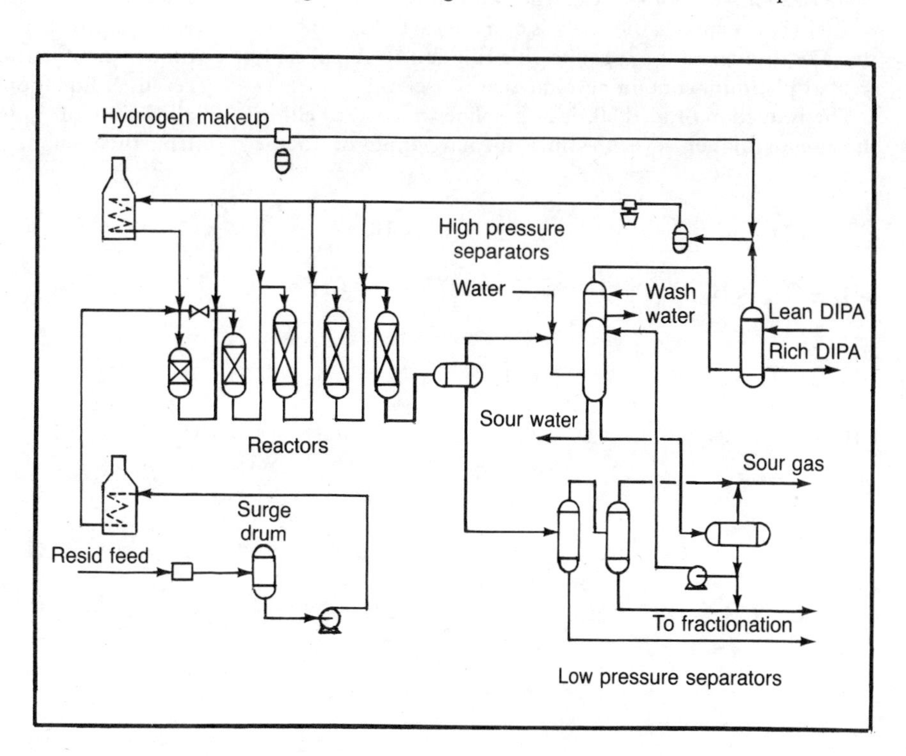

Fig. 3-16—Union Oil Company of California's resid desulfurization process—Unicracking/HDS.[58]

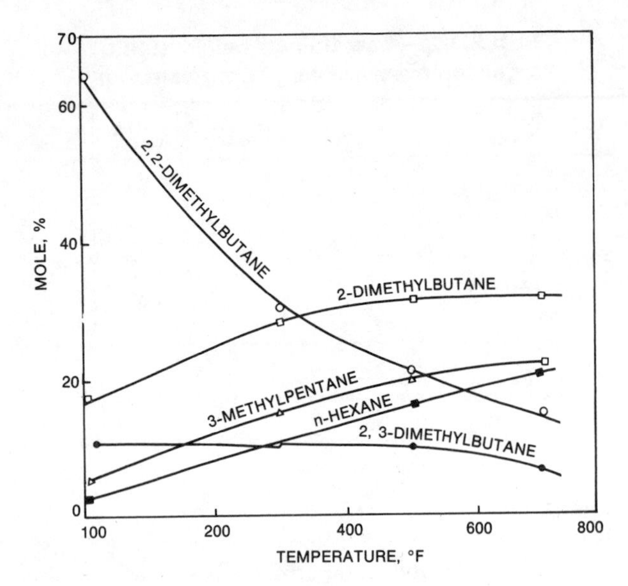

Fig. 3-17—Vapor phase isomer equilibrium for hexanes.[60]

TABLE 3-23—C_5/C_6 Feedstock and Product Analysis from IFP Isomerization Process[62]

	Feed, wt. %	Product, wt. % Based on Feed
H_2	0.3	0
C_{1-3}	....	1.9
C_4	0.4	0.5
Isopentane	21.3	36.4
n-Pentane	26.4	12.1
Cyclopentane	2.4	1.7
2.2-Dimethylbutane	1.7	13.1
2.3-Dimethylbutane	1.0	4.3
2-Methylpentane	9.6	12.7
3-Methylpentane	7.3	7.4
n-Hexane	21.6	4.7
Cyclohexane	3.3	2.7
Methylcyclopentane	2.7	2.7
Benzene	1.8	....
2.2-Dimethylpentane	0.2	0.1
2.4-Dimethylcyclopentane	0.3	
Total	100.3	100.3
Sp. gr.	0.657	0.651
RON	67	83.2
MON	65	81.5

catalysts. A typical catalyst is platinum on a zeolite base. This type of catalyst can be regenerated. These catalysts are isomerization selective and are used at temperatures and pressures in the range of 200-400° F and 200-500 psig in the presence of hydrogen. The isomerization product may go directly to a molecular sieve unit for extraction of unisomerized *n*-paraffins.[61]

The dual function catalysts are strongly acidic and produce carbonium ions by abstraction of a hydride ion, H^-, from the paraffinic hydrocarbon to produce a carbonium ion intermediate. It is also possible that dehydrogenation of the normal paraffin takes place on the platinum site followed by addition of a proton at the Brönsted-Lowery acid site to produce the carbonium ion intermediate. A rearrangement can then take place by a methide, CH_3, shift similar to isomerization reactions in catalytic reforming.

Table 3-23 contains the analysis of a typical C_5/C_6 feed and the **IFP** isomerization product distribution.[62] A specially treated platinum-containing catalyst was used.

The reduction of lead alkyls in gasoline will lead to either the use of higher severities in reforming units or to the addition of other components to upgrade the pool octane number and ensure satisfactory volatility characteristics. The isomerization of C_5/C_6 cuts and/or dimerization of propylene to isohexanes may be the answer.

Alkylation

The term *alkylation* generally applies to the acid catalyzed reaction between isobutane and various light olefins. The product is highly branched paraffin hydrocarbons, *alkylate,* used for blending to improve the octane number of gasoline. Alkylate is the best of all possible motor fuels, having both excellent stability and high octane number. In addition, aklylate has a high heat of combustion per pound, a low vapor pressure and a desirable boiling range.

The alkylation reaction involves the addition of a proton, H^+, to the double bond of an olefin to form a carbonium ion. This ion then adds to another olefin molecule to form a new carbonium ion which abstracts a hydride ion, H^-, from isobutane to become a paraffin hydrocarbon. At the same time a new carbonium ion is produced to continue the reaction. Equations 1, 2, and 3 illustrate this sequence in the alkylation of isobutylene by isobutane to produce 2,2,4-trimethylpentane, "isooctane."

$$\underset{(1)}{CH_3-\overset{\overset{CH_3}{|}}{C}=CH_2} + H^+ \rightarrow CH_3-\overset{\overset{CH_3}{|}}{C^+}-CH_3 \quad (1)$$

$$CH_3-\overset{\overset{CH_3}{|}}{C^+}-CH_3 + CH_3-\overset{\overset{CH_3}{|}}{C}=CH_2 \rightarrow CH_3-\underset{\underset{CH_3}{|}}{\overset{\overset{CH_3}{|}}{C}}-CH_2-\overset{\overset{CH_3}{|}}{C^+}-CH_3 \;\; (2) \quad (2)$$

$$CH_3-\underset{\underset{CH_3}{|}}{\overset{\overset{CH_3}{|}}{C}}-CH_2-\overset{\overset{CH_3}{|}}{C^+}-CH_3 + CH_3-\underset{\underset{H}{|}}{\overset{\overset{CH_3}{|}}{C}}-CH_3 \rightarrow CH_3-\underset{\underset{CH_3}{|}}{\overset{\overset{CH_3}{|}}{C}}-CH_2-\underset{\underset{H}{|}}{\overset{\overset{CH_3}{|}}{C}}-CH_3 + CH_3-\overset{\overset{CH_3}{|}}{C^+}-CH_3 \quad (3)$$

TABLE 3-24—The Ranges of Operating Conditions for H_2SO_4 and HF Alkylation[64]

Process catalysts	H_2SO_4	HF
Temperature °C	2–16	16–52
Isobutane/olefin feed	3–12	3–12
Olefin space velocity, vo/hr./vc	0.1–0.6	...
Olefin contact time, min.	20–30	8–20
Catalysts acidity, wt. %	88–95	80–95
Acid in emulsion, vol. %	40–60	25–80

The initial hydrogen ion is furnished by an acid catalyst, either concentrated sulfuric acid or anhydrous hydrofluoric acid.

Other olefins that are alkylated commercially include propylene, 1-butene, 2-butene and pentenes. The alkylation of propylene by isobutane leads to three products: 2,2-trimethylbutane (60-80%), 2-methylhexane (10-30%) and 2,2,3-trimethylbutane, triptane, (7-11%). 2-Butene produces much higher octane alkylate than 1-butene. Clear research octane number of 2-butene alkylate is 98.5 where 1-butene alkylate is 92.5. Consequently, 1-butene is isomerized to 2-butene before alkylation. The ratio of 2-butene to 1-butene is 7 or 8 to 1, depending upon feed quality and operating conditions.[63]

Both sulfuric acid, H_2SO_4, and hydrofluoric acid, HF, catalyzed alkylations are low temperature processes. Table 3-24 gives the conditions used and the temperature is determined by the olefin being alkylated.[64] Sulfuric acid catalyzed alkylation of propylene, for example, requires a temperature range of 10 to 15°C. Higher temperatures are used with HF alkylation. Pressure has no significant effect provided it is sufficient to maintain the reactants in the liquid phase. The isobutane-olefin ratio has a pronounced effect on the reaction; increasing the ratio up to 12 to 1 improves both alkylate yield and the Octane number of the products for both processes.

Dimerization

The dimerization of propylene to isohexenes provides another process to upgrade the Pool Octane number and ensure satisfactory volatility of gasoline. Dimerization was first used (1935) to dimerize isobutylene to diisobutylene, "isooctene," 2,4,4-trimethyl-1-pentene (80%) and 2,4,4-trimethyl-2-pentene (20%). Both phosphoric acid and sulfuric acid were used as catalysts.

At present, the feedstock is either propylene or a propylene-propane mixture to give isohexenes or propylene-butene mixtures to yield isoheptenes. The main product from propylene alone is 2,3-dimethyl-2-butene. Table 3-25 contains an analysis of a typical propylene dimerization feed and the products from its IFP Dimersol dimerization.[62] The isononenes come from the reaction between a hexene and propylene, *trimerization*.

Phosphoric acid dimerization is reported to produce appreciable amounts of isononenes and isodecenes, *tetramerization*. The IFP process is a once-through operation with a homogeneous catalyst system and mild operating conditions (Fig. 3-18). Fig. 3-19 shows the Chevron Research Corporation's process for the dimerization of propylene to 4-methyl-1-pentene.[65] A dispersed and stabilized potassium metal catalyst in a heavy white oil is used.

TABLE 3-25—Typical Feed and Products from the Dimerization of Propylene[62]

	Vol. %	Total	Wt. %	Total
Feed				
Propylene	71	...	...	...
Propane	29	100	...	...
Products				
LPG				
Propylene	4.2	...	...	...
Propane	34.6	...	...	...
Isohexanes*	61.2	100	...	...
Isohexenes			92.0	...
Isononenes	...	...	6.5	...
Heavier	...	...	1.5	100
ASTM distillation (°F) IBP 133				
10 136				
50 140				
90 160				
95 320				
EP 370				

*"Dimersol isohexenes"

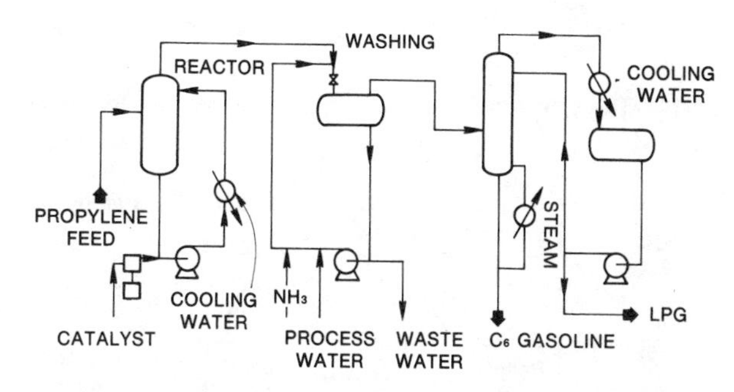

Fig. 3-18—IFP Dimersol process for dimerization of propylene to isohexenes.[62]

Catalytic dimerization is a source of byproduct propylene trimer and tetramer for use in **OXO** processes. Heptenes are produced by the dimerization of propylene and butenes and are also used as **OXO** feedstock.

REFINING FOR THE NEXT 20 YEARS

Axel R. Johnson states that "the petroleum industry is in a state of transition, with the HPI sector appearing to be relegated to a narrowing scope concerned with transportation fuels and petrochemicals."[66] However, he points out that the HPI should build now to have a wider role in fuel production. This will include entry into other energy markets such as utility fuels, gas supply, and industrial fuels. To do this it will be necessary to develop an expanded role for heavy crudes and coal in the refinery raw materials picture. The long term answer to the predicted shortfall in crude oil and natural gas is conversion of heavier fossil materials, particularly coal and shale into gas and liquids.

Several approaches are open to the industry. Gasification of residual oil using either air or oxygen technology is feasible, and the technology can be converted to coal gasification. The same applies to hydrogenation of coal to produce liquid fuels.

Proposed refinery-chemical complexes for 1980, 1990, and 2000 are illustrated in Figs. 3-20 through 3-22.[66] The future inputs for an example complex are given in Table 3-26; the products are given in Table 3-27.

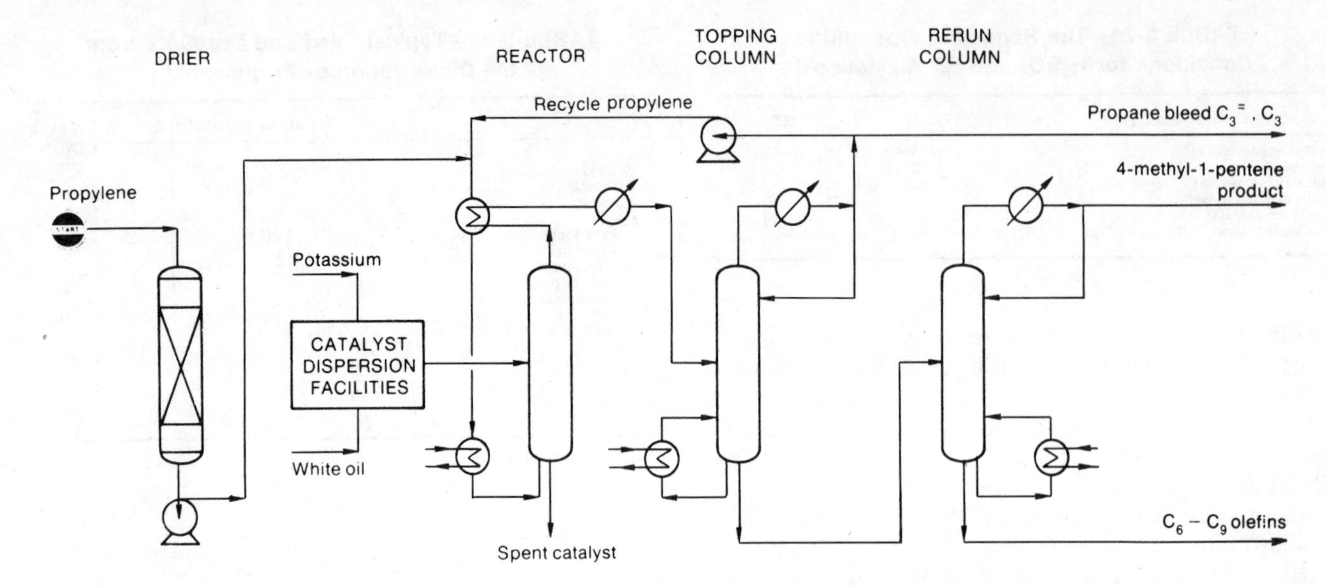

Fig. 3-19—The Chevron Research Corporations process for the dimerization of propylene to 4-methyl-1-pentene.[65]

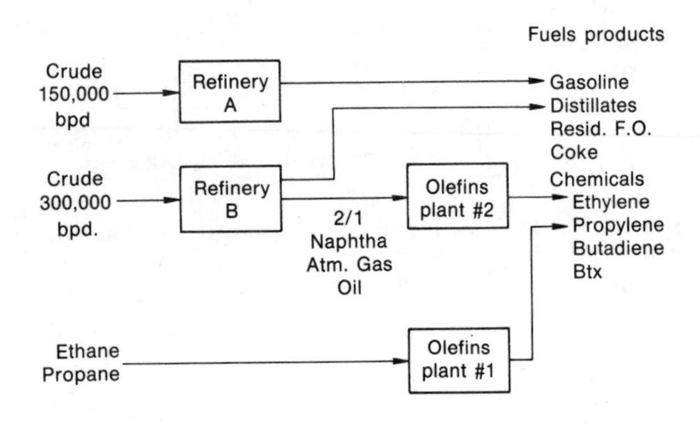

Fig. 3-20—Refinery-chemicals complex, 1980.[66]

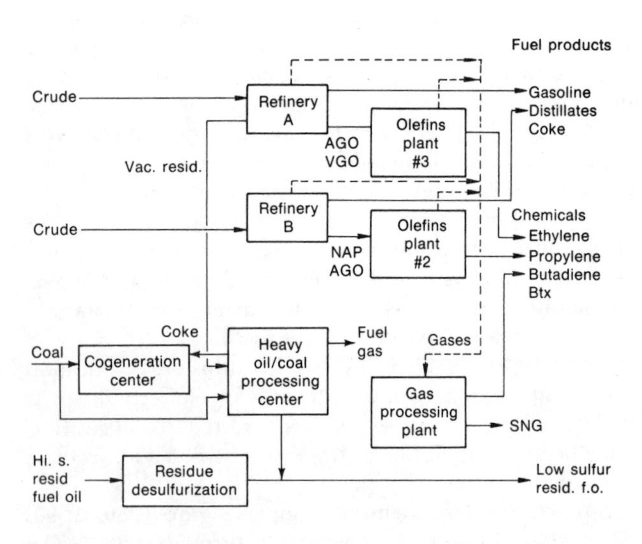

Fig. 3-21—Refinery-chemicals complex, 1990.[66]

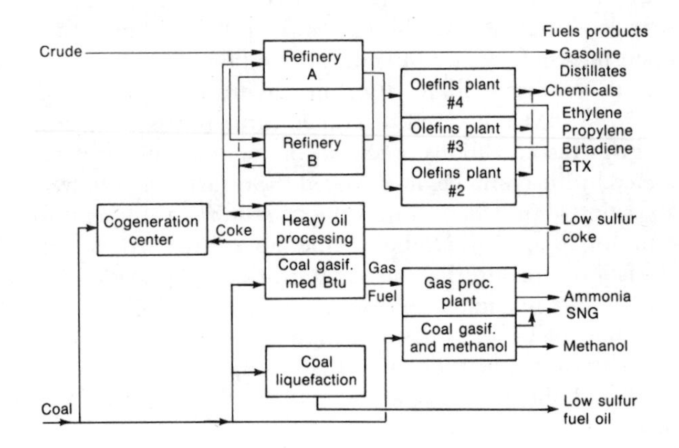

Fig. 3-22—Refinery-chemicals complex, 2000.[66]

TABLE 3-26—Future Inputs for an Example Complex[66]

Inputs	1980	1990	2000
Petroleum, bpd			
Sweet crude	350,000	170,000	100,000
Sour crude, lt	100,000	348,000	329,000
Sour crude, hvy	....	50,000	100,000
High S fuel oil	....	75,000	....
Total petroleum	450,000	643,000	529,000
Natural gas, MM SCFD	113	....	....
C_2/C_3, MM lb/yr	484	....	....
Coal, T/D	....	10,700	35,300
TOTAL (COE), bpd	471,000	687,000	675,000

TABLE 3-27—Future Products from an Example Complex[66]

Products	1980	1990	2000
SNG, MM scfd	....	57	212
LPG, bpd	19,000	17,000	14,000
Gasoline, bpd	207,000	216,000	156,000
Distillates, bpd	140,000	178,000	150,000
Resid. F. O., (1% S), bpd	19,000	117,000	48,000
Resid. F. O., (3% S), bpd	25,000	....	....
Low S. coke, T/D	640	630	650
Ethylene, Bill, lb/yr	1.89	2.66	3.66
Propylene, Bill, bbl/yr	0.54	1.00	1.60
Butadiene, Bill, lb/yr	0.18	0.38	0.59
BTX, gal/yr	55	110	146
Methanol, T/D	....	....	1,000
Ammonia, T/D	....	....	700

LITERATURE CITED

1. Stork, K., M.A. Abrahams and A. Rhoe. *Hydrocarbon Processing*, Vol. 53, No. 11, 1974, pp. 157-166.
2. Sinkar, S. (Ray). *The Oil and Gas Journal*, Feb. 23, 1976, pp. 103-105.
3. Eickerman, Richard. *The Oil and Gas Journal*, Mar. 26, 1979, pp. 113-116, 121.
4. Posey, L.G., P.E. Kelly and C.B. Cobb. *The Oil and Gas Journal*, Oct. 23, 1978, pp. 131-136.
5. *Hydrocarbon Processing*, Vol. 57, No. 4, 1978, p. 122.
6. Symoniak, M.F. and A.C. Frost. *The Oil and Gas Journal*, March 15, 1971, pp. 76-79.
7. Reber, R.A. and M.F. Symoniak. *Ind. Eng. Chem. Div.*, 169th ACS National Meeting, Paper 75, April 1975.
8. Grebbel, J. *The Oil and Gas Journal*, April 14, 1975, pp. 85-94.
9. Chopra, S.J. and P.K. Mukhopadhyay. *Hydrocarbon Processing*, Vol. 57, No. 2, 1978, pp. 113-117.
10. Brenken, H. and F. Richter. *Hydrocarbon Processing*, Vol. 58, No. 1, 1979, pp. 127-129.
11. Rhoe, A. and C. deBligniere. *Hydrocarbon Processing*, Vol. 58, No. 1, 1979, pp. 131-138.
12. Notarbartolo, M., C. Meneqazzo and J. Kuhn. *Hydrocarbon Processing*, Vol. 58, No. 9, 1979, pp. 114-118.
13. Manzanilla-Sadilla, F., O. Moreno L., M.C. Sze and W.V. Bauer. *Hydrocarbon Processing*, Vol. 58, No. 3, 1979, pp. 97-102.
14. Foulkes, P.S. and M.D. Harber. *The Oil and Gas Journal*, Mar. 20, 1978, pp. 85-86, 88.
15. Royoichi, Takahoshi and Washimi Koichi. *Hydrocarbon Processing*, Vol. 55, No. 11, 1976, p. 93.
16. Aalund, Leo R. *The Oil and Gas Journal*, Mar. 26, 1979, pp. 79-80.
17. Burk, Donal P. *Chemical Week*, Mar. 28, 1979, pp. 42-54.
18. Hughes, T.R., *et al. The Oil and Gas Journal*, May 17, 1976, pp. 121-124, 129-130.
19. *The Oil and Gas Journal*, April 5, 1976, p. 93.
20. *The Oil and Gas Journal*, April 19, 1976, p. 74.
21. *The Oil and Gas Journal*, May 3, 1976, p. 252.
22. Brabard, M., H.O. Braun and C.C. Bate. *The Oil and Gas Journal*, Nov. 15, 1976, p. 86.
23. Pollitzer, E.L., J.C. Hayes and V. Haensel. *The Chemistry of Aromatics Production via Catalytic Reforming*, Refining Petroleum for Chemicals, Advances in Chemistry Series, No. 97, American Chemical Society, 1970, pp. 20-23.
24. Kugelman, A.K. *Hydrocarbon Processing*, Vol. 55, No. 1, 1976, p. 95.
25. *Hydrocarbon Processing, 1977 Petrochemicals Handbook*, p. 132.
26. Nelson, W.L. *The Oil and Gas Journal*, Aug. 2, 1971, p. 76.
27. *The Oil and Gas Journal*, Feb. 28, 1972, p. 60.
28. Haensel, V. and G.E. Addison. *Advances in Catalytic Reforming*, 7th World Petroleum Congress, Vol. 4, 1967, pp. 113-123.
29. McKenna, W.L., G.H. Owen, and G.R. Mettick. *The Oil and Gas Journal*, Vol. 62, No. 20, 1964, p. 106.
30. Archibald, R.C., B.S. Greensfelder, G.R. Holzman, and D.H. Powel. *Ind. Eng. Chem.*, Vol. 52, 1960, p. 745.
31. Ryland, L.B., M.W. Tamele and J.N. Wilson. *Catalysis*, Vol. 7, New York: Reinhold Publishing Co., 1960.
32. Hall, W.K., *et al. J. Catalysis*, Vol. 2, 1963, p. 506.
33. Ward, J.W. *J. Catalysis*, Vol. 9, 1967, p. 396.
34. Baker, R.W., and J.J. Blazek. 31st Midyear Meeting, API Division of Refining, Houston, May 10, 1966.
35. Oblad, A.G. *The Oil and Gas Journal*, March 27, 1972, pp. 84-106.
36. Ebel, R.H. *The Oil and Gas Journal*, April 1, 1968, p. 116.
37. Rheaume, L., R.E. Ritter, J.J. Blazek, and J.A. Montgomery. *The Oil and Gas Journal*, May 17, 1976, pp. 103-110.
38. Magee, J.S., R.E. Ritter and L. Rheaume. *Hydrocarbon Processing*, Vol. 58, No. 9, (1979) 123-130.
39. *Hydrocarbon Processing*, Vol. 53, No. 9, 1974, p. 19.
40. *The Oil and Gas Journal*, Nov. 16, 1964, pp. 190-191.
41. Carter, C.P. *Hydrocarbon Processing*, Vol. 58, No. 9, 1979, p. 103-108.
42. Billon, A., J. Peries, E. Fehr and E. Lorenz. *The Oil and Gas Journal*, Jan. 24, 1977, p. 43.
43. Edison, R.R., J.O. Siemssen and G.P. Mosoloites. *Hydrocarbon Processing*, Vol. 55, No. 5, 1976, p. 133.
44. *Chemical Engineering*, Nov. 22, 1976, p. 86.
45. *The Oil and Gas Journal*, March 20, 1978, p. 94.
46. Burke, D.P. *Chemical Week*, Sept. 13, 1978, p. 27.
47. Gearhart, L.G. *Hydrocarbon Processing*, Vol. 55, No. 5, 1976, p. 127.
48. Dale, G.H. and D.L. McKay. *Hydrocarbon Processing*, Vol. 56, No. 9, 1977, p. 102.
49. *Hydrocarbon Processing*, Vol. 53, No. 9, 1974, pp. 127, 131.
50. Ward, J.W. *Hydrocarbon Processing*, Vol. 55, No. 9, 1976, p. 101-106.
51. Alcock, L., *et al. The Oil and Gas Journal*, July 8, 1975, pp. 102-110.
52. Sikonia, J.G., W.L. Jacobs and S.A. Gemibcki. *Hydrocarbon Processing*, Vol. 57, No. 5, 1978, p. 119.
53. Billon, A., J.P. Franck, J.P. Peries, E. Fehr, E. Galleie, and E. Lorenz. *Hydrocarbon Processing*, Vol. 57, No. 5, 1978, p. 124.
54. Billon, A., J.P. Franck and J.P. Peries. *Hydrocarbon Processing*, Vol. 54, No. 9, 1975, pp. 139-144.
55. Oblad, A.G. *The Oil and Gas Journal*, March 27, 1972, pp. 84-106.
56. Scott, J.W. and A.G. Bridge. *Origin and Refining of Petroleum*, #7, Washington, D.C.: American Chemical Society, 1971, p. 116.
57. Bridge, A.G., J.W. Scott, and E.M. Reed. *Hydrocarbon Processing*, Vol. 54, No. 5, 1975, p. 74-81.
58. Richardson, R.L. and F.C. Riddick. *The Oil and Gas Journal*, May 28, 1979, pp. 80, 85-86, 88, 93-94.
59. *The Oil and Gas Journal*, June 14, 1976, pp. 83-86.
60. Lawrance, P.A. and A.A. Rawlings. Proceedings, 7th World Petroleum Congress, 1967, p. 137.
61. Cartwright, C.W. and R.J. Stock. *The Oil and Gas Journal*, Sept. 18, 1978, pp. 142-145.
62. Andrews, J.W., *et al. Hydrocarbon Processing*, Vol. 54, No. 5, 1975, pp. 69-73.
63. Rogers, C.L. *The Oil and Gas Journal*, Nov. 8, 1971, pp. 60-61.
64. Lafferty, W.L. and R.W. Stokeld. *Origin and Refining of Petroleum*, Advances in Chemistry Series 103, ACS Washington, D.C., 1971, p. 134.
65. *Hydrocarbon Processing, 1979 Petrochemical Handbook*, Vol. 58, No. 11, p. 196.
66. Johnson, Alex R. *Hydrocarbon Processing*, Vol. 58, No. 9, 1979, pp. 109-112.

Nonhydrocarbon Products

The primary aim of refinery processes is to get the most out of a barrel of crude oil. The main thrust is to produce fuels with lesser emphasis on peripheral areas such as the production of petrochemical feedstocks. Even less attention is given to the production of byproduct elements and compounds. These elements and compounds, however, are valuable petrochemicals in their own right and have a place in refinery operations.

The byproduct elements are hydrogen, sulfur and carbon and the non-hydrocarbon compounds are cresylic acids and naphthenic acids. While all are byproducts, some are produced during major refining processes such as catalytic reforming and hydrogen processing while others are from the bottom of the barrel. Still others are indigenous to petroleum and are extracted from various high boiling refinery streams. Aromatics (**BTX**) are major petrochemicals from catalytic reforming and are the subject of Chapter 10.

HYDROGEN

Hydrogen is produced in large quantities during catalytic reforming. Formerly this hydrogen was utilized as a fuel or in the production of ammonia. This latter use accounts for petroleum refiners becoming involved in the nitrogen fertilizer business. The advent of extensive hydrogen requirements for modern refinery processes has put a new perspective to hydrogen and its value. Currently, reformer hydrogen is used to a large extent for hydrogen treating processes and hydrocracking. Hydrogen for ammonia synthesis comes primarily from methane reforming which is discussed in Chapter 5. Fig. 4-1 shows the past and predicted demand for hydrogen by the large-consuming industries.[1]

SULFUR

Aldous Huxley once said "Chemists cheer for the flag and H_2SO_4." And why not? Sulfuric acid is Number 1 on the list of industrial chemicals with an annual U.S. production of over 40 million tons and has a myriad of uses. A nation's industrial development can be measured by its sulfuric acid consumption, and sulfuric acid production accounts for about 85 percent of sulfur utilization. The United States alone uses sulfur at an annual rate of more than 100 pounds per capita. From where does it all come?

The Frasch process for obtaining elemental sulfur from its underground deposits is the largest source of sulfur in the United States. Other sources are the chemical conversion of sulfur dioxide and hydrogen sulfide to sulfur. The sulfur dioxide is obtained as a byproduct of sulfide ore smelting and from antipollution installations associated with the use of high sulfur fuels. In general, it is more economical to oxidize sulfur dioxide to sulfur trioxide for conversion to sulfuric acid than to reduce it to sulfur. The roasting of pyrites, FeS_2, is an important world-wide source of sulfur dioxide, but not in the United States.

The principal world source of hydrogen sulfide is sour natural gas and associated gas. An increasingly important source is refinery hydrogen processes which convert the

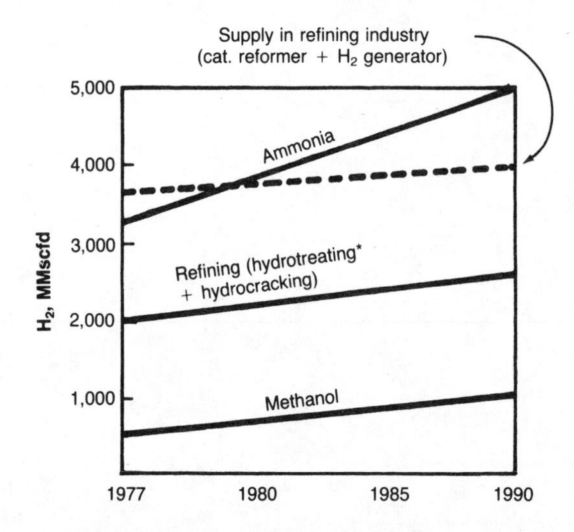

Fig. 4-1—The demand for hydrogen by the large consuming industries between 1977 and 1990.[1]

TABLE 4-1—World Sulfur Supply Forecast[3]

Source	1976	1978	1980	1985
	MM Metons			
Frasch & Native	17	18	19	21
Byproduct brimstone	15	17	19	25
Pyrites	12	12	13	16
Other forms	8	9	10	12
Locked in	(2)	(1)	(1)	(1)
Total available supply	50	55	60	73
Location	MM Metons			
United States	11	12	13	14
Canada	7	7	7	7
Western Europe and Africa	8	9	9	10
Asia/Oceania	4	5	7	9
L.A.	3	3	3	4
Western World	33	36	39	44
Locked in	(2)	(1)	(1)	(1)
Comecon	19	20	22	30
Total world supply	50	55	60	73

sulfur compounds present in the stream being processed into hydrogen sulfide. Approximately 47 percent of U.S. sulfur production is from hydrogen sulfide. The source of this hydrogen sulfide is equally divided between natural gas and the hydrogen sulfide produced as a byproduct of refinery processes. With the increasing use of hydrotreating and hydrocracking, refinery byproduct hydrogen sulfide will become more important. In Japan, for example, nearly all of the sulfur is currently being produced from petroleum refinery hydrogen sulfide. It is predicted that by 2000 the current reserves of Frasch sulfur in the United States and Mexico will be exhausted.[2] A world-wide sulfur supply forecast is given in Table 4-1 by both source and location.[3]

From Hydrogen Sulfide. Commercial processes for the recovery of sulfur from hydrogen sulfide involve vapor-phase oxidation of the hydrogen sulfide by sulfur dioxide—the *Claus reaction*. The sulfur dioxide is obtained by oxidation of hydrogen sulfide and at the same time some elemental sulfur is formed. The hydrogen sulfide oxidation to sulfur dioxide is exothermic ($-\Delta H$) and spontaneous at high temperatures and rapid at low temperatures in the presence of a catalyst.

In a typical Claus plant the incoming acid gas feed, H_2S, is subjected to a high temperature (up to 1,450°K) non-catalytic free flame combustion with air in stoichiometric proportions according to Equation 4-1.[4]

$$3\ H_2S + 3/2\ O_2 \rightarrow 3H_2O + 3/x\ S_x$$
$$\Delta H = -145 \text{ to } -173 \text{ kcal} \qquad (4\text{-}1)$$

Free sulfur is produced directly in yields of as much as 70 percent. Sulfur dioxide is also produced, Equation 4-2.

$$H_2S + 3/2\ O_2 \rightarrow H_2O + SO_2$$
$$\Delta H = -124 \text{ to } -138 \text{ kcal} \qquad (4\text{-}2)$$

Sulfur dioxide then reacts with unconverted hydrogen sulfide over a catalyst to produce more sulfur, Equation 4-3.

$$2\ H_2S + SO_2 \rightarrow 2\ H_2O + 3/x\ S_x$$
$$\Delta H = -21 \text{ to } -35 \text{ kcal} \qquad (4\text{-}3)$$

The reaction gas, containing a mixture of sulfur, sulfur dioxide and unreacted hydrogen sulfide, is cooled and the sulfur removed. The rest of the stream enters a catalytic converter where the hydrogen sulfide reacts with the sulfur dioxide to form sulfur and water.

The *catalysts* used in a Claus converter are activated alumina, bauxite and Porocel. Evaluation of these catalysts has shown that the ratio between pore volume and particle size is an important physical property related to catalyst activity.[5] A good catalyst is one which causes a high hydrogen sulfide conversion, a high carbonyl sulfide, COS, and carbon disulfide, CS_2, reconversion, and has a low loss of activity due to catalyst poisoning. Carbon-sulfur compounds found in the tail gas are produced by side reactions in the front-end furnace due to the presence of hydrocarbons in the feed gas. They can be reconverted to hydrogen sulfide by the use of an appropriate catalyst.[6]

Claus catalysts are deactivated by aging or poisoning with sulfation of the catalysts the most serious cause of loss of activity.[7] It is stated that the higher the catalyst activity toward oxidation, the higher its activity toward sulfate formation.

Small amounts of *hydrogen* are formed in the Claus process. This has been attributed to dissociation of hydrogen sulfide in the hot hydrogen sulfide preflame region. The heat needed for this dissociation is furnished by the exothermic oxidation of the hydrogen sulfide. The dissociation reaction is endothermic, Equation 4-4.

$$H_2S \rightarrow H_2 + S$$
$$\Delta H = +38 \text{ kcal} \qquad (4\text{-}4)$$

Current research is directed toward the development of a commercially viable process for the recovery of hydrogen and sulfur by cracking hydrogen sulfide. The preliminary studies show that the sulfur produced by such a process may be more expensive than that produced by the Claus process.[8]

Readers wishing more extensive information about the Claus process are referred to the following articles:

"GPA Panelist Outlines Claus Process Improvement in Sulfur Recovery."[9]

"Advances in Claus Technology," by P. Grancher.[10]

"Sulfur Costs Vary with Process Selection," by H. Fischer.[11]

Utilization. The use of sulfur for the production of sulfuric acid so dominates the statistics of sulfur utilization that its many other uses are obscured. Sulfur is a versatile element with uses ranging from dusting powder for roses to rubber vulcanization to sulfur-asphalt pavements. Flower-sulfur is used in match production and in certain pharmaceuticals. "Insoluble" sulfur is used as an additive to cutting oils to prevent welding of small metal particles to the cutting tools. Sulfur is also used in high pressure lubricants and as a coating for urea used as a fertilizer.

Sulfur can replace 30-50% of the asphalt used in making a blend for road construction. This is done by using simple mixing equipment. The sulfur-asphalt blend conserves hydrocarbons and provides a market for sulfur. For the user, it provides road surfaces that have nearly double the strength of conventional pavement and can thus be thinner. The roads are also said to be more resistant to climatic changes. If this use proves out, road construction may become a large new market for elemental sulfur. The impregnation of concrete with molten sulfur is another potential large sulfur

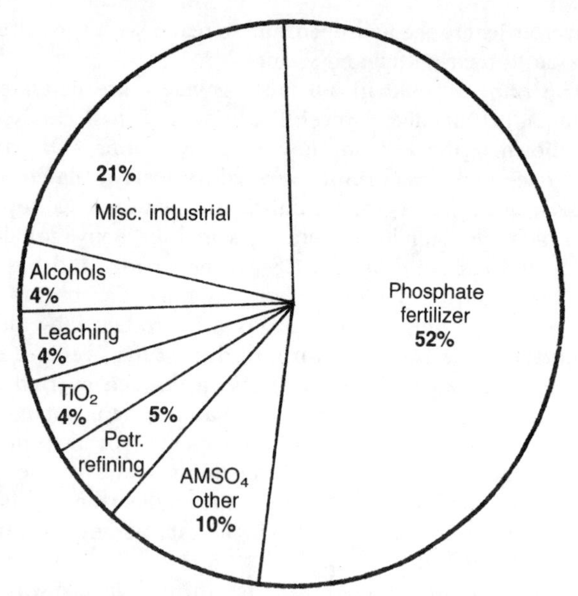

Fig. 4-2—Western world 1976 sulfur consumption by weight percent end use.[3]

use. Concrete impregnated with sulfur has increased compressive and tensile strength and its corrosion resistance is tripled. Potential applications are sewer pipes and bridge decks.

Sulfur is used to produce industrial compoúnds other than sulfur dioxide for sulfuric acid production. For example, sulfur is used to produce *phosphorus pentasulfide* used for the preparation of zinc dithiophosphates which, in turn, are used as corrosion inhibitors in lubrication oils. *Carbon disulfide* is produced commercially from sulfur and petroleum hydrocarbons, especially methane, or by heating sulfur with charcoal. Besides being a good solvent, carbon disulfide has many industrial uses such as the manufacture of rayon and as an intermediate for the production of thiourea. Sulfur is also used in the synthesis of dyes and in the manufacture of enamels.

Sulfur may enter a new field—the production of polymeric sulfur nitride, *polythiazyls*, $(\mathbf{SN})_x$.[12] These polymers have optical and electrical properties of metals and are malleable. Very thin epitaxial films can polarize light and, when deposited on different plastic substrates, may be potentially useful in optical communications equipment. This is not a potential tonnage use of sulfur but it does illustrate that research in sulfur chemistry is not dead. Fig. 4-2 shows the 1976 sulfur consumption by wt.% end-use.[3]

The present status of sulfur with its surplus resources, especially in the form of hydrogen sulfide, will encourage greater research effort toward the utilization of elemental sulfur. Sulfur has been known since antiquity. It is the brimstone of the Bible and reputedly fuels the fires of Hell. If the current over-production of sulfur continues, some producers may wish to consign it there.

INDUSTRIAL CARBONS

There is an important but little known group of highly useful materials which can be described as high purity *industrial carbons*.[13] These carbons serve as key components in the manufacture of such end-use items as tires, plastic building materials, amorphous and graphitized electrodes and many, many essential small volume uses. Industrial carbons have many things in common, not the least of which is their petroleum source, either as petroleum byproducts or made from refinery byproducts.

A detailed classification of industrial carbons has been made by Stokes.[13] Included are (1) carbon black, (2) selected petroleum coke (green or calcined), (3) needle coke (green or calcined), (4) electrode pitch, (5) pitch for structural carbon manufacture including graphite shapes for atomic reactors, (6) carbonaceous reducing agents for the new Alcoa primary aluminum process, (7) activated carbon. This is not necessarily an all inclusive list but is representative of the many different industrial carbons.

Various *petroleum pitches* are derived from cracking processes and isolated as distillation fractions. They are used in road surfacing, roofing materials and various adhesive and sealant formulations. They are also used as binders in the fabrication of composite structures which can subsequently be pyrolyzed to carbon. The properties and uses of petroleum pitches have been discussed extensively in the *Symposium on Petroleum Derived Carbon*[14] as have other carbons in Stokes' classification of industrial carbons.[13] Petroleum pitches are not pure substances or simple mixtures of relatively simple compounds.

Of the industrial carbons, we will note only carbon black and coke, both of which are essentially pure carbon. They represent a sizeable portion of the half billion dollar-a-year industrial carbon business.

Carbon Black

Carbon black can be defined as elemental carbon made under conditions permitting control of its particle size, degree of particle agglomerization and surface chemistry. The product is our most important industrial carbon and is the seventh largest dollar value chemical sold in the United States. The world-wide market is approximately seven billion pounds annually.[15] By 1980 the total value of the carbon black will reach the $1 billion level on a world basis.[13]

The main constituent of carbon black is elemental carbon with a varying amount of volatile material and ash. The volatile material is a mixture of high boiling hydrocarbons and oxygen containing compounds. These oxygen compounds are acidic and affect the acidity properties of the black. There are several types of carbon blacks and their characteristics depend upon both the particle size and the method of production (Table 4-2).[16]

Properties. The three most important properties of carbon blacks are pH, and specific surface and particle structure. These properties are closely related to the process used to produce the black and to the characteristics of the feedstock.

The *pH* of a carbon black varies from acidic, $\mathrm{pH} < 7$, for channel blacks to basic, $\mathrm{pH} > 7$, for furnace blacks which are water cooled. The basic character of furnace blacks is caused by the presence of evaporation deposits from the water quench. The pH of the black has a pronounced influence on the vulcanization time of rubber. Acidic pH acts as a cure retarder while alkaline pH activates the vulcanization process.

TABLE 4-2—Selected Physical and Chemical Properties of Several Carbon Blacks and Their Uses[16]

Carbon Black Type	Average particle Diameter Å	Surface Area sq.m/g (Electron Microscope Method)	Surface Area sq.m/g (Nitrogen Adsorption Method)	Volatile Matter wt. %	pH	Important Uses
Color and ink channel	100–300	100–218	110–1000	5–18	2.7–5.5	Automotive enamels & paints, printing ink, etc.
Rubber grade channel						
Conductive (CC)	250	110	225	4.5	4.3	Used in sandblast hoses, conductive soles & heels.
Hard Processing (HPC)	260	105	140	5.0	4.0	Used in tire treads.
Medium Processing (MPC)	280	106	120	5.0	4.0	
Easy Processing (EPC)	300	95	100	5.0	5.0	
Gas furnace blacks						
Semi reinforcing (SRF)	700	25	25	1	9.8	Used in tire carcass & side walls.
High Modules (HMF)	500	40	35	1	9.5	
Fine Furnace (FF)	400	60	75	1	9.5	Used in truck tire carcass.
Oil furnace blacks						
General purpose (GPF)	550	40	25	0.9	9.1	Used in tire carcass, tread base, side walls.
Fast extruding (FEF)	400	60	40	0.9	9.0	
High Abrasion (HAF)	280	75	75	1.6	9.0	Used in tire treads, heels and soles.
Intermediate Abrasion (ISAF)	240	120	130	1.0	9.3	
Conductive (CF)	190	120	220	1.6	8.2	Used in conductive rubber goods.
Thermal blacks						
Fine Thermal (FT)	1850	16	16	0.5	8.9	Used in natural rubber inner tubes, footwear.
Medium Termal (MT)	5200	6	6	0.5	8.5	Used in wire insulation, mechanical goods, footwear, etc.

The *specific surface* is the dominate property in respect to reinforcement properties of carbon black. A high specific surface produces better abrasion resistance to the rubber. Both processing time and mixing power requirements increase rapidly with decreasing average particle size. The surface area of blacks is measured by use of liquid nitrogen adsorption[17] and by electronmicroscopy (Table 4-2). The grades used in the rubber industry range from 9 to 153 m^2/gram.

Surface activity is related to the surface and to the amount of surface oxygen on the carbon black particle. It apparently influences adhesion of the black to the elastomer and also is related to reduction of heat build-up.

Structure is a concept related to the degree of agglomeration of the carbon black. The structure of the black is directly related to the process used for its production. Thermal blacks have large, individual particles and have *low structure.* Oil-based furnace blacks have chain-like particles and a *high structure.* High structure contributes to modules, extrudability and hardness but process safety (scorch time) is reduced.

The trend is toward blacks combining the reinforcing action of the high surface with greater ease of dispersion and processing, faster extrusion and durability of the resulting product, all of which is a characteristic of high structure.[18] When used with rubber, carbon black does not dissolve or undergo chemical reactions but remains as a discrete phase.

Although bonding of carbon black to polymers may be related to the surface structure of the black, a complete understanding of this phenomena is lacking.

Production. Carbon black is produced by pyrolysis, partial oxidation or by complete oxidation of a portion of the feedstock and pyrolysis of the remainder. The exothermic oxidation furnishes the energy for the endothermic pyrolysis. The following reactions illustrate this process with methane as the feedstock.

$$CH_4 + 2O_2 \rightarrow CO_2 + 2H_2O \quad \Delta H = -191 \text{ kcal}$$

$$CH_4 \rightarrow C + 2H_2 \quad \Delta H = +20.3 \text{ kcal}$$

Almost any hydrocarbon material can be used as a feedstock to produce carbon black. Natural gas and petroleum distillates, especially those rich in aromatics, are used. Tars produced from catalytic cracking units and from naphtha and gas oil cracking units are excellent feedstocks and usually give a 92 weight percent carbon and are essentially free of sulfur. Still other feedstocks are being investigated. For example, cokes produced from fluid or delayed coker units with low sulfur and ash content are possible substitutes for cokes from distillate and natural gas.[15]

The nucleation of carbon black has been studied and it appears that in the case of thermal blacks polyaromatic molecules are first formed which then condense into liquid droplets which are the precursors of the carbon black particles.[19]

The *channel process* (channel black) is a process of more historical than economic interest. Not more than 5 percent of carbon black is made by this process. The process depends upon burning natural gas in less than sufficient air for complete combustion. Fine particles of carbon are formed and deposit on cold steel channels which move slowly over the flames. The particle size is changed by varying the distance of the channels from the flames and by the extent of combustion regulated by the gas to air ratio.

The *thermal process* (thermal black) depends upon the pyrolysis of natural gas in preheated furnaces containing a checkerwork of hot bricks. The natural gas is cracked to carbon and hydrogen, and the cooled bricks are then reheated by combustion of natural gas in the regenerative cycle. This process is more efficient than the channel process but the quality of carbon black is less. The average particle size is large and, therefore, thermal black is used in inner tube formulations rather than in tire tread which requires abrasion resistance.

The *furnace process* (furnace black) is a partial combustion process. Two types of reactors are in use depending upon whether the feed is a gas or a liquid. The hot gases and carbon from the reactor are quenched with a water spray and then further cooled by heat exchange with the air used for the partial combustion of the hydrocarbons. Fig. 4-3 is a flow diagram of an oil furnace black process.[20] The type of black produced from this process depends upon the feed and reactor temperature.

Although most of the furnace processes now use oil distillates and catalytic cracking byproduct (decant oil-fractionator bottoms) rather than natural gas (the process is more costly with natural gas), a new economic evaluation of the two feedstocks in light of the new prices of crudes and natural gas might be required.

A slow but steady growth lies ahead for the U.S. carbon black market, production, and price as long as we continue to travel on rubber tires. Approximately 92 percent of carbon black goes into rubber and there is no price-competitive substitute which can provide comparable strength and abrasion resistance to tire and other products.[21]

Petroleum Coke

Practically all of the raw materials used in the production of carbon are byproducts of some other industry. Petroleum coke is no exception for it comes from the material left at the tail end of the refining process. The primary purpose of coking has been to upgrade heavy residuals or bottom-of-the-barrel materials into valuable distillate products. Consequently, some refiners coked only refinery resids that brought the least amount of profit or were difficult to sell because coke was considered a "no-value" product.

In recent years, however, the emphasis is to produce a quality product for anodes used by the aluminum industry and for the specialty market. Coke can no longer be considered a byproduct nor delayed coking a garbage disposal process.[8]

The quality of petroleum coke varies from oil field to oil field and from refinery to refinery. It even may differ depending upon the mix of products from the refinery. In any petroleum coking process essentially all of the non-volatiles, metals and much of the sulfur of the feed remain in the coke.

Production. There are two major processes, delayed coking and fluid coking. The delayed coking plants are about 14 percent less expensive than the fluid coking plants.[22]

Delayed coking utilizes one of the simplest processes in a refinery (Fig. 4-4).[23] The reactor system consists of a short contact time heater coupled to a large drum in which the preheated feed "soaks" on a batch basis. Coke is gradually formed in the drum. A unit has at least one pair of drums; one drum is filling while the other is being decoked. When

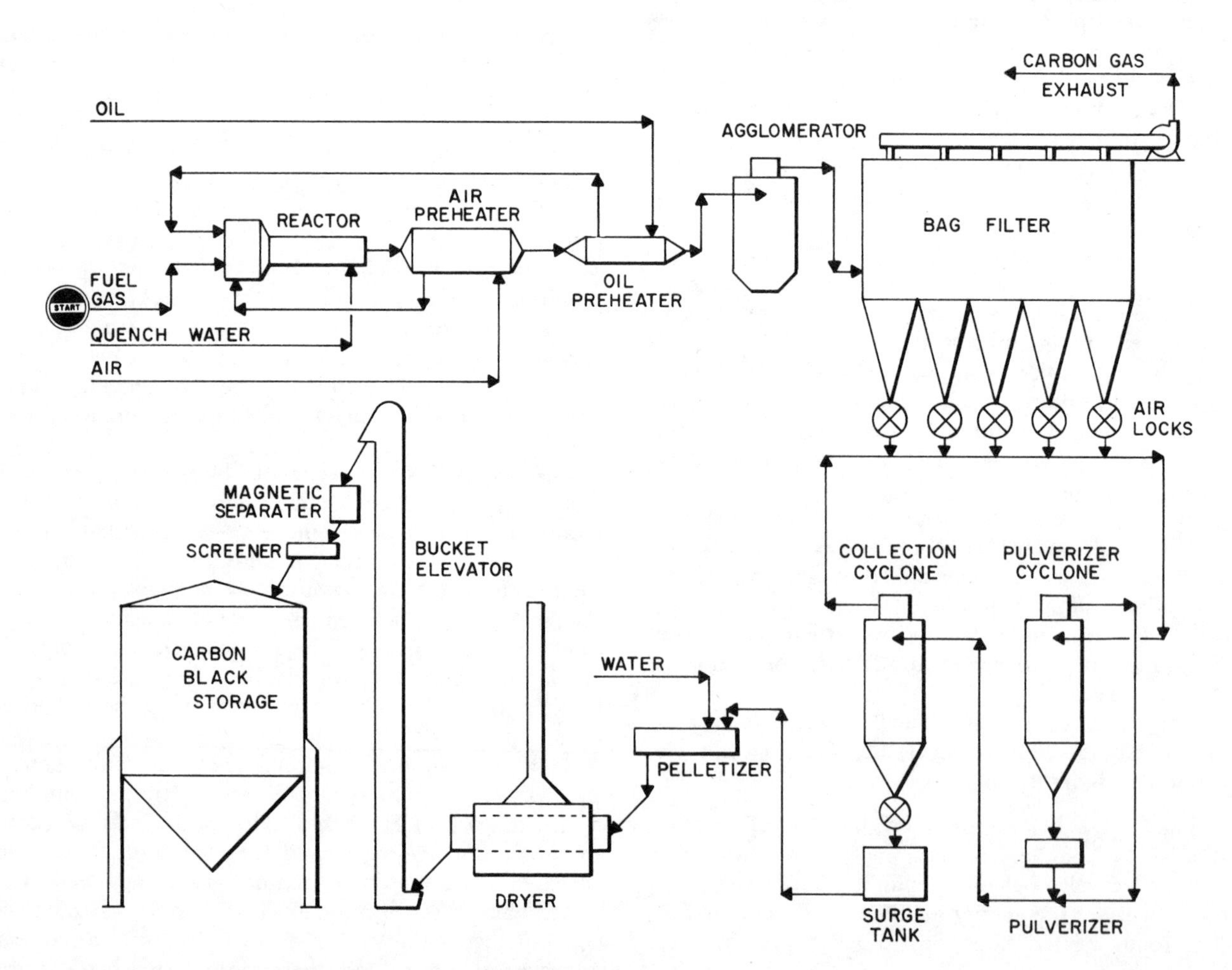

Fig. 4-3—Carbon black (oil black by furnace process of Ashland Chemical Co.).[20]

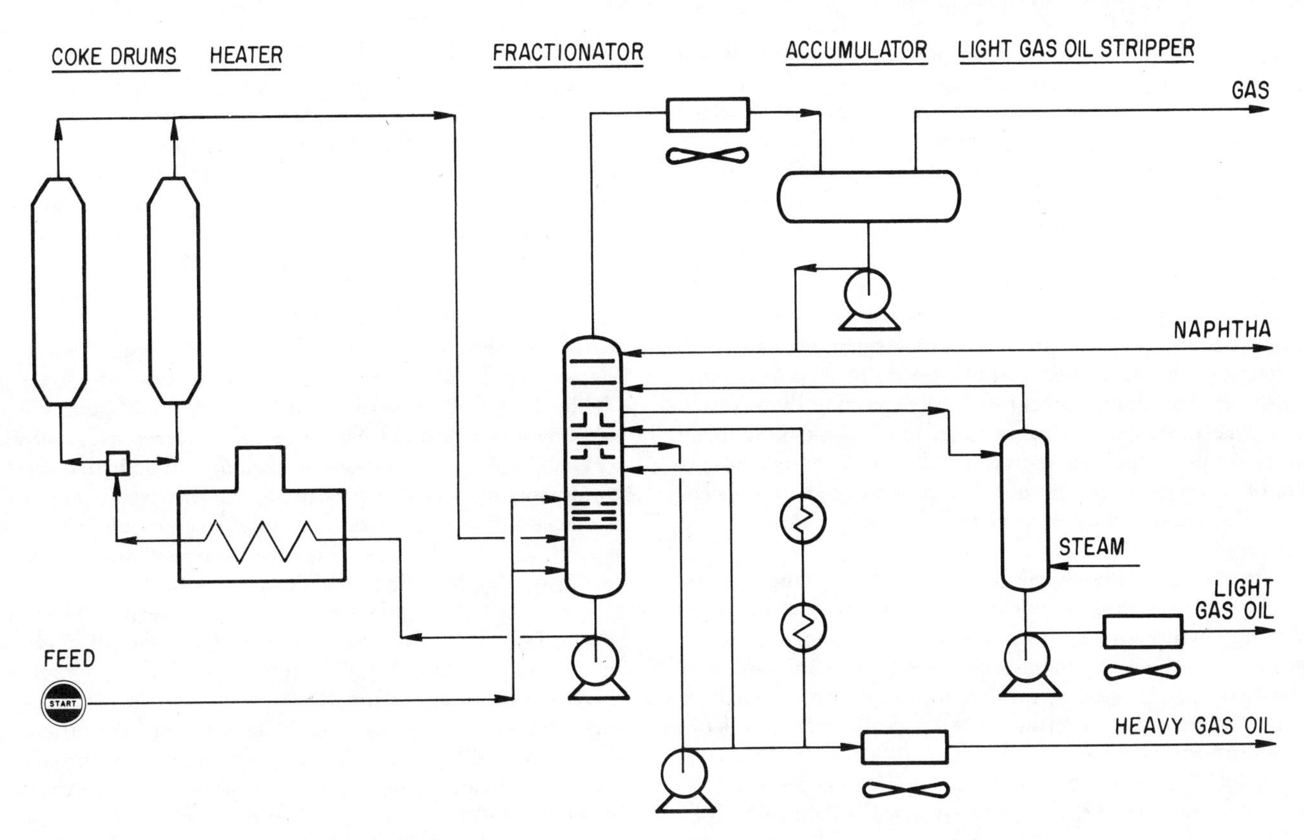

Fig. 4-4—A typical delayed coking installation.[23]

the coke reaches a predetermined level in one drum, flow is diverted to the other drum so that the process is continuous. The unconverted hydrocarbons in the full drum are removed by steam stripping and the drum is decoked by water jets under at least 3,000 psig.

Operating conditions for delayed coking are: pressure 25-30 psig at a temperature of 910-930°F and a recycle ratio of about 0.25 based on equivalent feed.[23] Too low a temperature produces soft, spongy coke. Coking conditions of higher severity as to pressure, temperature and recycle produce greater yields of coke having lower sulfur, lower VCM and lower coefficient of thermal expansion. The quality of coke from delayed coking is primarily related to the quality of the feed.

Raw petroleum coke, *green coke,* contains 5-30 percent volatile matter. This must be reduced to 0.5 percent or less for the coke to be used as industrial carbon. This reduction in volatiles is accomplished by calcining the green coke in rotary kilns or hearths at 1,200-1,400°C. The *calcined coke* has a density of 2.08-2.13 g/cc and 0.2 percent volatile matter (Table 4-3).[24]

The aluminum industry provides the largest single outlet for petroleum coke. The specifications for aluminum electrodes require low sulfur and metal content. Desulfurization and demetallization of coke feed may be a logical combination for high sulfur, high metals crude used to produce coke with reduced metal and sulfur content. Table 4-4 compares the composition of delayed coking vacuum residues of three different crudes—first without treatment and then after desulfurization of the feed.[32] Hydrodesulfurization of the

TABLE 4-3—Typical Properties of Petroleum Coke Used in Carbon and Graphite Manufacture[24]

Property	Raw Coke	Coke Calcined To 1300°C
Aluminum, ppm*	15–100	15–100
Boron, ppm	0.1–0.5	0.2–1.5
Calcium, ppm	25–500	25–500
Fixed carbon, %	87–97	97–99
Hydrogen, %	3.0–4.5	0.1
Iron, ppm	50–2000	50–2000
Manganese, ppm	2.0–100	2.0–100
Magnesium, ppm	10–250	10–250
Nickel, ppm	10–100	10–100
Nitrogen, %	0.1–0.5	0.1
Silicon, ppm	50–300	50–300
Sulfur, %	0.2–2.5	0.2–2.5
Titanium, ppm	2.0–60	2.0–60
Vanadium, ppm	5.0–500	5.0–500
Moisture, %	0.5–2.0	Negligible
Volatile matter, %	5.0–15	0.2
Real density, g/cc	1.6–1.8	2.08–2.13
Ash, %	0.1–1.0	0.2–1.5

*Parts per million

TABLE 4-4—Delayed Coking Vacuum Residuum[32]

Item	Arabian Light		Arabian Heavy		North Slope	
	Untreated	VRDS	Untreated	VRDS	Untreated	VRDS
Feed						
Sulfur, wt. %	4.1	0.8	5.7	0.9	2.3	0.4
Coke						
Sulfur, wt. %	6.7	1.5	7.7	2.0	3.0	1.0
Vanadium, ppm	230	60	500	150	300	40
Yield, wt. % of vac. resid	34	20	37	10	29	10
Coker feed						
Relative rate	1.0	0.84	1.0	0.43	1.0	0.50

feed substantial reduces the quantity of coke produced, but yields a higher quality coke. Hydroconversion also reduces the coker feed rate. To use the Arabian heavy crude which has a high asphaltene content, severe **VRDS** hydrotreating is required to reduce the sulfur content of the feed to 0.9%. This level is required to produce green coke having about 2 wt.% sulfur and 150 ppm vanadium. The coker feed rate is reduced by about 50%.

Fluid coking is a fluid solids process that thermally cracks the feed to gaseous and liquid products plus coke. The process utilizes the heat produced by the burning of about 25 percent of the coke produced to provide process heat (925-1,000°F). The fluid coke is formed by spraying the resid on hot coke particles. The conversion into coke is immediate with complete disorientation with the crystallites in the recycle of product coke. The burning process tends to concentrate the metals but does not reduce the sulfur content of the coke.

Fluid coke has several characteristics which make it undesirable for most petroleum coke markets. These characteristics are: high sulfur content, low volatility, poor crystalline structure and low grindability index.[15] The sulfur content of fluid coke varies between 5 and 8 percent which makes it undesirable for both fuel and metallurgical uses. The volatile matter in fluid coke is low (*ca.* 3 percent) because the coke is partially calcined in the burner portion of the process (1,125-1,175°F). Because of the disoriented crystal structure of fluid coke, its grain density (1.9-2.0 g/cc) is lower than the grain density of delayed coke (2.1 g/cc). The low grindability index (Hardgrove scale) of fluid coke (20-30) is related to its extreme hardness. Delayed coke has a grindability index of *ca.* 100 which is the same as bituminous coal.

Flexicoking integrates conventional fluid coking with coke gasification.[25, 26, 27] Fig. 4-5 is a diagrammatic representation of the process.[28] Flexicoking gasification produces a substantial concentration of metals in the coke product and also causes significant desulfurization of the residual coke.[27] This metal concentration may represent a potential source of vanadium and nickel. Both the quantity and quality of the coke can be varied independent of the resid feedstock. Typical yields for processing various resids are given in Table 4-5.[26] The gaseous products are hydrogen, water, carbon monoxide and carbon dioxide, referred to as *coke gas*.

The liquid products from flexicoking can be hydrotreated to give many desirable products. One of these is naphtha, which is suitable for bimetallic catalytic reforming jet fuel and diesel fuel. The gas oil fraction can be hydrotreated to satisfy fuel oil blending specifications.[29]

Needle coke is a so-called crystalline coke preferred for graphitized electrodes. It is produced from very pure petroleum tars by a variation of the delayed coking process. Typical feedstocks are slurry and decant oils from cat cracking and thermal cracking tars.[30] High aromaticity of the feed is the key to the process. Know-how plays a very important role and operating techniques are closely guarded. A recent development links a needle coke unit to an ethylene plant.[31] Needle coke is usually calcined before use. In the calciner, the remaining moisture and the volatiles are removed.

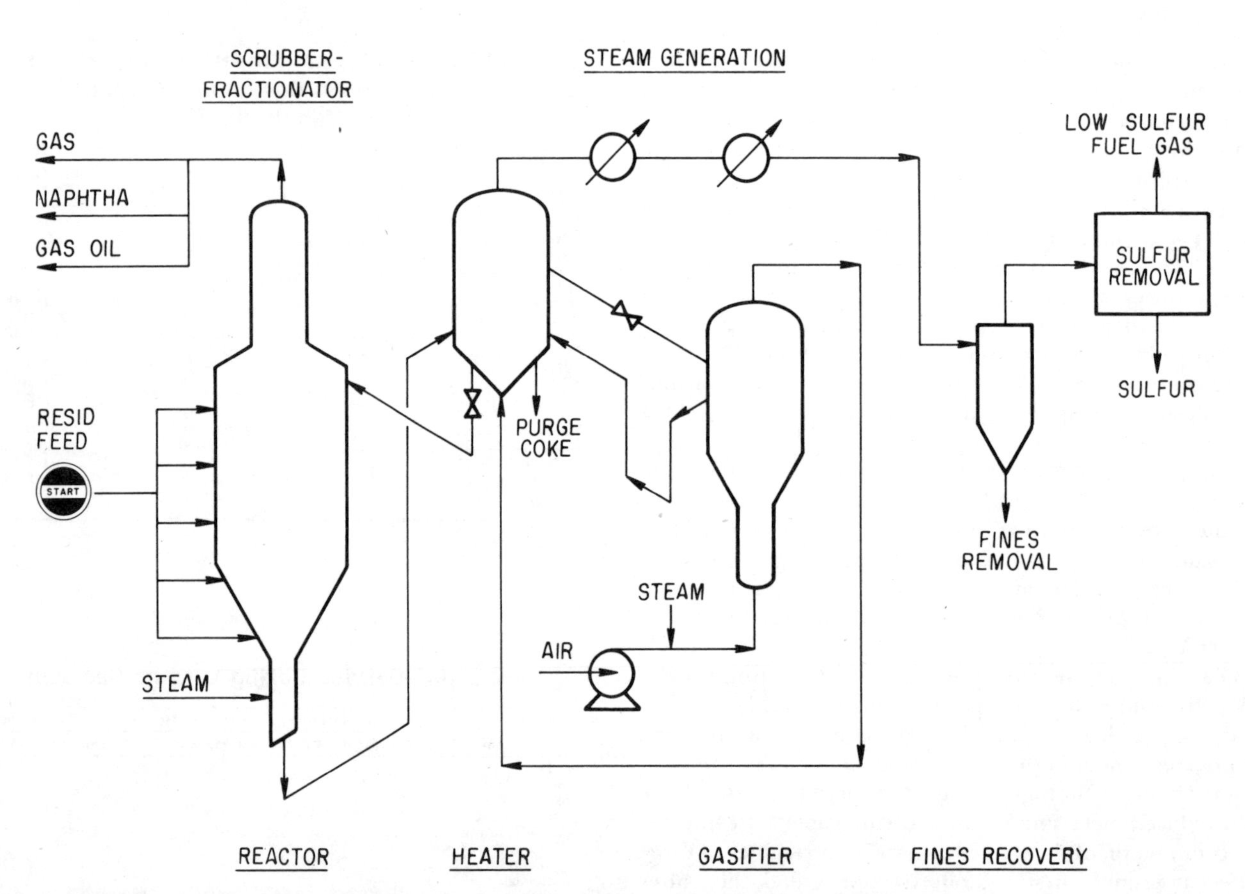

Fig. 4-5—Diagram of the Flexicoking process to produce gases, liquids, and coke from various feedstocks.[28]

TABLE 4-5—Flexicoking Yields[26]

Inspection Data	Arabian Heavy	Prudhoe Bay	South Louisiana Mix
Feed properties			
Cut range, °F	1050+	1050+	1050+
Gravity, °API	3.0	5.5	8.7
Sulfur, wt%	6.0	2.5	1.7
Nitrogen, wppm	4800	6600	4600
Conradson carbon residue, wt %	27.7	20.0	17.9
Metals, wppm	269	150	40
Yields based on fresh feed			
H_2S, wt %	1.45	0.53	0.36
C_3^-, wt %	9.62	8.85	8.53
C_4, wt %	0.67	0.58	0.57
C_5/370°F, LV %	15.0	16.0	16.1
370-650°F, LV %	16.7	20.4	21.6
650-975°F, LV %	28.9	36.9	39.1
Purge coke, wt %	0.69	0.48	0.43
Gas heat content (LHV)			
Btu/scf	127	113	108
× 10^3 Btu/bbl fresh feed	1370	900	760

Utilization. The uses of petroleum coke are somewhat confused by including coke on spent cracking catalyst as a fuel (Fig. 4-6). This coke is burned in the regeneration step to carbon monoxide and carbon dioxide and, of course, never reaches the open market. In fact, it is never even seen.

The utilization of green petroleum coke is given in Table 4-6.[24] The major use of amorphous type coke is for non-graphitized electrodes. This type of electrode is used as the anodes in the manufacture of primary aluminum. On a green coke basis, about 0.55-0.60 pounds of coke are consumed for 1 pound of aluminum. On a calcined basis, prebaked aluminum anodes require 0.45-0.55 pounds of carbon per pound of aluminum. On the same basis, silicon carbide requires 1.4 lbs., phosphorous 1.18 lbs., calcium carbide 0.69 lbs. and graphite 1.25 lbs.

Table 4-7 contains typical specifications for calcined petroleum coke for electrodes used in aluminum production.[13] These specifications are, perhaps, more theoretical than actual. No one knows the exact degree of deviation. The

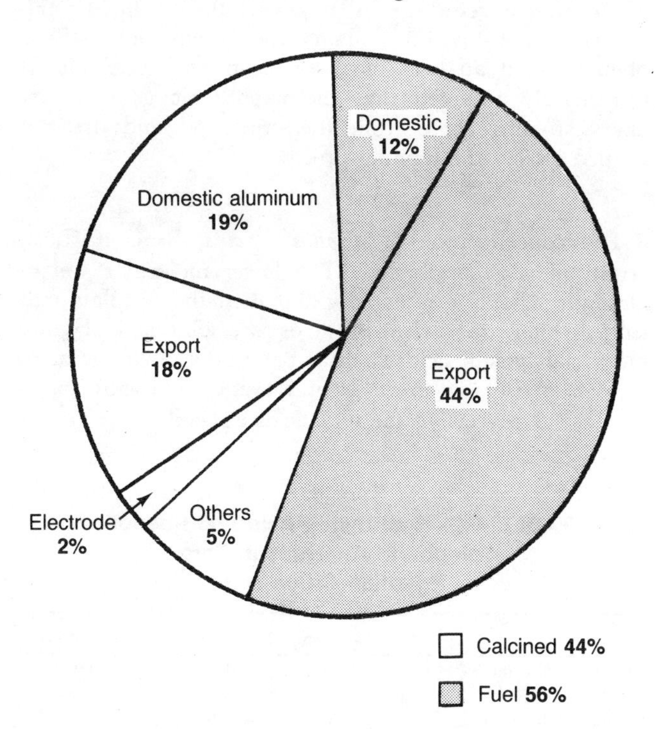

Fig. 4-6—The petroleum coke markets in 1978.[38]

TABLE 4-6—The Uses of Raw (Green) Petroleum Coke[24]

Low Ash Fuel	High Purity Reactant	Ferrous Metallurgy	Calcination
Power plants	Calcium carbide	High density foundry coke	Calcined petroleum coke
Domestic fuel	Silicon carbide	Blast furnace coke	
Cement kilns	Other carbides		

TABLE 4-7—Typical Specifications for Calcined Petroleum Coke for Aluminum Production[13]

Volatile matter	2.5% max.
Ash	0.05% max.
Metals	
Iron	0.02%
Vanadium	0.001%
Nickel	0.001%
Manganese	0.001%
Nonmetals	
Silicon	0.02%
Sulfur	1%*

*Or less when EPA rulings are enforced

world consumption of coke for aluminum in 1985 is predicted to be 17.5 million tons per year. The green coke requirements for graphitized electrodes and other baked and structural carbons are projected to be 4.4 million tons for a total industrial carbon use of 22 million tons per year of green coke. The demand for needle coke on a green coke basis is estimated to be 1 million tons in 1985. The 1978 petroleum coke demand, in million tons and by type, was: delayed, 12.50; fluid, 1.50; needle, 0.36; and inventory, 0.16.[38]

CRESYLIC ACID

In a broad sense, the term *cresylic acid* includes cresols and xylenols and even phenol (Table 4-8) which are obtained from coal tar and refinery streams. In a narrower use of the term, only the cresols are included.

Production. The production of cresylic acid from petroleum sources illustrates the far reaching, all pervading, effect of pollution control. It also illustrates the impact of changing technology in the petroleum industry. In the past, about 50 percent of cresylic acid production was petroleum based. It was obtained as a byproduct of refinery processes that produced spent caustic waste streams. These streams were the product of caustic treating to remove mercaptan sulfur. The production of low lead and lead-free gasoline requires hydrodesulfurization, a process that requires no caustic. Cresols are now synthesized at an appreciable increase in price.

Cresylic acid, however, is still obtained to some extent from petroleum. When crude oil is distilled at atmospheric pressure, phenol and cresols are distributed in the gasoline fraction. The percentage of these phenols in cracked gasoline is more than in straight run gasoline.[33]

Phenol and the cresols are weak acids which form water soluble salts by the action of bases. This permits their extraction by use of an alkaline solution, usually sodium hydroxide. The aqueous layer contains, in addition to sodium phenate and cresylate, a small amount of sodium

TABLE 4-8—Properties of Phenol, Cresols and Xylenols

Name	Formula	MP(°C)	BP(°C)	Density 20/4°C	pKa	$K_a \times 10^{-10}$
PHENOL	OH	42.5	182	1.0722	10.0	1.1
CRESOLS						
O-CRESOL	OH, CH_3	31	191	1.02734	10.2	0.63
M-CRESOL	OH, CH_3	11	202	1.0336	10.01	0.98
P-CRESOL	OH, CH_3	35.5	202	1.0178	10.17	0.67
XYLENOLS						
2,4-DIMETHYLPHENOL	OH, CH_3, CH_3	26	211	0.9650		
2,5-DIMETHYLPHENOL	OH, CH_3, CH_3	75	212			
3,4-DIMETHYLPHENOL	OH, CH_3, CH_3	62.5	225	0.9830		
3,5-DIMETHYLPHENOL	OH, CH_3, CH_3	68	219.5	0.9680		

naphthenate and sodium mercaptans. Before the advent of hydrodesulfurization, this process was used to sweeten refinery streams and provided, as a byproduct, cresylic acids. The mercaptans in the extract can be oxidized to disulfides that are insoluble in the water layer and are separated from the cresolate solution by decantation.[34]

Free cresylic acid is obtained by treating the solution with a weak acid or sulfuric acid. Refinery flue gases can be used as an inexpensive source of carbon dioxide to spring the cresylic acid. The solution is passed counter current to flue gas and the freed cresylic acid is extracted with a solvent such as a mixture of naphtha and methanol. The cresylic acid is devolatilized from hydrocarbons and then distilled to separate phenol and *o*-cresol, a mixture of *m*- and *p*-cresol, or a mixture of cresols and xylenols. Isolation of pure components of the cresol and xylenol fractions is accomplished by selective precipitation and by sulfonation.

Phenol and cresols can be extracted by a physical-chemical adsorption process using high surface adsorbent resins. Aqueous streams with low phenol or cresylic acid concentrations are passed through one or more beds of the resin. The beds are regenerated by a 1 percent sodium hydroxide solution or by use of a solvent.[35] The cresylic acid can be obtained from the caustic wash by sulfuric acid. Although this method of removal of phenols was developed to clean-up waste streams, it has possibilities in the cresylic acid production from petroleum processes.

Utilization. Cresylic acid is used as a degreasing agent and as a disinfectant in the form of a stabilized emulsion in a soap solution. The cresols are also used as flotation agents and as wire enamel solvents. In general, however, cresylic acid and the cresols are not used in the acid form because of their acidity. A mixture of *o*- and *m*-cresols is used in the production of phenol-formaldehyde resins. Tricresyl phosphates are produced by the reaction between a mixture of cresols and phosphorus oxychloride. The esters are used as plasticizers for vinyl chloride polymers. Tricresyl phosphate esters are also used as a gasoline additive. Their function is to reduce carbon deposits in the combustion chamber.

NAPHTHENIC ACIDS

Naphthenic acids, the primary acidic portion of crude oil, are straight-chain alkyl cycloparaffin carboxylic acids. The low molecular weight naphthenic acids contain 8 to 12 carbon atoms and are cyclopentane derivatives.

$$\begin{array}{l} \quad CR_2 \qquad\qquad\qquad O \\ \;\;/\;\;\backslash \qquad\qquad\qquad\; \| \\ R_2C \quad CR(CH_2)_n - C - OH \\ \;\;| \qquad\;\; | \\ R_2C - CR_2 \qquad R = H \text{ or an alkyl group (usually } -CH_3) \end{array}$$

Cyclohexane-based naphthenic acids are also present in the low molecular weight fraction but in small amounts. High-boiling naphthenic acids from lube oils are monocarboxylic acids, $C_{14} - C_{19}$, with an average of 2.6 rings.

Naphthenic acids constitute about 50 percent by weight of the total acidic compounds in crude oil. Naphthene-base crudes contain a higher percentage of naphthenic acids than paraffin-base crudes. Consequently, it is more economical to isolate these acids from naphthene-base crudes.[33] Naphthenic acids occur in relatively large amounts (3 percent) in Russian, Romanian and Polish crudes while the percentage is low (0.1-0.3%) in American crudes. Table 4-9 contains the naphthenic acid and phenolic acid analyses of three Egyptian crude oils. It is generally accepted that the naphthenic acids in refinery streams come from the crude oil and are not formed during the refining process.

The concentration of *aliphatic acids,* $R-\overset{O}{\overset{\|}{C}}-OH$, in crude oil is generally low. The lower molecular weight aliphatic acids are essentially absent in the distillate cuts from which naphthenic acids are obtained. Higher molecular weight acids such as palmitic and stearic acids have been isolated from several crudes and may be as high as 7.7 percent of the total acids isolated.[36]

TABLE 4-9—Naphthenic Acid and Phenolic Compound Analyses in Three Egyptian Crude Oils

	Morgan	Amer	Yiedma
Organic acidity, mg KOH/g..	0.344	0.714	0.0508
Naphthenic acid content, wt. %..................	0.184	0.403	0.0015
Phenolics, wt. %...........	0.124	0.0082	......

Production. Although the separation of naphthenic acids from hydrocarbon mixtures is a relatively simple process, their direct extraction from crude oil is not feasible. The naphthenic acids of commercial importance are concentrated in the refinery streams boiling between 400 and 700° F. They are extracted from light gas oil and kerosine fractions by using a 7 to 10 percent caustic solution.

The presence of other acidic compounds which form sodium salts can create isolation problems. Phenols and cresols (cresylic acids), however, do not constitute a particular problem. They are weaker acids than the naphthenic acids and are also concentrated in the gasoline streams. In addition, they require a more concentrated caustic to form sodium salts. Esterification may be used for the separation of these two types of acidic compounds. The methyl esters of the naphthenic acids boil about 50°F lower than the unreacted phenols.[37]

The use of dilute caustic solution for the separation of naphthenic acids is preferred since the naphthenic acid salts are emulsifying agents. When strong caustic solutions are used the solubility of hydrocarbon oils in the salted-out sodium naphthenates increases. The aqueous layer, which contains the naphthenates, is separated from the hydrocarbon layer and treated with dilute mineral acids to spring the acids. The free acids are separated from the aqueous layer, dried and distilled. When purity is not critical, the acids are purified by treatment with strong sulfuric acid. Table 4-10 contains typical specifications for naphthenic acids used to produce driers (Type A) and those used to produce corrosion inhibitors and emulsifiers (Type B).[39]

Utilization. Free naphthenic acids are corrosive and are mainly used as their salts and esters. Their sodium salts are emulsifying agents for the preparation of agricultural insecticides, additives for cutting oils and emulsion breakers in the oil industry.

Other metal salts of naphthenic acids have many and varied uses, especially where an oil-soluble metal compound is required. Calcium naphthenate is a lubricating oil detergent additive, lead naphthenate is widely used as an extreme pressure agent for lubricating oils and zinc naphthenate is an antioxidant. Lead, zinc and barium naphthenates are wetting agents used in pigment grinding and as dispersion agents for paints. Metal naphthenates were extensively used as driers in oil-based paints before this type of paint was replaced for general use by water-based paints. Among these driers are zinc, cobalt and lead naphthenates which are soluble in oil paints, possess good heat stability and are non-corrosive.

Copper naphthenates are used for mildew-proofing sandbags, rope for use at sea and other wood, cotton, jute and hemp products. Copper naphthenate is an especially effective fungicide for cotton textiles.[40] The fungicidal action is associated with the copper in the copper naphthenate. Copper naphthenates are preferred over copper soaps because the naphthenates are taken up by the fiber from the oil soluble solution more readily than copper soaps.

Among the many diversified uses for naphthenates is the use of aluminum naphthenates as a gelling agent for gasoline flame throwers (Napalm). Naphthenates are used as catalysts for various reactions such as the OXO reaction. Manganese naphthenates are oxidation catalysts used to produce carboxylic acids from petroleum waxes. Metal naphthenates are also used as surface detackifiers and carrying agents for polyester resins.

TABLE 4-10—Properties of Two Types of Naphthenic Acids[38]

	Type A*	Type B**
Density (d_4^{20})	0.972	0.987
Viscosity SU/210, °F	40.1	159.0
Pour point, °F	−30	40
Refractive index (n_4^{20})	1.476	1.503
Average molecular weight of deoiled acids	206	330
Unsaponifiable matter (wt. %)	12.5	6.3
Acid number, mg KOH/g	235	0.987

*Used to produce driers
**Used to produce inhibitors and emulsifiers.

LITERATURE CITED

1. Posey, L.G., P.E. Kelly and C.B. Cobb. *The Oil and Gas Journal*, Oct. 23, 1978, pp. 131-136.
2. *Chemical Week*, Aug. 2, 1978, pp. 19-20.
3. Newton, B.F. *Hydrocarbon Processing*, Vol. 57, No. 1, 1978, pp. 181-184.
4. Raymont, M.E.D. *Hydrocarbon Processing*, Vol. 54, No. 5, 1975, pp. 177-179.
5. Burns, R.A., R.B. Lippert, and R.K. Kerr. *Hydrocarbon Processing*, Vol. 53, No. 11, 1974, pp. 181-186.
6. Bechtold, E. *Ber. Bunsen. Physik. Chem.*, Vol. 69, No. 4, 1965, p. 328.
7. Pearson, M.J. *Hydrocarbon Processing*, Vol. 52, No. 2, 1973, p. 81.
8. Raymont, M.E.D. *Hydrocarbon Processing*, Vol. 54, No. 7, 1975, pp. 139-142.
9. *The Oil and Gas Journal*, Aug. 7, 1978, pp. 92-94, 99.
10. Grancher, P. *Hydrocarbon Processing*, Vol. 57, No. 7, 1978, pp. 155-160.
11. Fischer, H. *Hydrocarbon Processing*, Vol. 58, No. 3, 1979, pp. 125-129.
12. *Chemical and Engineering News*, May 26, 1976, pp. 18-19.
13. Stokes, C.A. *Preprints*, Division of Petroleum Chemistry, A.C.S., Vol. 20, No. 3, 1975, pp. 690-701.
14. *Preprints*, Division of Petroleum Chemistry, A.C.S., Vol. 20, No. 2, (1975) 312-464; Vol. 20, No. 3, 1975, pp. 681-710.
15. Gotshall, W.W. *Preprints*, Division of Petroleum Chemistry, A.C.S., No. 20, No. 3, 1975, pp. 702-705.
16. Smith, W.R. *Petroleum Products Handbook*, Vol. 4, New York: McGraw-Hill Book Company, 1960; *Encyclopedia of Chemical Technology*, Vol. 4, Sec. 15, 1969.
17. Emmet, H.P. and T. DeWitt. *Ind. Eng. Anal. Ed.*, Vol. 13, No. 28, 1941.
18. Hahn, A.V.G. *The Petroleum Industry*, New York: McGraw-Hill Book Co., 1970, p. 570.
19. Lahaye, J. and G. Prado. *Preprints*, Division of Petroleum Chemistry, A.C.S., Vol. 20, No. 2, 1975, p. 389.
20. *Hydrocarbon Processing, Petrochemicals Handbook*, November 1975, p. 122.
21. *The Oil and Gas Journal*, April 19, 1976, p. 35.
22. Nelson, W.L. *The Oil and Gas Journal*, May 24, 1975, pp. 60-62.
23. *Hydrocarbon Processing, 1974 Refining Processes Handbook*, September 1974, p. 124.
24. Mantell, G.L. *Preprints*, Division of Petroleum Chemistry, A.C.S., Vol. 20, No. 2, 1975, pp. 312-320.
25. Matula, J.P. and H.N. Weinberg. *The Oil and Gas Journal*, Sept. 18, 1972, pp. 67-71.
26. *The Oil and Gas Journal*, March 10, 1975, pp. 53-56.
27. Metrailer, W.J., R.C. Royle and G.C. Lahn. *Preprints*, Division of Petroleum Chemistry, A.C.S., Vol. 20, No. 3, 1975, pp. 681-689.
28. *Hydrocarbon Processing, 1974 Refinery Processes Handbook*, September 1974, p. 125.
29. Busch, R.A., J.J. Kociscin, H.F. Schroeder, and G.N. Shah. *Hydrocarbon Processing*, Vol. 58, No. 9, 1979, pp. 136-142.
30. Rose, K.E. *Hydrocarbon Processing*, Vol. 50, No. 7, 1971, pp. 85-92.
31. Albers, Barend and D.P. Zwartbol. *The Oil and Gas Journal*, June 4, 1979, pp. 137-141.
32. Rossi, W.J., B.S. Deighton and A.J. MacDonald. *Hydrocarbon Processing*, Vol. 56, No. 5, 1977, p. 107.
33. Lochte, H.L., and E.R. Littman. *Petroleum Acids and Bases*, New York: Chemical Publishing Company, Inc., 1955, p. 124.
34. *Chem. Eng.*, Vol. 66, 1962, pp. 68-70.
35. Fox, C.R. *Hydrocarbon Processing*, Vol. 54, No. 7, 1975, pp. 109-111.
36. Tanaka, Y. and C.A. Kuwata, *C.A.*, Vol. 23, 1929, p. 4051.
37. Nenitzescu, C.P., D.A. Isacescu and T.A. Volrap. *Ber.*, Vol. 71B, 1938, p. 2056.
38. Matson, John A. *The Oil and Gas Journal*, Mar. 24, 1980, pp. 93-94.
39. Jolly, S.E. *Encyclopedia of Chemical Technology*, Vol. 13, New York: Interscience Publishers, John Wiley & Sons, Inc., 1967, p. 727.
40. Mari, A., S.A. Fam, A.A. Abou-Zeid, and M.I. Khalil. Second Arab Conference on Petrochemicals, Paper No. 15, (P-3), Abu Dhabi, March 15-22, 1976.
41. Matson, John A. *The Oil and Gas Journal*, Mar. 24, 1980, pp. 93-94.

5

Petrochemicals from Methane

Petrochemical processes begin with relatively few basic raw materials, expand into a complex network of chemicals and converge to materials that serve specific functions as consumer products.[1] The raw material base for the petrochemical industry primarily depends upon the types of intermediates and final products required by industry and the consumer. Almost all petrochemicals are derived from three sources:

1. Carbon monoxide/hydrogen (synthesis gas, syn gas) from reforming natural gas (methane)
2. Olefins from pyrolysis of ethane, propane-butane (LPG, LP-gas) or distillates
3. Aromatics from catalytic reforming.

The three main sources for petrochemicals lead to products which are marketable items in their own right, as well as raw materials for a great many other petrochemicals used both as intermediates and as finished products. One man's product is often another man's raw material. For example, a producer of basic petrochemicals could consider methane (natural gas) as his only raw material and synthesis gas (CO/H_2), after conversion to methanol, as his finished product. An intermediate producer uses the merchant methanol as raw material to produce formaldehyde as a finished product while a resin manufacture would see the formaldehyde as a basic raw material for the production of phenolformaldehyde resins.

How does methane, CH_4, that unassuming little molecule, fit into this picture? In a very large way—provided it isn't burned to death in our mad scramble for energy! In spite of its small size, methane is the precursor of a wide variety of compounds as shown in Fig. 5-1. It is not easy, however. First the molecule must be rent assunder and then reformed into useful products such as methanol, formaldehyde, ammonia, and urea. These had an overall United States production in 1978 of 53.3 billion pounds. The transformation of methane into these compounds is effected through synthesis gas, which may come from many sources other than methane.

SYNTHESIS GAS

Synthesis gas, *syn gas*, is a general term used to designate various mixtures of carbon monoxide and hydrogen. These mixtures are used as such, and they are also sources of pure hydrogen and pure carbon monoxide.[2] The mixtures of CO/H_2 can be produced from almost anything containing carbon and hydrogen—from methane to manure,[3] from coal[2] to crude oil residues.[4]

Production. Two general types of reactions are used for the production of syn gas: partial oxidation and steam reforming. Steam reforming is the more important process when methane (natural gas) is the carbon-hydrogen source. This process is extensively documented in the literature. Partial oxidation is primarily used for heavy fuel and resid. Table 5-1 gives the composition of gaseous products from both sources at thermodynamic equilibrium.[2]

Typical conditions for the steam reforming of methane to produce a 3 to 1 ratio of hydrogen to carbon monoxide are

$$CH_4 + H_2O \rightarrow 3\ H_2 + CO \quad \Delta H_{800\,°C} = +\,54.2\ \text{kcal}$$

Temperature:	830-850° C
Pressure:	400-500 psig
Catalyst:	Promoted nickel based

Partial oxidation of methane

$$2\ CH_4 + O_2 \rightarrow 2\ CO + 4\ H_2 - \Delta H$$

is a non-catalytic process operating in a temperature range of 1,300-1,500°C at pressures of 200-2,000 psig.[5] When heavy resids are partially oxidized by oxygen and steam at

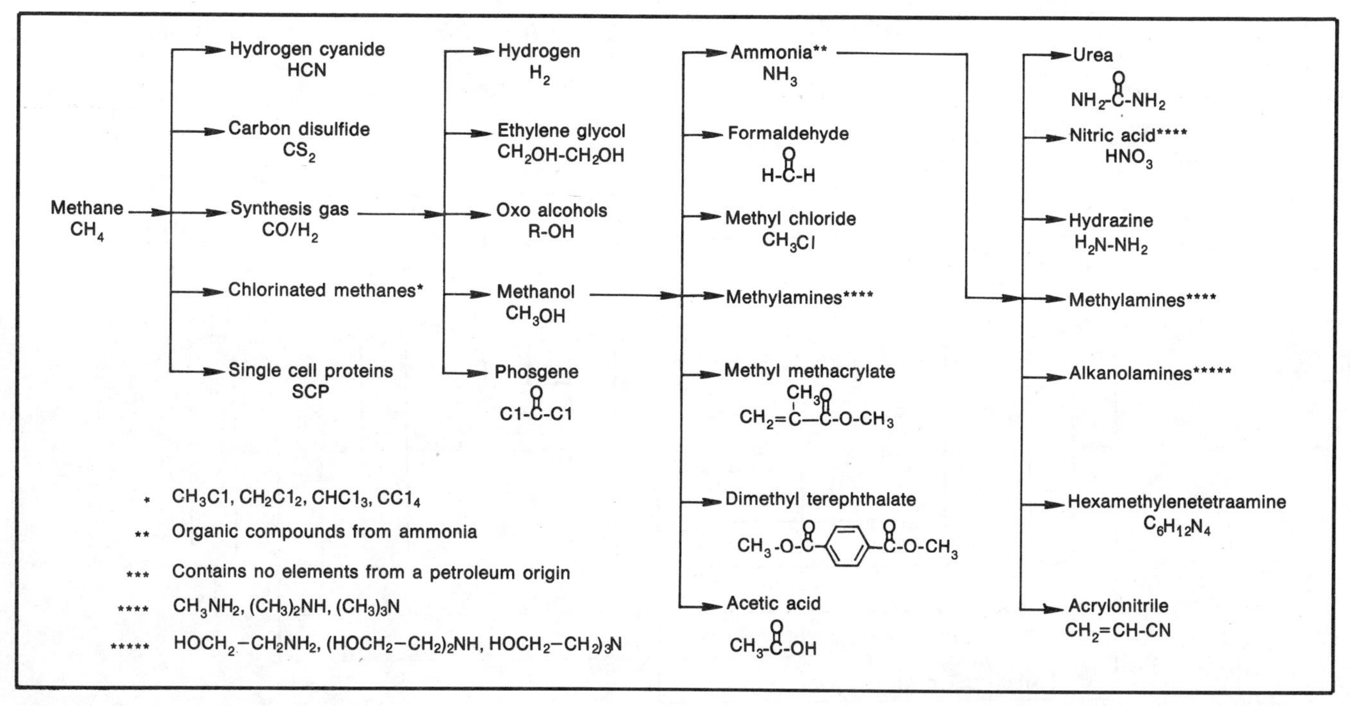

Fig. 5-1—Major compounds derived directly or indirectly from methane. Many of these compounds also have other raw material sources.

TABLE 5-1—Composition of Reformed Gas and Partial Oxidation Gas[2]

	Volume % dry sulfur free				
Production process	CO	H_2	CO_2	$N_2 + A$	CH_4
Steam-natural gas reforming...	15.5	75.7	8.1	0.2	0.5
Partial oxidation-heavy fuel oil..	47.5	46.7	4.3	1.4	0.3

1,400-1,450°C and 55-60 atm pressure the product gas consists of equal parts of hydrogen and carbon monoxide,[6] Fig. 5-2.[7]

The shift reaction

$$CO + H_2 \rightleftarrows CO_2 + H_2$$

is used to produce pure hydrogen.[8] Typical reaction temperatures are 425-500°C over an iron oxide catalyst promoted with chromium oxide. Carbon monoxide is purified by a copper liquid absorption method and by cryogenic separation.[2]

Chemicals from Syn Gas

Pure hydrogen is produced from Syn Gas. Hydrogen combined with atmospheric nitrogen is used for the production of ammonia. Ammonia is the parent compound of many chemicals—especially compounds used as fertilizers. Syn Gas with the appropriate ratio of carbon monoxide to hydrogen is used to produce methanol, oxo alcohols, ethylene glycol, and other chemicals. Methanol is a precursor for a host of other important chemicals such as formaldehyde, acetic acid, methyl chloride, and methylamines (see Fig. 5-1).

Carbon monoxide from Syn Gas is used for carbonylation reactions and in the hydroformylation reactions (oxo reactions). The latter reactions are used for the production of oxo alcohols. Carbonylation and hydroformylation reactions will be discussed when they are applied to a specific raw material such as methanol and propylene.

Ammonia. Direct synthesis of ammonia, *azane,* from hydrogen and atmospheric nitrogen is a classic heterogeneous catalytic reaction. Ammonia is one of the most important inorganic chemicals in the market place and is exceeded in production only by sulfuric acid and lime. The United States produced 33.9 billion pounds of ammonia in 1978. The world demand is predicted to reach 300 billion pounds by 1990.[9] The main consumer of ammonia is the fertilizer industry (about 80%). Anhydrous ammonia is utilized as a fertilizer by direct application to soils, and in the form of various compounds such as ammonium nitrate, urea, and ammonium phosphate. Ammonia is also used to produce plastics, resins, fibers, and explosives.

Production. The Haber Process is the most widely used process for ammonia production. A multi promoted iron oxide catalyst accelerates the reaction. The reaction is very slow at normal room temperatures, though conversion is theoretically high.

$$N_{2(g)} + 3\ H_{2(g)} \rightleftarrows 2\ NH_{3(g)} \quad \Delta H_{25°C} = -22.08 \text{ kcal}$$

By LeChatelier's Principle, high pressures and low temperatures favor the equilibrium production of ammonia. Promoted iron oxide, Fe_3O_4, with other oxides accelerate the reaction rate so that good yields can be obtained in a reasonable time at moderate temperatures. Fig. 5-3 shows the Pullman-Kellogg Process for ammonia production.[11]

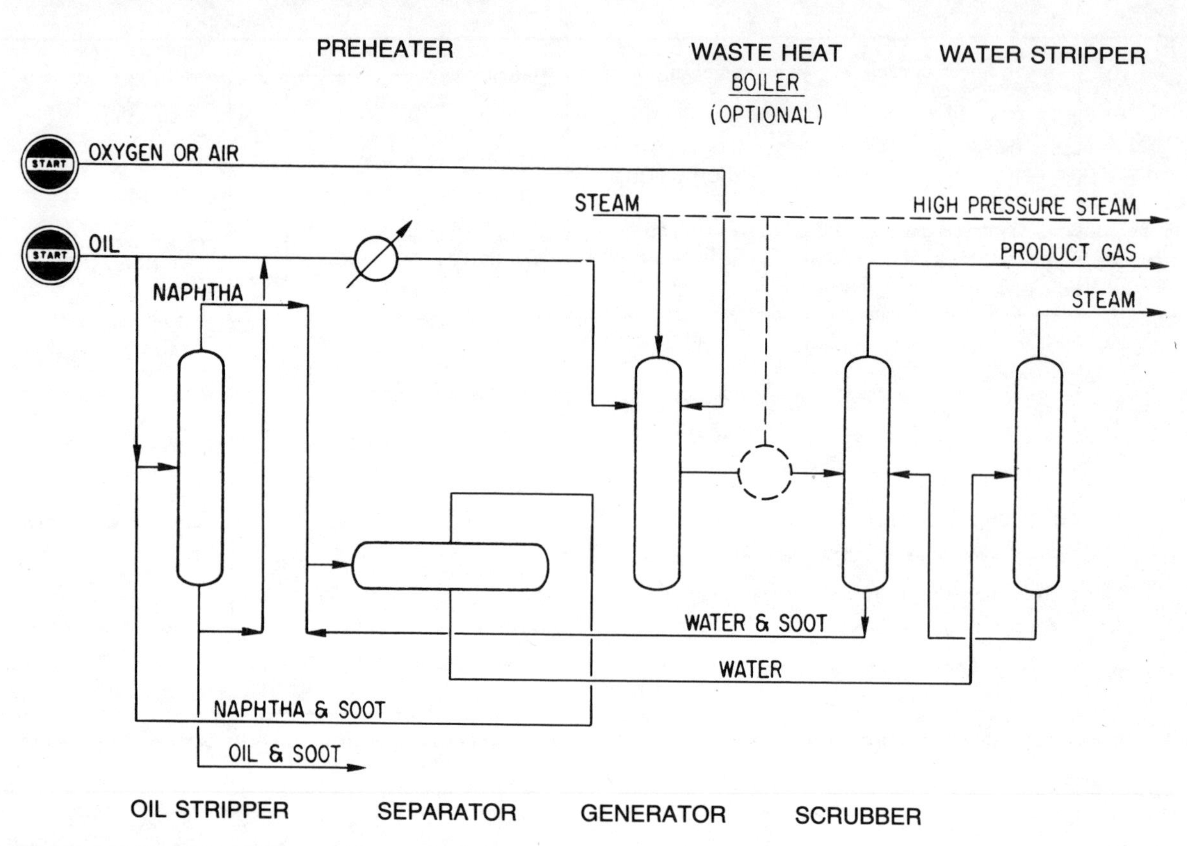

Fig. 5-2—The Texaco Development Corporation process for the production of synthesis gas.[7]

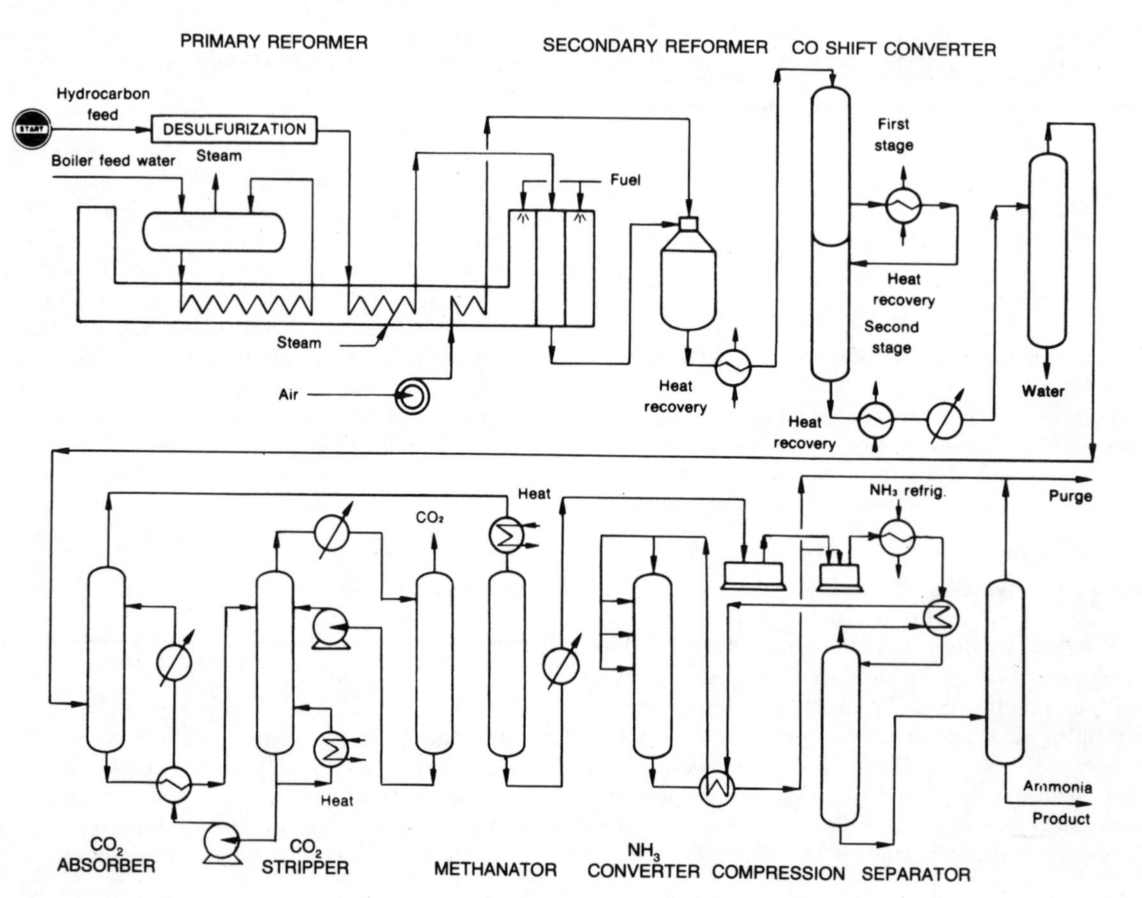

Fig. 5-3—The Pullman Kellogg process for the production of ammonia.[11]

Synthetic ammonia pressure ranges from 2,000-4,500 psig for large units. In practice, hydrogen and nitrogen in the ratio of 3:1 are prepared in the secondary reformer where air is introduced to supply the nitrogen. The combustion heat of partially reformed gas supplies the energy to reform the remainder of the gas.

Urea. A byproduct from ammonia production—carbon dioxide—reacts with ammonia to produce urea. Urea is an important solid fertilizer containing about 45% nitrogen. The fertilizer use of urea accounts for about 75% of its production. Other urea uses are animal feeds (10%) and adhesives, plastics, and resins (15%).[10]

Production. The production of urea is based on the two-step reaction of carbon dioxide and ammonia. In the first step, ammonium carbamate is formed by an exothermic reaction.

$$CO_{2(g)} + 2\ NH_{3(g)} \rightleftarrows NH_2COONH_{4(s)}$$

Equilibrium production is favored at high pressures and low temperatures. High operating pressures are desirable for the separation absorption which results in a higher carbamate solution concentration. Reactor temperature ranges between 170-190°C and the pressure is about 150 atmospheres. The NH_3/CO_2 mole ratio is about 3:1.

The second step is the decomposition of the carbamate to urea and water at temperatures of about 200°C. The reaction is endothermic.

$$NH_2COONH_{4(s)} \rightleftarrows NH_2CONH_{2(s)} + H_2O_{(g)}$$

$$\Delta H_{25^\circ C} = 6.32 \text{ kcal}$$

Unreacted ammonia is sent to the carbamate decomposer as a stripping agent. The urea solution leaving the carbamate decomposer is expanded by heating at low pressures; the ammonia is recycled. Decomposition in the presence of ammonia inhibits corrosion.

In the urea plant, corrosion is one of the most serious problems—especially with the recycle pumps which handle the hot, highly corrosive ammonium carbamate solution. Either reciprocating pumps with a modified cylinder design (for large plant capacities) are used, or high pressure centrifugal carbamate solution pumps that are made of high chromium two-phase stainless steel alloy (for small capacity plants). Titanium is used for the reactor lining. The reactor operates at high NH_3/CO_2 ratios and high temperatures and pressures.[12] Fig. 5-4 shows the Snamprogetti urea process.[13]

Nitric Acid. Nitric acid is commerically produced by oxidizing ammonia with air over a platinum-rhodium wire gauze catalyst. The reactions involved are highly exothermic.

$$4\ NH_{3(g)} + 5\ O_{2(g)} \rightarrow 4\ NO_{(g)} + 6\ H_2O_{(g)}$$

$$\Delta H_{25^\circ C} = -21 \text{ kcal}$$

$$2\ NO_{(g)} + O_{2(g)} \rightarrow 2\ NO_{2(g)}$$

$$\Delta H_{25^\circ C} = -27 \text{ kcal}$$

$$3\ NO_{2(g)} + H_2O_{(l)} \rightarrow 2\ HNO_{3(aq)} + NO_{(g)}$$

$$\Delta H_{25^\circ C} = -32 \text{ kcal}$$

Although high pressures favor the second reaction, the reverse is true for the first reaction. Therefore, the preferred reaction conditions are atmospheric pressure and a temperature of about 900°C. The atmospheric pressure conversion gives low platinum loss and high yield. Nitric acid is mainly used for the production of ammonium nitrate. It is also used as a nitrating agent for paraffins and aromatic compounds.

Hydrazine, NH_2-NH_2. Hydrazine, *diazane,* is produced by the Rashing Process. Ammonia is oxidized using sodium hypochlorite to produce chloramine, NH_2Cl, which further reacts with ammonia to produce hydrazine.

$$2\ NH_3 + NaOCl \rightarrow NH_2\text{-}NH_2 + NaCl + H_2O$$

The hydrazine is then evaporated from the sodium chloride solution and sold as the hydrate, $N_2H_4 \cdot H_2O$.

Hydrazine can also be produced by the Puck Process which uses hydrogen peroxide as the oxidizing agent.

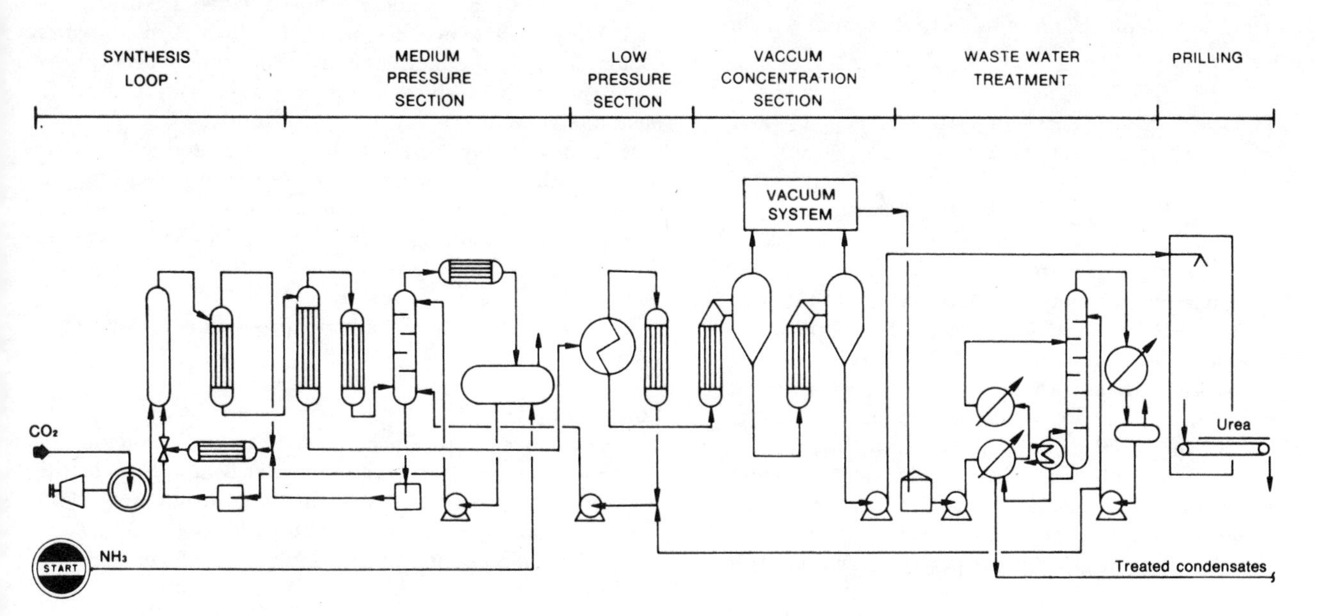

Fig. 5-4—The Snamprogetti process for the production of urea from liquid ammonia and carbon dioxide.[13]

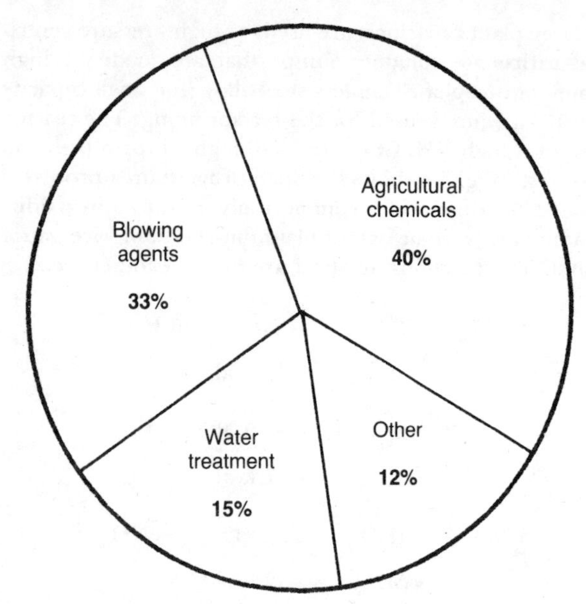

Fig. 5-5—The world-wide markets of hydrazine.[14]

$$2\ NH_3 + H_2O_2 \rightarrow NH_2\text{-}NH_2 + 2\ H_2O$$

Newer plants now use a ketone route for the production of hydrazine. The current world production is 40,000 tons/year.

Anhydrous hydrazine is known for its use as a rocket fuel. Hydrazine-based explosives are being evaluated for fracturing gas-bearing formations. However, hydrazine also has many other uses as shown in Fig. 5-5. It is used for the production of amino cresols, for azodicarbonamide in pesticides, and in synthetic rubber as an anti-corrosion inhibitor.[14, 15]

Hydrazine is an efficient reducing agent. It scavenges oxygen from steam boilers and can selectively reduce a nitro group without reducing the double bond.

$$O_2N\text{-}C_6H_4\text{-}CH{=}CH\text{-}COOH + NH_2\text{-}NH_2 \rightarrow H_2O + NO_2 + H_2N\text{-}C_6H_4\text{-}CH{=}CH\text{-}COOH$$

METHANOL

Methanol has many claims to distinction in addition to being synthesized from syn gas, the feedstock of many different sources. In 1978, the United States produced 6.36 billion pounds of methanol, which ranks fourth behind urea, ethylene dichloride, and vinyl chloride among the nonhydrocarbon chemicals. Of these compounds, methanol is far more versatile in its utilization. The world-wide production capacity for 1985 is estimated at 5.86 billion gallons.[16] Although most of the current applications for methanol are mature, the 6.5 percent growth rate per year for the 1977-1990 period is caused by continued growth in most existing applications, and additional demand in acetic acid, methyl tertiary butyl ether (**MTBE**), single cell protein (**SCP**), gasoline blending, and peak power generation for utilities. In 1979, the world's effective methanol production capacity was about 3.5 billion gallons.

Production. The production of methanol is represented by the equation

$$2\ H_2 + CO \rightarrow CH_3OH \qquad \Delta H^\circ_{298} = -30.6\ \text{kcal/mole}$$

This reaction is effected by either a high pressure process or a low pressure process. The high pressure process operates at 400°C over a copper or zinc-chromium oxide ($Zn\text{-}CrO_3$) catalyst at pressures between 4,000 and 6,500 psig.

The low pressure synthesis of methanol is currently the process of choice with all new capacity being of this type. Reaction conditions for the Lurgi process (Fig. 5-6)[6, 18] are

Temperature:	250-260°C
Pressure:	725-1,176 psig
Catalyst:	Cu + Small amounts of Zn + a "component which will increase its resistance to aging."

Zinc oxide is also used in place of zinc.

The Imperial Chemical Industries (ICI) low pressure process operates in a temperature range of 250 to 300°C at pressures between 750 and 1,500 psi over a copper-aluminum-zinc synthesis catalyst.[19]

The synthesis reaction takes place over a bed of heterogeneous catalyst arranged in either sequential adiabatic beds or placed within heat transfer tubes. This may change, however, with the introduction of Chem Systems three phase fluidized bed reactor (Fig. 5-7).[20] The synthesis gas is passed upward into the reactor concurrent with an inert liquid hydrocarbon. The catalyst is fluidized by the hydrocarbon which also absorbs the heat of reaction thus giving excellent temperature control. Separation among the solid, liquid and gas phases occurs at the top of the reactor. The catalyst remains in the reactor and the liquid hydrocarbon is recycled. The gas phase, which contains 15-20 vol.% methanol, is cooled to obtain the methanol. The conventional two phase system normally produces a reactor exit gas with a 5-6 vol.% methanol content.

Methanol production is usually an integrated, continuous process with gaseous or liquid hydrocarbons furnishing the carbon monoxide and hydrogen by either steam reforming or partial oxidation (Fig. 5-6).[6, 18] The relationship between the cost of methanol production and the cost of the raw material for the syn gas is shown in Fig. 5-8.[21]

Approximately 50% of methanol production is oxidized to formaldehyde.

Formaldehyde

Formaldehyde is produced by the catalyzed oxidation of methanol and (along with various other compounds) by the non-catalytic oxidation of propane-butane mixtures. This latter method, however, accounts for only about 8% of the annual 1.6 billion pound production of 100% formaldehyde.

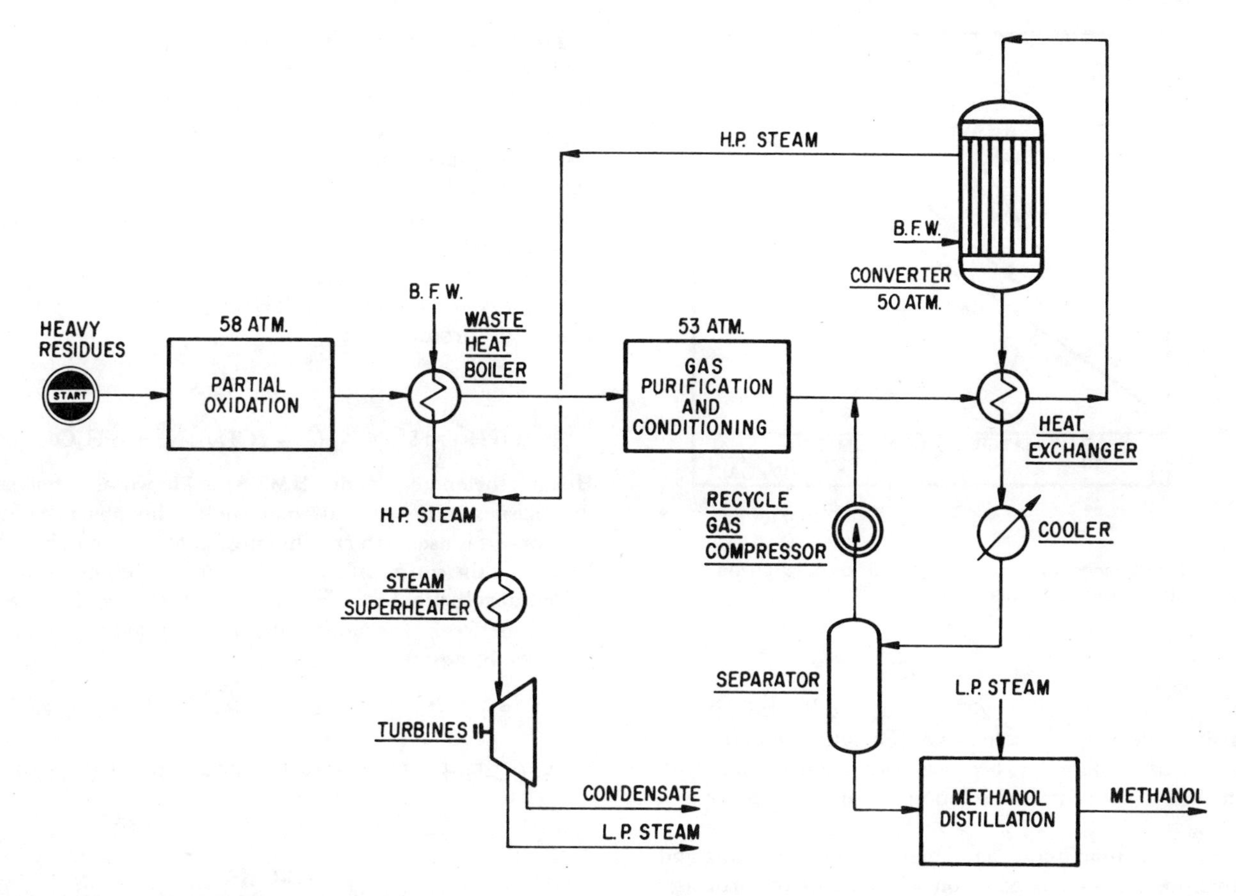

Fig. 5-6—Schematic diagram of the Lurgi low-pressure process for the production of methanol from syn gas with partial oxidation of heavy residues as the source of the syn gas.[6]

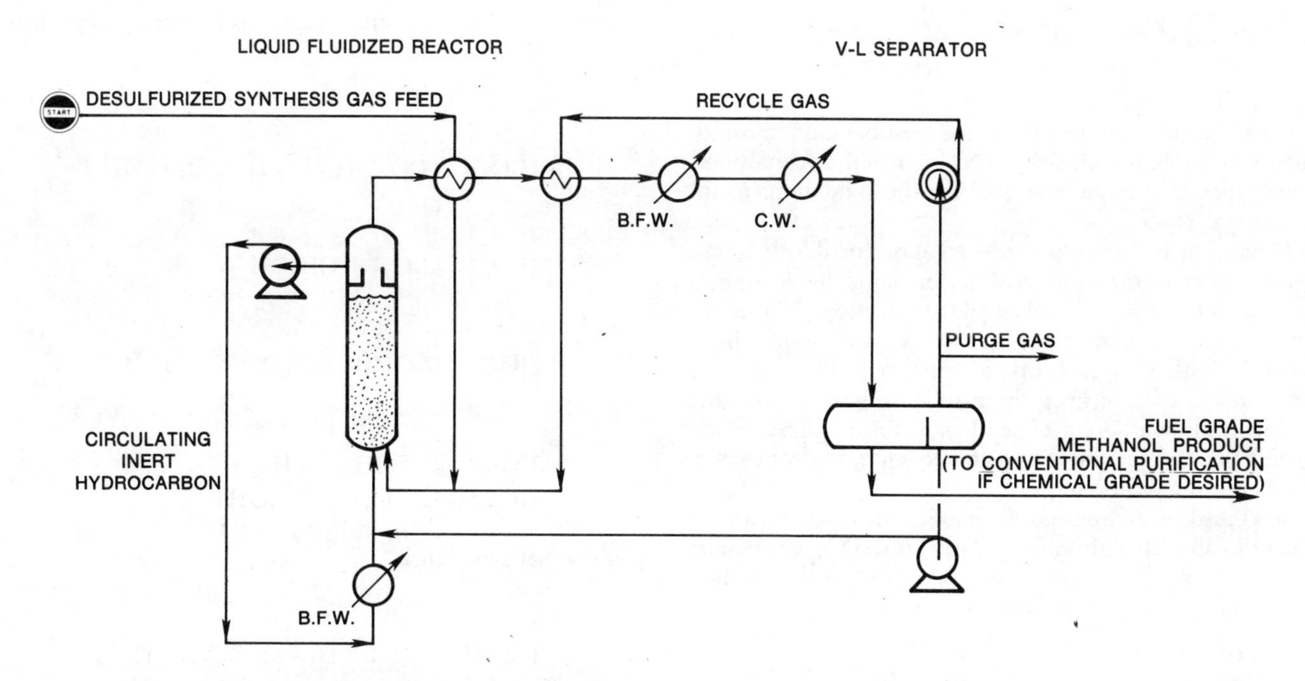

Fig. 5-7—The Chem Systems three-phase methanol synthesis process.[20]

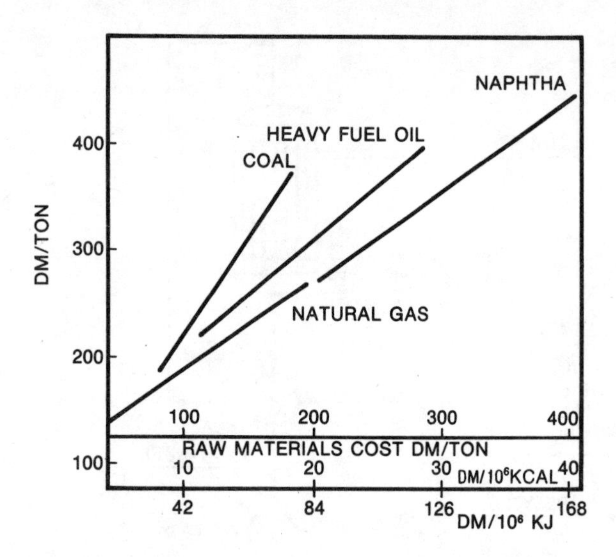

Fig. 5-8—The dependency of methanol production costs on the cost of the raw material used to produce syn gas.[21]

Formaldehyde is usually sold as a 37% solution in water with methanol as a stabilizer because formaldehyde tends to polymerize in concentrated solution and in the absence of a solvent.

There is no unanimous agreement on the reactions and mechanisms for the air oxidation of methanol to formaldehyde. Both reactions and mechanisms depend upon operating conditions and catalyst composition. There are, however, two principal reactions:[22]

$$CH_3OH \rightarrow H\text{-}\overset{\overset{\displaystyle O}{\|}}{C}\text{-}H + H_2 \qquad \Delta H^\circ_{298} = +20.4 \text{ kcal}$$

$$2\,H_2 + O_2 \rightarrow 2\,H_2O \qquad \Delta H^\circ_{298} = -57.8 \text{ kcal}$$

At least eight side reactions are possible and most of them are probable under any set of reaction conditions. Two types of catalysts are used for this oxidation: metal and metal oxide.

Oxidation by air over a silver gauze (or 0.5 to 3 mm grains) is of historic interest but commercially represents only about 8% of formaldehyde production. The silver catalyzed process operates at 600° C and atmospheric pressure with a large excess of methanol. The excess of methanol is used to keep the reaction mixture above the upper explosion limits of methanol/air mixtures which is 36.5 vol. % methanol. As a consequence, the conversion is low but the yield is high.

5-9 is based on an iron-molybdenum oxide catalyst with the active catalyst being iron molybdate, $Fe(MoO_4)_3$. Chromium or cobalt oxides are sometimes used to dope the catalyst. General operating conditions are:

Temperature:	400-425° C
Pressure:	Atmospheric
Catalyst:	$Fe(MoO_4)_3$
Air/methanol ratio:	Large excess of air
Yield (conversion):	98%

The large excess of air is used to keep the air/methanol ratio below the lower explosion limits.

Utilization. Formaldehyde has many and varied uses, but its primary use is in the production of urea-formaldehyde resins and glues. The glues are used in the production of particle boards and plywood. Phenol-formaldehyde resins and glues and melamine resins also take a substantial portion of formaldehyde production.

One of the major synthesis uses of formaldehyde is *hexamethylenetetramine,* $(CH_2)_6N_4$ the reaction with ammonia to produce

$$6\,H\text{-}\overset{\overset{\displaystyle O}{\|}}{C}\text{-}H + 4\,NH_3 \rightarrow (CH_2)_6N_4 + 6\,H_2O$$

Hexamethylenetetramine, **HMTA**, is known as *hexamine* in the plastics industry and *urotropin* in the pharmaceutical industry. It is used as a crosslinking agent for phenolic resins where it is the source of formaldehyde to effect crosslinking.

Pentaerythritol, $C(CH_2OH)_4$, is prepared by the reaction between formaldehyde and acetaldehyde in the presence of a base.

$$4\,H\text{-}\overset{\overset{\displaystyle O}{\|}}{C}\text{-}H + CH_3\text{-}\overset{\overset{\displaystyle O}{\|}}{C}\text{-}H + NaOH \rightarrow C(CH_2OH)_4 + H\text{-}\overset{\overset{\displaystyle O}{\|}}{C}\text{-}ONa$$

Pentaerythritol is used in the production of alkyd resins and rosin resins.

One of the processes for the preparation of *ethylene glycol* is the reaction of formaldehyde with carbon monoxide and water to form glycolic acid. The glycolic acid is esterified with methanol; then the ester is hydrogenated to ethylene glycol and methanol.

$$H\text{-}\overset{\overset{\displaystyle O}{\|}}{C}\text{-}H + CO + H_2O \rightarrow HOCH_2\text{-}\overset{\overset{\displaystyle O}{\|}}{C}\text{-}OH$$

$$HOCH_2\text{-}\overset{\overset{\displaystyle O}{\|}}{C}\text{-}OH + CH_3OH \rightarrow HOCH_2\text{-}\overset{\overset{\displaystyle O}{\|}}{C}\text{-}OCH_3 + H_2O$$

$$HOCH_2\text{-}\overset{\overset{\displaystyle O}{\|}}{C}\text{-}OCH_3 + 2\,H_2 \rightarrow HOCH_2\text{-}CH_2OH + CH_3OH$$

The net reaction is

$$H\text{-}\overset{\overset{\displaystyle O}{\|}}{C}\text{-}H + CO + 2\,H_2 \rightarrow HOCH_2\text{-}CH_2OH$$

All of the reactants in the net reaction come directly or indirectly from syn gas. The intermediate, glycolic acid, is used in chelating formulations for iron.

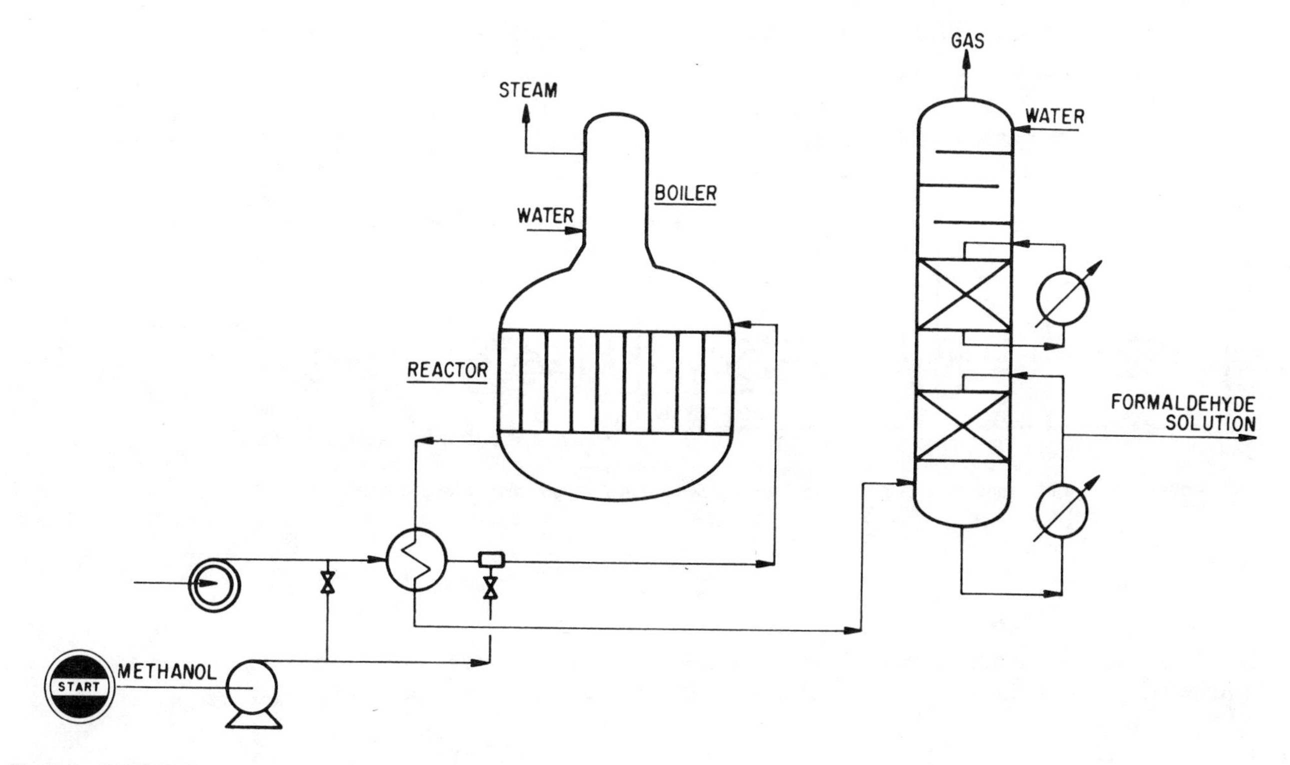

Fig. 5-9—IFP-CdF Chimie iron-molybdenum oxide catalyzed oxidation of methanol to formaldehyde utilizes this type of process arrangement.[22, 23]

Acetic Acid

Acetic acid is produced commercially by three major processes: oxidation of acetaldehyde, oxidation of *n*-butane and carbonylation of methanol. The liquid phase carbonylation of methanol has been developed by Monsanto within the last decade.

The Monsanto process can be represented by the simple reaction

$$CH_3OH + CO \rightarrow CH_3\text{-}\overset{\overset{\displaystyle O}{\|}}{C}\text{-}OH$$

The carbonylation reaction uses a monomeric rhodium iodide complex in a homogeneous catalytic cycle. The reaction conditions are:

Temperature:	200°C
Pressure:	215 psig
Catalyst:	Rhodium promoted by iodine
Yield:	99%

The reaction is reported to take place by a five-step mechanism.[24] Fig. 5-10 gives a flow diagram of the process.[25] The main byproduct comes from the water-gas reaction.

$$CO + H_2O \rightarrow CO_2 + H_2$$

This process for the production of acetic acid has an advantage over the other commercial processes because both the methanol and the carbon monoxide come from the same source—syn gas. Liquid phase carbonylation of methanol will very likely account for essentially all new world-wide acetic acid capacity.

Methyl Chloride

Methyl chloride is produced by the chlorination of methane (*ca* 35%) and by the reaction between methanol and hydrogen chloride (vapor phase) or hydrochloric acid (liquid phase).

$$CH_3OH + HCl \rightarrow CH_3Cl + H_2O$$

Vapor Phase

Temperature:	340-350° C
Pressure:	Substantially atmospheric
Catalyst:	Various*
Yield:	95%

*Ignited alumina gel, zinc chloride on pumice, cuprous chloride or activated carbon.

Liquid Phase

Temperature:	100-150° C
Catalyst:	Zinc chloride

Methyl chloride also can be produced from methanol and hydrochloric acid which is generated by the reaction between sodium chloride and sulfuric acid.

$$2\,CH_3OH + H_2SO_4 + 2\,NaCl \rightarrow 2\,CH_3Cl + Na_2SO_4 + 2\,H_2O$$

Methyl chloride is primarily an intermediate for other chemicals. The production of silicones represents about 60% of methyl chloride utilization with the remainder

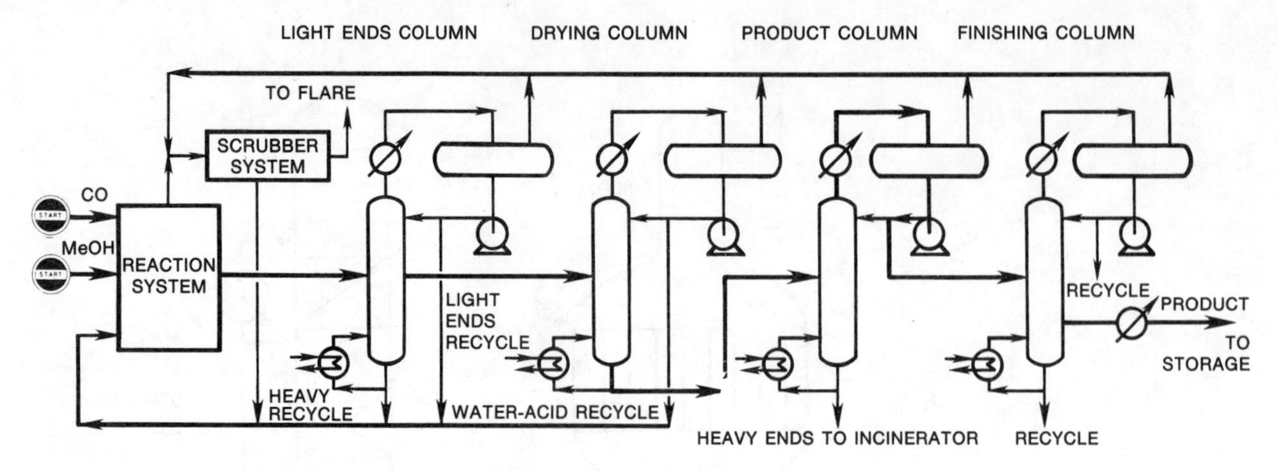

Fig. 5-10—Arrangement of Monsanto's methanol carbonylation process for the production of acetic acid.[25]

distributed among tetraethyllead and methyl cellulose production. The 1978 production of methyl chloride was 454 million pounds.

Methylamines

The synthesis of the methylamines from methanol and ammonia presents an interesting problem of product distribution.

$$CH_3OH + NH_3 \rightarrow CH_3NH_2 + H_2O$$
$$CH_3OH + CH_3NH_2 \rightarrow (CH_3)_2NH + H_2O$$
$$CH_3OH + (CH_3)_2NH \rightarrow (CH_3)_3N + H_2O$$

Temperature: 380-450° C
Pressure: 200 psi
Catalyst: Alumina-gel
Yield: 95% (on methane), 97% (on ammonia)
Mole ratio NH_3:CH_3OH 2:1

Ratio of products (mole %) at equilibrium (450° C):

Monomethylamine (**MMA**)	43
Dimethylamine (**DMA**)	24
Trimethylamine (**TMA**)	33

Unfortunately dimethylamine is the most widely used of the three methylamines. The relative yield of **DMA** is increased by recycling the monomethylamine.

The primary use of *monomethylamine* is in the synthesis of Sevin (Carboryl), a widely used insecticide. **MMA** is also used in the production of other pesticides and pharmaceuticals. *Dimethylamine* has many different uses, mainly in the synthesis of solvents (dimethylformamide and dimethylacetamide) for acrylic and polyurethane fibers and in the synthesis of various surface-active agents and rubber chemicals as well as unsymmetrical dimethylhydrazine which is used as a rocket fuel. *Trimethylamine* has only one major use and that is in the synthesis of choline salts which are high-energy additives for poultry feed. **TMA** is used to some extent to make anionic ion-exchange resins.

Methyl Methacrylate

Methanol is used in the synthesis of many methyl esters. One of the most interesting is methyl methacrylate because of its novel synthesis. It all starts with hydrogen cyanide and acetone in the presence of a base.

$$CH_3\text{-}\overset{\overset{\displaystyle O}{\|}}{C}\text{-}CH_3 + HCN \xrightarrow{OH^-} CH_3\text{-}\underset{\underset{\displaystyle OH}{|}}{\overset{\overset{\displaystyle CH_3}{|}}{C}}\text{—}CN$$

With sulfuric acid this becomes the sulfuric acid salt of the amide of methacrylic acid.

$$CH_3\text{-}\underset{\underset{\displaystyle OH}{|}}{\overset{\overset{\displaystyle CH_3}{|}}{C}}\text{—}CN + H_2SO_4 \rightarrow CH_2 = \overset{\overset{\displaystyle CH_3}{|}}{C}\text{—}\overset{\overset{\displaystyle O}{\|}}{C}\text{-}NH_3^+HSO_4^-$$

The amide salt is esterified by methanol to give methyl methacrylate.

$$CH_2 = \overset{\overset{\displaystyle CH_3}{|}}{C}\text{——}\overset{\overset{\displaystyle O}{\|}}{C} - NH_3^+ HSO_4^- + CH_3OH \rightarrow$$

$$CH_2 = \overset{\overset{\displaystyle CH_3}{|}}{C}\text{—}\overset{\overset{\displaystyle O}{\|}}{C}\text{-}OCH_3 + NH_4HSO_4$$

The *hydrogen cyanide* used in this synthesis is produced directly from methane, and indirectly by a process that utilizes carbon monoxide, ammonia, and methanol. The methanol is recovered and recycled.

$$CO + CH_3OH \rightarrow H\text{-}\overset{\overset{O}{\|}}{C}\text{-}OCH_3$$

$$H\text{-}\overset{\overset{O}{\|}}{C}\text{-}OCH_3 + NH_3 \rightarrow H\text{-}\overset{\overset{O}{\|}}{C}\text{-}NH_2 + CH_3OH$$

$$H\text{-}\overset{\overset{O}{\|}}{C}\text{-}NH_2 \rightarrow HCN + H_2O$$

The net reaction is

$$CO + NH_3 \rightarrow HCN + H_2O$$

Dimethyl Terephthalate

Methanol is used for the esterification of terephthalic acid to dimethyl therephthalate (**DMT**). This ester is subsequently transesterified with ethylene glycol to make polyester fibers and film.

$$2\ CH_3OH + HO\text{-}\overset{\overset{O}{\|}}{C}\text{-}C_6H_4\text{-}\overset{\overset{O}{\|}}{C}\text{-}OH \rightarrow$$

$$CH_3O\text{-}\overset{\overset{O}{\|}}{C}\text{-}C_6H_4\text{-}\overset{\overset{O}{\|}}{C}\text{-}OCH_3 + 2\ H_2O$$

$$n\ CH_3O\text{-}\overset{\overset{O}{\|}}{C}\text{-}C_6H_4\text{-}\overset{\overset{O}{\|}}{C}\text{-}CH_3 + n\ HOCH_2\text{-}CH_2OH \rightarrow$$

$$\left[-OCH_2\text{-}CH_2O\text{-}\overset{\overset{O}{\|}}{C}\text{-}C_6H_4\text{-}\overset{\overset{O}{\|}}{C}\text{-}\right]_n + 2n\ CH_3OH$$

This is an example of methanol playing an important but transitory role in the production of consumer products.

Methanol's Future.

Methanol may have a more important role as a basic petroleum building block in the future because of the multisources of syn gas. When the oil and gas run their course there will still be coal and after coal there will be wood—all of these can be converted into syn gas and from there to methanol.

It is now possible to convert natural gas and coal to methanol and then to gasoline with a research octane number (RON) between 90 and 100 (Fig. 5-11).[26]

$$n\ CH_3OH \rightarrow (CH_2)_n + n\ H_2O$$

Reactions conditions are:

Temperature:	453°C
Pressure:	25 psi
Catalyst:	Shape selective zeolite
Recycle ratio:	9:1
Yields (wt.%):	Hydrocarbons 43.4 Water 56

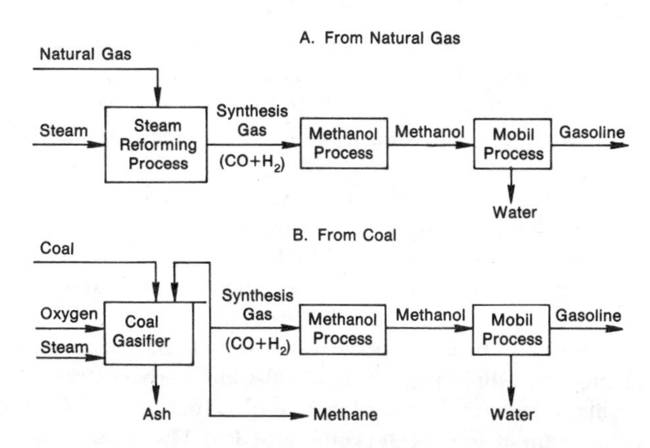

Fig. 5-11—The Mobil methanol-to-gasoline process starting with either natural gas or coal.[26]

Of the hydrocarbons, 89.7% are in the gasoline range with a research octane number of 92.[27] A flow diagram of the gasoline producing unit is shown in Fig. 5-12.[28]

The reaction between methanol and isobutylene to produce methyl *tert*-butyl ether (**MTBE**), a gasoline additive, is another utilization with appreciable potential. The reaction is noted in Chapter 9. And another example of the synthesis potential of methanol is the production of aromatics by passing methanol over a high selica-zeolite catalyst at 260°C.[29] The conversion is quantitative with a selectivity of 40-50 percent. Still another example is the production of

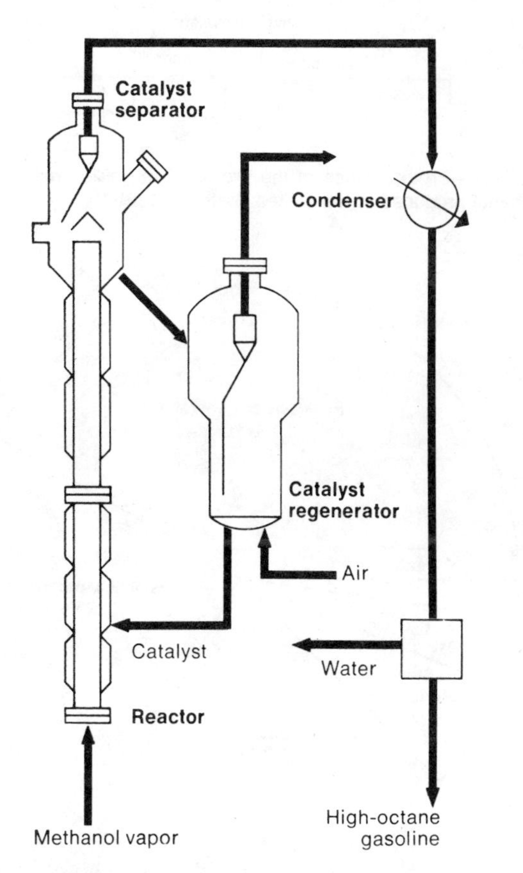

Fig. 5-12—The reactor unit of the Mobil process for producing gasoline from methanol.[28]

ethylene. In this process methanol and syn gas are passed over a cobalt catalyst to produce ethanol. The ethanol is then dehydrated to ethylene. While this process is currently not attractive, it could become a viable one in the not too distant future.

It has been proposed that methanol serve as a hydrogen carrier. Methanol can be readily transported from the areas of production such as the Middle East to the areas where the hydrogen will be utilized. The methanol is then decomposed into hydrogen and carbon dioxide (Fig. 5-13).[30]

There are two potential uses of methanol that could change the whole picture of world-wide methanol requirements. These are the use of methanol as fuel[31, 32] and for the production of single cell proteins, **SCP**.[21] The acceptance of either or both of these uses could increase methanol production by several orders of magnitude. Distribution of the predicted 1985 world-wide 68 million metric ton production is given in Fig. 5-14.[17] This production estimate is considered to be conservative. A prediction of 200 million metric tons has been made with the major differences being in the much larger predicted methanol demand for motor fuel and by the steel industry.[16]

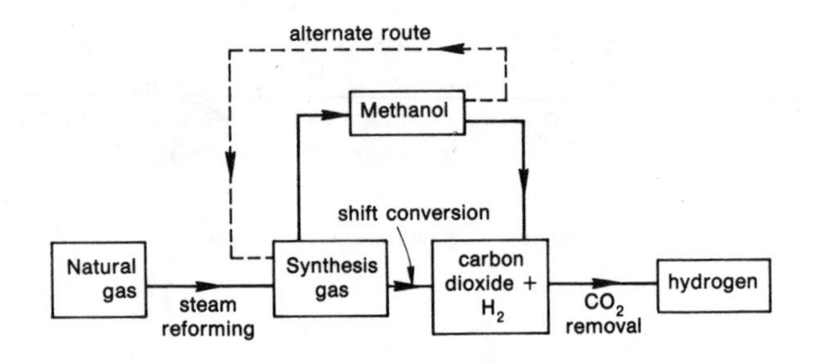

Fig. 5-14—Predicted use of the projected 68 million metric tons of methanol produced in the United States in 1985.[17]

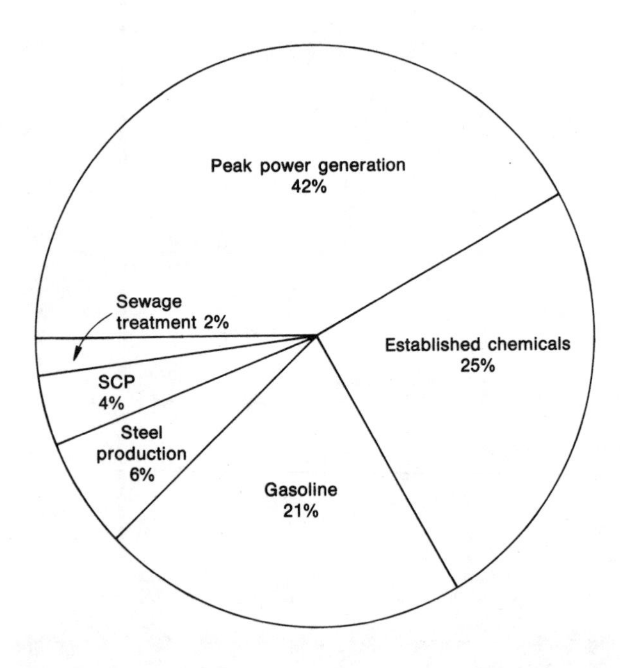

Fig. 5-13—A proposed method for the transportation of hydrogen by way of methanol.[30]

Predicting the future of methanol is as difficult as predicting the future of a newborn baby. There are too many "ifs."

Extensive development work is in progress on the synthesis of ethylene glycol from syn gas.

$$3\,CO + 5\,H_2 \rightarrow HOCH_2CH_2OH + CH_3OH$$

Temperature:	200°C
Pressure:	8000 psi
Catalyst:	$Rh(CO)_2 + C_sO(CO)H$

The amount of byproduct methanol varies with the reaction conditions.

The commercialization of this process could make ethylene glycol available from a wide variety of raw materials. It would be independent of petroleum and natural gas.

CHLOROMETHANES

The chloromethanes

methyl chloride	chloromethane	CH_3Cl
methylene chloride	dichloromethane	CH_2Cl_2
chloroform	trichloromethane	$CHCl_3$
carbon tetrachloride	tetrachloromethane	CCl_4

are produced by the chlorination of methane. This process presents much the same type of problem as the production of methylamines from methanol: consecutive reactions taking place under the same conditions. This leads to a mixture of products. The ratio of these products can be regulated within fairly wide limits by varying the mole ratio of methane to chlorine.

The relative demand for the first three chlorinated methanes is about 47% methylene chloride, 29% methyl chloride and 24% chloroform. The demand for methylene chloride, however, is increasing at a more rapid rate than for the other two chloromethanes. Carbon tetrachloride is in a class by itself and with an uncertain future because of its relationship with the fluorocarbons.

Production. Methane can be chlorinated both thermally and photochemically with thermal chlorination being the important commercial process.

$$CH_4 + Cl_2 \rightarrow CH_3Cl \xrightarrow{+Cl_2} CH_2Cl_2 \xrightarrow{+Cl_2}$$
$$CHCl_3 \xrightarrow{+Cl_2} CCl_4 + 4\,HCl$$

Temperature:	350-370° C	
Pressure:	Atmospheric +	
Mole ratio: CH_4:Cl_2	1.7:1.0	
Yields: on methane	75%	
on chlorine	95%	
Product distribution:	Methyl chloride	58.7%
	Methylene chloride	29.3%
	Chloroform	9.7%
	Carbon tetrachloride	2.3%

A yield of about 80% methyl chloride can be obtained with a methane to chlorine ratio of 10:1 at 450° C over a partially reduced cupric chloride on pumice catalyst. A high ratio of chlorine to methane gives a high yield of carbon tetrachloride. One of the commercial processes for the production of carbon tetrachloride along with coproduct perchloroethylene, $CCl_2 = CCl_2$, and lesser amounts of hexachloroethane, $CCl_3 - CCl_3$, hexachorobenzene, C_6Cl_6, and phosgene, $COCl_2$, is by the high temperature chlorolysis of miscellaneous chloroparaffins. These are swing plants and the ratio of carbon tetrachloride to perchloroethylene can be varied as much as 50%. A small percentage of carbon tetrachloride is produced by the chlorination of carbon disulfide.

It is more difficult to obtain a high yield of the desirable methylene chloride. The Vulcan process gives a good yield along with lesser amounts of chloroform and carbon tetrachloride, by the chlorination of methyl chloride (Fig. 5-15).[33] The methyl chloride is produced from methanol and the hydrogen chloride formed by the chlorination of the methyl chloride. Excess methylene chloride can be converted to chloroform and carbon tetrachloride by liquid phase mercury-lamp photochemical chlorination.

Utilization of Methylene Chloride. Methylene chloride has a wide variety of markets, some of which have great potential. One of these is in aerosol propellant mixtures which utilize about 13% of the present methylene chloride production. Because of its nonflammability, it is in a position to replace chlorofluorocarbons in mixtures that contain hydrocarbons. Its use as a blowing agent for urethane foams is another growth area. It is also used for solvent degreasing and paint stripping. Paint removal takes about 25-30% of current methylene chloride production. Exports account for another 17% with the remaining production being divided among many small uses. Methylene chloride has no important uses as an intermediate in the preparation of other compounds. The 1979 production of methylene chloride was 625 million pounds.

Chloroform. By far the largest use of chloroform is in the production of chlorodifluoromethane, $CHClF_2$, (Fluorocarbon 22).

$$CHCl_3 + 2\ HF \rightarrow CHClF_2 + 2\ HCl$$

This compound is used as a refrigerant and as an aerosol propellant. It is also used to synthesize tetrafluoroethylene, $CF_2 = CF_2$ (**TFE**).

$$2\ CHClF_2 \rightarrow CF_2 = CF_2 + 2\ HCl$$

Tetrafluoroethylene is polymerized to highly heat and chemical resistant polymers.

The growth potential of chloroform is to some extent dependent on the future use of chlorofluorocarbons as aerosol propellents.

Carbon Tetrachloride. Carbon tetrachloride, even more than chloroform, has been intimately associated with the fate of chlorofluorocarbons as aerosol propellants, Fig. 5-16.[34] About 90% of the past 1 billion pounds of carbon tetrachloride produced was used in the production of trichlorofluoromethane (CCl_3F, Fluorocarbon 11), and dichlorodifluoromethane (CCl_2F_2, Fluorocarbon 12).

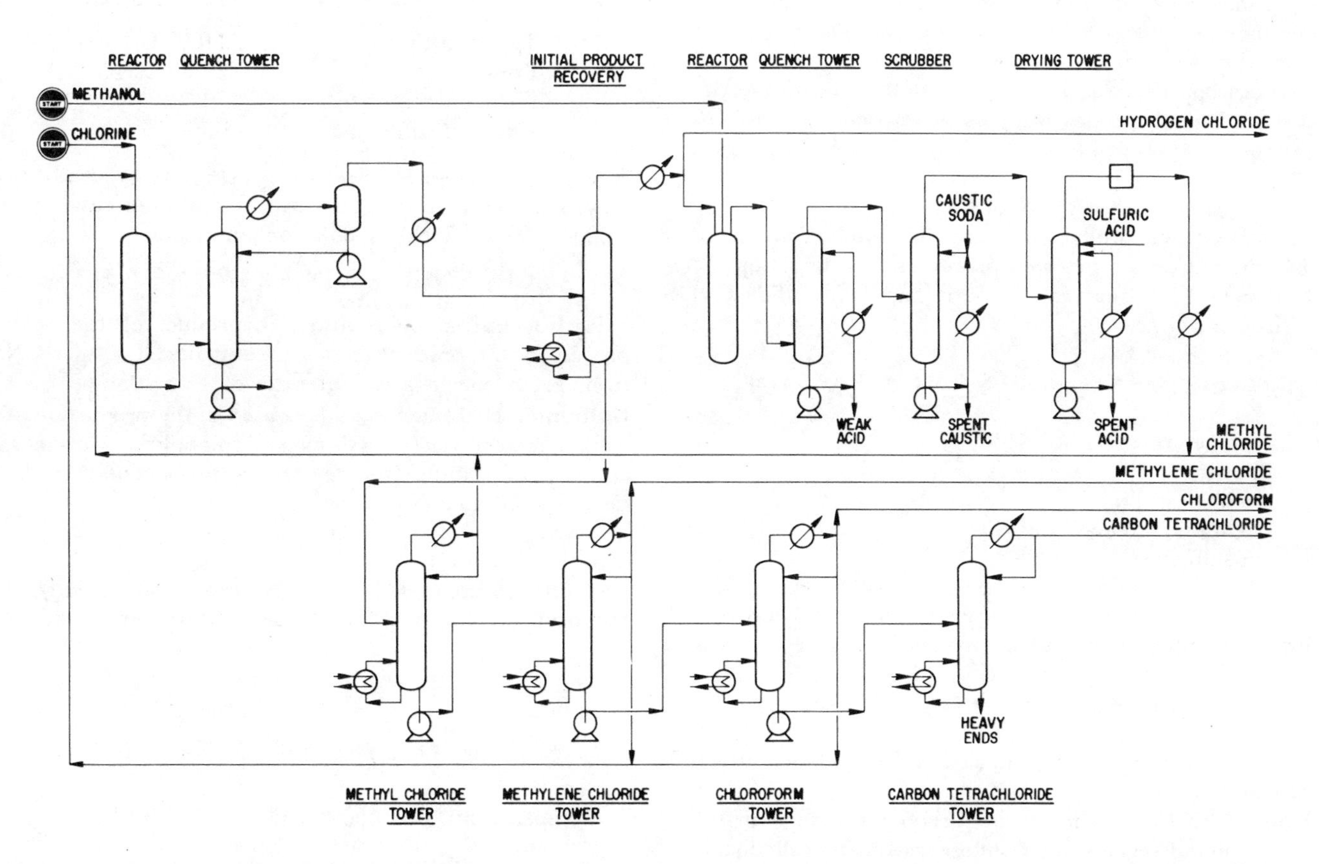

Fig. 5-15—The Vulcan process to produce polychlorinated methanes by the chlorination of methyl chloride produced from methanol.[33]

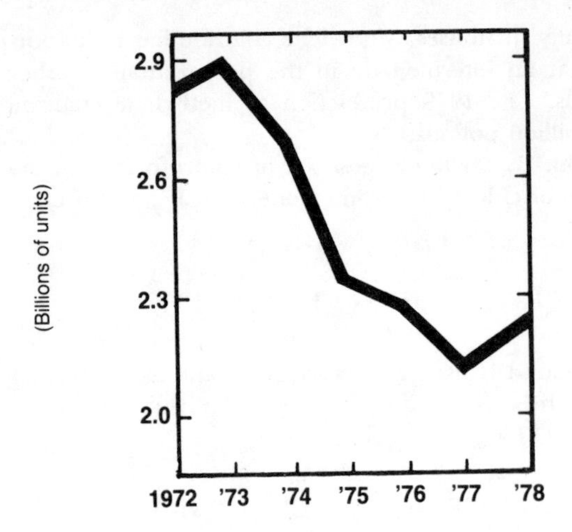

Fig. 5-16—The United States production of aerosols.[34]

$$CCl_4 + HF \rightarrow CCl_3F + HCl$$
$$CCl_4 + 2\ HF \rightarrow CCl_2F_2 + 2\ HCl$$

Temperature:	65-100°C
Pressure:	Atmospheric
Catalyst:	Molten $SbCl_5$
Yield:	A mixture of "11" and "12"

Approximately one half of these chlorofluorocarbons went into personal product aerosols such as colognes and perfumes, deodorants, hair sprays and shave creams. From there they went into the stratosphere. Fluorocarbon 11 and fluorocarbon 12 are also used in air conditioning and refrigeration. Even these uses face possible restrictions. The 1978 production of carbon tetrachloride was 748 million pounds.

CARBON DISULFIDE

Production. The production of carbon disulfide from methane and sulfur is a two step process. In the first step the methane and sulfur react to form carbon disulfide and hydrogen sulfide.

$$CH_4 + S \rightarrow CS_2 + 2\ H_2S\ 4\ \Delta H = +\ 19.5\ \text{kcal}$$

Temperature:	675°C
Pressure:	20-30 psig
Catalyst:	Activated Al_2O_3 or clay
Yield on methane:	85-90%
on sulfur:	92%

The second step is sulfur recovery from the coproduct hydrogen sulfide by the Claus process. The net reaction is

$$CH_4 + 2\ S + O_2 \rightarrow CS_2 + 2\ H_2O$$

Carbon disulfide is also produced by the reaction between carbon and sulfur.

Utilization. The major use of carbon disulfide is in the production of regenerated cellulose: rayon and cellophane. A secondary use is for the production of carbon tetrachloride. This process consists of two steps with the first step being the production of carbon tetrachloride (60%) and sulfur monochloride (40%).

$$CS_2 + 3\ Cl_2 \rightarrow CCl_4 + S_2Cl_2$$

Temperature:	30° C
Pressure:	Atmospheric
Catalyst:	Iron powder

The second step is the reaction between carbon disulfide and sulfur monochloride to produce more carbon tetrachloride and to recover the sulfur.

$$2\ S_2Cl_2 + CS_2 \rightarrow CCl_4 + 6\ S$$

The net reaction is

$$CS_2 + 2\ Cl_2 \rightarrow CCl_4 + 2\ S$$

The sulfur is used to produce more carbon disulfide.

Other uses of carbon disulfide include the production of perchloromethyl mercaptan, $Cl_3C\text{-}S\text{-}Cl$, used as an intermediate in the synthesis of the fungicide "Captan."

Xanthates, R-O-C(=O)-S-Na, are produced for use as ore flotation agents and ammonium thiocyanate, $NH_4\text{-}S\text{-}CN$, is a corrosion inhibitor in ammonia handling systems.

HYDROGEN CYANIDE

Production. Hydrogen cyanide is produced from methane, ammonia and air (oxygen) according to the following reaction (Andrussow Process):

$$2\ CH_4 + 2\ NH_3 + 3\ O_2 \rightarrow 2\ HCN + 6\ H_2O$$

Temperature:	1100° C
Pressure:	20-30 psig
Yield on methane:	70%
on ammonia:	60%

A process developed in Germany (Degussa Process) uses only methane and ammonia over a platinum catalyst at 1,200-1,300°C. Yield is 83% on ammonia

$$CH_4 + NH_3 \rightarrow HCN + 3\ H_2 \quad + \Delta H$$

Hydrogen cyanide is also a byproduct of the Sohio process for the production of acrylonitrile, $CH_2 = CHCN$, from propylene in a weight ratio of 1 to 10.

Utilization. Hydrogen cyanide is used in the production of methyl methacrylate, acrylonitrile, adiponitrile, acrylates, chelates and sodium cyanide. Annual production is in the range of 400 million pounds.

Oxamide. Oxamide, $H_2N\text{-}C(=O)\text{-}C(=O)\text{-}NH_2$, can be produced from hydrogen cyanide by a simple catalytic oxidation (Fig. 5-17).[35]

$$4\ HCN + O_2 + 2\ H_2O \rightarrow 2\ H_2N\text{-}\overset{\overset{O}{\|}}{C}\text{-}\overset{\overset{O}{\|}}{C}\text{-}NH_2$$

Temperature:	50-80° C
Pressure:	Atmospheric
Catalyst:	$Cu(NO_3)_2$ in aqueous acetic acid
Yield:	99%

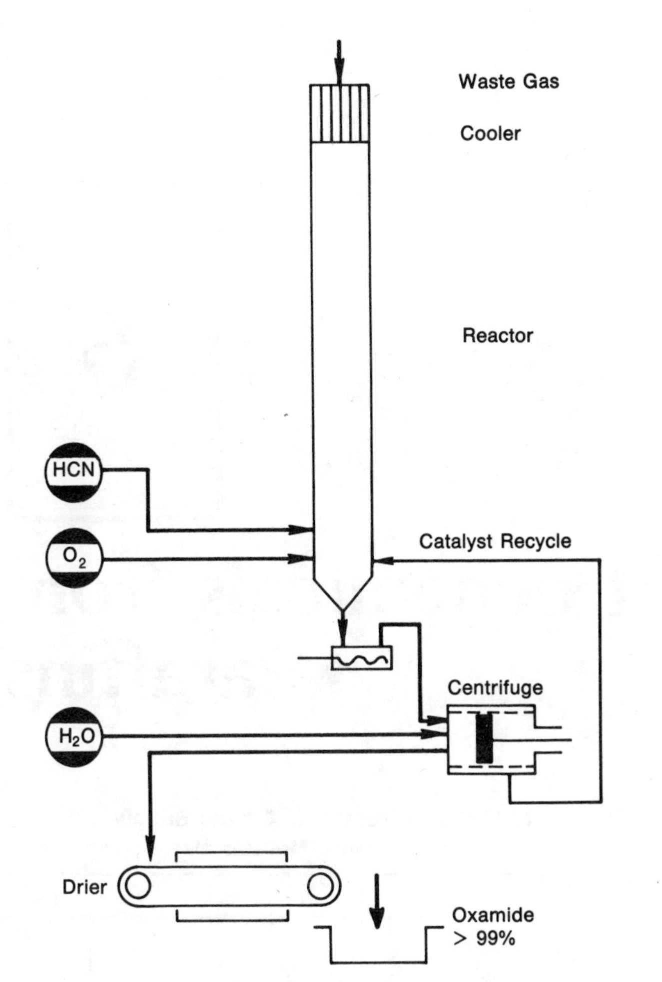

Fig. 5-17—Flow diagram for the Hoechst AG continuous process for the preparation of oxamide.[35]

Cyanogen, NC-CN, is produced when the aqueous acetic acid catalyst is replaced by acetonitrile, CH_3CN, containing 5-10% water. The yield is more than 90%.

Cyanoformamide, $NC\text{-}\overset{\overset{\displaystyle O}{\|}}{C}\text{-}NH_2$, is formed at temperature below 45° C in an acetic acid/water system.

Oxamide is a long-lived fertilizer. It will steadily release nitrogen in a useful form over the entire vegetation period. Oxamide may also be used as a stabilizer for nitrocellulose preparations. Its diacetyl derivative when added to detergents is a perborate activator.

Cyanogen is a well established industrial chemical with many applications.

Cyanoformamide is a new industrial chemical with large-scale uses yet to be developed.

LITERATURE CITED

1. Ross, James W. *Hydrocarbon Processing*, Vol. 54, No. 11, 1975, p. 76D, 76VV.
2. Foo, K.W. and I. Shortland. *Hydrocarbon Processing*, Vol. 55, No. 5, 1976, pp. 149-152; Vol. 55, No. 7, 1976, p. 67.
3. Engler, C.R., W.P. Walawender and Fan Liang-tseng. *Environmental Science and Technology*, Vol. 9, No. 13, 1975, pp. 1152-1157.
4. *Hydrocarbon Processing, 1975 Petrochemical Handbook*, Vol. 54, No. 11, p. 206.
5. Kuhre, C.J. and C.J. Shearer. *The Oil and Gas Journal*. Sept. 6, 1971, pp. 85-90.
6. *Hydrocarbon Processing, 1979 Petrochemical Handbook*, Vol. 58, No. 11, p. 162.
7. *Hydrocarbon Processing, 1977 Petrochemical Handbook*, Vol. 56, No. 11, p. 130.
8. *Hydrocarbon Processing, 1973 Petrochemical Handbook*, Vol. 52, No. 11, p. 137.
9. *Chemical Age*, Mar. 17, 1978, p. 4.
10. *Chemical and Engineering News*, Jan. 22, 1979, p. 11-12.
11. *Hydrocarbon Processing, 1979 Petrochemical Handbook*, Vol. 58, No. 11, p. 131.
12. Otsuka, Eiji, I. Shigeru and T. Jajima. *Hydrocarbon Processing*, Vol. 55, No. 11, 1976, p. 164.
13. *Hydrocarbon Processing, 1979 Petrochemical Handbook*, Vol. 58, No. 11, p. 248.
14. *Chemical Week*, Sept. 20, 1979, pp. 34-35.
15. *Chemical Age*, Jan. 20, 1978, p. 2.
16. *Hydrocarbon Processing*, Vol. 58, No. 7, 1979, p. 19.
17. Savage, P.R. *Chemical Engineering*, Nov. 22, 1976, pp. 74H-74L.
18. *The Oil and Gas Journal*, March 17, 1975, pp. 112, 114, 118.
19. *The Oil and Gas Journal*, March 12, 1973, p. 85.
20. Sherwin, M.B. and M.E. Frank. *Hydrocarbon Processing*, Vol. 55, No. 11, 1976, pp. 122-124.
21. Dimmling, W. and R. Seipenbusch. *Hydrocarbon Processing*, Vol. 54, No. 9, 1975, pp. 169-172.
22. Chauvel, A.R., P.R. Courty, R. Maux, and C. Petitpas. *Hydrocarbon Processing*, Vol. 52, No. 9, 1973, pp. 179-184.
23. *Hydrocarbon Processing, 1975 Petrochemical Handbook*, Vol. 54, No. 11, p. 149.
24. *Chemical and Engineering News*, Aug. 30, 1971, p. 19.
25. Grove, H.D. *Hydrocarbon Processing*, Vol. 51, No. 11, 1972, pp. 76-78.
26. Meisel, S.L., J.P. McCullough, C.H. Lechthaler, and P.B. Weisz. *CHEMTECH*, Vol. 6, No. 2, 1976, pp. 86-89.
27. Stinson, S.C. *Chemical and Engineering News*, April 2, 1979, pp. 28-30.
28. *Chemical and Engineering News*, Jan. 30, 1978, pp. 26, 28.
29. Belgian Patent 818708 to Mobil Oil Co.
30. Jonchere, J.P. "Methanol, A Privileged Hydrogen Carrier," Second Arab Conference on Petrochemicals, United Arab Emirates, Abu Dhabi, March 15-22, 1976, *The Oil and Gas Journal*, June 14, 1976, pp. 71-73.
31. Humphreys, G.C. "Associated Gas Utilization via Methanol," International Seminar on Petrochemical Industries, Baghdad, Iraq: Oct. 25-30, 1975.
32. Burk, D.P. *Chemical Week*, Sept. 24, 1975, pp. 33-42.
33. *Hydrocarbon Processing, 1975 Petrochemicals Handbook*, Vol. 54, No. 11, p. 127.
34. *Chemical Week*, May 23, 1979, pp. 32-33.
35. Riemenschneider, W. *CHEMTECH*, Vol. 6, No. 10, 1976, pp. 658-661.

6

Petrochemicals from *n*-Paraffins

The chemical utilization of paraffin hydrocarbons, *alkanes*, is markedly restricted by their lack of chemical reactivity. Those with lower molecular weight are oxidized to various products such as acetic acid, maleic anhydride and acetaldehyde, and chlorinated to perchloroethylene. By far the most important use of the lower molecular weight hydrocarbons, however, is to produce olefins. Higher molecular weight *n*-paraffins are chlorinated, sulfonated or can be fed to microorganisms to produce single cell proteins, **SCP**.

The lower molecular weight paraffin hydrocarbons in the United States are recovered from natural gas and liquefied petroleum gas (**LPG**, **LP**-gas). In 1976, only about 20% of the ethane in natural gas was recovered and used as petrochemical feedstock. The remaining 80% was burned as a constituent of natural gas. By 1985 the recovery will be 30% from essentially the same quantity of natural gas (Table 6-1).[1] Table 6-2 gives the total U.S. ethane supply and demand in 1985.[1]

Propane's recovery from natural gas is currently 72%, but only 13% of this is available for petrochemicals. By 1981 the propane recovery from natural gas is estimated to be 79% and the 1986 recovery will be 88%.[2] The availability of this propane for chemical use will be 9% and 7%, respectively. The drop in propane availability for petrochemicals reflects the increasing use of **LP**-gas for domestic and commercial heating. In the Middle East ethane and propane are increasingly being recovered from associated gas.

TABLE 6-1—U.S. Natural Gas Production and Ethane Supply[1]

	Natural gas production, trillion cubic feet (excluding Alaska)	Contained ethane, million barrels	Ethane recovered from natural gas	
			Percent	MM bbls.
1974	21.6	637	19	118
1975	20.1	583	20	118
1980	19.1	530	26	137
1985	22.3	505	30	151

TABLE 6-2—Total U.S. Ethane Supply and Demand[1] MM Barrels

	1974	1975	1980	1985
Supply				
Natural gas	118	118	137	151
Refining	6	6	8	8
Total	124	124	145	159
Demand				
Ethylene	123	123	144	158
SNG	1	1	1	1
Total	124	124	145	159

The high molecular weight *n*-paraffins, C_8-C_{30}, are extracted from the appropriate refinery streams by molecular sieve processes. The melting points and boiling points of *n*-paraffins from methane through *n*-eicosane plus the densities at 20°C of those that are liquids are given in Table 6-3.

ETHANE

Ethane, like methane, has its relationship with petrochemicals primarily through a single product. With methane it is synthesis gas, $CO + H_2$, and with ethane, it is ethylene, $CH_2 = CH_2$. Ethane is the most desirable feedstock for ethylene production when a minimum amount of byproduct is desired. The direct conversion of ethane to vinyl chloride is the only other process currently in use for the production of a petrochemical from ethane.

The amount of ethane available for petrochemical use in 1976 was 335,000 bpd.[2] The figures for 1981 and 1986, respectively, are 367,000 bpd and 392,000 bpd. The remaining recovered ethane is for refinery and non-petrochemical use.

TABLE 6-3—Selected Physical Properties of the *n*-Paraffins, *n*-Alkanes, from Methane through *n*-Eicosane (C_{20})

Name	Formula	M.P., °C	B.p., °C	Density (at 20°C)
Methane	CH_4	−183	−162	
Ethane	CH_3CH_3	−172	−88.5	
Propane	$CH_3CH_2CH_3$	−187	−42	
Butane	$CH_3(CH_2)_2CH_3$	−138	0	
Pentane	$CH_3(CH_2)_3CH_3$	−130	36	0.626
Hexane	$CH_3(CH_2)_4CH_3$	−95	69	0.659
Heptane	$CH_3(CH_2)_5CH_3$	−90.5	98	0.684
Octane	$CH_3(CH_2)_6CH_3$	−57	126	0.703
Nonane	$CH_3(CH_2)_7CH_3$	−54	151	0.718
Decane	$CH_3(CH_2)_8CH_3$	−30	174	0.730
Undecane	$CH_3(CH_2)_9CH_3$	−26	196	0.740
Dodecane	$CH_3(CH_2)_{10}CH_3$	−10	216	0.749
Tridecane	$CH_3(CH_2)_{11}CH_3$	−6	234	0.757
Tetradecane	$CH_3(CH_2)_{12}CH_3$	5.5	252	0.764
Pentadecane	$CH_3(CH_2)_{13}CH_3$	10	266	0.769
Hexadecane	$CH_3(CH_2)_{14}CH_3$	18	280	0.775
Heptadecane	$CH_3(CH_2)_{15}CH_3$	22	292	
Octadecane	$CH_3(CH_2)_{16}CH_3$	28	308	
Nonadecane	$CH_3(CH_2)_{17}CH_3$	32	320	
Eicosane	$CH_3(CH_2)_{18}CH_3$	36		

Vinyl Chloride. Vinyl chloride production is in the 6-billion-pound-per-year class. Any compound with a production of this magnitude is likely to be produced by several different processes. Vinyl chloride is no exception. Oxychlorination, hydrochlorination, dehydrochlorination and combinations of these reactions are used commercially. The various processes that utilize ethylene as a feedstock are noted in Chapter 8.

Vinyl chloride from ethane, the *Transcat Process*, utilizes a combination of chlorination, oxychlorination and dehydrochlorination in a molten salt reactor (Fig. 6-1).[3, 4]

Temperature:	310-640°C
Pressure:	Nominal
Catalyst:	Copper oxychloride
Yields: On ethane	80%
On chlorine	90-99%

During this reaction the copper oxychloride is converted to copper(I) and copper(II) chlorides. These are air oxidized to regenerate the copper oxychloride. The copper oxychloride acts as an oxygen carrier and is also involved in the hydrogen chloride recovery.

$$2\,CuCl + O_2 \rightarrow 2\,CuO \cdot CuCl_2$$
$$2\,CuCl + Cl_2 \rightarrow 2\,CuCl_2$$
$$4\,CuCl + 4\,HCl + O_2 \rightarrow 4\,CuCl_2 + 2\,H_2O$$

This process is not restricted to ethane. The feedstock can be partially or wholly replaced by ethylene and hydrogen chloride or by chlorohydrocarbon wastes and byproducts. The process can also be used to produce trichloroethylene, $CHCl{=}CCl_2$, perchloroethylene, $CCl_2{=}CCl_2$ and chloromethanes.

PROPANE

Propane is more versatile than ethane as a feedstock for petrochemicals. This is at least partly caused by the presence of the more reactive secondary hydrogen atoms in the molecule.

$$CH_3 - \underset{\displaystyle H}{\overset{\displaystyle H}{\overset{|}{\underset{|}{C}}}} - CH_3$$

The two types of hydrogen, however, cause isomer problems in substitution reactions as illustrated in the production of nitropropanes.

Nitropropanes. The *nitration* of propane has been limited industrially primarily because the product is a complex mixture. Not only are both possible substitution products formed but also carbon-carbon fission products are formed as well.

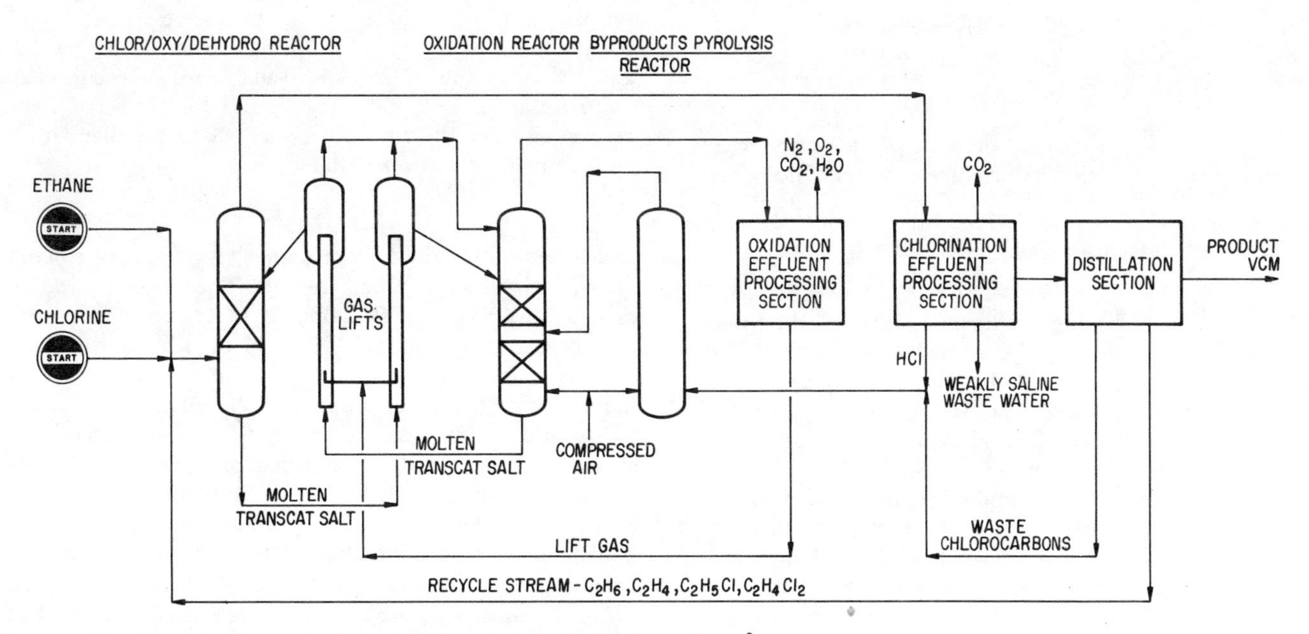

Fig. 6-1—The Transcat Process for the production of vinyl chloride from ethane.[3]

$$CH_3-CH_2-CH_3 \xrightarrow{HNO_3} \begin{cases} CH_3-CH_2-CH_2NO_2 \\ CH_3-CHNO_2-CH_3 \end{cases} \text{55-65 wt. \%}$$

$$\rightarrow CH_3-CH_2NO_2 \quad \text{20-25 wt. \%}$$

$$\rightarrow CH_3NO_2 \quad \text{10-30 wt. \%}$$

Temperature:	390-440° C
Pressure:	100-125 psig
Catalyst:	None
Mole ratio: $C_3H_8:HNO_3$	25:1 (over-all 5:1)
Yields: On HNO_3	30%

The nitropropanes are good solvents and nitromethane has been used as an additive in fuels used in racing cars.

Nitropropanes react with formaldehyde to produce nitroalcohols.

$$CH_3-CH_2-CH_2NO_2+HCHO \rightarrow CH_3-CH_2-\underset{}{\overset{NO_2}{\overset{|}{C}H}}-CH_2OH$$

These difunctional compounds are versatile solvents but expensive.

Hydrolysis of primary nitroparaffins with sulfuric acid produces the hydroxylamine salt and the corresponding carboxylic acid.

$$CH_3-CH_2-CH_2NO_2+H_2SO_4+H_2O \rightarrow NH_2OH \cdot H_2SO_4+CH_3-CH_2-\overset{O}{\overset{||}{C}}-OH$$

Perchloroethylene. Chlorination of propane at 480-640°C with an excess of chlorine yields a mixture of perchloroethylene (Perchlor) and carbon tetrachloride.

$$CH_3-CH_2-CH_3+8\,Cl_2 \rightarrow \underset{\text{Perchlor}}{CCl_2=CCl_2}+CCl_4+8\,HCl$$

The yield is 95% based on chlorine. The carbon tetrachloride can be recycled to produce more perchloroethylene.

$$2\,CCl_4 \rightarrow CCl_2=CCl_2+Cl_2$$

Other methods for the production of perchloroethylene are noted in Chapter 5. The annual U.S. production of perchloroethylene is in the 700-million-pound range.

Processes of this general type, with or without a catalyst, frequently utilize the chlorinated byproducts of other processes such as the heavy ends from vinyl chloride, chloromethanes, methylchloroform and ethylene dichloride plants. For example, trichloroethylene and perchloroethylene are produced by using byproduct ethylene dichloride without yielding hydrogen chloride as a byproduct (Fig. 6-2).[5]

Propane is not chlorinated to produce monochloropropanes.

Oxidation. The *catalytic oxidation* of propane has received some attention but no commercial processes have been developed. Dewing, *et al*, report that an antimonytin oxide catalyst (46% Sb + 25% Sn) at 490°C can catalyze the oxidation of propane to acrolein in a 30% yield.[6]

Propane has also been proposed as an alternate for propylene in the production of acrylonitrile.[7] The equation for the reaction is

$$CH_3-CH_2-CH_3+NH_3+2\,O_2 \rightarrow CH_2=CH-CN+4\,H_2O$$

Temperature:	500°C
Pressure:	Atmospheric
Catalyst:	Antimony-uranium oxide + CH_3Br (Sb:U 5:1)
Yield: Conversion:	85%
Selectivity:	71%

Propylene is probably an intermediate.

Noncatalytic oxidation of propane produces propylene oxide, $CH_3\text{-}CH\overset{O}{-}CH_2$, industrially on a very small scale. The reaction conditions are 400°C and 20 bar with a wt.% yield on propane of 32; other pertinent data are in Table 6-4.[8] Obviously, this process has an extremely elaborate separation operation. The organic-laden wastewater (about 12 tons per ton of propylene oxide produced) is a considerable burden under today's environmental constraints. It is doubtful that this process will be a serious factor in overall propylene oxide production.

Noncatalytic oxidation of propane is associated with the oxidation of butane and is noted in the next section.

The competition between petrochemical and fuel uses for propane is one-sided because propane is committed politically to home-heating and farm markets.[9] Added suplies of propane must come from imports because essentially all of the available propane is already being extracted from natural gas.[10] The amount of propane available for petrochemicals in 1976 was 127,000 bpd.[2] The figures for 1981 and 1986, respectively, are 105,000 bpd and 90,000 bpd. Other supply and demand data are in Table 6-5.[1]

n-BUTANE

n-Butane is obtained from natural gas, refinery streams (from hydrocrackers), and imports. The major utilization of *n*-butane, over 500,000 bpd, is to control vapor pressure of gasoline. About 41,000 bpd of *n*-butane in 1976 was used to produce ethylene. *n*-Butane has been the main feedstock for the production of butadiene by dehydrogenation processes. This source is being replaced by steam cracking for ethylene which produces considerable butadiene as a byproduct. The amount of *n*-butane utilized by the chemical industry during the period 1977-1987 will be only a small portion of the total *n*-butane available.

The chemistry of *n*-butane is more varied than that of propane, partly because *n*-butane has four secondary hy-

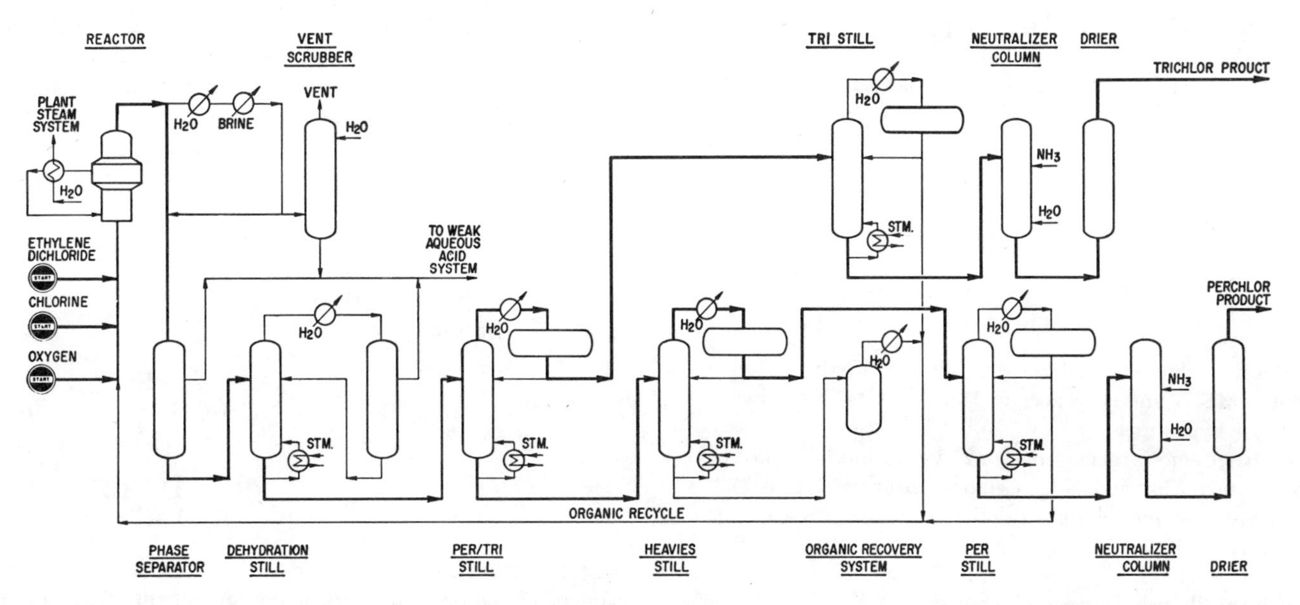

Fig. 6-2—The production of trichloroethylene and perchloroethylene by the oxychlorination-oxydehydrochlorination of various chlorine-containing feedstocks.[5]

TABLE 6-4—Input and Output for the Noncatalytic Oxidation of Propane[8]

Input/kg PO			Output/kg PO		
Propane	3.11	kg	Acetaldehyde	0.77	kg
Oxygen	4.17	kg	Formaldehyde	0.66	kg
Process water	10.2	kg	CO_2/CO	2.43	kg
Lime	?	kg	Miscellaneous	0.10	kg
Cooling water	0.3	m^3	Waste-water	12.	kg
Steam	7.5	kg			
Electricity	1.1	Kwh			

TABLE 6-5—Total United States Propane Supply and Demand[1] MM Barrels

	1974	1975	1980	1985
Demand				
Fuel markets	245	275	342	406
Chemicals	65	37	56	52
Others	5	4	2	2
Total	315	316	400	459
Supply				
Natural gas	207	203	172	172
Refining	92	95	137	171
Canada/Venezuela imports	22	23	32	31
	321	321	341	374
Shortfall	(6)	(5)	59	85

drogen atoms and three carbon-carbon bonds that can be broken.

$$CH_3-\underset{H}{\overset{H}{C}}-\underset{H}{\overset{H}{C}}-CH_3$$

Other than for the production of ethylene and butadiene, the chemical utilization involves its oxidation to various organic acids, aldehydes, ketones, alcohols and esters; usually in complex mixtures.

The oxidation of hydrocarbons in general proceeds by a free radical-chain reaction. Non-catalyzed reactions of this type are characterized by:

- A multiplicity of products
- Low selectivity to any single product
- Low conversions
- Are highly exothermic.[11]

Improved selectivity has been obtained by use of homogeneous catalysts such as soluble salts of cobalt or manganese. The most striking difference between catalytic and non-catalytic oxidation of *n*-butane is that catalytic oxidation produces acids while non-catalyzed oxidations produce aldehydes as the main products.

Acetic Acid. Acetic acid is the most important carboxylic acid produced industrially. The annual production in the United States is over 2.5 billion pounds per year with this figure predicted to increase to over 3 billion in 1980.

As with most compounds produced on a large scale, acetic acid has several different industrial processes. The process with the brightest future is the carbonylation of methanol. This process was discussed in Chapter 5. The production of acetic acid by the oxidation of acetaldehyde will be noted in Chapter 8. The third method is the oxidation of **LP**-gas and of *n*-butane. In Europe the preferred feedstock is light naphtha (pentane and hexane).

The catalytic liquid phase oxidation, **LPO,** of *n*-butane is currently the most important process in the United States for the production of acetic acid.

$$CH_3-CH_2-CH_2-CH_3 \xrightarrow{O_2}$$

$$CH_3-\overset{O}{\overset{\|}{C}}-OH + \text{byproducts} + H_2O$$

Temperature:	150-225°C	
Pressure:	*ca* 800 psi	
Catalyst:	Cobalt (or manganese) acetate	
Yield: wt. %	$CH_3-\overset{\overset{O}{\|}}{C}-OH$	75-80
	$H-\overset{\overset{O}{\|}}{C}-OH$	6
	CH_3-CH_2OH	6
	CH_3OH	4
	Other	9-4

"Other" includes acetaldehyde, acetone, methyl ethyl ketones, (**MEK**), and ethyl acetate. If desired, **MEK** can become a major byproduct (*ca* 13 wt.%).[11] When a manganese catalyst is used, more formic acid (*ca* 25 wt.%) and less acetic acid (61 wt.%) is formed. The Celanese process (Fig. 6-3) has been summarized, along with other processes for acetic acid production, by Lowry and Aguilo.[12] Liquid phase oxidation of hydrocarbons in general, and of n-butane in particular, has been discussed by Saunby and Kiff.[11]

Acetaldehyde. The *non-catalytic vapor phase oxidation* of n-butane illustrates the non-selectivity of non-catalyzed oxidation of hydrocarbons. A complex mixture of simple compounds is formed.

Temperature:	360-450°C	
Pressure:	100 psi	
Catalyst	None	
Yields: wt.% on 100 lbs. of n-butane	$CH_3-\overset{\overset{O}{\|}}{O}-H$	31
	$H-\overset{\overset{O}{\|}}{C}-H$	33
	CH_3OH	20
	$CH_3-\overset{\overset{O}{\|}}{C}-CH_3$	4
	Mixed solvents	12

Maleic Anhydride. Maleic anhydride is produced by the catalytic oxidation of benzene (Chapter 10) and the *oxidative*

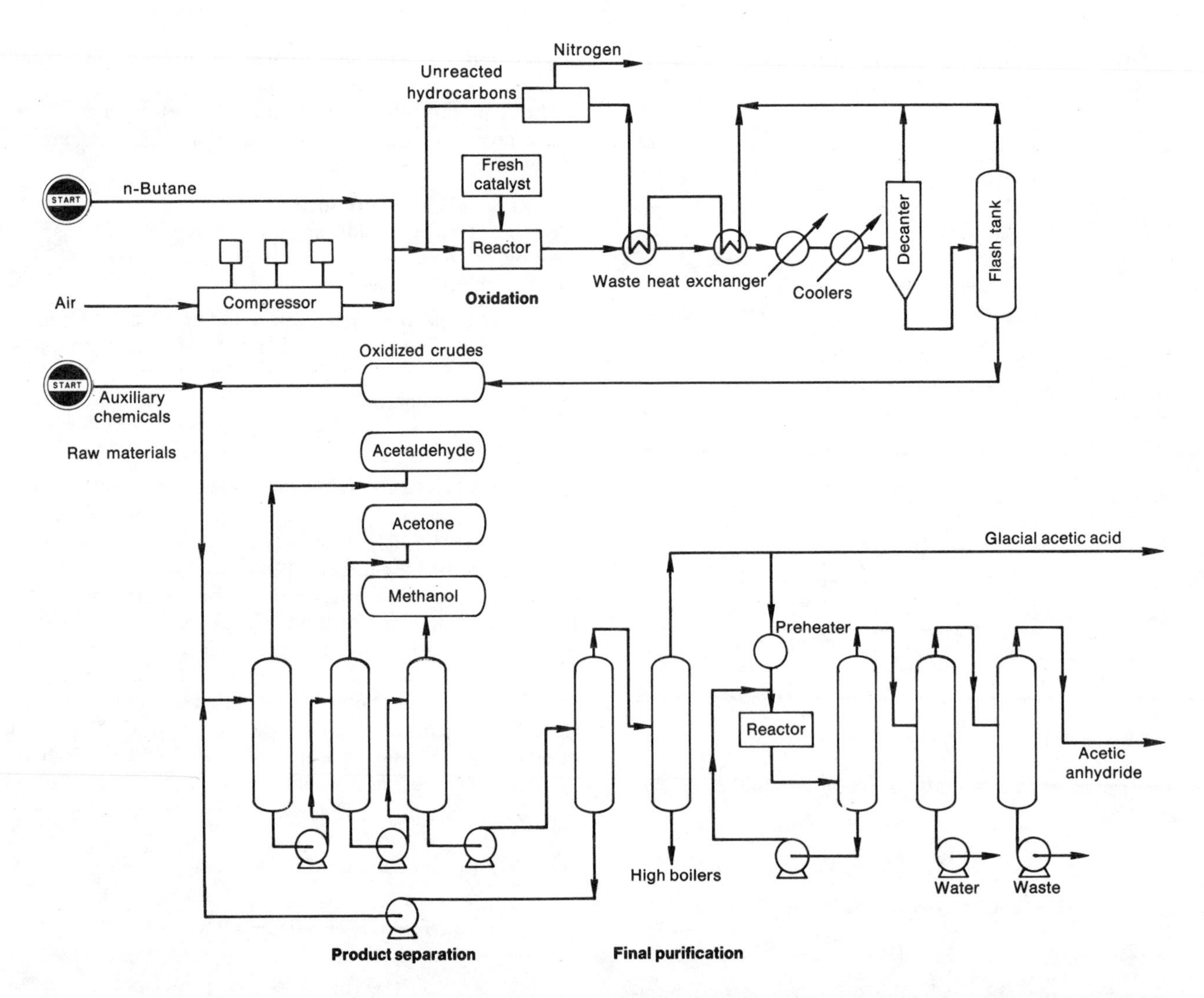

Fig. 6-3—The Celanese liquid phase oxidation, LPO, process for oxidation of n-butane to acetic acid and recovery of by-products.[11]

dehydrogenation-oxidation of butane. This process is currently of considerable interest.

$$2\ CH_3-CH_2-CH_2-CH_3 + 7\ O_2 \rightarrow 2\ \begin{matrix} H-C-C(=O) \\ \| \quad\quad\ \ \ \ \ \ >O \\ H-C-C(=O) \end{matrix} + 8\ H_2O$$

Temperature:	500°C
Pressure:	Atmospheric
Catalyst:	Iron + V_2O_5 -P_2O_5 on silica-alumina
Yield: Conversion	18.6%
Selectivity	76.5%

Maleic anhydride has also been obtained by the oxidation of *n*-butane at 490°C over a cerium chloride Co-Mo oxides catalyst supported on silica with a hydrogen chloride-*n*-butane mole ratio of 0.59:1.00.[13] A 63 wt.% yield was reported. The function of the cerium chloride is to dehydrogenate the butane to butene. The butenes are then oxidized to maleic anhydride over the Co-Mo oxide catalyst. The use of a phosphorus-vanadium catalyst has been reported to give a 55-60 percent yield.[14]

The kinetics of the oxidation of *n*-butane have been reported by Kuz'michev and Skarchenko.[15]

Maleic anhydride *capacity* increased from 375 million pounds in 1975, to 501 million pounds in 1979. Most of this new capacity uses butane as feedstock, and benzene-based capacity is being shifted to butane, primarily because of economics. Butane is also more environmentally desirable than benzene.[31] The *demand* for maleic anhydride, however, was 365 million pounds in 1979.[16]

The principal *use* of maleic anhydride is in the preparation of unsaturated polyester resin (Fig. 6-4).[17] These resins are used to fabricate glass-fiber-reinforced materials for such widely divergent uses as glass reinforced boats and bathroom fixtures. The automobile industry is increasing its use of glass-fiber-reinforced materials to reduce automobile weight. Other uses include the production of fumaric acid, pesticides and alkyd resins. Maleic acid esters are used as plasticizers and lubricants.

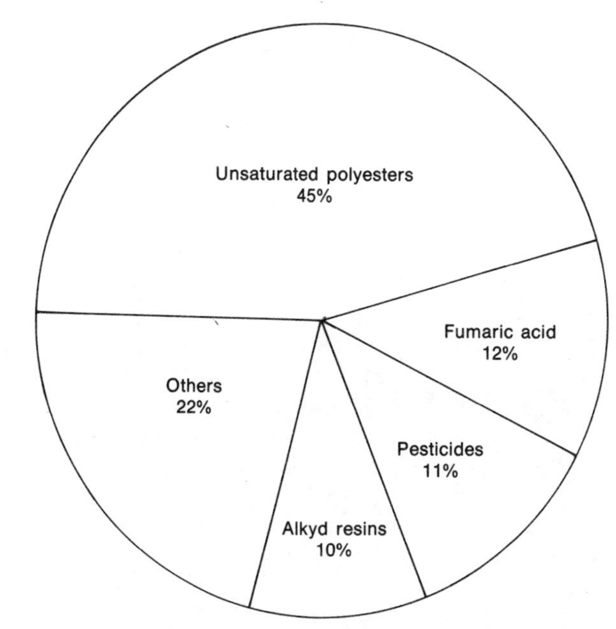

Fig. 6-4—Distribution of maleic anhydride end uses for 1975.[17]

LIGHT NAPHTHA (C_5-C_7)

Light naphtha containing hydrocarbons in the C_5-C_7 range is the preferred feedstock in Europe for **LPO** oxidation to produce acetic acid (Fig. 6-5).[18]

Temperature:	170-200°C
Pressure:	750 psig
Catalyst:	Manganese acetate
Yield: (wt.%)	40
	Plus a mixture of acids, esters, alcohols, aldehydes, ketones, etc.

As many as 13 distillation columns are used to separate this complex mixture of products. The number of products can be reduced by recycling most of them to extinction. The final products are predominately acetic acid with about 10% formic acid and propionic acid and 2% succinic acid.[19]

Rouchaud and Lutete have made an indepth study of the **LPO** catalytic oxidation of *n*-hexane.[20]

Temperature:	160°C
Pressure:	380 psi
Catalyst:	I Manganese naphthenate (0.64 wt% Mn)
	II Manganese naphthenate (0.64 wt% Mn) + 40 wt.% Na_2CO_3

	I	II
Conversion: (mole %)	53%	45%
Yield: (Relative to *n*-hexane)	83%	79%
Distribution of acids: (mol. %)		
$CH_3-C(=O)-OH$	65	28
$H-C(=O)-OH$	18	5
$CH_3-CH_2-C(=O)-OH$	12	32
$CH_3-(CH_2)_2-C(=O)-OH$	4	28
$CH_3-(CH_2)_3-C(=O)-OH$	1	7

Kinetics of the accumulation of acids show that the distribution of acids result simultaneously from the oxidation of the *n*-hexane and its alcoholic and ketonic deg-

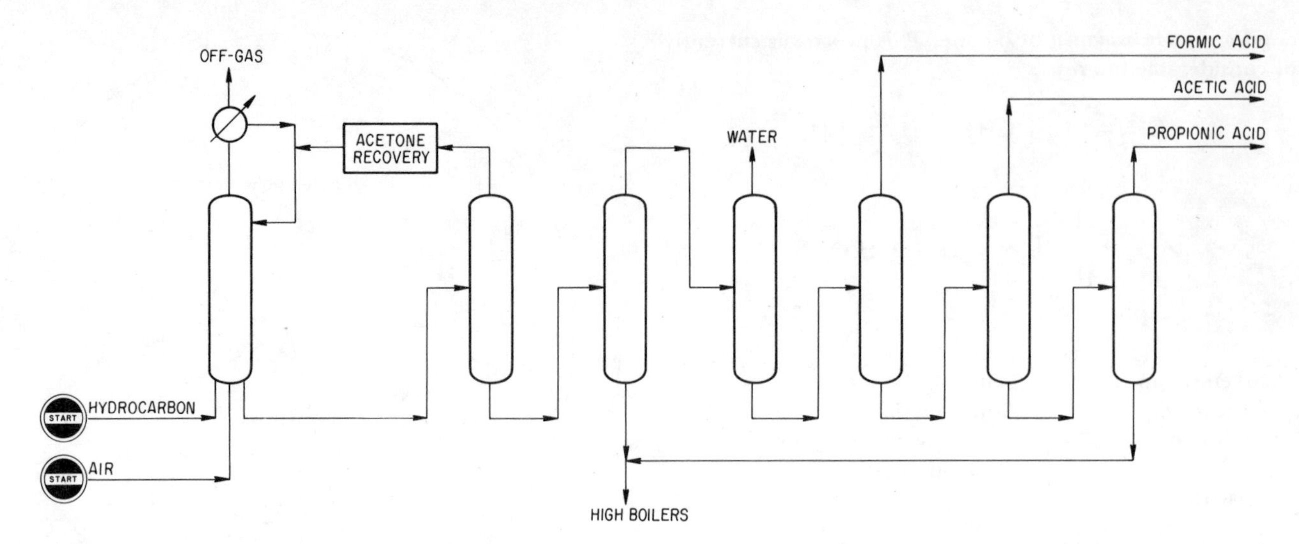

Fig. 6-5—The BP Chemicals Ltd. process for production of acetic acid by liquid phase oxidation, LPO, of a light hydrocarbon distillate.[18]

radation derivatives, and from the decarboxylative oxidation of the acids.

HIGH MOLECULAR WEIGHT *n*-PARAFFINS (C_8-C_{30})

The high molecular weight paraffins are obtained from the appropriate refinery streams by several different extractive processes. These are noted in Chapter 3. The various mixtures of *n*-paraffins are oxidized, chlorinated, dehydrogenated, sulfonated and fermented.

Fatty Acids. The *oxidation* of C_{18}-C_{30} *n*-paraffins to produce fatty acids for the soap industry is of some industrial interest in Europe. The over-all reaction can be expressed as[21]

$$R_1(CH_2)_n - CH_2 - CH_2 - (CH_2)_n R_2 + 5/2\, O_2 \rightarrow$$

$$R_1(CH_2)_n - \overset{\overset{\displaystyle O}{\|}}{C} - OH + R_2(CH_2)_n - \overset{\overset{\displaystyle O}{\|}}{C} - OH + H_2O$$

Temperature:	105-120°C	
Pressure:	200-900 psi	
Catalyst:	Manganese salts	
Yield: (wt.%)	Fatty acids for soap	60
	Shorter fatty acids	25
	$CO_2 + CO$	10
	Other	5

Under periods of food shortage, these acids may be used as food.

Fatty Alcohols. An extensive study has been made on the air oxidation of *n*-paraffin fractions rich in C_{12}-C_{14} *n*-paraffins to produce fatty alcohols.[22] Typical operating conditions for the oxidation of a fraction containing 95% *n*-paraffins are:

Temperature:	120-130°C	
Pressure:	Atmospheric	
Catalyst:	Boron trioxide, butyl borates (0.44% B)	
Conversion: (wt.%)	30.5	
Yield: (wt.%)	Acids:	8.9
	Alcohols:	76.2
	Other:	14.9

tert-Butyl hydroperoxide (0.5%) was used to initiate the reaction.

The alcohols were mainly secondary with the same number of carbon atoms and the same structure per molecule as the parent paraffin hydrocarbon.

n-Paraffins also can be oxidized to alcohols by a dilute oxygen stream (3-4%) in the presence of a mineral acid. The acid converts the alcohol to an ester which prohibits further oxidation of the alcohol to an organic acid. Essentially all of the alcohols are secondary. Alcohols of this type are of commercial importance as plasticizers and for the production of monoolefins, polymers, lube oils and detergents.

Chloroparaffins. The *liquid phase monochlorination* of *n*-paraffins has been investigated using *n*-octane with both thermal and catalytic (PCl_3) processes.[23] Selectivity of about 75% for monochlorination can be obtained at 30°C for thermal chlorination and at 60-70°C for catalytic chlorination. Substitution of secondary hydrogens predominate. At 50°C with thermal chlorination the ratio is 2.6 to 1 with an 18% conversion. With catalytic chlorination, the ratio is 2.1 to 1.0 with a 15% conversion. A high ratio of hydrocarbon to chlorine favors monochlorination.

Monochloroparaffins are used to some extent for the production of detergents and related products.

Polychlorination up to a chlorine content of 70% can be readily carried out on the whole range of *n*-paraffins from C_{10} to C_{30} (Fig. 6-6).[24]

Temperature:	80-120°C
Pressure:	Atmospheric
Catalyst:	None (or light)
Yield: On chloride	100%

Polychloroparaffins are used as cutting oil additives, plasticizers and fire retardant chemicals.

Sulfonated *n*-Paraffins. *Secondary linear paraffin sulfonates* (secondary alkane sulfonates, **SAS**) are produced by the reaction between sulfur dioxide, oxygen and *n*-paraffins, (C_{15}-C_{17}).[25]

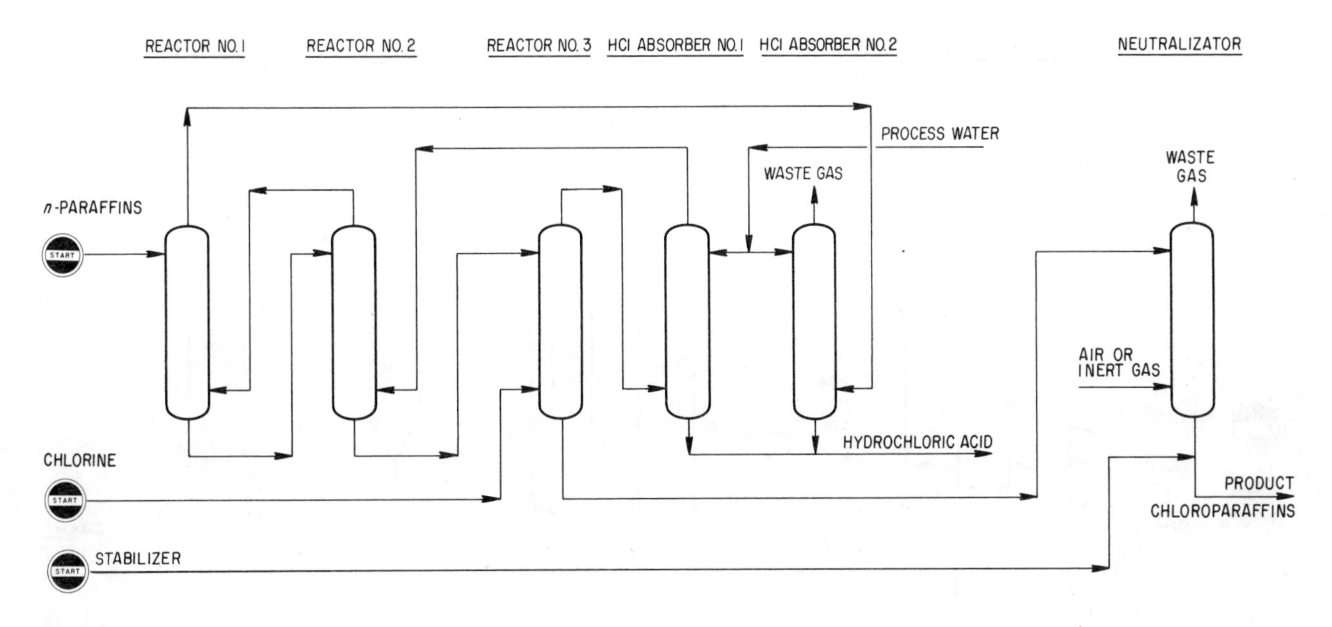

Fig. 6-6—A process for the production of polychlorinated paraffin hydrocarbons.[24]

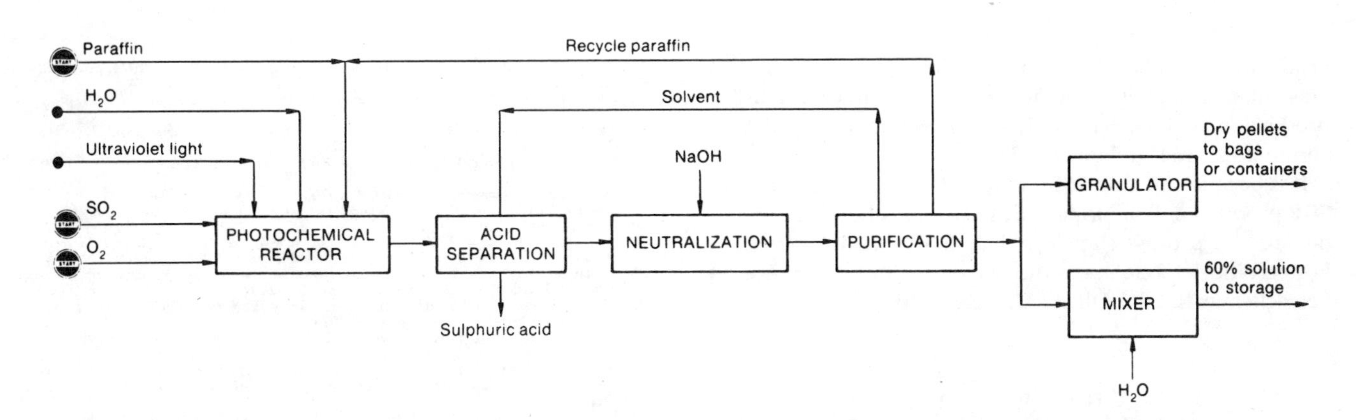

Fig. 6-7—The ATO Chimie photochemical process for the production of linear paraffin sulphonates from paraffins, SO_2 and O_2.[25]

$$RH + 2\ SO_2 + O_2 + H_2O \rightarrow RSO_3H + H_2SO_4$$

The temperature and pressure are slightly above ambient. The reaction is catalyzed by ultraviolet light with a wavelength between 3,300-3,600 Å and a power level of several kw. Both polysulfonates and monosulfonates in a constant 9.5 to 1 ratio are produced by this process (Fig. 6-7).

These sulfonates are nearly 100% biodegradable, soft and stable in hard water and have good washing properties. However, there is a tendency for the powdered form to pick up moisture and become tacky.[26] Commercial production, as well as use, is currently limited to Europe.

Single Cell Protein, SCP. The production of *single cell protein*, **SCP**, is not generally considered a synthetic process and the proteins are not individual compounds in the sense that acetic acid is. The production of **SCP** is microbial farming in which the substrate serves in place of the sunlight and fertilizer of conventional agriculture. The present state of the art permits the production of **SCP** in forms nearly as acceptable as the natural product and from every conceivable feedstock from gas oil to manure. The current substrates of choice are *n*-paraffin hydrocarbons, methane, methanol and ethanol.

Single cell protein is not a protein but an organism containing protein and a variety of other constituents, some of which limit the usefulness of **SCP** for human food. This is especially true with *n*-paraffins as the substrate. **SCP** is protein only in the sense that fish, cattle or chickens are protein. However, half a ton of steer will produce about one half a pound of protein per day whereas half a ton of yeast will produce 2.5 tons of protein in the same period of time.[27]

The reliability and economics of producing high quality *n*-paraffins is a critical factor in the use of *n*-paraffins for the production of **SCP**. The *n*-paraffins, C_{10}-C_{30}, used in **SCP**

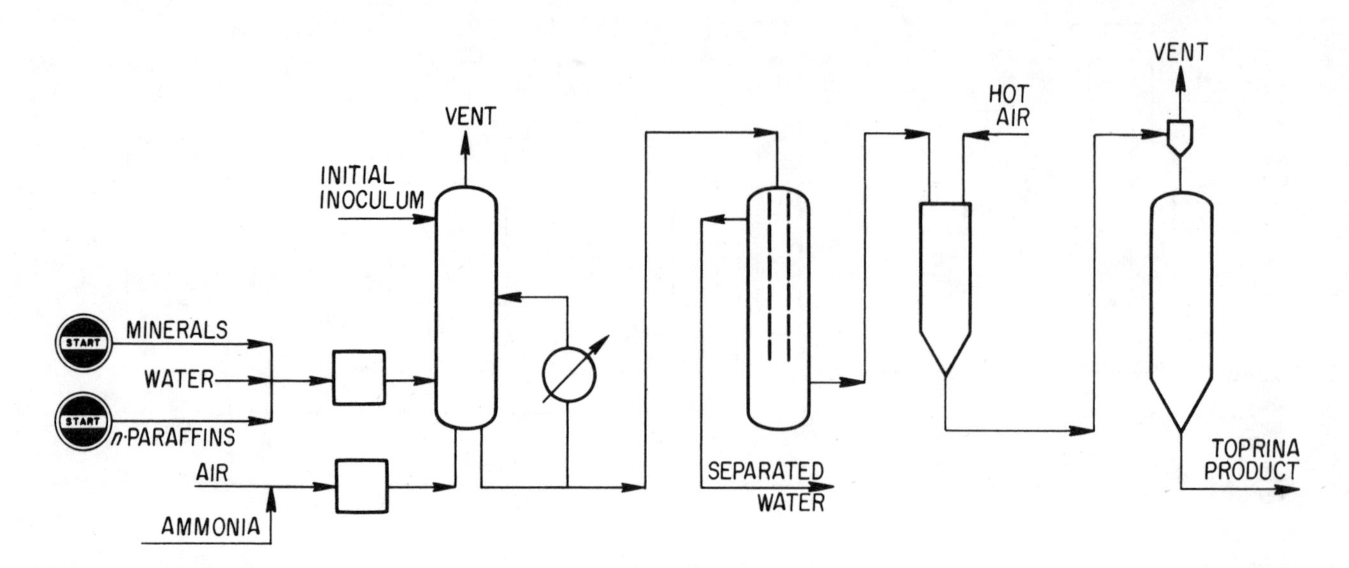

Fig. 6-8—A schematic drawing of a continuous process for the production of single cell protein, SCP, from *n*-paraffins obtained from feedstock boiling in the 175-350°C range.[29]

production are obtained from kerosine and gas oil by 5-**A** molecular processes. The world-wide theoretical feasible production in 1980 will be 124 million tons of *n*-paraffins and in 1985 154 million tons.[28]

It is beyond the scope of this book to discuss the details of **SCP** production from *n*-paraffins. A comprehensive description of the process shown in Fig. 6-8[29] was presented by Foster Wheeler Italiana at the Second Arab Conference on Petrochemicals, Abu Dhabi, U.A.E., 1976.[30]

LITERATURE CITED

1. Luger, C.R. *Hydrocarbon Processing*, Vol. 55, No. 6, 1976, pp. 105-106.
2. Minet, R.C., F.W. Tsai. *The Oil and Gas Journal*, Mar. 21, 1977, pp. 135-141.
3. *Hydrocarbon Processing, 1973 Petrochemical Handbook*, Vol. 52, No. 11, p. 92.
4. *Chemical and Engineering News*, Jan. 1, 1973, pp. 18-19.
5. *Hydrocarbon Processing, 1975 Petrochemical Handbook*, Vol. 54, No. 11, p. 169.
6. Dewing, J., C. Barret and J.J. Rooney. Ger. Offen. 1,903,619 (1969).
7. DuPont, U.S. Patent 3,846,474 (Nov. 5, 1974); U.S. Patent 3,818,067 (June 18, 1974).
8. Simmrock, K.H. *Hydrocarbon Processing*, Vol. 57, No. 11, 1978, pp. 105-113.
9. *The Oil and Gas Journal*, Newsletter, Mar. 28, 1977.
10. *Chemical Week*, Aug. 6, 1975, pp. 23-24.
11. Saunby, J.B. and B.W. Kiff. *Hydrocarbon Processing*, Vol. 55, No. 11, 1974, pp. 247-252, 113.
12. Lowry, R.P., and Adolfo Aquilo. *Hydrocarbon Processing*, Vol. 53, No. 11, 1974, pp. 103-113.
13. Agasiev, R.A., *et al. Azerb. Khim. Zh.*, No. 5, (1969) 128; *Chem. Abstr.*, No. 73, p. 24875.
14. *Chemical Engineering*, Vol. 83, No. 3, 1975, p. 57.
15. Kuz'michev, S.P. and V.K. Sharchenko. *Kinet. Katal* (Eng. Transl.), Vol. 11, 1970, p. 652.
16. *Chemical and Engineering News*, Oct. 4, 1976, p. 9.
17. Browstein, A.M. *Trends in Petrochemical Technology*, Tulsa: Petroleum Publishing Co., 1976, p. 259.
18. *Hydrocarbon Processing, 1979 Petrochemical Handbook*, Vol. 58, No. 11, p. 121.
19. Millidge, A.F. *Education in Chemistry*, Sept. 1970, pp. 189-200.
20. Rouchaud, Jean and Bernard Lutete. *Industrial and Engineering Chemistry*, Product Research Division, Vol. 7, No. 4, 1968, pp. 266-270.
21. Dumas, Theodore and Walter Bulani. *Oxidation of Petrochemicals, Chemistry and Technology*, New York: John Wiley and Sons, 1974.
22. Marer, A. and M.M. Hussain. Second Arab Conference on Petrochemicals, United Arab Emirates, Paper No. 21 (P-3), Abu Dhabi, March 15-23, 1976.
23. Hassan, B.-E. M., A.A. El-Basousi and M.I. Roshdy. Second Arab Conference on Petrochemicals, United Arab Emirates, Paper No. 6 (P-4), Abu Dhabi, March 15-23, 1976.
24. *Hydrocarbon Processing, 1975 Petrochemical Handbook*, Vol. 54, No. 11, p. 128.
25. *Hydrocarbon Processing, 1979 Petrochemical Handbook*, Vol. 58, No. 11, p. 186.
26. Kerfoot, O.C. and H.R. Flammer. *Hydrocarbon Processing*, Vol. 54, No. 3, 1975, pp. 74-78.
27. Laine, B.M. *Hydrocarbon Processing*, Vol. 53, No. 11, 1974, pp. 139-142.
28. Dimmling, W. and R. Seipenbusch. *Hydrocarbon Processing*, Vol. 54, No. 9, 1975, pp. 169-172.
29. *Hydrocarbon Processing, 1975 Petrochemical Handbook*, Vol. 54, No. 11, p. 203.
30. Foster Wheeler, Italiana, Second Arab Conference on Petrochemicals, United Arab Emirates, Paper No. 14 (P-3), Abu Dhabi, March 15-23, 1976.
31. *Chemical Week*, Dec. 5, 1979, p. 42.

Production of Olefins

The three most important olefins used for the production of petrochemicals are **ethylene**, **propylene** and **butadiene**, and their production is inseparable. All three are produced in various ratios by the cracking of feedstocks as different as ethane and crude oil (Table 7-1).[1-4]

Ethylene is the dominant olefin with a production of 28 billion pounds in 1978. The past and predicted sources of ethylene in the United States are shown in Fig. 7-1.[5] In Europe about 90% is obtained by cracking naphtha, Fig. 7-2.[5] It has also been predicted that in 1990 the sources of ethylene in Europe will be: naphtha (67%), gas oil (20%), **LPG** (9%), and ethane (4%).[84] In both the United States and Europe, heavier feedstocks are playing a greater role in ethylene production as supplies of traditional feedstocks (ethane, **LPG**, and naphtha) fail to keep pace with ethylene demand.

As feedstocks progress from ethane through heavier fractions with lower hydrogen content the yield of ethylene decreases and feed per pound of ethylene increases markedly. Heavy gas oil requires over three times the amount of feed per pound of ethylene as for ethane cracking. The total

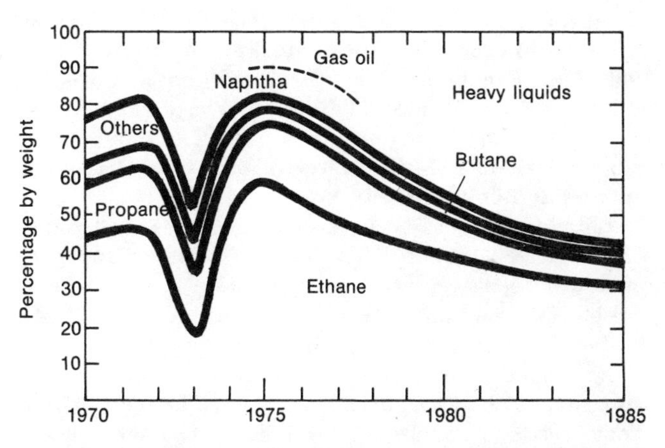

Fig. 7-1—The past, present, and future sources of ethylene in the U.S.[5]

TABLE 7-1—Typical Yields from Various Feedstocks (Including Ethane Recycle)

Feedstock	Products wt. %				
	Ethylene	Propylene	Butadiene	Aromatics (BTX)	Other
Ethane[1]	84.0	1.4	1.4	0.4	12.8
Propane[1]	44.0	15.6	3.4	2.8	34.2
n-Butane[1]	44.4	17.3	4.0	3.4	30.9
Light naphtha[2]	40.3	15.8	4.9	4.8	34.2
Full range naphtha[2]	31.7	13.0	4.7	13.7	36.9
Reformer raffinate[2]	32.9	15.5	5.3	11.0	35.3
Light gas oil[2]	28.3	13.5	4.8	10.9	42.5
Heavy gas oil[2]	25.0	12.4	4.8	11.2	46.6
Waxy distillate[2]	28.3	16.3	6.4	4.5	44.5
Crude resid[3]	21	7	2	11	59
Crude oil[4]	32.8	4.4	3.0	14.4	45.4

Note: Superscript denotes literature cited.

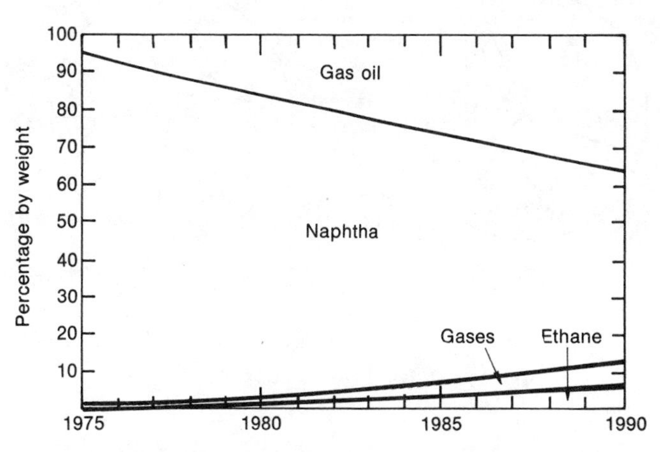

Fig. 7-2—The Western Europe sources of ethylene—past and future.[5]

amount of propylene, butadiene and BTX also increases with increase in molecular weight of the feedstock.

The feedstock of choice is determined by the olefin and BTX distribution desired. Of special interest is the decrease in the ethylene to propylene weight ratio as the feedstock becomes heavier. With ethane the ratio is 1 to 0.017, with propane the ratio is 1 to 0.35, with butane 1 to 0.39, with full range naphtha 1 to 0.41 and with heavy gas oil the ratio is 1 to 0.50. Operation at a lower severity will increase the propylene to ethylene ratio. The yield of propylene as well as that of ethylene can be an important consideration in feedstock selection. An in depth study by Janakievski on alternate feedstocks for olefin production indicates that the feedstocks to be utilized in the Middle East, especially Saudi Arabia, should be light hydrocarbons from associated gas.[6] Until 1976, ethylene production from heavy liquids was less than 2% of the total. The major impetus for this change in feedstock was the anticipated decrease in natural gas production. This change in feedstock increases the availability of propylene for petrochemical conversions. It is expected that the ratio of propylene to ethylene will rise from the current 25:100 to around 50:100 by 1990.

It has been predicted, however, that the consumption of ethane as a fuel will remain relatively stable in the United States for the next few years; there is even a possibility that ethane will show a small increase in usage by 1985. The use of propane is expected to rise gradually from 1979 through 1985. Butane usage probably will remain constant, at least until the LP gas from Middle East associated gas hits the market. A relatively large source of ethylene comes from refinery gas. An informal survey of this source has been reported by Bill R. Minton.[8]

Utilization of feedstocks in existing operations has become a fine science wrought by order-of-magnitude changes in costs. Wilkinson's review of ethylene price trends and the possible ethylene supply/demand balance reveals a high degree of uncertainty as to feedstock trends.[9]

The United States name plate ethylene capacity as of January 1, 1979, was 32.49 billion pounds. This increased to about 38 billion pounds/year by January 1, 1980, will increase to 40.5 billion pounds/year by January 1, 1981, and 42.6 billion pounds/year by January 1, 1982.[10] World-wide ethylene capacity on January 1, 1980, was about 52 million tons/year. The world-wide geographical production of ethylene in 1976 and the predicted production in 1990 are shown in Fig. 7-3.[11] Figure 7-4 shows the distribution of the 1979 noncommunist 43.2 million metric ton ethylene capacity.[12]

Propylene is the second most important olefin. Its growth rate is the fastest of any petrochemical raw material, appreciably higher than the growth rate of ethylene. The demand for petrochemical propylene is now almost equal to its fuel uses and it will soon be the larger of the two. In 1980 the petrochemical utilization is predicted to represent 54% and fuel products 46%. The fuel products are alkylate, polymer gasoline, refinery fuel and a component in LP-gas.

In 1980, 37% of petrochemical propylene will come from refinery gases, and 63% as a coproduct from ethylene production.[13] Coproduct propylene capacity as of June 30, 1979, was about 5.00 million metric tons per year. On the other hand, by 1985 the percentages are predicted to be equal. Other experts feel that more propylene will be drawn from the refinery pool, while still others say that the supply of propylene from refineries for chemical use is only moderately expandable. Obviously, predictions made in respect to propylene sources will be, at best, educated guesses. This is frequently the situation when the compound is both an important coproduct of a dominate main product and also a byproduct of a large and important refinery process. In addition, propylene has no production of its own. The changing ethylene feedstock picture compounds the problem as does the phasing out of lead in gasoline. The constant increase in the need for alkylate and polymer gasoline to maintain octane number will divert refinery propylene from chemical to fuel uses.

Butadiene runs a poor third in the production and demand race with ethylene and propylene. As with propylene, butadiene is a coproduct with ethylene. It is produced at a rather constant 4-5% when the feedstock is butane or higher. The overall ratio of ethylene to butadiene is 1 to

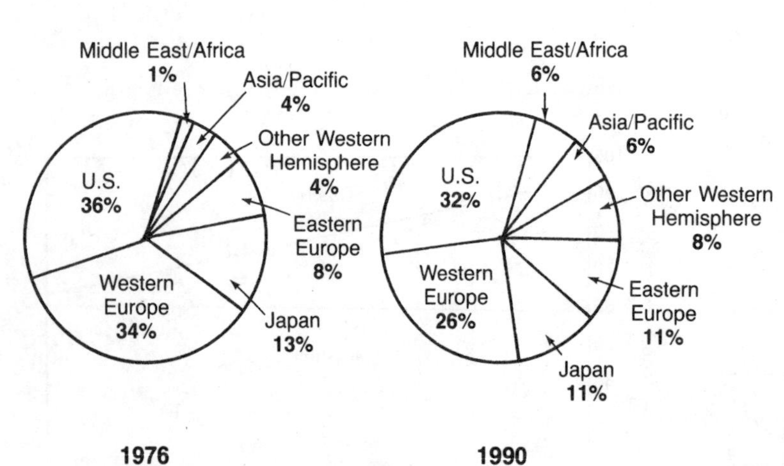

Fig. 7-3—World-wide geographical distribution of ethylene production.[11]

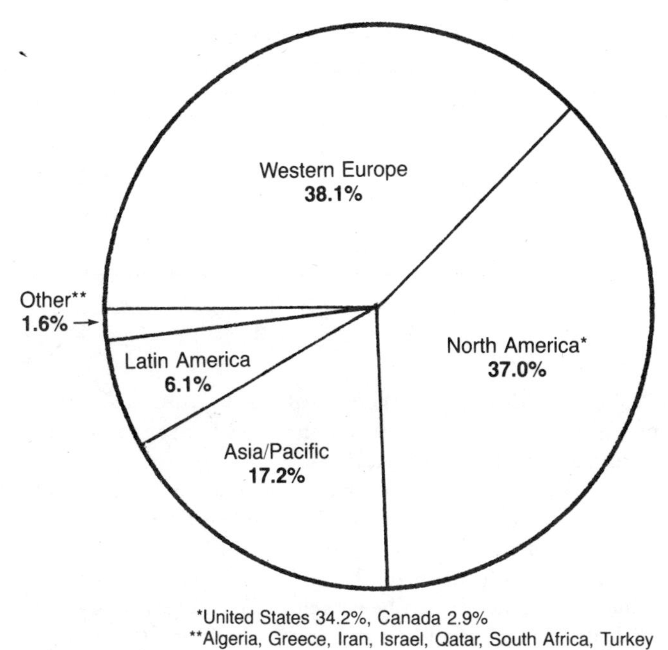

Fig. 7-4—Distribution of the 1979 noncommunist 43.185 million metric ton ethylene capacity.[12]

0.08. Coproduct butadiene capacity as of June 3, 1978, was about 0.93 million metric tons per year. The 1978 United States production of butadiene was 3.52 billion pounds, the world-wide production excluding the U.S.S.R. and China was 9.63 billion pounds, and the 1979 world-wide production was about 9.94 billion pounds.[80]

Butadiene, in addition to being an ethylene coproduct, is produced by butane and butene dehydrogenation and is also obtained from refinery off gas. Coproduct production accounted for 58% of the 1978 production and this percentage is predicted to increase to 67% in 1985. This increase is a reflection of the expected ethylene shift to heavier feedstocks. Ethylene crackers will be the leading source of butadiene in the future. Dehydrogenation of butane and the butenes is currently being phased out as a butadiene source. However, Welch, Croce, and Christmann have made an in-depth study of an oxidative dehydrogenation process for butene to butadiene and find that conversion and overall energy costs show this process to be best for butadiene production.[14]

It is also predicted that by 1981 butadiene demand should be about 4.5 billion pounds and at least 3.6 billion will be coproduct butadiene. Most of the remaining 1.7 billion pounds will be imported.

OLEFIN FEEDSTOCKS

The choice of feedstock for olefin production is progressively growing more complicated. The U.S. natural gas production has peaked out, and the price of all feedstocks has increased dramatically along with construction costs. Because of the changing feedstock picture, a plant now must be capable of handling a wide variety of raw materials to insure a supply position. However, Luger of the Phillips Chemical Division of Phillips Petroleum stated, "Unfortunately, you cannot afford to design a plant to run on everything from ethane to gas oil. The answer to the feedstock is the key to the whole project. It involves looking at the future availability of each feedstock, the pricing of each feedstock, and future coproduct values of propylene, butadiene, benzene and fuel oil. Heavier feedstocks such as naphtha, gas oil, and resids are becoming important precursors for petrochemicals."[15] Currently only 5-10% of the United States ethylene plants can handle natural gas liquids, **NGL,** and heavier feeds.[87]

The plants utilizing heavier feeds are more costly than those designed for ethane but the heavier feeds also produce appreciably more propylene, butadiene and **BTX** as coproducts. Their separation is the major factor in the added production cost of the ethylene. This cost is to some extent compensated for by the coproduct credits. In this article we will be concerned primarily with actual production processes rather than in the economics of the processes.

For the over-all picture one must keep in mind that regardless of the feedstock, olefin production is a gigantic destroyer of energy, an enormous heat-sink. Olefin production is very energy intensive.[16]

With all feedstocks, the main effort is to obtain high selectivity—maximum production of olefins and aromatics and a minimum of methane and C_5+. The main factors influencing the selectivity are the types of compounds present, the reaction temperature and/or residence time, film effect, partial pressure[16] and surface reactions.[17]

In general, the higher the molecular weight of the feedstock, the easier it cracks. For example, the typical furnace outlet temperature for the cracking of ethane is *ca* 800° C while the temperature for the cracking of naphtha or gas oil is *ca* 675-700° C.

Paraffin hydrocarbons, **alkanes,** are the easiest to crack. Straight chain paraffins, ***n*-alkanes,** give high ethylene yields, while branched chain paraffins, **isoparaffins,** such as isobutane and isopentane, give high yields of propylene. Naphthenes, **cycloparaffins,** are more resistant to cracking because the ring must be opened first. At low temperatures they tend to dehydrogenate to aromatic compounds. Higher temperatures produce olefins and diolefins. Aromatic rings, **benzene rings,** tend to pass through the cracking furnace unchanged. Aromatic compounds with side chains such as toluene and the xylenes, lose these side chains by hydrodealkylation and become benzene.

$$C_6H_5CH_3 + H_2 \rightarrow C_6H_6 + CH_4$$

This accounts for the relatively high benzene concentration in pyrolysis gasoline.

In respect to **residence time,** conversion in the fluid (bulk conversion) has to be distinguished from conversion in the laminar layer near the tube wall. This is the so-called **film effect.** The film effect is a function of fluid velocity, tubewall temperature, and coil diameter. Olefins are formed by primary reactions and, therefore, require a low residence time.

Low velocity produces a high film effect. A high film effect will cause a high conversion in the laminar layer at the expense of the primary products. This is especially true for the less refractory products such as propylene and butadiene.

The **reactor surface** is also of importance because it influences the yields of both desired and undesired products. The effect of reactor materials is shown in Table 7-2 on the pyrolysis of ethane to ethylene.[17]

Partial pressures of the hydrocarbons strongly influence secondary reactions such as the production of aromatics at the expense of olefins. Maximum ethylene yield will be obtained at low residence times. For maximum olefin yields, a compromise must be made between residence time, film effect, partial pressure of the hydrocarbon and reaction temperature.

Cracking for the simultaneous production of olefins and aromatics is controversial. Olefin production requires short residence times; aromatic yields increase with residence time and partial pressure. In general, the feedstock and process conditions can be selected to produce a wide range of product requirements.

An ethylene plant consists of two main sections: the pyrolysis section where the feedstock is thermally cracked and the separation section which separates the specific products.

GAS FEEDSTOCKS

The gaseous feedstocks for ethylene production are ethane, propane and *n*-butane and various mixtures of these.

Ethane. The advantage of ethane as a feedstock is a high ultimate ethylene yield combined with a minimum of coproducts. At a 60% per pass conversion level, the ultimate yield of ethylene is about 80 wt %, based on the ethane being recycled to extinction. A conversion of 60% on a single-pass basis is considered optimum. This gives an acceptable ultimate ethylene yield and furnace run length without excessive recycle.

The analysis of a typical ethane feedstock is

Acetylene	1.9 wt %
Ethane	94.2
Propylene	3.0
Propane	0.9

The unsaturates come with the recycled ethane.

Typical operating conditions are

Temperature	750-850° C	
Catalyst	None	
Pressure	1-1.25 atms (bars)	
Steam diluent	0-50%	
Conversion	60%	
Yields: wt %	Hydrogen and methane	12.9
	Ethylene	80.9
	Propylene	1.8
	Butadiene	1.9
	Other*	2.5

*Other: Propane 0.3; butenes 0.4; butane 0.4; C_5 and heavier 1.4.

TABLE 7-2—The Effect of Surface Material on Ethane Pyrolysis to Ethylene[50]

	Ethylene Yields	
	30% conversion	60% conversion
304 stainless steel	84%	69%
Incoloy 8000	89	79.5
Vycor	93.5	89
"Wall-less" reactor	97.5	93

These yields are the theoretical yields when acetylene, methylacetylene, and propadiene are hydrogenated to ethane and propane, respectively.[18]

A typical ethane cracker has several identical pyrolysis furnaces in which fresh ethane feed and recycle ethane are cracked with steam as a diluent (Fig. 7-5).[19,20,21] The outlet temperature is usually in the 850°C range. The furnace effluent is quenched in a heat exchanger and further cooled by direct contact in a water quench tower where the diluent steam is condensed. This water is recycled to the pyrolysis furnace. The cracked gas is compressed, acid constituents are removed, and the purified gas dried.

Hydrogen and methane are removed from the pyrolysis products in the demethanizer. The product stream is hydrogenated to remove acetylene, or the acetylene is separated as a product. Ethylene is separated in the ethylene tower from the unreacted ethane and higher boiling products. The ethane is recycled to extinction. The other products are separated and either sold, burned as fuel, or absorbed into a refinery operation.

Propane. The supply of ethane can be increased by deeper ethane extraction from natural gas, but the supply picture for propane to be used for olefin production is not encouraging.[8, 22] The maximum propane extraction from natural gas is essentially quantitative and the utilization of this propane is primarily as a fuel.

Propane cracking for olefin production is somewhat more complicated than ethane cracking. The processing sequence up to the demethanizer is the same. Essentially the same reaction conditions are used and the reaction furnaces are also the same. However, the ease of cracking increases with molecular weight. For example, the same conditions of temperature and residence time that give a 60% conversion for ethane give a 90% conversion for propane.[23]

Propane gives a lower ethylene yield and a larger quantity of coproducts (Table 7-1). Propane produces a two-and-a-half fold increase in residual gas, and over a ten-fold increase in propylene, twice as much butadiene, and significantly more aromatic pyrolysis gasoline. Because of these coproducts, the separation section is more extensive than that used for ethane.

The ratio of coproducts and the degree of conversion is determined by the severity of the reaction conditions. This is illustrated in Fig. 7-6.

Ethane-Propane Mixtures. The cracking of mixtures of ethane and propane is an alternative to cracking each separately. *Cocracking* viability has been investigated by Minet and Hammond.[24] They found that cocracking has the following effect, relative to separate cracking:

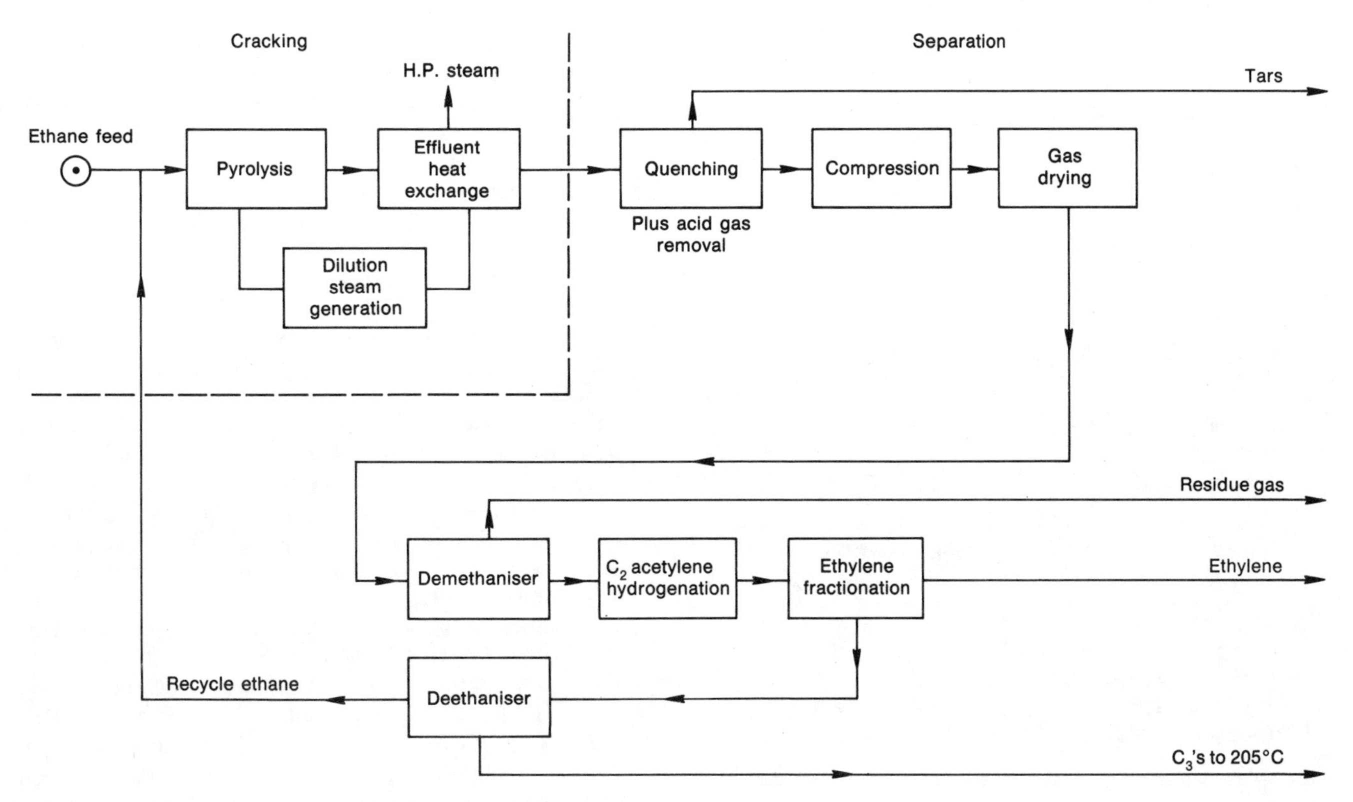

Fig. 7-5—Basic plant for the production of ethylene from ethane.

Reduced yields	Increased yields
Ethylene	Hydrogen and methane
Propylene	Ethane
Propane	C_5+
Mixed C_4's	

The extent of these effects is shown in Table 7-3. The cocracking effect on ethylene yield is small compared with the effect on the propane-propylene yields. Overall, cracking ethane and propane separately increases the yield of propylene with a small decrease in the feedstock requirement for constant ethylene production.

TABLE 7-3—The Effect of Cracking Ethane and Propane Separately and as a Mixture[24]

Products*	Separate Cracking**	Cocracking 50/50 mixture**
Hydrogen + Methane	20.5	22.8
Ethylene	61.6	60.9
Propylene	9.8	6.9
Butadiene	2.2	1.9
Butenes	1.2	0.8
C_5+	4.7	6.7

*Weight percent
**Ethane and propane recycled to extinction.

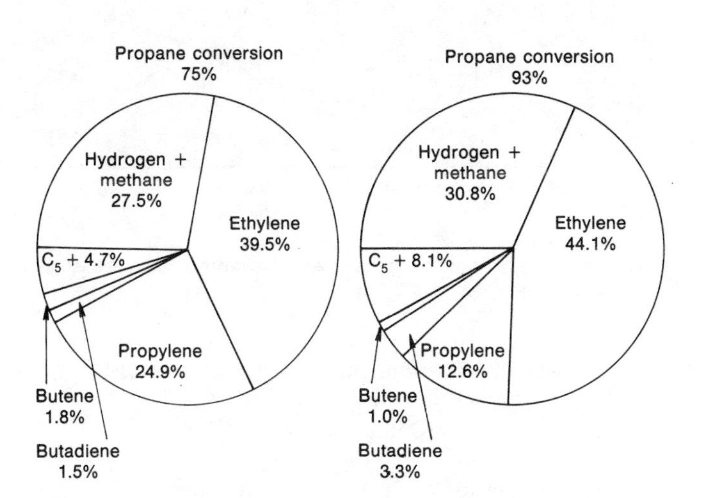

Fig. 7-6—The influence of conversion severity on the theoretical product yield for the cracking of propane. Acetylene, methylacetylene, and propadiene are hydrogenated and both ethane and propane are recycled to extinction. (wt %)[18]

n-Butane. *n*-Butane is a minor source of ethylene. A cracker designed for *n*-butane feedstock is substantially the same in concept as one used for propane cracking. *n*-Butane is cracked at the highest conversion level because any unconverted butane will be contained in the C_4+ products. This makes recovery of the butadiene and butenes more difficult and expensive. Fig. 7-7 gives the products for one-pass *n*-butane cracking at a 90% conversion level.[6]

LPG. In the early 1980's there will be a marked world-wide increase in available **LPG**, especially from the Middle East. By 1985 the world surplus will be comparable to the entire 1979 United States market.[85]

Ethylene plants in the United States have been designed to utilize ethane as the feedstock. With the expected increase in the availability of **LPG**, thought is now being given to the design or modification of ethane-based ethylene plants which will accommodate **LPG** as well as ethane. Fig. 7-8

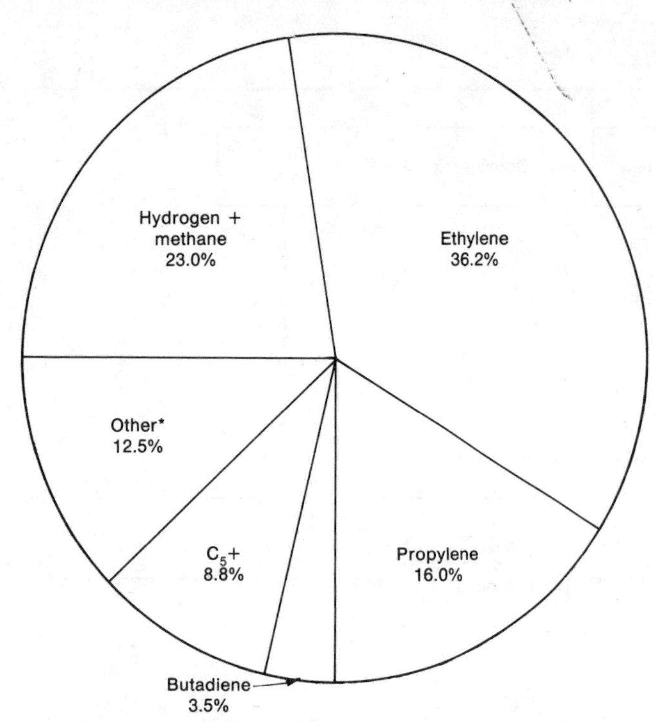

Fig.7-7–One-pass cracking of *n*-butane with 90% conversion.(wt%)[6]

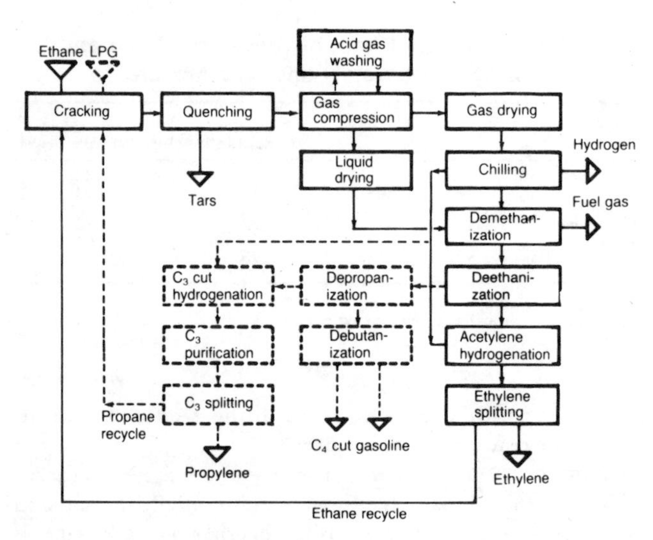

Fig. 7-8—Modifications required to convert an ethylene plant from ethane to LPG feed. The dotted lines and equipment are added for LPG.[86]

indicates the modification required to make this conversion from ethane to **LPG**.[86] The plant should be able to give 75-80% plant capacity. It has been estimated that the petrochemicals industry can convert 15-20% of its naphtha feedstock base to **LPG**.[88]

LIQUID FEEDSTOCKS

The major liquid feeds for olefin production are light virgin naphtha (**LVN**), full range naphtha (**FRN**), reformer raffinate, atmospheric gas oil (**AGO**), vacuum gas oil (**VGO**), resids and crude oil. The feedstocks are usually cracked with lower residence times and higher temperatures and with higher steam dilution ratios than is used for gas feedstocks. The reaction section of the plants is essentially the same as with gas feedstocks but the design of the convection and quenching sections are different. This is required because of the wider variety and greater quantity of coproducts. Feedstock governs the plant design.[25] For example, provision must be made for the processing of *pyrolysis gasoline*, pyro gasoline.[26] Fig. 7-9 shows the relationships among feedstock, propylene/ethylene ratios, and pyrolysis gasoline yields.[27] Table 7-4 gives the types of compounds in a typical pyrolysis gasoline.[27]

An olefin plant which utilizes a liquid feedstock requires an additional pyrolysis furnace for cracking coproduct ethane and propane and an effluent quench exchanger. This is followed by an oil quench and a primary fractionator for fuel oil separation. In contrast, a gas cracker requires a simple direct-contact water quench tower off the cracking unit. A liquid feed cracker also contains a propylene tower and a methylacetylene removal unit. A unit for first stage hydrotreating of pyrolysis gasoline may be included.

High olefin yields are favored by low hydrocarbon partial pressure, pressure drop and residence time.[28] This is detrimental to **BTX** production. Thus, it is not possible to maximize conditions for both high olefin yield and high **BTX** yield. A compromise to operate at a longer residence time (0.3-0.5 sec) and a lower outlet temperature. For naphthas and gas oils, high-severity, short residence-time furnaces are required. This will maximize ethylene yield and minimize coking.

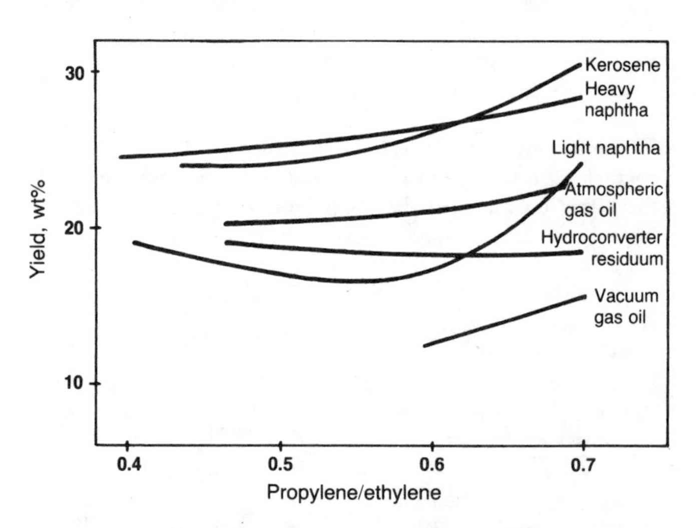

Fig. 7-9—The pyrolysis gasoline yield from various liquid feedstocks vs. propylene/ethylene ratio.[27]

TABLE 7-4—Composition of the C_5 and C_6-C_8 Fraction of Pyrolysis Gasoline by Compound Type*[27]

Group of Compounds	Fraction of Pyrolysis Gasoline Total	C_5	C_6-C_8
Paraffins, %	1.2	4.7	0.8
Mono-olefins, %	1.2	5.7	0.6
Diolefins, %	13.2	89.3	3.1
Cycloparaffins, %	0.5	0.3	0.5
Aromatics, %	83.9	...	95.0

*Feed stock: heavy naphtha; propylene/ethylene ≈0.5

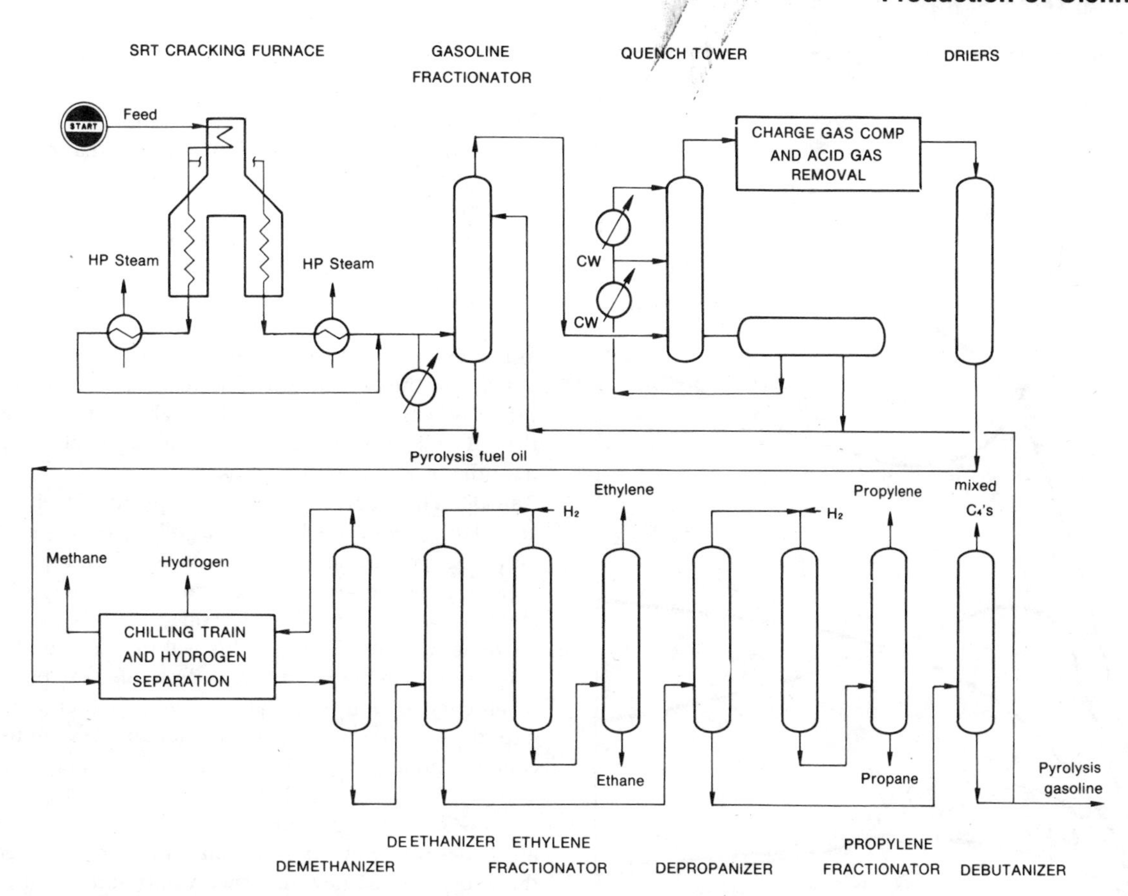

Fig. 7-10—The C-E Lummis process for the cracking of naphtha or gas oil for the production of ethylene.[30]

As previously noted in this chapter, the idea of designing an ethylene plant which will process feedstock from ethane to gas oil is technically feasible but economically unsound. However, Western European ethylene producers are quickly moving to improve their feedstock situation by building more dual-feed crackers—ethylene plants that are able to utilize both naphtha and gas oil.[29]

Naphtha. One of the advantages of naphtha over gas feedstocks is the wider spectrum of coproducts—provided, of course, you wish to obtain a variety of coproducts.

Fig. 7-10 represents a typical plant for naphtha cracking.[30] Conventional operating conditions are[28]

Outlet temperature:	840°C	
Catalyst:	None	
Pressure:	Atmospheric +	
Steam dilution: kg/kg HC	0.6	
Residence time: sec	0.35	
Yields, wt%	Ethylene	31.0
	Propylene	14.7
	Butadiene	4.4
	BTX	14.3
	Other	35.6

Table 7-5 and Fig. 7-11 contain data that indicate the influence of severity on product distribution.[31, 32] An increase in severity increases the production of ethylene, at the expense of propylene and the butenes. Both methane and **BTX** increase. The naphtha feedstock required for high severity cracking is 15-20% higher than that for moderate severity cracking. Pyrolysis gasoline accounts for about two-thirds of this additional naphtha feed. The additional pyrolysis gasoline accounts for the additional **BTX** formed.

High severity cracking of Kuwait full-range naphtha compared to propane feed shows propylene about 29% greater than low-severity cracking and the C_4 mixture is up 280%.[25] The butadiene is up 218%. Pyrolysis gasoline has increased seven-fold and contains 67% aromatics. Benzene accounts for 54% of this gasoline.

The overall product distribution is determined by the naphtha characteristics—and there is no typical naphtha.

TABLE 7-5—Typical Once-Through Yields from Naphtha Feedstock*[31,32]

	Cracking Severity	
Products**	Low	High
Methane	10.3	15
Ethylene	25.8	31.3
Propylene	16.0	12.1
Butadiene	4.5	4.2
Butenes	7.9	2.8
BTX	10	13
C5+	17	9
Fuel oil	3	6
Other***	5.5	6.6

*Sp. gr. 60/60°F 0.713
Boiling range °C 32-170
Aromatics 7
**Weight percent
***Ethane (3.3 and 3.4%), acetylene, methylacetylene, propane, hydrogen.

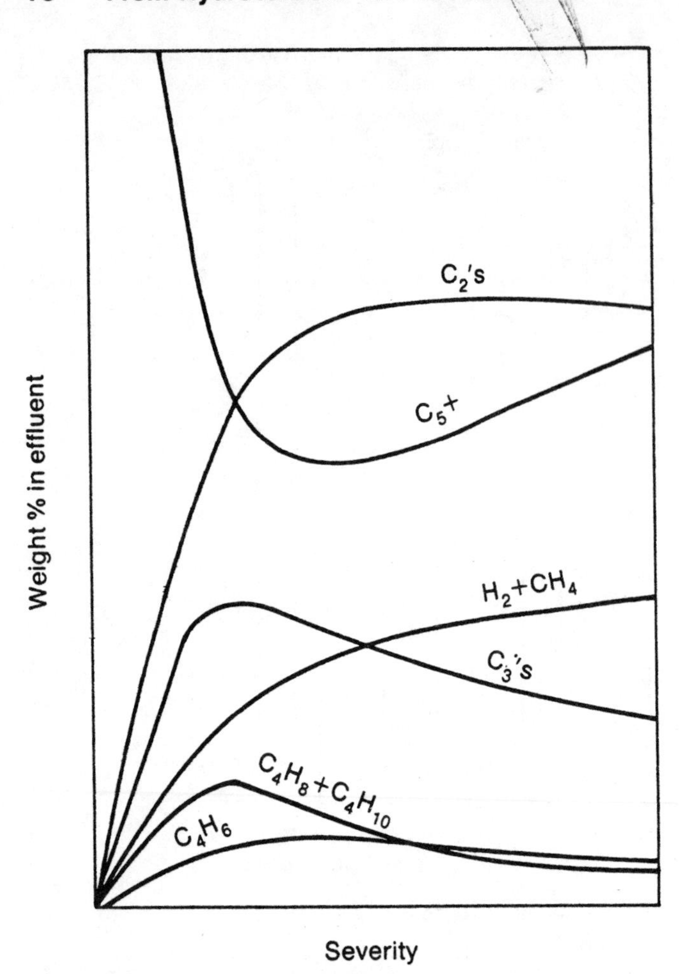

Fig. 7-11—Naphtha cracking—typical variation of product distribution with severity.[32]

Naphthas are normally characterized by their specific gravity, ASTM boiling range, the C/H ratio and PONA analysis which provides the distribution of *n*-paraffins and isoparaffins, olefins, naphthenes and aromatics. *n*-Paraffins produce more ethylene, while branched chain paraffins (isoparaffins) will yield more propylene.

Unwanted components in the product stream are often removed by a chemical reaction, usually hydrogenation. The place of hydrogenation in the production of olefins by hydrocarbon cracking has been reviewed by Watson.[33] The catalyst is palladium, Pd, on alumina. The compounds hydrogenated are:

C_2	Acetylene	$HC \equiv CH$
C_3	Methylacetylene	$CH_3 - C \equiv CH$
	Propadiene	$CH_2 = C = CH_2$
C_4	Ethylacetylene	$CH_3 - CH_2 - C \equiv CH$
	Dimethylacetylene	$CH_3 - C \equiv C - CH_3$
	Vinylacetylene	$CH_2 = CH - C \equiv CH$
	Butadiene	$CH_2 = CH - CH = CH_2$
C_5^+	Dienes and acetylenes,	C_5/C_9 fractions.

The hydrogenation of acetylene is a gas phase reaction. C_3 and C_4 hydrogenations can be either gas phase or liquid phase while C_5^+ constituents are hydrogenated only in the liquid phase. The primary purpose of hydrogenation is to improve the quality of specific products—for example, to upgrade ethylene from chemical to polymer specifications.

Reformer Raffinate. Raffinate from aromatic extraction units contains C_6 to C_8 hydrocarbons and has essentially the same cracking requirements as naphtha. Because raffinates are high in isoparaffins, they produce relatively less ethylene and more propylene than virgin (straight run) naphtha.

Currently only about 8,000 bpd of the 135,000 bpd raffinate production is used for olefin production.[34] Because of the low octane number of raffinate, more will become available in the future because less raffinate will be used in gasoline as the production of no-lead gasoline increases. The raffinate will be replaced by the higher octane virgin naphthas. The raffinate will then, to some extent, replace the naphtha as ethylene feedstock.

Gas Oil. The cracking of gas oils to produce olefins is an old art—dating from the 1930's. Gas oils were quickly replaced by the more desirable and available ethane, propane and butane. Now we are turning again to gas oils. World-wide ethylene capacity in 1977 based on gas oil was 6% of the total ethylene capacity; in 1985, this will increase to 18-22%.[35]

Gas oils, in general, are not as desirable as naphtha for the production of ethylene. They have a higher density than naphtha, a lower hydrogen content and a higher sulfur content and a higher concentration of aromatic compounds. The wt % aromatics goes from 11% for light gas oil to 48% for **VGO**. Heavy gas oil has about 33% aromatics. These aromatics constitute an important difference between naphtha and gas oil. They influence yields and running time of the system. Gas oils produce lower yields of ethylene and appreciably more heavy fuel oil. Gas oil plants also cost more for the same ethylene production.[36]

Gas oils are characterized as either atmospheric gas oil, **AGO**, or vacuum gas oil, **VGO**. They differ in their boiling range with the atmospheric gas oil having the lower range, 232-327°C, while vacuum gas oil has a range of 299-538°C (Table 7-6).[32] Atmospheric gas oil contains about 10 times more sulfur than full-range naphtha. Vacuum gas oil contains about three times as much sulfur as **AGO** (Table 7-7).[37] Because of this sulfur, gas oil feedstocks must be hydrodesulfurized before cracking. High sulfur oils can be cracked but the resulting high levels of sulfur in the products cause processing and environmental problems. Between 50 and 60% of the sulfur in the AGO and VGO feedstock appears in the pyrolysis fuel oil. High sulfur concentrations in the feedstock also may cause severe corrosion problems.

The same type of cracking plant (Fig. 7-12) and conditions are used for gas oils as those for naphtha. The gas oil throughput is about 20-25% higher than that for naphtha.[29] The ethylene capacity for **AGO** is about 15% lower than for

TABLE 7-6—Characteristics of Typical Atmospheric Gas Oil, AGO, and Vacuum Gas Oil, VGO[32]

Properties	Gas Oil	
	Atmospheric AGO	Vacuum VGO
Specific gravity, °API	38.6	30.0
Specific gravity, 15/15°C	0.832	0.876
Boiling range, °C	232-327	299-538
Hydrogen, wt%	13.7	13.0
Aromatics, wt%	24.0	28.0

TABLE 7-7—The Sulfur Content of Atmospheric Gas Oil, AGO, and Vacuum Gas Oil, VGO, from Various Grade Oils[37]

Crude origin	Sulfur, wt% AGO	Sulfur, wt% VGO
Kuwait	0.7	2.7
Kirkuk	1.0	2.5
Gach Saron	0.82	1.6
Agha Jari	0.33	1.5
Kharsaniyah	1.0	3.0
Safaniyah	1.03	3.0
Abu Dhabi	0.7	1.3
Nigerian	0.1	0.28
Libyan	0.27	0.4
Ekofisk	0.06	0.19
Brunei	0.05	0.15

TABLE 7-8—Typical Once-Through Data for Atmospheric Gas Oil, AGO, and Vacuum Gas Oil, VGO[31,32,39]

Products*	AGO Severity Low	AGO Severity High	VGO Severity Low	VGO Severity High
Methane	8.0	13.7	6.6	9.4
Ethylene	19.5	26.0	19.4	23.0
Ethane	3.3	3.0	2.8	3.0
Propylene	14.0	9.0	13.9	13.7
Butadiene	4.5	4.2	5.0	6.3
Butenes	6.4	2.0	7.0	4.9
BTX	10.7	12.6	18.9	16.9
C_5–205°C**	10.0	8.0		
Fuel oil	21.8	19.0	25.0	21.0
Other***	1.8	2.5	1.4	1.8

* Weight %.
** Other than **BTX**.
*** Acetylene, methylacetylene, propane, hydrogen.

naphtha.[38] There must be a careful balance between furnace residence time, hydrocarbon partial pressure and other factors in order to avoid problems inherent in cracking **AGO**. These problems include fouling in the hot end of the plant, both in the furnaces and the transfer line exchangers.[29]

Table 7-8 contains the results of a typical once through cracking of **AGO** and **VGO** at low and high severity.[31, 32, 39] Fig. 7-13 shows product distribution with respect to severity for a typical gas oil feedstock.[40] The relationship between reaction temperature and olefin product distribution is shown in Fig. 7-14.[41]

Table 7-9 contains a comparison of products from the high severity cracking of light virgin naphtha, **LVN**, (43°-84° C) from a light Arabian crude with the products from **VGO** which is the 370°-470°C fraction of Kuwait crude and contains 1.9 wt % sulfur.[42, 43] The liquid products from the **VGO** feed are approximately 45 wt % on feed or about twice the liquid product yield from the **LVN** feed. The heavy cracked oil fraction contains 5 wt % sulfur.

Pyrolysis fuel is high in aromatic polynuclear material that contains unsaturated gum formers and possibly a small amount of carbon. It can be blended to only a limited degree and can present combustion and handling problems. Cracked fuel oil also can be used for carbon black production and as a feedstock for specialty coke.

A general survey of olefin production by gas oil cracking has been published by Zdonik and various authors.[40, 44-47] The emphasis is on feedstock composition, furnace and infrastructure design, and process variables. Lassman and Wernicke have a method to predict with excellent reliability the product yields from steam-cracking gas oils.[48]

Resid is cracked to produce ethylene, propylene and pitch as the major products[31, 49] by a different process from the production of acetylene as the major product.[50] The *acetylene* intensive process cracks the feedstock by an oxygen-fed

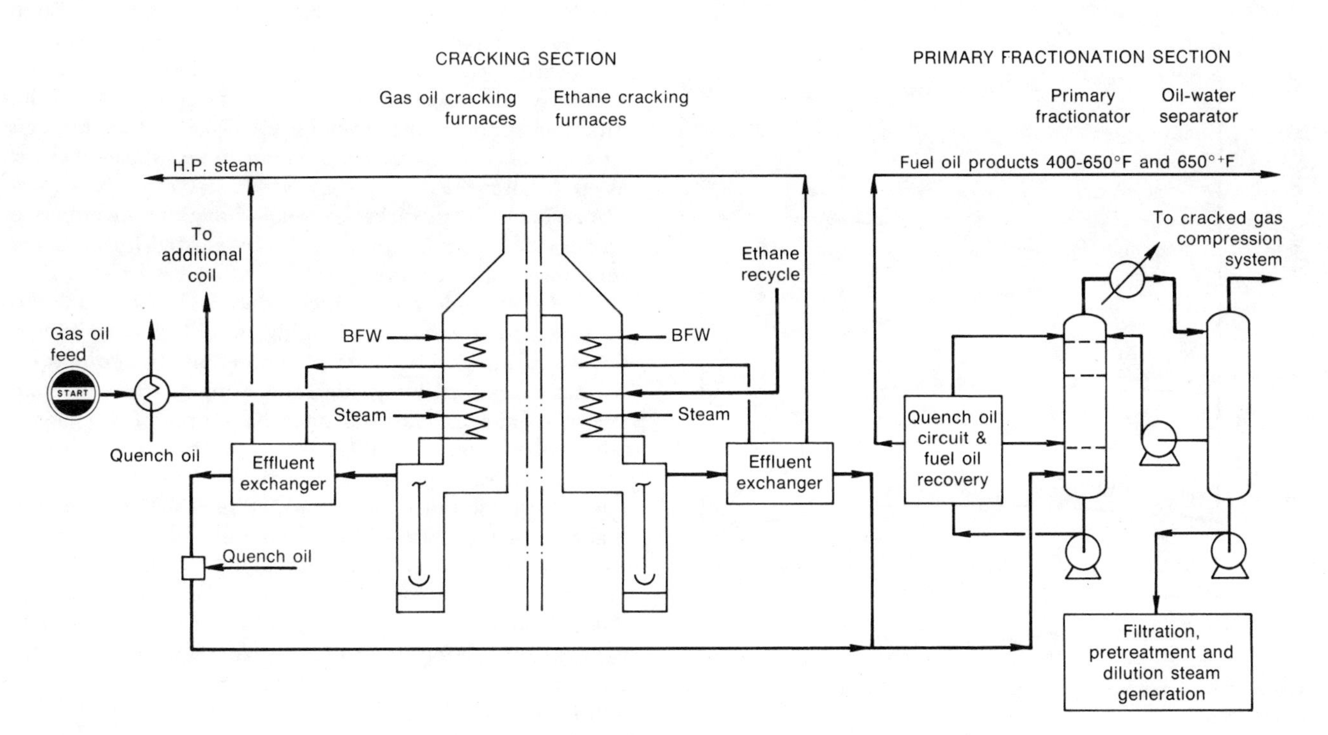

Fig. 7-12—A simplified version of the cracking and primary fractionation section of a gas oil cracking plant.[42]

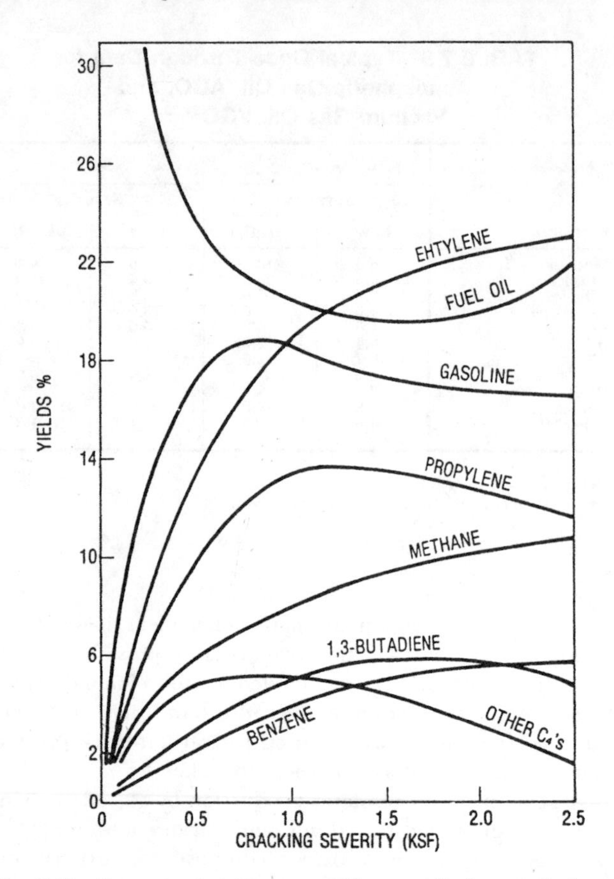

Fig. 7-13—Component yields vs. cracking severity for a typical gas oil.[40]

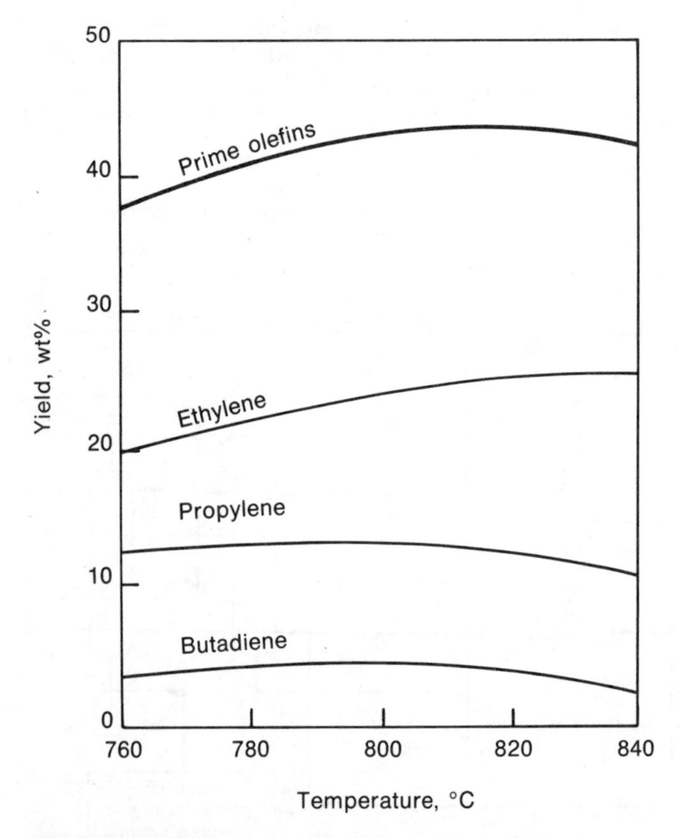

Fig. 7-14—The influence of temperature on olefin production from gas oil cracking.[37]

TABLE 7-9—Furnace-Outlets Yields for the Steam Cracking of Light Virgin Naphtha, LVN, and Vacuum Gas Oil, VGO[42,43]

Products	(% w) LVN	VGO
Hydrogen	1.0	0.4
Methane	16.7	9.1
Acetylene	0.6	0.2
Ethylene	31.3	16.6
Ethane	4.3	4.4
C_3 and C_4 acetylenes	1.1	0.5
Propylene	16.2	13.3
Propane	0.5	1.0
Butadienes	4.7	4.1
Isobutene	2.1	1.9
n-Butenes	1.9	3.2
Butanes	0.2	0.1
C_4 and lighter S compounds	....	0.6
Steam cracked naphtha (C_5 to 190°C)	17.9	17.3
Heavy cracked oil (190°C +)	1.5	27.3
Total	100.0	100.0
Properties of heavy cracked oil (190°C +)		
Specific gravity	1.08	1.11
Sulfur (% w)	0.08	5.0
Viscosity at 90°C (SUS)	10	300

flame burning under the surface of the resid (Fig. 7-15). The following wt % yields are obtained:

Acetylene	29%
Ethylene	33%
Propylene	13%
C_4+	25%

Appreciable lean gas, $CO:H_2$ = 1.65:1.00, is also produced. This process can utilize crude oil as well as fuel oil and vacuum resid.

Crude resid is the feedstock for a cyclic thermal cracking process to produce ethylene, propylene and **BTX**.[31, 49] Pitch for use as an electrode bonder in aluminum cell anode manufacture is an interesting coproduct (Table 7-10). The process flow diagram is shown in Fig. 7-16.[49] Vertical refractory-lined reactors operate in pairs in a cyclic reaction-regeneration pattern. The atomized feedstock is sprayed downward on a stacked bed of hot, 1200°C+, refractory checker-work of bricks. Each complete cycle is 4 minutes with the endothermic cracking cycle being 1.25 minutes. This process can crack natural tar sands and shale oil as well as crude oil resids.

A recent process cracks heavy liquid vacuum-tower bottom feedstock in a fluidized bed of coke particles at a cost considered to be competitive.

Crude Oil. Chemical companies are always looking backward in respect to integration. The ultimate of backward integration is cracking crude oil to obtain olefins. In addition this ultimate backward integration will eliminate the uncertainties of ethane, LP-gas and naphtha supply. The yields of ethylene, acetylene and **BTX** are greater than from naphtha and gas oil cracking. However, coproducts include fuel oil, pitch, naphtha and tar fractions, all of which are not normally marketed by chemical companies.

Two processes have been developed for the cracking of crude oil. In one process the crude oil is sprayed directly into 2,000°C superheated steam inside the reactor. The yield patterns vary widely according to changes in residence time and temperature. In a typical process a C_2 yield of 35.9 lbs/100 lbs of feed (Arabian light crude) was obtained.[51] The

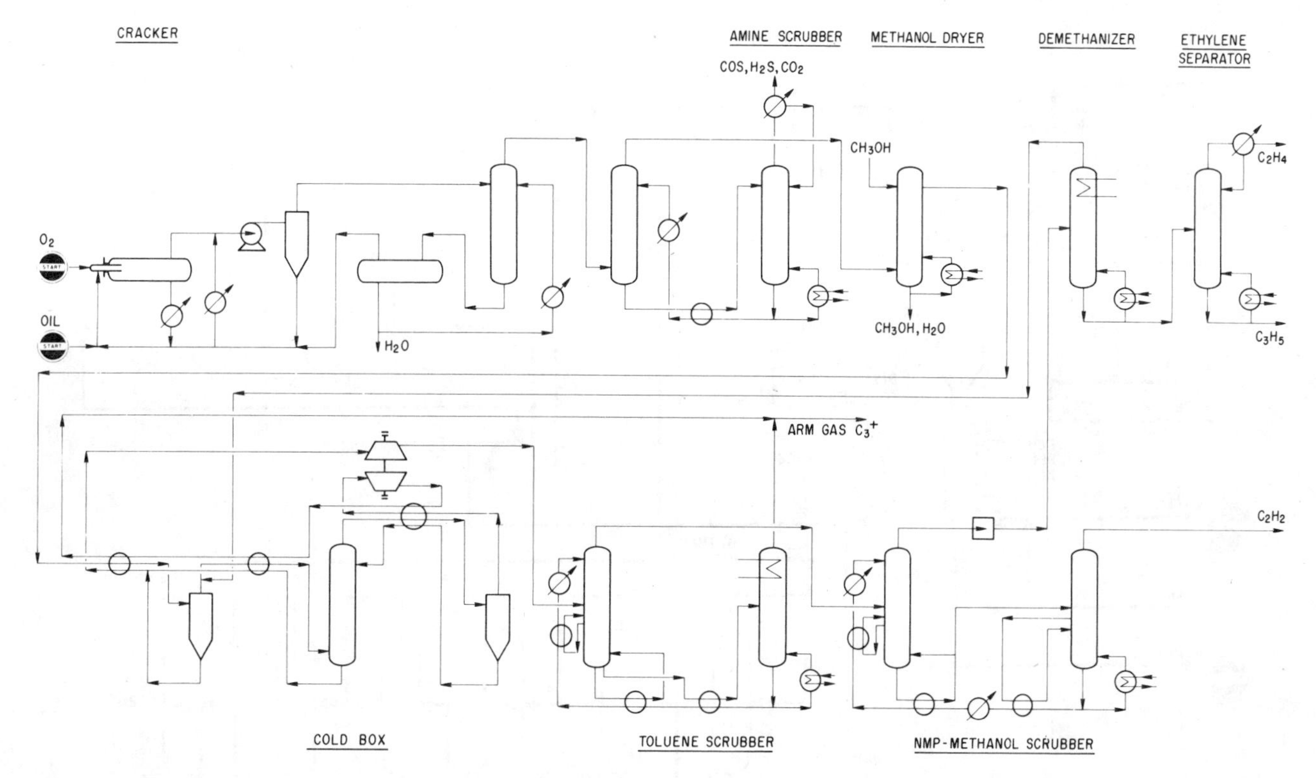

Fig. 7-15—An authothermic cracking process for the production of acetylene, ethylene, propylene, and lean gas, $CO+H_2$, from either resid or crude oil.[50]

TABLE 7-10—Product Yields from the Cracking of Various Crude Resid[49]

Characterization	Australia Gippsland	Indonesia Minas	Middle East Iranian	FCC Cycle oils
	Feedstock Source			
Sp. Gr.	0.88	0.89	0.975	0.92
UOP "K" factor	12.2	12.3	11.6	11.3
C/H wt ratio	6.4	6.3	7.7	7.3
Products*				
Hydrogen	1	1	1	1
Dry-fuel gas**	22	21	16	18
Ethylene	21	20	15	16
Propylene	7	7	5	5
Butadiene	2	2	1	1
Butenes	2	2	1	1
C_5/C_6 unsaturates	2	2	2	2
Benzene	7	7	5	6
Toluene	3	3	2	3
Xylene	1	1	1	1
Aromatic distillates	11	13	15	18
Pitch	13	13	22	18
Carbon	8	8	14	10

*Weight %.
** Methane, ethane, propane.

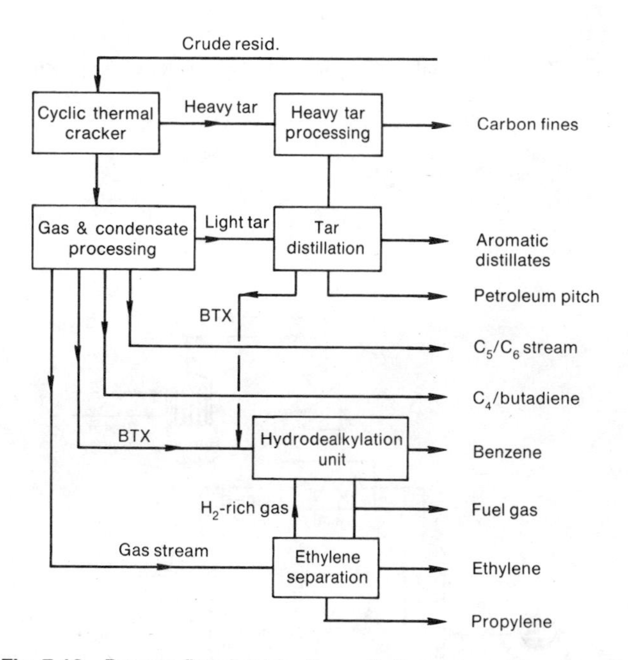

Fig. 7-16—Process flowsheet for the cyclic thermal cracking of crude resid.[49]

ethylene to acetylene weight ratio was 7.54 to 1.00. This same technique can be used for light naphtha or gas oils.

The other process is direct cracking by a fluidized bed reactor.[52] Fig. 7-17 gives a flow diagram of a process of this type. It uses partial combustion as a source of heat and inorganic oxide particles as the solid particles in the fluidized bed. Other processes use carbon particles. The residence time is 0.2-0.3 sec and the ethylene, propylene, butene-butadiene ratio is 1:0.4:0.3.

A similar process which operates at 900-925°C with a residence time of less than 0.10 sec is said to provide substantial improvements in yields and feedstock utilization.[53] It is possible to get 65-70% of the crude oil into basic chemicals. The Advanced Cracking Reactor (ACR) process, Fig. 7-18, represents a significant advance in olefin technology.[54] The process is reported to obtain maximum C_2 and olefin yields from most petroleum distillates and selected crude oils.

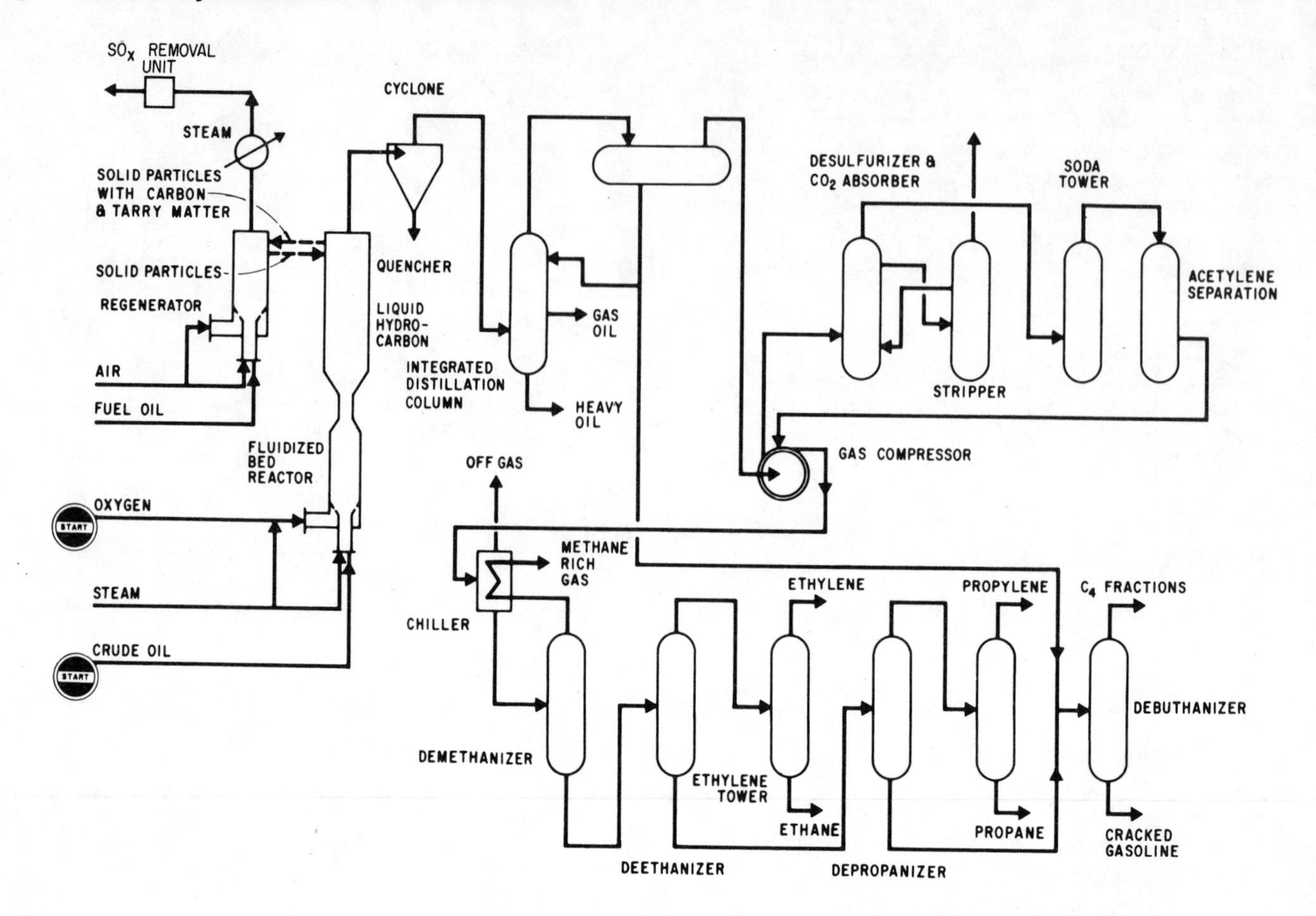

Fig. 7-17—Fluidized bed cracking of crude oil for olefin production.[52]

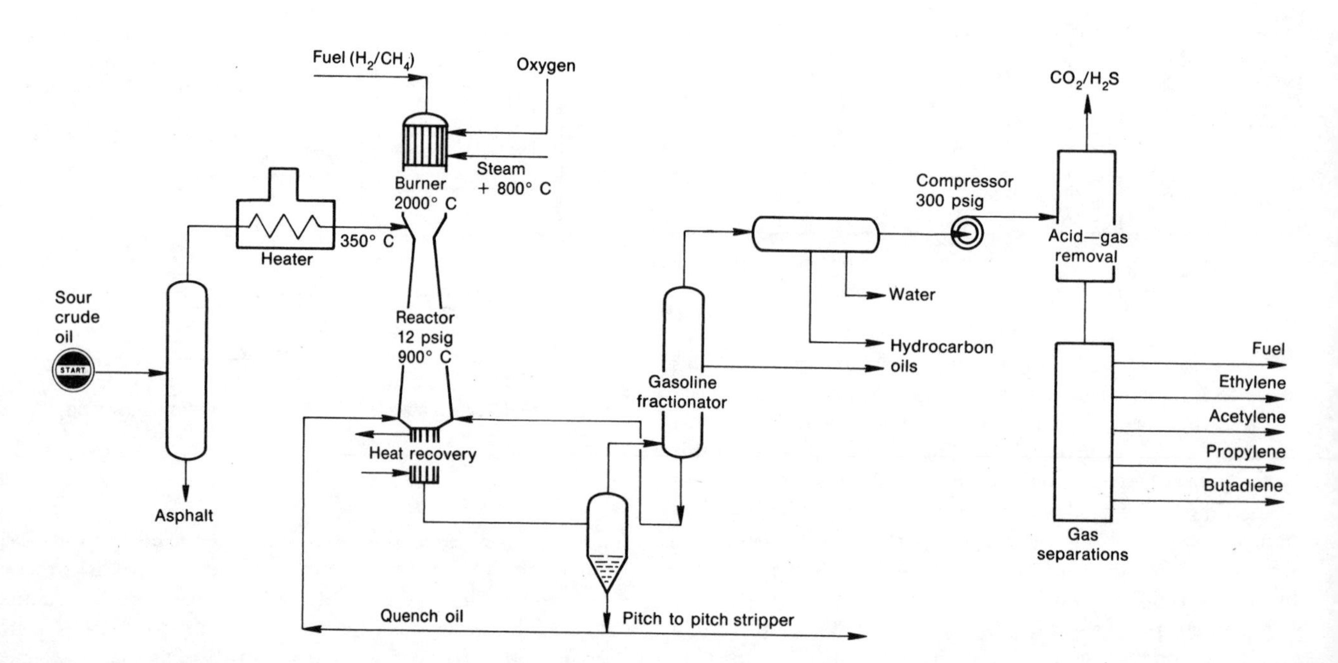

Fig. 7-18—The Advanced Cracking Reactor (ACR) for producing olefins from crude oil.[54]

OTHER PROCESSES

There is a continuing, never ending, search and research for better processes to produce ethylene, other olefins and **BTX**. These efforts are illustrated by the development of hydropyrolysis and the Millisecond Furnace. Concomitant with these developments is research on coproduct treatment to keep ahead of the changing coproduct distribution brought about by the changing feedstock and process variables.

HYDROPYROLYSIS

Hydropyrolysis is a cracking process characterized by operating in the presence of hydrogen under pressure. It is basically a hydrocracking process, but differs from conventional hydrocracking by the absence of a catalyst, higher operating temperatures and a shorter processing time. Fig. 7-19 is a schematic drawing of a hydropyrolysis unit for ethylene production.[55] Reaction conditions are

Temperature:	800-900°C
Catalyst:	None
Pressure:	10-30 atms (bars)
Reaction time:	"far less than 0.1 sec"

Yields vary with feedstock and recycling. With *naphtha* feedstock, a maximum ethylene production of 44-45% is obtained by propylene recycle and separate steam cracking of the produced ethane. Methane production reaches 34% and a 20% yield of highly aromatic gasoline. Only 2-3% of heavy fuel is produced. *Gas oil,* 178-375°C, feedstock, however, yields 35% ethylene, 25-30% methane and 13% heavy fuel.

Conoco has patented a process that operates at 700-750°C at 5-100 atms (bars) for 0.017 sec.[56] The effluent is rapidly heated and expanded to form products at 1-3 atms and 800-900°C.

One important characteristic of hydropyrolysis is the capability of recycling certain undesirable coproducts. This contributes to increased yields of desired products, especially ethylene. By recycling the coproducts to extinction, yields of 35% ethylene and 18% propylene can be obtained.

Increased cracking severity increases the ethylene yield but beyond about a 35% severity the increase is at the expense of coproduct olefins as shown in Fig. 7-20.[55] The increase in methane production is due to methylation reactions. Aromatics also increase with an increase in severity.

A catalyst has been developed for a hydroconversion process to upgrade heavy feedstock to obtain maximum olefin output.[88] The catalyst selectively cracks and hydrogenates polyaromatics and hetrocyclic compounds, but keeps the saturation of monoaromatic compounds and isomerization activity suppressed. Fig. 7-21 is a flow analysis of a combined hydroconverter and steam cracker.[88]

The recovery of large quantities of hydrogen gas per unit of product is reported to be a distinct disadvantage to the hydropyrolysis process.[57] In steam cracking the steam is easily condensed and removed from the gaseous products after cracking.

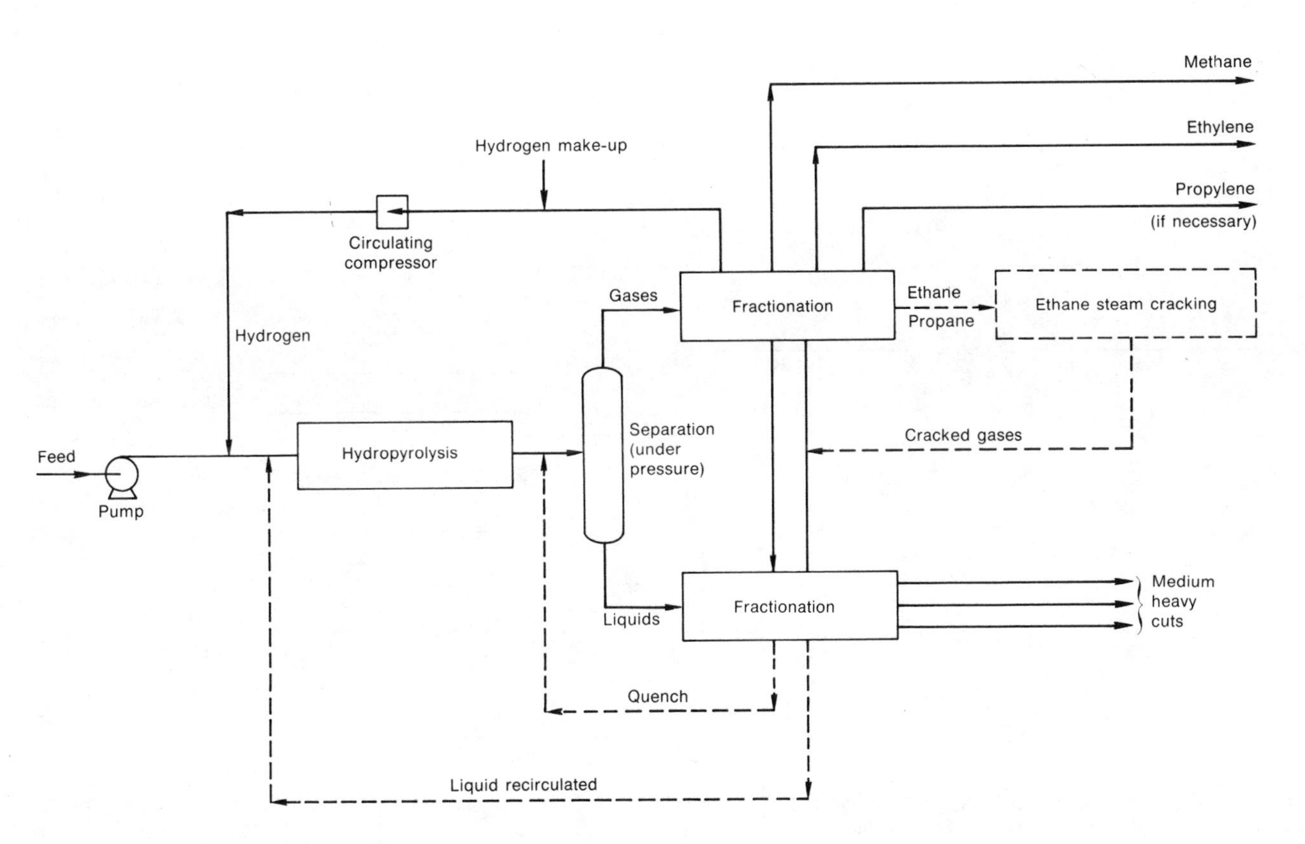

Fig. 7-19—Schematic arrangement of a hydropyrolysis for ethylene production.[55]

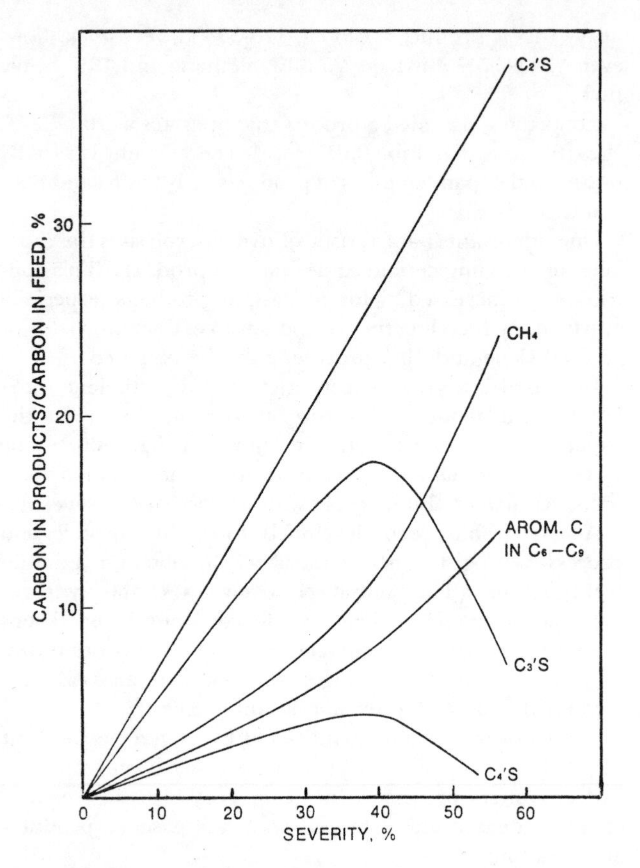

Fig. 7-20—With rising severity, more ethylene production is at the expense of coproduct production in gas oil hydropyrolysis and no recycle.[55]

The fundamental role of hydrogen in hydropyrolysis has been delineated by Barre, et al:[55]

- Contrary to other diluents, hydrogen participates actively in the initiation of decomposition reactions and encourages cracking, even under pressure.
- It opposes the formation of heavy products and greatly reduces the coking tendency by stabilizing the unsaturated products.
- By participating in the reaction, it can also make up for any possible lack of hydrogen in the charge. This permits treating, without difficulty, those raw materials whose H:C ratio is low, e.g., heavy petroleum cuts, various cracked olefin fractions, alkylaromatics, etc.
- Because of the highly exothermic nature of the hydrogenation reaction and the ability to control the development of such reactions by modifying the processing time, pressure and excess hydrogen, the thermal cracking conditions can be controlled and exothermic conditions achieved, if necessary, to obtain the very high temperatures which are difficult to achieve otherwise.

Millisecond Pyrolysis

Another breakthrough in olefin pyrolysis appears to be the development of the Millisecond Furnace.[58, 59, 60] This furnace operates between 0.03 and 0.1 sec and in the outlet temperature range of 870°C (moderate severity) and 925°C (high severity). Under these conditions the ethylene yields increase 10 to 20%. Single-pass ethylene yields of over 33 wt% can be obtained from naphtha while methane yields are

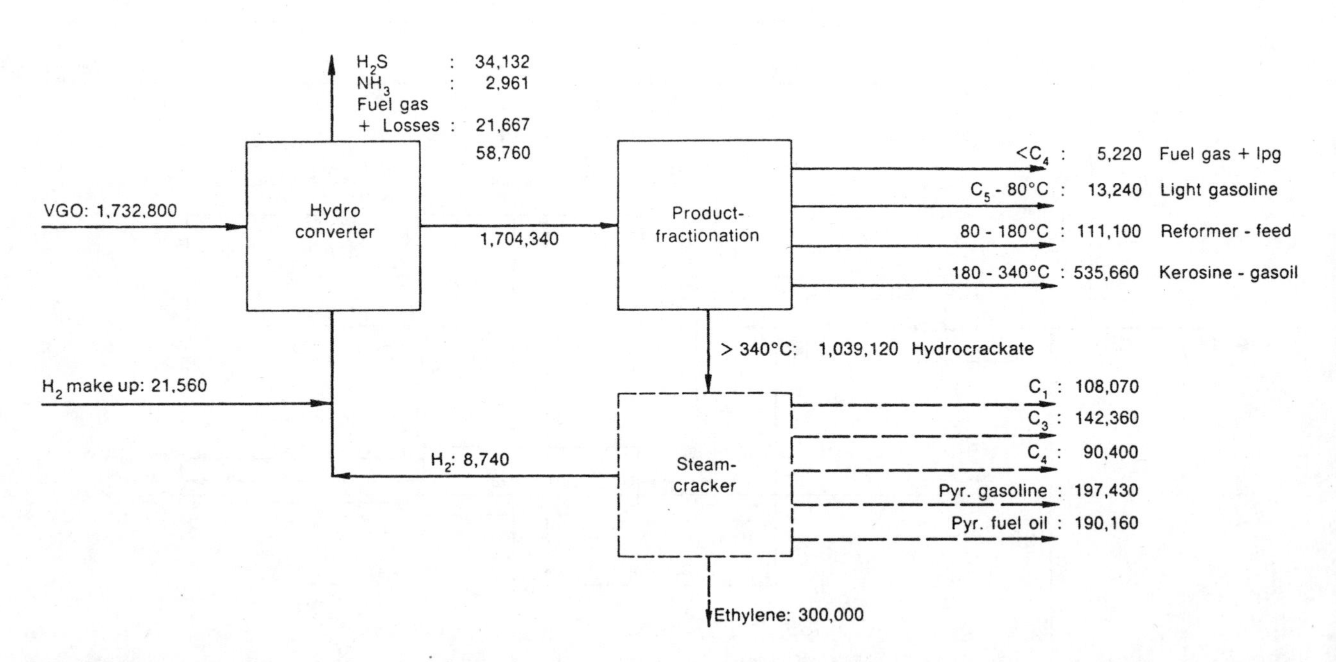

Fig. 7-21—Flow analysis of combined steam cracker and Linde hydroconverter based on feed of VGO for 300,000 metric tons/year of ethylene. Basis: Arabian Light, $Gravity_{15}=0.91$, Cut 340-550°C.[88]

below conventional furnace processes. "The Millisecond Furnace probably represents the last important step that can be taken with respect to this critical pyrolysis variable because contact times below the 0.01 sec range lead to the production of acetylenes in large quantities."[56]

The relationship between pyrolysis yields and contact time are shown in Fig. 7-22.[55] Table 7-11 gives pyrolysis data for full-range naphtha under conventional high severity conditions and under millisecond conditions at both moderate and high severity. High severity millisecond conditions give a higher ethylene yield than conventional cracking, with or without ethane recycle. Propylene remains the same, butadiene is appreciably higher and tailgases are significantly lower.

The process has been demonstrated as capable of the same yield enhancement for mixed C_5's, kerosine, reformer raffinate and light gas oil.

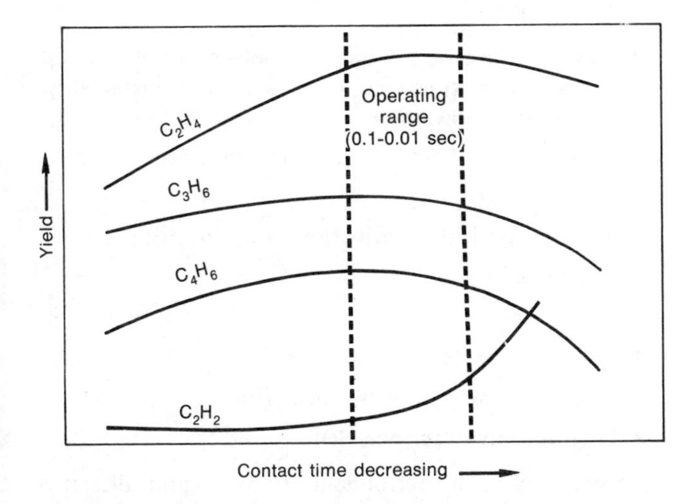

Fig. 7-22—Relationship between contact time and yield of acetylene, ethylene, propylene, and butadiene for Millisecond Furnace pyrolysis of naphtha.[58]

TABLE 7-11—Typical Data for Pyrolysis of Full-Range Naphtha[58]

	Furnace Type		
Severity	**Conventional High**	**Millisecond Moderate**	**Millisecond High**
Products			
Hydrogen	1.0	1.0	1.2
Methane	17.0	12.8	15.2
Acetylene	0.7	0.7	1.3
Ethylene	28.5	29.0	31.8
Ethane	3.8	3.2	2.8
Methylacetylene	0.6	1.0	1.2
Propylene	11.6	15.0	11.6
Propane	0.3	0.4	0.3
Butadiene	3.7	5.4	4.7
Butenes	2.7	4.5	2.2
Butane	0.2	0.2	....
Fuel oil	29.9	26.8	27.7
Product ratios			
$H_2 + CH_4/C_2H_4$	0.631	0.476	0.516
C_3H_6/C_2H_4	0.407	0.517	5.365
C_4H_6/C_2H_4	0.130	0.186	0.148
Big three yield			
$C_2H_4 + C_3H_6 + C_4H_6$	43.8	49.4	48.1
Ultimate C_2H_4 yield**	32.2	32.2	35.2

*Weight %.
** Ethane recycle.

Thermal Regenerative Cracking

Thermal Regenerative Cracking (**TRC**) is a new process that is said to make ethylene from about any feedstock at a 20% reduction in cost. It is reported to have several advantages over conventional coil cracking: lower capital costs, energy needs, plant emissions, a broad range of cracking severities, and therefore, product distribution.[61]

Cracking Oil by Steam and Molten Salts

Cracking Oil by Steam and Molten Salts (**COSMOS**) uses an externally heated tubular furnace to which can be applied the conventional naphtha-cracking technology that is entirely different from the internally heated processes:[62]

- While gas oil used to be the heaviest possible heavy feedstock for the conventional naphtha cracker, it has been possible to crack crude oil or heavy oil by the use of molten salts which act as the catalyst for water-gas reaction.
- Plugging of the reactor tube can be prevented by means of forming a wet wall with molten salts on the internal surface of the reactor tube. Specifically, the wet wall serves to wash down tarry matters and causes selective water-gas reaction of such material.
- Suitable feedstock for COSMOS Process is paraffin-rich low-sulfur crude oil such as Minas or Taching.
- When the typical product pattern of thermal cracking of Minas by the COSMOS Process is compared with that of conventional naphtha cracking, the COSMOS Process gives a comparable ethylene yield, while its propylene yield is somewhat lower. COSMOS Process produces more H_2, CO and CO_2 inasmuch as tarry matter is positively gasified.
- It is possible to achieve a cost reduction by some 20 yen/kg-ethylene when the price of Minas crude oil is lower than that of naphtha by 10 yen/kg.

The thermal cracking of heavy oil, such as crude oil, is accompanied with production of large amounts of carbon and pitch, which give rise to plugging problems. The carbon and pitch produced are treated with the molten salt (inside the tubes) which acts as the catalyst for the water-gas reaction of these materials. Molten salts used are a ternary eutectic mixture consisting of Na_2CO_3, K_2CO_3 and Li_2CO_3 (hereafter called "Melt"). The reason for using Melt is: (1) the melting point is as low as 394°C, (2) its thermal stability is relatively high, and (3) it is nontoxic. Fig. 7-23 gives a flow diagram of the COSMOS Process.[62]

Coproduct Treatment

The cracking of hydrocarbons for olefins is predominately for the production of ethylene. Ethylene demand, therefore, governs the amount of coproduct propylene, butadiene and **BTX**. The type of feedstock and operating conditions also influence the ratio of these products. Market conditions will necessitate simultaneous adjustment of ethylene and propylene. The trend toward heavier feedstocks will increase C_4 production.

The problem has been investigated by Bonnifay, *et al*, and they have proposed the following five key processes to provide greater product flexibility.[63]

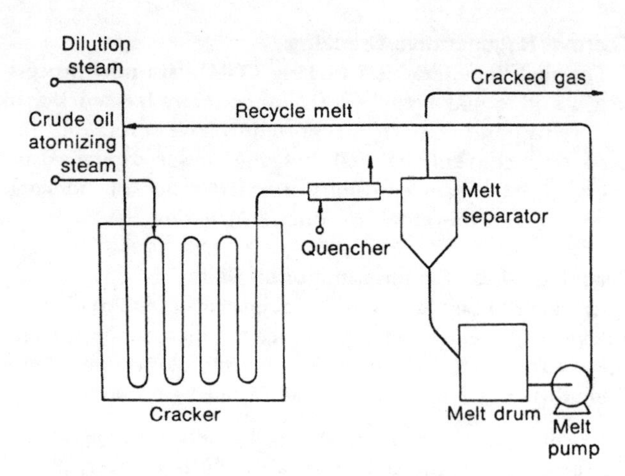

Fig. 7-23—Schematic flow diagram of COSMOS Process.[62]

1. *Hydrogenation of raw C_3's*
   This removes methylacetylene and propadiene to give polymer grade propylene.
2. *Total hydrogenation*
   This results in the C_3 and C_4 cuts being totally hydrogenated.
3. *Dimerization*
   Once through liquid-phase dimerization of propylene produces isohexanes.
4. *Acetylene removal*
   The raw C_4 cuts are selectively hydrogenated with minimum loss of butadiene.
5. *Hydroisomerization*
   Hydroisomerization is used to isomerize 1-butene (−6.3° C) to 2-butene (*cis* −3.7° C, *trans* +0.9° C) to make the separation of isobutylene (−6.6° C) and the *n*-butenes feasible.

These processes either improve the marketability of the coproducts, ease their separation or purify other coproducts and prepare the less desirable coproducts for recycle to increase the ethylene yield.

Acetylene

Acetylene technology has passed through a series of developments using different feedstocks: first from calcium carbide, which is becoming uneconomical because of currently high energy costs; next from partial oxidation of natural gas and hydrocarbon cracking processes suitable for heavy feedstocks, including crude oil. As ethylene production moves deeper into the barrel, more acetylene will become available. Presently, this source accounts for about 10% of acetylene production from hydrocarbon sources. An economic analysis of the transfer price of ethylene and acetylene as a function of raw material type, price, and type of technology has been made by Verde, Riccardi and Moreno.[64]

Acetylene yields of olefin plants expressed as a wt% of ethylene production are listed in Table 7-12.[65] The low acetylene production in the U.S. is the result of high consumption of ethane as the ethylene feedstock. A 1.3 billion lb/yr naphtha-based plant will produce between 13 million and 32 million lb/yr coproduct acetylene. This acetylene is removed by either hydrogenation or extraction. *Hydrogenation* converts the acetylene to ethane; consequently, this is not a method for obtaining acetylene from an ethylene plant.

TABLE 7-12—Olefins Plant's Acetylene Yields, Acetylene as Wt. % of Ethylene[65]

Cracking feedstock	Wt. % of Ethylene
Ethane	0.5
Propane	1.3-1.6
Butane	1.1-1.5
Naphtha	1.0-2.5
Gas oil	1.0-1.5

Extraction removal is applied to the pyrolysis C_2 fraction, either before or after final fractionation, to produce product ethylene. The preferred extraction solvents have the following characteristics:[65, 66]

- Low vapor pressure to minimize solvent loss, but a high enough vapor pressure, when heated, to enable stripping of dissolved gases
- Stability and non-reactivity with the process gases
- Low freezing point
- Miscibility with water, without reaction with it
- Low toxicity
- Low viscosity
- Low cost and ready availability
- High dissolving power for acetylene
- High selectivity for acetylene.

Acetone, N-methylpyrrolidone (NMP) and dimethyl formamide (DMF) are some of the better known acetylene extraction solvents.

OTHER SOURCES OF ETHYLENE

Because of the importance of ethylene as a petrochemical, many interesting and ingenious methods for its production have been proposed. Among these are from propylene, by the Wulff furnace process, from lower olefins *via* Fischer-Tropsch synthesis, from the dehydration of ethanol, from coal by way of synthesis gas and by the pyrolysis of steer manure.

Fischer-Tropsch Synthesis

The Fischer-Tropsch synthesis, **FT**, was developed to produce gasoline from carbon monoxide and hydrogen—synthesis gas. With the growing importance of ethylene and other olefins, the **FT**-synthesis has been reexamined. The production of lower olefins via **FT** is the subject of a definitive article by Büssemeir, Frohning and Cornils.[67]

The main advance in adapting the **FT**-synthesis for the production of C_2-C_4 olefins is the development of a new, selective catalyst. The catalyst of choice is a doped iron catalyst. The following activators are especially suitable: oxides of titanium, vanadium, molybdenum, tungsten and manganese. They reduce alkane formation and limit primary product distribution to C_5 or C_6.

TABLE 7-13—Reaction Conditions, Yields, and Product Distribution Using Newly Developed FT Catalysts[67]

Catalyst*	I	II	III
Product yield (g/Nm^3 syn gas introduced)			
CH_4	22.1	20.2	21.1
C_2H_4	11.5	16.2	21.2
C_3H_6	14.3	25.3	31.7
C_4H_8	11.5	16.2	16.9
Σ C_2-C_4 alkenes	37.3	57.7	69.8
Σ C_2-C_4 alkanes	8.1	8.3	8.5
Reaction conditions**			
Temperature, °C	360	340	280
Pressure, bars	10	10	10
H_2/CO ratio	1	1	1
Total yield, g/Nm^3 (Without CH_4, C_5 and higher hydrocarbons included)	178.4	132	103
Conversion rate, %	87.6	63.1	49.5

* Catalyst composition not given.
** Gas phase, fixed-bed synthesis.

TABLE 7-14—Reaction Conditions, Yields and Product Distribution for a Liquid Phase FT Synthesis Using Newly Developed Catalysts[67]

Catalyst*	IV	V
Product yield, g/Nm^3		
CH_4	36	28
C_2H_4	3.1	24
C_3H_6	10.8	27
C_4H_8	15.2	17
C_5H_{10}	8.4	8
Σ C_2-C_5 alkenes	37.5	76
Σ C_2-C_5 alkanes	58	26
Reaction conditions		
Temperature, °C	290	350
Pressure, bars	12	20
H_2/CO ratio	0.67	1
Total yield, g/Nm^3 (Without CH_4, C_6- and higher hydrocarbons included)	95.5	127
Conversion rate, %	91	83

* Catalyst composition not given.

Tables 7-13 and 7-14 contain reaction conditions, yields and product distribution by newly developed catalysts.[67] Both gas phase, fixed-bed synthesis and liquid phase synthesis are represented. The compositions of the five different catalysts used were not given.

The Fischer-Tropsch synthesis may be one of the viable sources of ethylene, propylene and naphtha in the future, with synthesis gas coming from coal.

Wulff Furnaces

Wulff furnaces in conjunction with conventional steam cracking furnaces have been used with naphtha feedstock to increase the acetylene yields.[68, 69]

Propylene Disproportionation

The disproportionation of propylene to ethylene and 2-butenes plus C_5+ products is an interesting process.[70] It not only can be used to convert the less desirable propylene to ethylene but also produces alkylate components for unleaded gasoline. This type of reaction is noted in Chapter 9.

Dehydration of Ethanol

The production of ethylene from ethanol would appear to be highly unsound from an economic standpoint. However, there are ethylene plants that use ethanol as the feedstock. They are economical under special circumstances such as a country with a large amount of fermentation ethanol available and an ethylene demand that will not require a world-competitive 1.3 billion lb/yr hydrocarbon-based plant.

The dehydration process is simple. Ethanol vapors are preheated over a catalyst of especially treated activated alumina for dehydration.[71] A yield of 94% is obtained in one pass. Fig. 7-24 is a flow diagram of a typical ethanol dehydration process.

The reaction takes place in two stages.

$$2\ CH_3\text{-}CH_2OH \rightarrow CH_3\text{-}CH_2\text{-}O\text{-}CH_2\text{-}CH_3 + H_2O$$

$$CH_3\text{-}CH_2\text{-}O\text{-}CH_2\text{-}CH_3 \rightarrow 2CH_2{=}CH_2 + H_2O$$

The production of ethylene from ethanol has been discussed in depth by Winter and Eng, especially in respect to the competitive position of this process.[71] Brant has compared ethanol and hydrocarbons as feedstock for ethylene and concluded ethanol could become competitive within the next 40-50 years.[72]

Manure

A chemical engineer at Texas Tech University has developed a manure pyrolysis process said to produce 180 pounds of ethylene per ton of the dry feedstock.[73] This equates to only 1.3×10^8 head of cattle to equal the U.S. annual ethylene demand.

1980 AND BEYOND

Many factors influence the choice of feedstock, all of which must be considered when assessing the future feedstock trends. The following factors have been summarized by Shinshin.[74]

(1) the non-stop efforts for improving the heavy oil cracking technology,

(2) the decline in natural gas reserves in the United States,

(3) United States refinery pattern changes expected from the demand for lead-free gasoline,

(4) increase in U.S. dependence on Eastern Hemisphere oil suppliers and possible product import changes,

(5) the expectation that the oil product demand in the Eastern Hemisphere will swing toward the lighter end of the barrel, thus reducing the availability of naphtha for the chemical industry,

(6) the improvement in the transportation facilities for LP-gas will help its movement to market places from the Eastern Hemisphere where it is available in abundant quantities, and

(7) the creation of petrochemical industries in OPEC member countries in the Middle East and Africa, depending on natural gas as a feedstock will influence the gas liquids supply to the international market.

The complexity of these influences gives an uncertainty to any prediction of future feedstock trends.

Future trends in ethylene production appear to be rather clear cut however. In the *U.S.* the trend will be away from ethane and LP-gas toward naphtha, gas oil

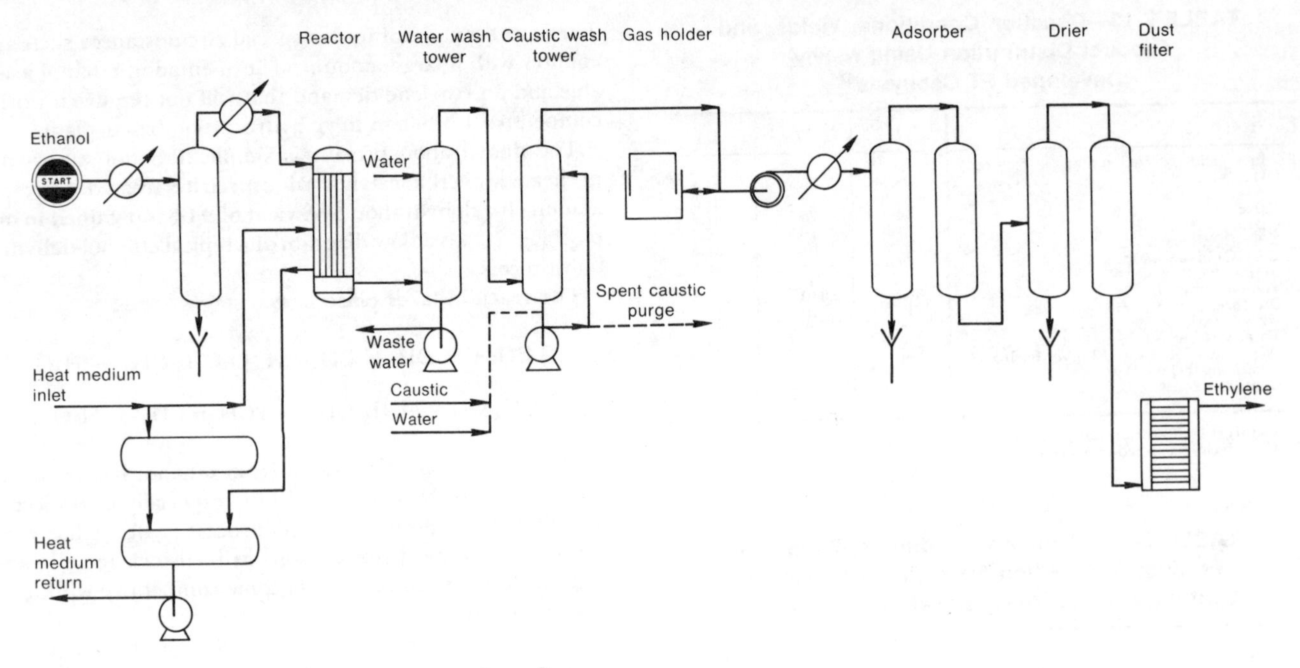

Fig. 7-24—Typical ethylene from ethanol process scheme.[71]

and heavier cuts. The new capacity will have greater flexibility to meet fluctuating feedstock availability. A large share of the industry will continue to depend on ethane for the foreseeable future. The long term trend toward liquid feedstock will be indicated by new capacity. This new capacity will have the capability to handle not only naphtha but also deeper cuts. Flexibility will be the design factor and will prevail although at a cost. This cost has been estimated at 8% more than for a naphtha only plant.[75] It is predicted that all new steam crackers projected through 1985 will use naphtha or gas oil. The high yield of energy coproducts mandates that only oil companies, singly or jointly with chemical companies can undertake an ethylene plant based on heavier feedstock.[76] For this reason, the petrochemical industry will continue to move closer to the oil industry, both technologically and financially.

The trend in *Europe* is away from naphtha in both directions. As new gas supplies from the Norwegian sector of the North Sea become available, ethane will be used. As more naphtha is required for fuel uses, the trend will be toward heavier feedstock, especially gas oils.[77] The entire feedstock situation in Europe is clouded by depressed refinery operations.[78]

The *Middle East* will use ethane as the predominant feedstock because of the abundance of associated gas, especially in Saudi Arabia. Ethane also simplifies coproduct marketing. The Middle East will account for about 4% of the total worldwide ethylene capacity in 1980 and for about 7.5% by 1990.[76] The impact of the Middle East countries will not be as great as many have suggested. The impact on the individual countries will be appreciable.

Japan is moving away from naphtha toward **LP**-gas and butane. This is a reflection of the increased availability of **LP**-gas from both the Middle East and Australia.

The geographical location of new ethylene plants and their impact on ethylene supply in 1980 has been reviewed by Hyde.[79] New countries will provide 24% of the new capacity with 8% being in the Middle East.

DIOLEFINS

Butadiene

Butadiene, CH_2-$CH=CH$-CH_2, is the raw material for the most widely used synthetic rubber, **SBR**—a copolymer between butadiene and styrene. *cis*-Polybutadiene is the polymer of butadiene which is widely used for tire production. Copolymers of butadiene, styrene, and acrylonitrile are important plastics. Production of **SBR** and *cis*-butadiene are discussed in Chapter 14; and of the copolymers of butadiene, styrene and acrylonitrile for plastics are discussed in Chapter 12. About 7% of the butadiene production is used for chemicals, and these are discussed in Chapter 8. The world-wide (less Russia and China) production of butadiene was 9.63 billion pounds; the 1978 United States production was 3.52 billion pounds.

Production. Most butadiene is produced as a by-product of ethylene production. The dehydrogenation of butane to butenes to butadiene is currently the most important "on purpose" source of butadiene.

This dehydrogenation takes place in one stage. Dehydrogenation catalysts such as alumina impregnated with chromic oxide, iron oxide, metal halides, and iodine are used in the presence of air and steam.

$$CH_3\text{-}CH_2\text{-}CH_2\text{-}CH_3 \rightarrow \text{Mixture of butenes}$$
$$\rightarrow CH_2=CH\text{-}CH=CH_2$$

The Phillips process uses an oxidative-dehydrogenation catalyst. The feed is a mixture of normal butenes, air, and steam. This mixture is passed over the catalyst bed at 900-1100°C. They hydrogen which is released from the dehydrogenation reacts with oxygen, thus removing it from the equilibrium mixture and shifting the reaction towards the formation of more butadiene. Conversion ranges between 75-80% and selectivity to butadiene is about 88-92%.[81] Continuous catalyst regeneration takes place *in situ*, and steam fed to the reactor controls reactor temperature.

Other Methods for the Production of Butadiene

From Ethyl Alcohol. In some parts of the world, as in the U.S.S.R., fermented alcohol can serve as a cheap feedstock. Butadiene is then produced from it. The reaction takes place in the vapor phase under normal or reduced pressures over zinc oxide-alumina or a magnesia catalyst promoted with chromium or cobalt. The formation of acetaldehyde has been suggested as an intermediate. Two moles of acetaldehyde condense and form crotonaldehyde which reacts with ethyl alcohol to give butadiene and acetaldehyde. Acetaldehyde is separated from butadiene and recycled.

The Hydration of Acetylene. This obsolete process was used by Germans during World War II. The hydration reaction produces acetaldehydes, followed by an Aldol condensation which produces aldol. The aldol is hydrogenated to 1,3-butanediol which is dehydrated to butadiene.

$$HC{\equiv}CH + H_2O \rightarrow CH_3\text{-}\overset{\overset{\large O}{\|}}{C}\text{-}H$$

$$2\ CH_3\text{-}\overset{\overset{\large O}{\|}}{C}\text{-}H \rightarrow CH_3\text{-}\overset{\overset{\large OH}{|}}{CH}\text{-}CH_2\text{-}CHO$$

$$CH_3\text{-}\overset{\overset{\large OH}{|}}{CH}\text{-}CH_2CHO + H_2 \rightarrow CH_3\text{-}\overset{\overset{\large OH}{|}}{CH}\text{-}CH_2\text{-}CH_2\text{-}OH$$

$$CH_3\text{-}\overset{\overset{\large OH}{|}}{CH}\text{-}CH_2\text{-}CH_2\text{-}OH \rightarrow CH_2{=}CH\text{-}CH{=}CH_2 + 2H_2O$$

Butadiene produced from different processes is purified from other olefinic hydrocarbons. Distillation is not used because the boiling points of these olefins are close, and extractive distillation is used. The solvents used are furfural, dimethyl formamide, and *n*-methyl pyrilidone.

From Acetylene and Formaldehyde. The reaction between acetylene and formaldehyde in the vapor phase uses a freshly prepared copper acetylide catalyst and produces 1,4-butynediol. This is hydrogenated to 1,4-butanediol.

$$HC{\equiv}CH + 2\ H\text{-}\overset{\overset{\large O}{\|}}{C}\text{-}H \longrightarrow HOCH_2\text{-}C{\equiv}C\text{-}CH_2OH \xrightarrow{H_2} HOCH_2\text{-}CH_2\text{-}CH_2\text{-}CH_2OH$$

The dehydration of 1,4-butanediol produces butadiene.

Isoprene

Isoprene, $H_2C{=}\overset{\overset{\large CH_3}{|}}{C}{-}CH{=}CH_2$, 2-methyl 1,3-butadiene, is an important elastomeric raw material. *cis*-Polyisoprene is similar in its structure to natural rubber. Most isoprene is used for the production of *cis*-polyisoprene. It is sometimes used as a copolymer with butyl rubber (3%).

Production. There are different routes for the production of isoprene. The choice of one process over the other depends upon the availability of raw materials and the economics of the selected process.

Dehydrogenation of Tertiary Amylenes (Shell Process). Isoprene is produced by dehydrogenation of a mixture of tertiary amylenes (2-methylbutene-1 and 2-methylbutene-2). The *t*-amylenes are extracted from a C_5 fraction from cat-cracking units by the use of aqueous sulfuric acid. The recovered *t*-amylenes are fed with steam over a dehydrogenation catalyst into the reactor. The reaction cycle is 24 hours including one hour for regeneration. The overall conversion and recovery of *t*-amylenes is 70%. Fig. 7-25 shows the Shell Process.[82]

From Isobutylene and Formaldehyde (IFP Process). Isobutylene, which can be separated from a C_4 fraction from the cat-cracker, reacts with two moles of formaldehyde to give dimethyl dioxane.

$$CH_3\text{-}\overset{\overset{\large CH_3}{|}}{C}{=}CH_2 + 2\ H\text{-}\overset{\overset{\large O}{\|}}{C}\text{-}H \rightarrow \text{(dimethyl dioxane: } (CH_3)_2C\text{-}O\text{-}CH_2\text{-}O\text{-}CH_2\text{-}CH_2 \text{ ring)}$$

The pyrolysis of dioxane gives isoprene and formaldehyde. The formaldehyde is recovered and recycled to the fluid bed reactor.

$$\text{(dimethyl dioxane)} \rightarrow CH_2{=}CH\text{-}\overset{\overset{\large CH_3}{|}}{C}{=}CH_2 + H\text{-}\overset{\overset{\large O}{\|}}{C}\text{-}H + H_2O$$

From Isobutylene and Methylal (Sun Oil Process). In this process, methylal is used instead of formaldehyde. The advantage of using methylal over formaldehyde is its lower reactivity to butene-1 than formaldehyde, so allowing a mixed feedstock of C_4 to be used. Also, unlike formaldehyde, methylal does not decompose to CO and H_2.

The first step in this process is to produce methylal by the reaction of methanol and formaldehyde using an acid catalyst.

$$H\text{-}\overset{\overset{\large O}{\|}}{C}\text{-}H + 2\ CH_3OH \overset{H^+}{\rightleftharpoons} CH_3O\text{-}CH_2\text{-}OCH_3 + H_2O$$

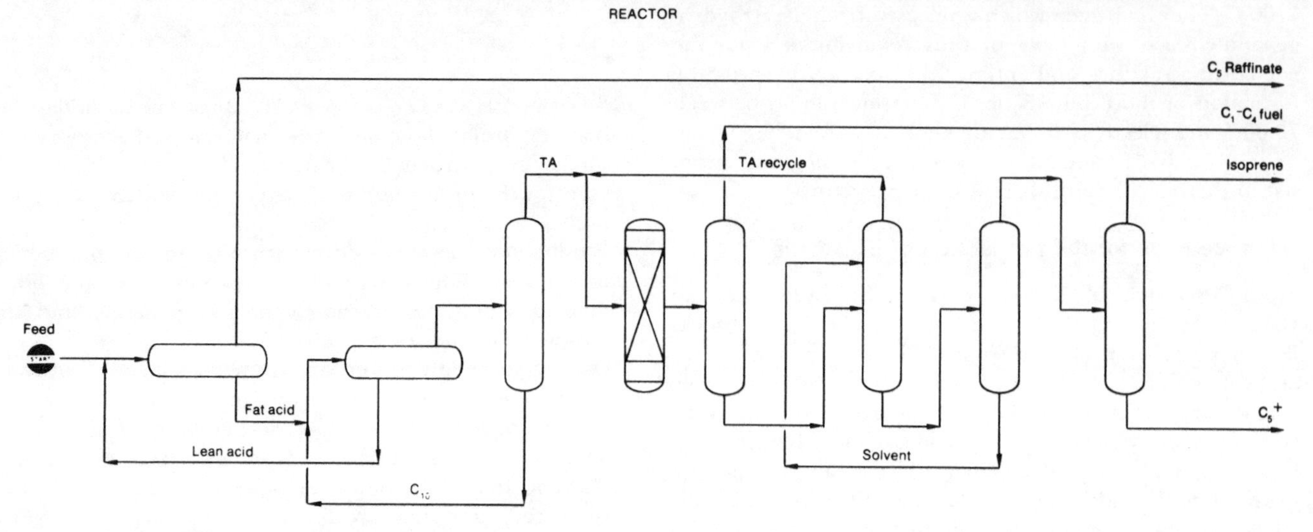

Fig. 7-25—The Shell process for the production of isoprene from tertiary amylenes.[82]

Methylal then reacts with isobutylene in the vapor phase to produce isoprene. 2-Butene in the C_4 mixture also reacts with methylal, but at a slower rate, to give isoprene. 1-Butene reacts very slowly to give 1,3-pentadiene.

$$CH_3\text{-}C(CH_3)=CH_2 + CH_3O\text{-}CH_2\text{-}OCH_3 \rightarrow CH_2=C(CH_3)\text{-}CH=CH_2 + 2\ CH_3OH$$

$$CH_3\text{-}CH=CH\text{-}CH_3 + CH_3\text{-}O\text{-}CH_2\text{-}OCH_3 \rightarrow CH_2=C(CH_3)\text{-}CH=CH_2 + 2\ CH_3OH$$

The reactor temperature is 250-350°C, the liquid hourly space velocity is 1-10 and olefin to methylal feed ratio is 6:1. Fig. 7-26 is a flow diagram for the Sun Process.[83]

From Propylene Dimer (Goodyear Process). When propylene is dimerized, 2-methyl-1-pentene is produced. Tripropyl aluminum in combination with nickel or platinum under about 200 atmospheres and 200°C are representative conditions.

$$2\ CH_3\text{-}CH=CH_2 \rightarrow CH_3\text{-}CH_2\text{-}CH_2\text{-}C(CH_3)=CH_2$$

$$CH_3\text{-}CH_2\text{-}CH_2\text{-}C(CH_3)=CH_2 \rightarrow CH_3\text{-}CH_2\text{-}CH=C(CH_3)\text{-}CH_3$$

$$CH_3\text{-}CH_2\text{-}CH=C(CH_3)\text{-}CH_3 \rightarrow CH_2=CH\text{-}C(CH_3)=CH_2 + CH_4$$

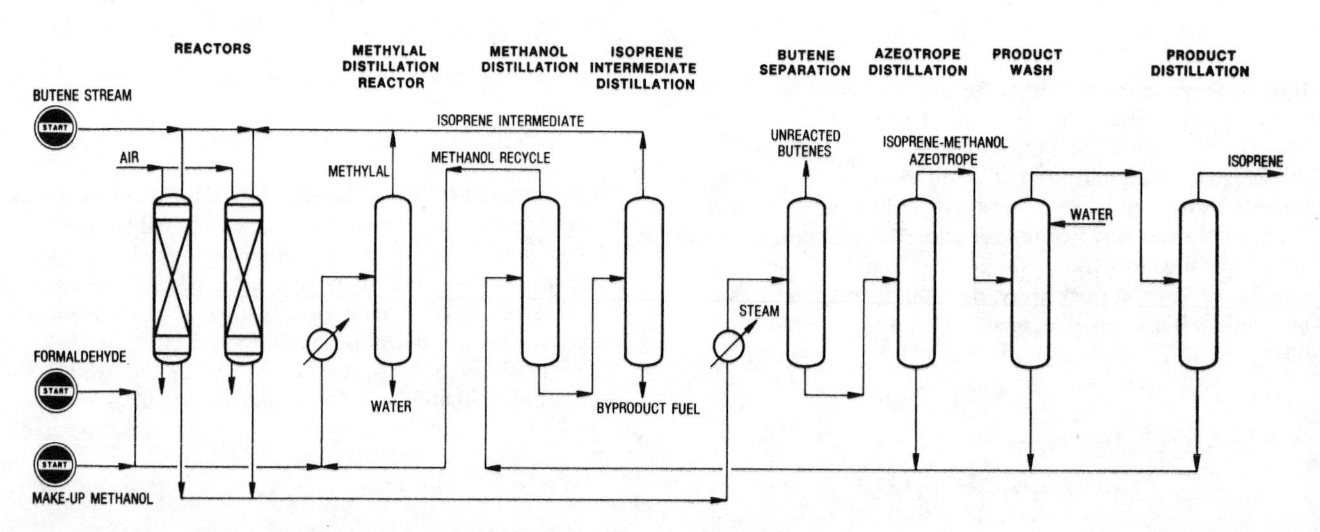

Fig. 7-26—The Shell Oil Company process for the production of isoprene from butylenes.[83]

An acid catalyst is used to isomerize 2-methyl-1-pentene to 2-methyl-2-pentene, which is subsequently pyrolyzed to butadiene and methane.

From C_5 Olefins. These olefins are produced by the reaction between ethylene and propylene over an acid catalyst to produce a mixture of C_5 olefins. These olefins are further dehydrogenated to isoprene.

$$CH_2{=}CH_2 + CH_3\text{-}CH{=}CH_2$$

$$\rightarrow \underset{\text{Main product}}{CH_3\text{-}\overset{\overset{\displaystyle CH_3}{|}}{C}{=}CH\text{-}CH_3} + CH_3\text{-}CH_2\text{-}\overset{\overset{\displaystyle CH_3}{|}}{C}{=}CH_2$$

LITERATURE CITED

1. *Hydrocarbon Processing, 1975 Petrochemical Handbook,* Vol. 54, No. 11, p. 141; *Hydrocarbon Processing, 1979 Petrochemical Handbook,* Vol. 58, No. 11, pp. 160-165.
2. *Hydrocarbon Processing, 1975 Petrochemical Handbook,* Vol. 54, No. 11, p. 143; *Hydrocarbon Processing, 1979 Petrochemical Handbook,* Vol. 58, No. 11, pp. 160-165.
3. Wett, Ted. *The Oil and Gas Journal,* Nov. 26, 1973, pp. 73-75.
4. Wilkinson, L.A. and S. Gomi. *Hydrocarbon Processing,* Vol. 53, No. 5, (1974) 109-111; S. Gomi and L.A. Wilkinson, *The Oil and Gas Journal,* June 17, 1974, pp. 59-62; O. Shinshin, International Seminar on Petrochemical Industries, Paper No. D4, Baghdad, Iraq, Oct. 25-30, 1975.
5. Piccioth, Marcello and V. Kaiser. *The Oil and Gas Journal,* April 2, 1979, pp. 112-115, 118, 120; *Hydrocarbon Processing,* Vol. 58, No. 6, 1979, pp. 99-105.
6. Janakievski, T.V. Second Arab Conference on Petrochemicals, Paper No. 6 (P-1), Abu Dhabi, Mar. 15-22, 1976.
7. Platzer, Norbert. *CHEMTECH,* Vol. 9, No. 1, 1979, p. 16-20.
8. Minton, Bill R. *The Oil and Gas Journal,* Oct. 22, 1979, pp. 62-63
9. Wilkinson, Loy, *Hydrocarbon Processing,* Vol. 58, No. 9, 1979, pp. 209-212.
10. *The Oil and Gas Journal,* Oct. 1, 1979, p. 49.
11. *Chemical and Engineering News,* Oct. 23, 1978, p. 12.
12. Wett, Ted. *The Oil and Gas Journal,* Sept. 3, 1979, pp. 59-64.
13. *The Oil and Gas Journal,* May 16, 1977, p. 50.
14. Welch, L.M., L.J. Croce and H.F. Christmann. *Hydrocarbon Processing,* Vol. 57, No. 11, 1978, pp. 131-136.
15. Fallwell, W.F. II. *Chemical and Engineering News,* April 26, 1976, p. 10-11.
16. Mol, A. and B. DeMoet. International Seminar on Petrochemical Industries, Paper No. 14, Baghdad, Iraq, Oct. 25-30, 1975.
17. *Chemical and Engineering News,* Dec. 8, 1975, pp. 29-30.
18. Nahas, R.S. and M.R. Nahas. Second Arab Conference on Petrochemicals, Paper No. 8 (P2), Abu Dhabi, Mar. 15-22, 1976.
19. Barnwell, J. and S.R. Martin. International Seminar on Petrochemical Industries, Paper No. A8, Baghdad, Iraq, Oct. 25-30, 1975.
20. Hayward, G.J. and P.N. Hoggett. *The Oil and Gas Journal,* July 28, 1975, pp. 81-84.
21. Hayward, G.L. and P.N. Hoggett. *Preprints,* Division of Petroleum Chemistry, ACS, Vol. 20, No. 3, 1975, pp. 659-668.
22. Luger, C.R. *Hydrocarbon Processing,* Vol. 55, No. 6, 1976, pp. 105-106.
23. Boyd, H.B., R. Orris, and T.A. Wells. *The Oil and Gas Journal,* March 7, 1977, pp. 98-102.
24. Minet, R.G. and J.D. Hammond. *The Oil and Gas Journal,* Aug. 4, 1975, pp. 80-82.
25. Baba, T.B. and J.R. Kennedy. *Chemical Engineering,* Jan. 5, 1976, pp. 116-128.
26. Dowsett, J.W. and J.R. Jones. *The Oil and Gas Journal,* Mar. 26, 1979, pp. 97-98, 100, 105, 108.
27. Lowr, B. and C. Schliebener. *The Oil and Gas Journal,* Mar. 26, 1979, pp. 82-85.
28. Goossens, A.G., R.F. Westerium and A. Mol. *The Oil and Gas Journal,* Aug. 25, 1975, pp. 92-95.
29. Smith, James, *Chemical Engineering,* Sept. 15, 1975, pp. 70C-70F.
30. *Hydrocarbon Processing, 1975 Petrochemical Handbook,* Vol. 54, No. 11, 1975, p. 141.
31. El Enany, M.M. and O.F. Abdel Rahman. Second Arab Conference on Petrochemicals, Paper No. 9 (P-2), Abu Dhabi, Mar. 15-22, 1976.
32. Barwell, J. and S.R. Martin. International Seminar on Petrochemical Industries, Paper No. 9 (P-2), Baghdad, Iraq, Oct. 25-30, 1975.
33. Watson, Allan. *The Oil and Gas Journal,* Nov. 8, 1976, pp. 179-182.
34. Minet, R.G. and F.W. Tsai. *The Oil and Gas Journal,* Mar. 21, 1977, pp. 135-141.
35. Lohr, B. and H. Dittman. *The Oil and Gas Journal,* July 4, 1977, pp. 53-58.
36. Tucker, W. and M.A. Abrahams, *The Oil and Gas Journal,* Apr. 11, 1977, pp. 81, 84.
37. Mol, A., J. Draaisma and B. DeMoet. *The Oil and Gas Journal,* Aug. 2, 1976, pp. 104-120.
38. Mol, A. and B. DeMoet, International Seminar on Petrochemical Industries, Paper No. 14, Baghdad, Iraq: Oct. 25-30, 1975.
39. Bassler, E.J. *The Oil and Gas Journal,* Mar. 17, 1975, pp. 93-96.
40. Zdonik, S.B., W.S. Potter and G.L. Hayward. *Hydrocarbon Processing,* Vol. 55, No. 4, 1976, pp. 161-166.
41. Greek, B.F. *Chemical and Engineering News,* Mar. 31, 1975, pp. 8-9.
42. *The Oil and Gas Journal,* May 26, 1975, pp. 103-108.
43. *Hydrocarbon Processing,* Vol. 54, No. 5, 1975, pp. 101-104.
44. Zdonik, S.B. and G.L. Hayward. *Hydrocarbon Processing,* Vol. 54, No. 8, 1975, pp. 95-98.
45. Green, E.J., S.B. Zdonik and L.P. Haller. *Hydrocarbon Processing,* Vol. 54, No. 9, 1975, pp. 164-168.
46. Zdonik, S.B., G.L. Hayward, S.H. Fishtine and J.C. Feduske. *Hydrocarbon Processing,* Vol. 54, No. 12, 1975, pp. 111-114.
47. Zdonik, S.B., G.L. Hayward, S.H. Fishtine and J.C. Feduske, *Hydrocarbon Processing,* Vol. 55, No. 1, 1976, pp. 149-154.
48. Lassmann, E. and H.J. Wernicke, *The Oil and Gas Journal,* Jan. 8, 1979, pp. 95-100.
49. Wett, Ted, *The Oil and Gas Journal,* Nov. 26, 1973, pp. 73-75.
50. *Hydrocarbon Processing,* Vol. 54, No. 11, 1975, p. 104.
51. *Chemical Week,* Apr. 16, 1975, pp. 35-39.
52. *Hydrocarbon Processing, 1975 Petrochemicals Handbook,* Vol. 54, No. 11, p. 166.
53. U.S. Patent 3,862,898 to M.W. Kellogg Co.
54. Ishikawa, T. and R.G. Keister. *Hydrocarbon Processing,* Vol. 57, No. 2, 1978, pp. 109-113.
55. Barre, C., E. Chahevkilian and R. Dumon. *Hydrocarbon Processing,* Vol. 55, No. 11, 1976, pp. 176-178.
56. Belgian Patent 840-343 to Continental Oil (Houston).
57. *Chemical Week,* July 23, 1975, pp. 29-30.
58. Ennis, B.P., H.B. Boyd and R. Orriss. *CHEMTECH,* Vol. 5, No. 11, 1975, pp. 693-699.
59. Prescott, J.H. *Chemical Engineering,* Vol. 82, No. 14, 1975, pp. 52-53.
60. Leftin, H.P. and D.S. Newsome. *Preprints,* Division of Petroleum Chemistry, Inc. ACS, Vol. 20, No. 3, 1975, pp. 669-680.
61. *Chemical Week,* Oct. 24, 1979, p. 38.
62. Yamaguchi, F., A. Sakai, M. Yoshitake and H. Saequsa. *Hydrocarbon Processing,* Vol. 58, No. 9, 1979, pp. 167-172.
63. Bonnifay, P., B. Cha, M. Derrien and J. Gaillard. *The Oil and Gas Journal,* Mar. 21, 1977, pp. 110-123.
64. Verde, L.R. Riccardi and S. Moreno. *Hydrocarbon Processing,* Vol. 5, No. 1, 1979, pp. 159-164.
65. Stork, K., J. Hanisian and J. Bac. *Hydrocarbon Processing,* Vol. 55, No. 11, 1976, pp. 151-154.
66. Miller, S.A. *Acetylene—Its Properties, Manufacture & Uses;* Vol. 1, 1965.
67. Büssememeier, B., C.D. Frohning and B. Cornils. *Hydrocarbon Processing,* Vol. 55, No. 11, 1976, pp. 105-112.
68. Wett, Ted. *The Oil and Gas Journal,* Sept. 4, 1972, pp. 103-110.
69. *The Oil and Gas Journal,* Mar. 12, 1973, p. 81.
70. Anderson, K.L. and T.D. Brown. *Hydrocarbon Processing,* Vol. 55, No. 11, 1976, pp. 119-122.
71. Winter, O. and M.T. Eng. *Hydrocarbon Processing,* Vol. 55, No. 1, 1976, pp. 123-133.
72. Brandt, D.E. *The Oil and Gas Journal,* Febr. 5, 1979, pp. 51-56.
73. "Concentrates", *Chemical and Engineering News,* Aug. 18, 1975, p. 19.
74. Wilkinson, L.A. and S. Gomi. *Hydrocarbon Processing,* Vol. 53, No. 5, 1974, pp. 109-111. Also Gomi, S. and Wilkinson. *The Oil and Gas Journal,* June 17, 1974, pp. 59-62; Shinshin, O. International Seminar on Petrochemical Industries, Paper No. D4, Baghdad, Iraq, Oct. 25-30, 1975.
75. Nahas, R.S. *Hydrocarbon Processing,* Vol. 54, No. 7, 1975, pp. 97-100.
76. *Hydrocarbon Processing,* Vol. 55, No. 11, 1976, p. 87.
77. Woodhouse, G.J. *The Oil and Gas Journal,* Aug. 2, 1976, pp. 96-98, 103.
78. Savage, P.R. *Chemical Engineering,* Nov. 8, 1976, pp. 72C-72G.
79. Hyde, M.C. *Hydrocarbon Processing,* Vol. 55, No. 4, 1976, pp. 53C-53F.
80. *The Oil and Gas Journal,* Apr. 30, 1979, p. 134.
81. Hutson, Tom, Jr., R.D. Skinner and R.S. Logan. *Hydrocarbon Processing,* Vol. 53, No. 6, 1974, p. 134.
82. *Hydrocarbon Processing, 1977 Petrochemical Handbook,* Vol. 56, No. 11, p. 175.
83. Peterson, H.J. and J.O. Turner, *Hydrocarbon Processing,* Vol. 53, No. 7, 1974, pp. 121-122.
84. *The Oil and Gas Journal,* Nov. 26, 1979, p. 26.
85. *Hydrocarbon Processing,* Vol. 58, No. 7, 1979, p. 21.
86. *The Oil and Gas Journal,* Jan. 7, 1980, pp. 106, 108, 110.
87. *Hydrocarbon Processing,* Vol. 58, No. 10, 1979, pp. 50-K, 50-L, 50-MM, 50-NN.
88. Goetzmann, S., W. Kreuter and H.J. Wernicke. *Hydrocarbon Processing,* Vol. 58, No. 6, 1979, pp. 109-112.

8

Petrochemicals from Ethylene

A basic reason why ethylene is a prime raw material for petrochemicals is that it is readily available, at low cost, and in high purity. Ethylene reacts by addition with low cost materials such as oxygen, chlorine, hydrogen chloride and water and the reactions take place under relatively mild conditions and usually with high yields. All of these reactions add weight to the ethylene molecule. And chemicals are sold by weight, not by the mole. Ethylene also reacts by substitution to produce vinyl monomers. It enters into the production of about 30% of all petrochemicals. Derivatives of ethylene are used for the production of plastics (65%), antifreeze (10%), fibers (5%), and solvents (5%).[1]

Ethylene and many ethylene derivatives are used for the production of polymers, which account for a very large percentage of ethylene utilization. The formation of polyethylene and ethylene-related polymers such as polystyrene, polyester, and polyvinyl chloride is the subject of Chapter 12. This chapter is restricted to individual compounds produced directly from ethylene, and those with the second degree of consanguinity such as ethylene glycol from ethylene oxide. These compounds, either directly or indirectly, account for a wide range of useful consumer products.

Ethylene oxide dominates the individual compounds produced from ethylene with ethylene dichloride, the precursor of vinyl chloride, next in quantity of ethylene utilized. Ethylbenzene for styrene production is third in quantity. These three account for 37% of the ethylene demand. Polyethylene accounts for 44% and the remaining 19% is scattered through a wide range of compounds—ethanol, linear alcohols, vinyl acetate, alpha olefins, and many others (Fig. 8-1).[2]

Many companies think that ethylene will continue to be the king of petrochemicals because of the actual and projected capacity and expanding demand as shown in Fig. 8-2.[3] The projected use for ethylene in 1982 is in: low-density polyethylene, 29%; high-density polyethylene, 18%; vinyl chloride, 12%; styrene, 9%; ethylene oxide, 17%; and others, 15%.[44]

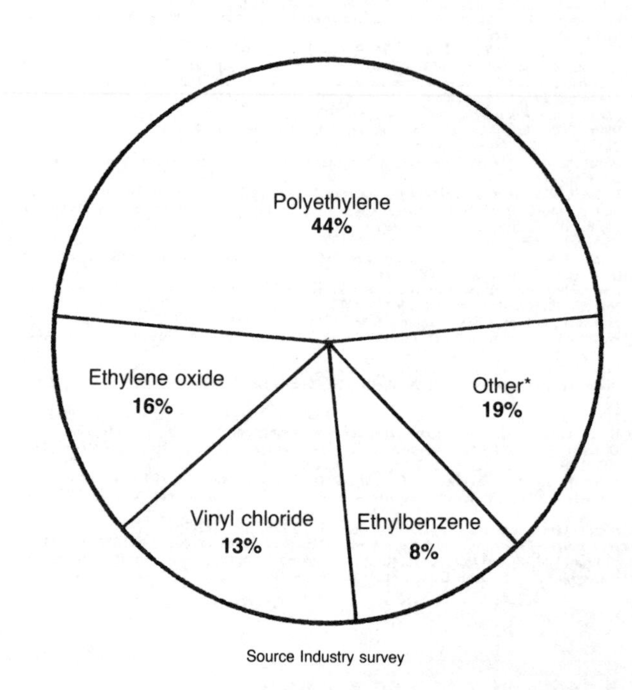

*Ethanol (3%), Acetaldehyde (3%), Vinyl acetate (2%), alpha olefins (2%), Ethylene dichloride, ethyl chloride, Ethyleneimine, Propionaldehyde.

Fig. 8-1—The 1979 demand pattern for ethylene.[2]

ETHYLENE OXIDE

Production. Ethylene oxide, EO, $CH_2\overset{O}{-}CH_2$ is produced by air or oxygen oxidation of ethylene over a silver catalyst (Fig. 8-3).[4]

$$2\ CH_2{=}CH_2 + O_2 \rightarrow 2\ \overset{O}{CH_2 - CH_2} \qquad \Delta H = -35.2\ \text{kcal}$$

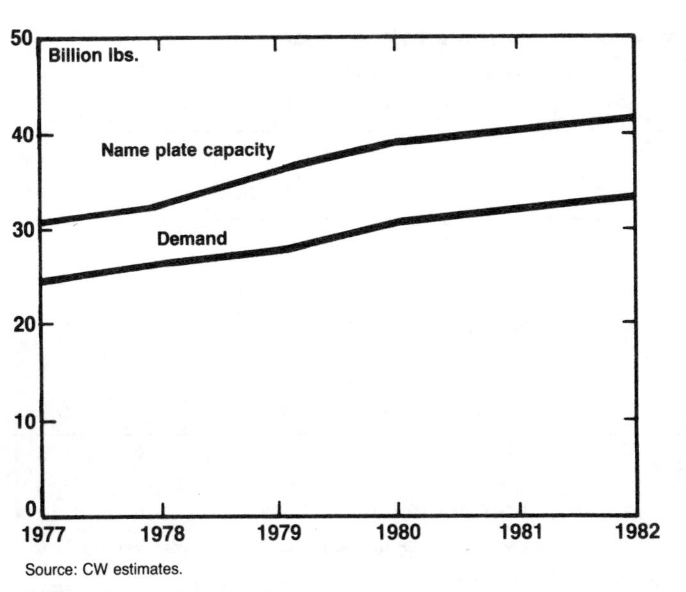

Fig. 8-2—The relationship between ethylene supply and demand through 1982.[3]

A concomitant reaction is

$$CH_2 = CH_2 + 3\ O_2 \rightarrow 2\ CO_2 + 2\ H_2O \quad \Delta H = -339.6 \text{ kcal}$$

Typical reaction conditions and selectivities are

Temperature:	200-300°C
Catalyst:	Ag_2O
Residence time:	1 sec.
Selectivity:	
Oxygen-based:	70-75 mole % Ethylene
Air-based:	66-72 mole % Ethylene

Selectivity is the ratio of moles of ethylene oxide produced per mole of ethylene reacted. Ethylene oxide selectivity is improved when the reaction temperature is lowered and the conversion of ethylene is decreased.

The use of high selectivity catalyst and control of temperature are key factors in successful production of ethylene oxide. Because of the low yield, appreciable research has gone into developing modified catalysts. In recent years, 78-84% selectivities have been obtained by incorporating alkali metal cations in, on, or under the silver particles on the alumina.[45] A selectivity of 88-94% has even been reported.[46] Table 8-1 gives a resumé of three different operating conditions patented for the oxidation of ethylene to ethylene oxide.[46]

Fig. 8-4 represents a process for both ethylene oxide and ethylene glycol. With air oxidation the reaction is carried out in two stages using main and purge reactors. Recycling of the unreacted gas is made in both stages to obtain the maximum utilization of ethylene. With oxygen oxidation the reaction is carried out in a single stage using the main reactor only. The oxidation reaction is controlled in a manner similar to that used for air oxidation. Most of the absorber outlet gas is recycled to the reactor and the rest is treated by potassium hydroxide solution to remove CO_2 and recycled to the reactor.

The use of oxygen in place of air for the production of ethylene oxide is somewhat controversial. Some of the variables are: fuel cost, nitrogen disposal, investment capital, flammability limits, pollution regulations, etc. The use of oxygen is currently being favored over air. In 1975, 57% of ethylene oxide was produced by an oxygen process. In 1965 the percentages were 18% by oxygen, 65% by air and 17% by the now obsolete chlorohydrin process.[6] The question of air

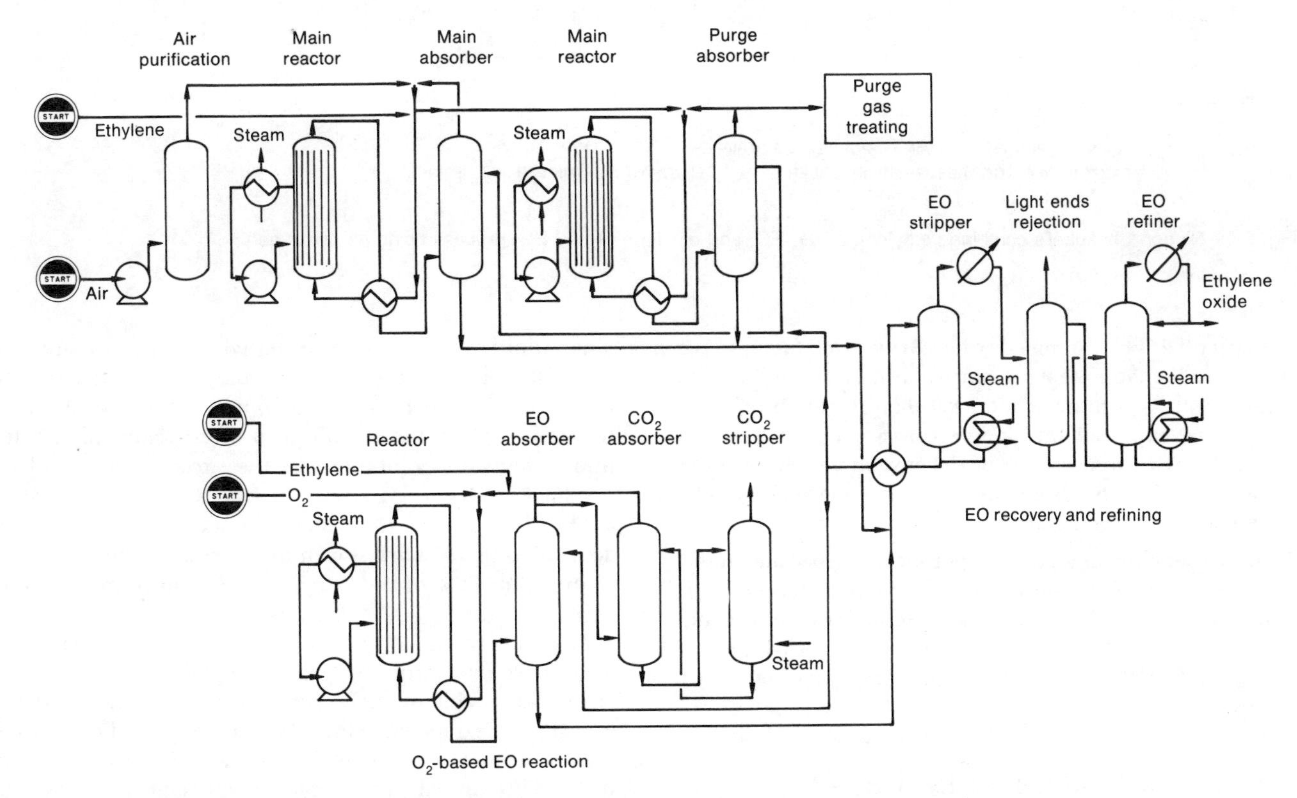

Fig. 8-3—Schematic of the SD/Halcon ethylene oxide process using either air or oxygen as the oxidant.[4]

TABLE 8-1—Conditions for the Oxidation of Ethylene to Ethylene Oxide[46]

Catalyst	Temp., °C	Press., atm	Feed composition $\%C_2^=$	Feed composition $\%O_2$	Feed composition Other	ppm Cl	GHSV h	$\%O_2$ conv.	Sel. to EO %	STY g EO/L × h
Shell 7.8% Ag 105 ppm Cs^+ on 5/16 in. alumina rings U.S. patent 3,962,136	252	15	30	8	bal. N_2	12	3300	52	81	277
Halcon 13.4% Ag 270 ppm Ba^{++} 53 ppm Cs^+ on 5 mm alumina spheres U.S. 4,066,575	241	21	15	7	4% CO_2 bal. N_2	.2	6000	–	78	–
ICI 8% Ag .3% K^+ .001 Rb^+ on 1 mm alumina particles W. German Offen. 28 20 170	240	16	30	8	bal. N_2	20	3000	9	94	–

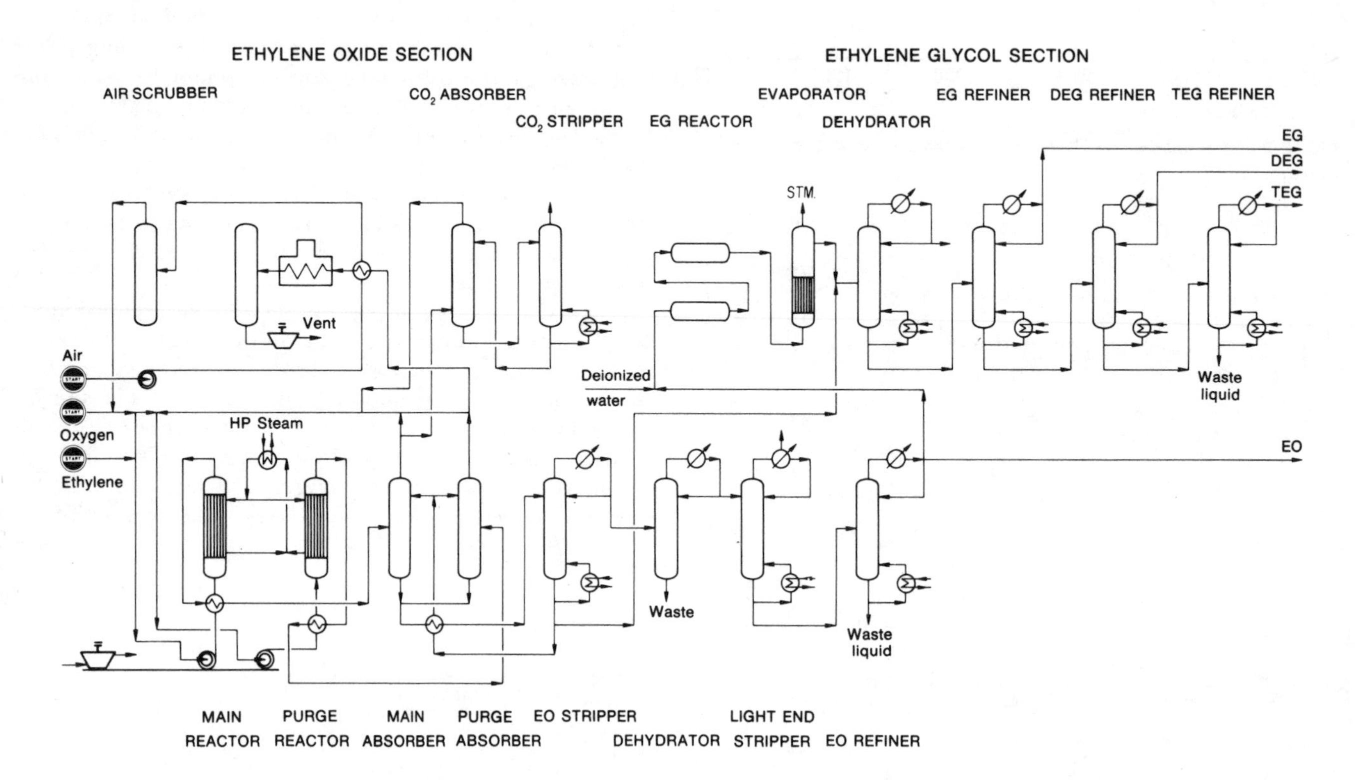

Fig. 8-4—Nippon Shokubai's combined ethylene oxide, EO, and ethylene glycol, EG, process using air or oxygen.[5]

vs. oxygen has been summed up by Brune De Maglie—"We can conclude that, except for particular local conditions, the oxygen process is more economical than the air process. It can also be expected that in the construction of new EO plants, the clear predominate of the oxygen process over the air process, which has been apparent for some years, will be more marked in the future."[7]

Uses. Ethylene oxide reacts exothermically, especially in the presence of a catalyst, with all compounds which have a labile hydrogen atom, such as water, alcohols, amines and organic acids. This reaction introduces the hydroxyethyl group, $-CH_2-CH_2OH$, into various types of compounds. For example:

$$R - CH_2OH + \overset{O}{\overbrace{CH_2 - CH_2}} \rightarrow R - CH_2O - CH_2 - CH_2OH$$

The addition of the hydroxyethyl group increases the water solubility of the resulting compound. Further reaction with ethylene oxide produces polyethylene oxide derivatives. The number of moles of ethylene oxide determines the water solubility and the surface activity of the product.

The uses of the 4.8 billion pounds of ethylene oxide demanded in 1978 are shown in Fig. 8-5.[47] Ethylene glycol accounts for 69% of this demand; surfactants are next with 13%.

There are two major types of these surfactants: (1) nonbiodegradable or "hard" surfactants; *alkylphenol ethoxylates* and (2) biodegradable or "soft" surfactants; *linear alcohol ethoxylates*. The linear alcohol ethoxylates are expected to have the highest growth rate of all EO derivatives—9 to 11% per year.[8] The annual growth rate of ethylene glycol will be lower than any of the other EO derivatives because of the

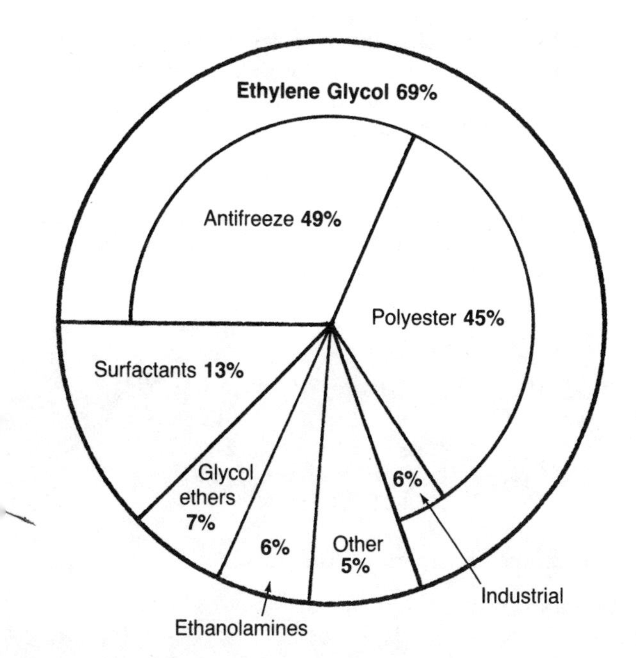

Fig. 8-5—1978 ethylene oxide demand and uses.[47]

predicted growth rate of only 3% for antifreeze, the single largest use of ethylene glycol.[9]

Other compounds with appreciable production from EO are the higher glycols, glycol ethers and the ethanolamines. Of lesser importance in respect to EO utilization are the tertiary alkyl mercaptoalcohols, glycol acetate and diacetate, *beta*-phenylethyl alcohol, and hydroxyethylcellulose. Ethylene oxide is used as a cold sterilant for bacteria, spores, and viruses, and with CO_2 for controlling weevils in nuts. It is an effective insecticide and is also used as an intermediate for other insecticides as well as for fungicides, explosives and resins.

All this adds up to an expected growth rate of 6% per year through 1985[10] with an annual rate of 7.6% through 1980.[11] Another estimate gives an annual growth rate of just under 5%, with an expected production of 6 billion pounds in 1983.[47] Others estimate 3.5 to 5.5%. The 1979 capacity was 6.6 billion pounds.

Ethylene Glycol

Essentially all *ethylene glycol*, **EG,** $CH_2OH\text{-}CH_2OH$, is produced by the hydration of ethylene oxide. It also can be produced directly from ethylene by the Oxirane process, *acetoxylation,* and the Teijin process, *oxychlorination.* The production of **EG** from *syn gas* by a direct synthesis process is predicted to be of major importance in the mid-1980s. At that time *syn gas* could offer a substantial cost advantage over ethylene.[12] Weismantel gives an interesting account of current practices in ethylene glycol production. The principal processes are compared as to their various merits and lack of merit.[48]

From Ethylene Oxide. The oxide ring, *epoxide ring,* is readily opened by water in the presence of hydrogen ions.

$$\overset{\;O\;}{\overbrace{CH_2 - CH_2}} + H_2O \xrightarrow{H^+} HOCH_2 - CH_2OH$$

A typical hydration process operates under the following general conditions:

Temperature	50-100°C
Catalyst:	0.5-1.0% H_2SO_4
Contact time:	30 min.
Ethylene oxide: water ratio:	1:10

Liquid phase processes operate in the 95°C and 200-300 psi range.

Di- and triethylene glycol ethers are formed to the extent of about 10% with the amount being determined by the ethylene-water ratio. The overall yield of useful products is essentially 100% based on ethylene oxide.

The formation of co-product di- and triethylene glycol ethers is not an economic burden on the monoglycol. They have many applications with the most important being water-based coatings. The tri-ethers are paramount in the brake fluid market.

From Ethylene by Acetoxylation. The production of ethylene glycol from ethylene by acetoxylation, the Oxirane process, is carried out in two steps.[13] The first step is the catalyzed liquid phase oxidation of ethylene in acetic acid to a mixture of mono- and diacetates of ethylene glycol.

$$2CH_2{=}CH_2 + 3CH_3 - \overset{\overset{\displaystyle O}{\|}}{C} - OH + O_2 \rightarrow CH_3 - \overset{\overset{\displaystyle O}{\|}}{C} - OCH_2 - CH_2OH + CH_3 - \overset{\overset{\displaystyle O}{\|}}{C} - OCH_2{-}CH_2O - \overset{\overset{\displaystyle O}{\|}}{C} - CH_3 + H_2O$$

The acetates are hydrolyzed to obtain the glycol and regenerate acetic acid for reuse.

$$CH_3 - \overset{\overset{\displaystyle O}{\|}}{C} - OCH_2 - CH_2OH + CH_3 - \overset{\overset{\displaystyle O}{\|}}{C} - OCH_2 - CH_2 - \overset{\overset{\displaystyle O}{\|}}{C} - CH_3 + 3\,H_2O \rightarrow 2\,HOCH_2 - CH_2OH + 3\,CH_3 - \overset{\overset{\displaystyle O}{\|}}{C} - OH$$

Net reaction:

$$2\,CH_2 = CH_2 + 2\,H_2O + O_2 \rightarrow 2\,HOCH_2 - CH_2OH$$

The following reaction conditions have been reported.[14]

Temperature	160°C
Pressure:	28 atm
Catalyst:	TeO_2 (promoted by Br compounds)
Conversion:	60%
Selectivity:	97%
Product distribution:	Diacetate 70%
	Monoacetate 25%
	Ethylene glycol 5%

Manganese acetate catalyst plus potassium iodide has also been developed.

The acetates are hydrolyzed to ethylene glycol at 107-130° C, 1.17 atm pressure, and a selectivity of 95%. The hydrolysis step is difficult to complete, and with the separation of the monoacetals and glycols, hard to accomplish. This part of the process requires almost half the capital investment and utilities of the project.[15] The mole % yield on ethylene is 94—appreciably higher than obtained by way of ethylene oxide. However, the economic advantage of the Oxirane process versus conventional silver catalyzed technology can be debated. "Ethylene efficiency improvement can be secured by the Oxirane process but it requires higher investment and energy costs. The silver catalyzed process should be viable for some time to come.[8] Corrosion is the major production problem with the Oxirane process because of the presence of acetic and formic acids. Further problems may be caused by the tellurium catalyst having a tendency to convert to the metal and plate out.[16] These problems apparently could not be overcome, as the 800 million pound-per-year plant utilizing this process was shut down.[52] The reason given was "continued unsatisfactory performance."

From Ethlyene by Oxychlorination. The *Teijin catalytic oxychlorination process* is a modern version of the obsolete chlorohydrin process for ethylene oxide production. In place of chlorine, concentrated (1*N*) hydrochloric acid is used and thallium (III) chloride, $TlCl_3$, is the "catalyst." The ethylene chlorohydrin may be hydrolyzed *in situ.*

$$CH_2{=}CH_2 + TlCl_3 + H_2O \rightarrow ClCH_2\text{-}CH_2OH + TlCl + HCl$$

$$ClCH_2\text{–}CH_2OH + H_2O \rightarrow HOCH_2\text{-}CH_2OH + HCl$$

The "catalyst" is regenerated by air or oxygen plus copper(II) chloride which gives the thallium(III) chloride the status of a catalyst.[17]

$$TlCl + 2\ CuCl_2 \rightarrow TlCl_3 + Cu_2Cl_2$$

$$2\ Cu_2Cl_2 + 4\ HCl + O_2 \rightarrow 4\ CuCl_2 + 2\ H_2O$$

The overall reaction is

$$2\ CH_2{=}CH_2 + 2\ H_2O + O_2 \rightarrow 2\ HOCH_2\text{–}CH_2OH$$

The reported reaction conditions are

Temperature:	60-250°C	
Pressure:	20 kg/cm²	
Catalyst:	$TlCl_3$	
Yield:	Ethylene glycol	89%
	Acetaldehyde	6%
	Other*	5%

*dioxane and diethylene glycol.

The acetaldehyde yield will be increased appreciably if the $Cl^-:Tl^{3+}$ ratio is less than *ca* 4:1. When the reaction temperature is above 120°C, the chlorohydrin is hydrolyzed *in situ*.

From Formaldehyde and Carbon Monoxide and From Synthesis Gas. These two sources of ethylene glycol are noted in Chapter 5.

Uses. Ethylene glycol is the 28th largest-volume chemical produced in the United States, with a production of 4.3 billion pounds in 1979. Plant capacity was 6.3 billion pounds on January 1, 1980.[52] The major uses if ethylene glycol are shown in Fig. 8-5.[47]

Ethanolamines. *Monoethanolamine,* **MEA,** *diethanolamine,* **DEA,** and *triethanolamine,* **TEA,** are produced as a mixture from the reaction of ethylene oxide with 25 to 50% aqueous ammonia.

$$\overset{O}{\overbrace{CH_2 - CH_2}} + NH_3 \rightarrow HOCH_2 - CH_2NH_2$$

$$\overset{O}{\overbrace{CH_2 - CH_2}} + (HOCH_2 - CH_2)_2NH \rightarrow (HOCH_2 - CH_2)_3N$$

Typical reaction conditions are

Temperature:	30-40° C	
Pressure:	10-20 psig	
Mole ratio:		
Ammonia: ethylene oxide 10:1		
Product distribution:	Monoethanolamine	75%
	Diethanolamine	21%
	Triethanolamine	4%

The relative proportions of mono-, di- and triethanolamine is dependent upon the ratio of ammonia to ethylene oxide (Table 8-2).[18] The ratio also varies with the temperature (50° to 275°) and pressure (15-1500 psi).

Fig. 8-6 is a flow diagram for the production of ethanolamines from ethylene oxide and ammonia.[19] Ethylene oxide/ammonia/ recycle **MEA** feed ratios are used to control the distribution of ethanolamines to accommodate varying market demands for each of the products.

Uses. The ethanolamines have unusually diverse industrial applications. The most important direct use for the ethanolamines is the sweetening of acid gases.[20] The most important indirect use is for the production of detergents (Fig. 8-7). The ethanolamines are also used as corrosion inhibitors and to stabilize chlorinated hydrocarbons by preventing decomposition in the presence of a metal or metallic compound.

The ethanolamines are used extensively for the production of ethanolamide detergents from fatty acids.

$$R - \overset{O}{\overset{\|}{C}} - OH + HOCH_2 - CH_2NH_2 \rightarrow R - \overset{O}{\overset{\|}{C}} - NH - CH_2 - CH_2OH + H_2O$$

TABLE 8-2—Weight Ratios of Ethanolamines as a Function of the Mole Ratios of the Reactants[18]

	Moles of Ethylene oxide/Moles of Ammonia		
	0.1	0.5	1.0
Monoethanolamine....	75—61	25—31	12—15
Diethanolamine.......	21—27	28—32	23—26
Triethanolamine......	4—12	37	65—59

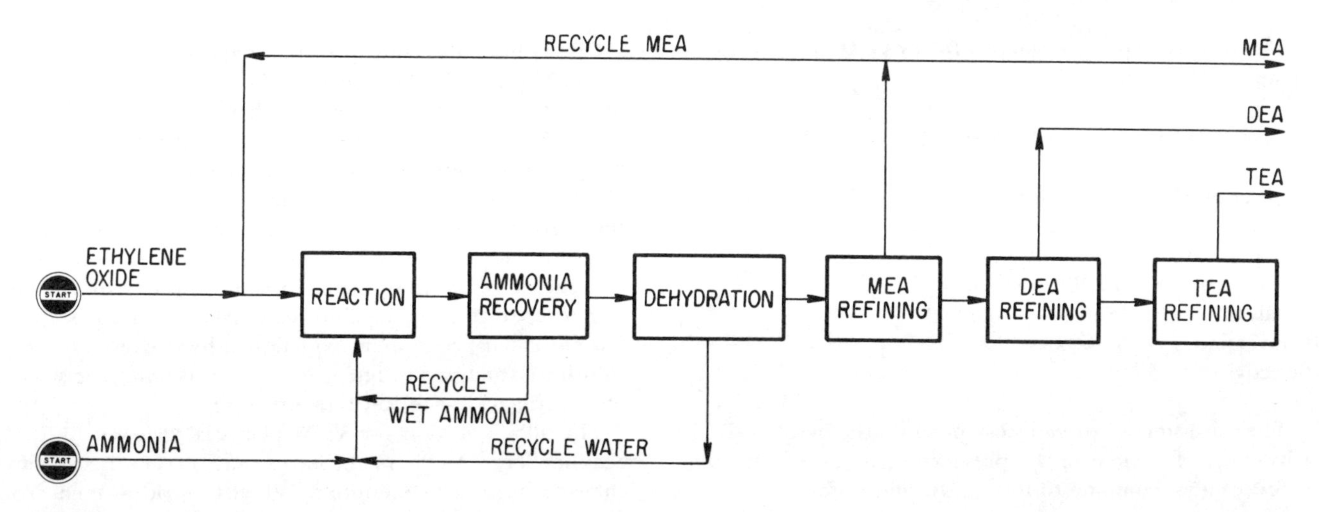

Fig. 8-6—Flow diagram for the manufacture of mono-, di-, and triethanolamine from aqueous ammonia and ethylene oxide.[19]

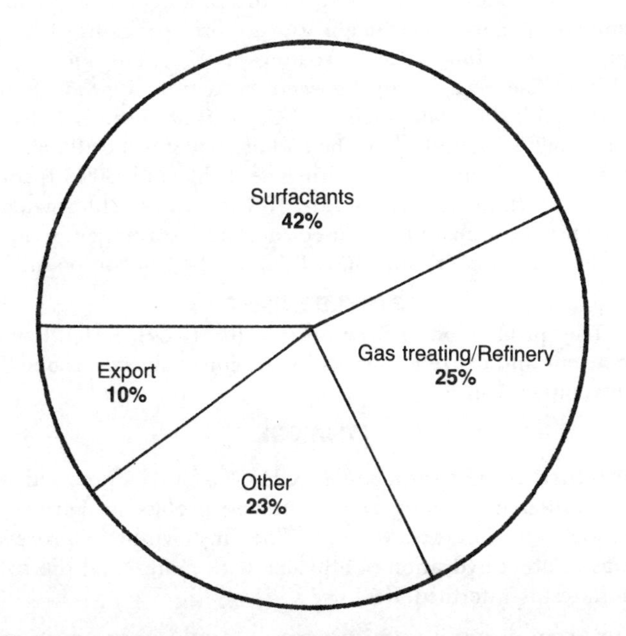

Fig. 8-7—Distribution of the projected 400 million pound production of ethanolamines in 1982.[21]

Lauric acid, $CH_3(CH_2)_{10}\text{-}\overset{O}{\overset{\|}{C}}\text{-}OH$, is the main fatty acid used. Monoethanolamides are used primarily in heavy-duty powder detergents as foam stabilizers, corrosion inhibitors and rinse improvers.

Ethanolamine soaps rank close behind the ethanolamides as important industrial products. They are formed by reaction of an ethanolamine with a fatty acid, a reaction similar to the formation of ethanolamide, but at a lower temperature and without a catalyst. The product is a salt rather than an amide. The fatty acids utilized are oleic, stearic and palmitic. These soaps are used extensively in cosmetic preparation.

A wide range of household and industrial specialties use ethanolamine soaps. These include "soluble" lubricating and cutting oils; floor, furniture, automobile and metal polishes; solvent cleaners, stain and paint removers and spotting soaps; and floor, rug, woodwork and paintbrush cleaners.

VINYL CHLORIDE

Production. In the United States, about 92% of the *vinyl chloride monomer,* **VCM,** $CH_2{=}CHCl$, is produced by the "balanced oxychlorination process." Three principal process steps are involved in this process in which ethylene and chlorine are converted to **VCM** (Fig. 8-8).[22, 23]

The first step is the liquid or vapor phase addition of chlorine to ethylene, *direct chlorination,* to produce ethylene dichloride, **EDC.**

$$CH_2{=}CH_2 + Cl_2 \rightarrow ClCH_2\text{-}CH_2Cl$$

This exothermal reaction may be catalyzed by ethylene bromide at 40-50° C or by iron(III) chloride, $FeCl_3$, in a temperature range of 53-65 or 109°C. The yield is about 90%.

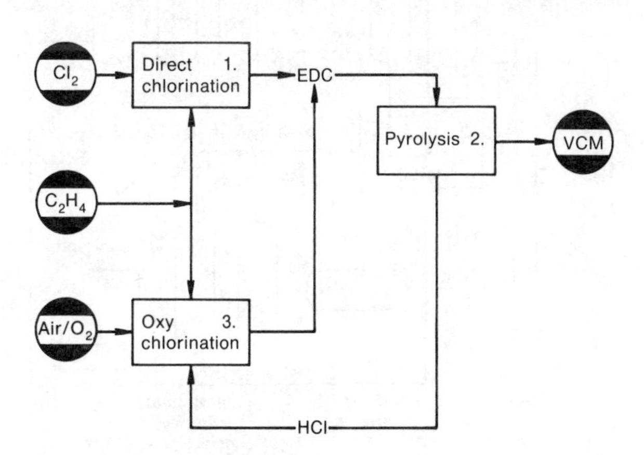

Fig. 8-8—The principal steps in the ethylene-based vinyl chloride monomer production.[23]

The second step is *pyrolysis* of **EDC** to **VCM** and hydrogen chloride, HCl.

$$CH_2Cl - CH_2Cl \rightarrow CH_2 = CHCl + HCl$$

Typical pyrolysis conditions are:

Temperature:	430-530°C.
Pressure:	*ca.* 400 psi
Catalyst:	Pumice or charcoal
Conversion:	50-60%
Selectivity:	95%

The third step is *oxychlorination* in which the pyrolysis HCl, ethylene and oxygen in the presence of modified Deacon-type[24] catalyst combine to form **EDC** and water.

$$2\ CH_2 = CH_2 + 4\ HCl + O_2$$
$$\rightarrow 2\ CH_2Cl - CH_2Cl + 2\ H_2O\ -\Delta H$$

The temperature is in the 225-235°C range for fluid bed and 230-290°C for fixed bed processes. The **EDC** is recycled to the pyrolysis unit.

The overall reaction is

$$4\ CH_2 = CH_2 + 2\ Cl_2 + O_2$$
$$\rightarrow 4\ CH_2 = CHCl + 2\ H_2O$$

and represents about a 90% yield based on ethylene. Fig. 8-9 shows a typical balanced vinyl chloride monomer process.[23]

The oxychlorination step is the heart of the process and has two major variants—reactor and oxidant. Either a fluidized bed or a fixed bed reactor is used—along with either oxygen or air. Reich has reported a definitive study of air vs. oxygen and has made a strong case for oxygen.[23]

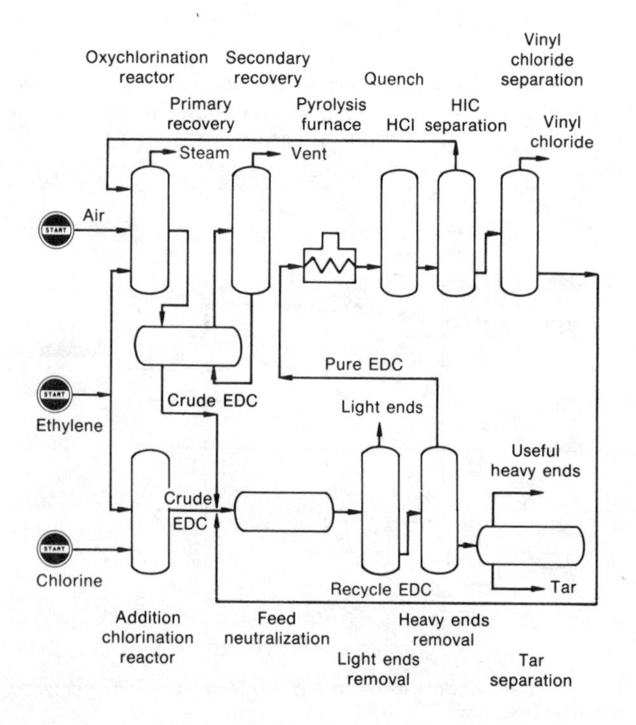

Fig. 8-9—Typical balanced vinyl chloride monomer process.[23]

Others have also concluded that oxygen is the best choice for **VCM** plants.[25] There seems to be little, if any, superiority of one type of reactor over the other.

Oxychlorination is an effective way to utilize the byproduct hydrogen chloride, but it is also the most costly process step. Many of the chlorinated compounds which have to be removed in the distillation section are byproducts of the oxychlorination step. An integrated process (Fig. 8-10) eliminates some of these problems because the oxychlorination step is separate from the chlorine-ethylene addition step. In this particular process, the oxidation of hydrogen chloride to chlorine is catalyzed by nitrogen oxide in a circulating stream of sulfuric acid—the KeChlor process.[26]

The alloy selection for VCM plants is reviewed by C.M. Schillmoller.[27] McPherson, Starks, and Fryar discuss vinyl chloride in an article entitled "Vinyl Chloride—What You Should Know."[28] Several corrections have been made for this article.[49]

Uses. The 1979 production of 7 billion pounds of vinyl chloride monomer was used for the production of homo- and copolymers. Their major uses are for extrusions such as pipe (55%), films (15%), coatings (10%), and moldings (10%). The relationship between capacity and production for VCM is shown in Fig. 8-11.[28] Capacity at the end of 1979 was 8 billion pounds.[50] At the end of 1980 it will be 9 billion pounds. The major use of ethylene dichloride (80%) is for the production of vinyl chloride. 10% is used as a chlorinated solvent; the remainder is used as a lead scavenger or for other purposes. Production is 1978 was 10.5 billion pounds.

ETHYLBENZENE

The production of *ethylbenzene* (C_6H_5-CH_2-CH_3) from ethylene and benzene, and its conversion to styrene, is noted in Chapter 10.

ETHANOL

Production. *Ethyl alcohol,* $CH_3 - CH_2OH$, production is considered by many to be the oldest profession. Fermentation was the process used. The first synthetic process was indirect hydration of ethylene with mono- and diethyl sulfates as intermediates.

$$3\ CH_2 = CH_2 + 2\ H_2SO_4 \rightarrow CH_3 - CH_2OSO_3H + (CH_3 - CH_2O)_2SO_2$$

Hydrolysis of these sulfates gave ethanol and regenerated the sulfuric acid.

$$CH_3 - CH_2OSO_3H + (CH_3 - CH_2O)_2SO_2 + 3\ H_2O \rightarrow 3\ CH_3 - CH_2OH + 2\ H_2SO_4$$

Synthetic ethyl alcohol is called *ethanol* to distinguish it from fermentation alcohol.

Indirect hydration has now been replaced by direct hydration of ethylene.

$$CH_2 = CH_2 + H_2O \rightarrow CH_3 - CH_2OH$$
$$\Delta H = -9.6\ kcal$$

Typical operating conditions are

Temperature:	325° C
Pressure:	1000 psi
Catalyst:	H_3PO_4 on diatomaceous earth
Conversion:	4-5%
Selectivity:	95-97%

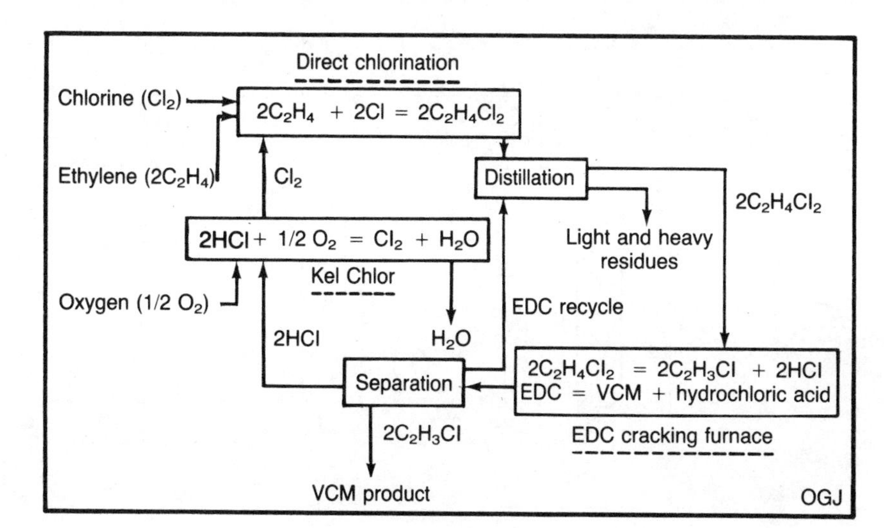

Fig. 8-10—The integrated vinyl chloride monomer process.[26]

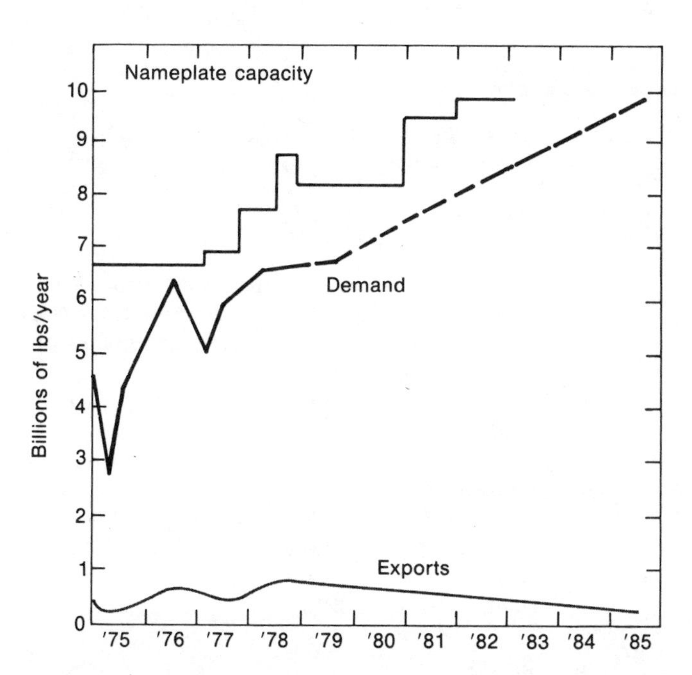

Fig. 8-11—The United States nameplate capacity vs. demand for vinyl chloride monomer.[28]

About 2-5% diethyl ether is the byproduct is this process. Other catalysts such as $Al(OH)_3$ gel and tungstic acid on silica gel have been reported.

Fig. 8-12 represents the flow diagram of a typical process for the hydration of ethylene to ethanol.[29] The initial product is 94.5-95% ethanol which can be dehydrated to anhydrous ethanol if desired. Anhydrous ethanol amounts to about 40% of production. Ethanol production was 196 million gallons (1.18 billion pounds) with a capacity of 270 million gallons (1.62 billion pounds).[51]

Uses. Ethanol has many and varied uses. These can be conveniently divided into solvent and chemical conversion uses. (Fig. 8-13)[30] The solvent uses of ethanol have been increasing at a faster rate than its use as an intermediate.[30, 31] This is partly caused by the rapid shift in the synthesis of acetaldehyde from ethanol based to direct oxidation of ethylene.

Ethanol is perhaps the most widely used intermediate. Among compounds synthesized from ethanol are ethyl chloride, ethyl ether, glycol ethyl ethers, ethyl vinyl ether, chloral, ethylamines, ethyl mercaptan, acetic acid and many different ethyl esters. It has been proposed as a future source of ethylene.[32] The ethanol would come from fermentation of organic material. The use of ethanol as a fuel is both controversial and beyond the scope of this book.

ACETALDEHYDE

Production. Historically, the production of *acetaldehyde,* $CH_3-\overset{\overset{O}{\|}}{C}-H$, was by the silver catalyzed oxidation of ethanol or by the chromium activated copper catalyzed dehydrogenation of ethanol. The more sophisticated processes used a combination of oxidation-dehydrogenation. The exothermic oxidation provided the heat required for the endothermic dehydrogenation.

$$\text{Oxidation: } 2\ CH_3-CH_2OH + O_2 \rightarrow 2\ CH_3-\overset{\overset{O}{\|}}{C}-H + 2\ H_2O$$

$$\text{Dehydrogenation: } CH_3-CH_2OH \rightarrow CH_3-\overset{\overset{O}{\|}}{C}-H + H_2$$

Currently acetaldehyde is produced directly from ethylene by use of a liquid phase homogeneous catalyst. This type of catalyst has several advantages. One distinct advantage is high selectivity which conserves the increasingly expensive raw materials. There is also a significant energy saving because of the low temperatures and pressures possible with homogeneous catalysts.

The homogeneous catalyst system used for the oxidation of ethylene to acetaldehyde consists of an aqueous solution of copper(II) chloride, $CuCl_2$ and a small quantity of palladium(II) chloride, $PdCl_2$. In the course of the oxidation, the palladium ion of the $PdCl_2$ is reduced to metallic palladium Equation 1. The palladium is reoxidized to palladium(II) ion (Pd^{2+}) by the copper(II) ion (Cu^{2+}) which becomes copper(I) ion (Cu^{+}), Equa-

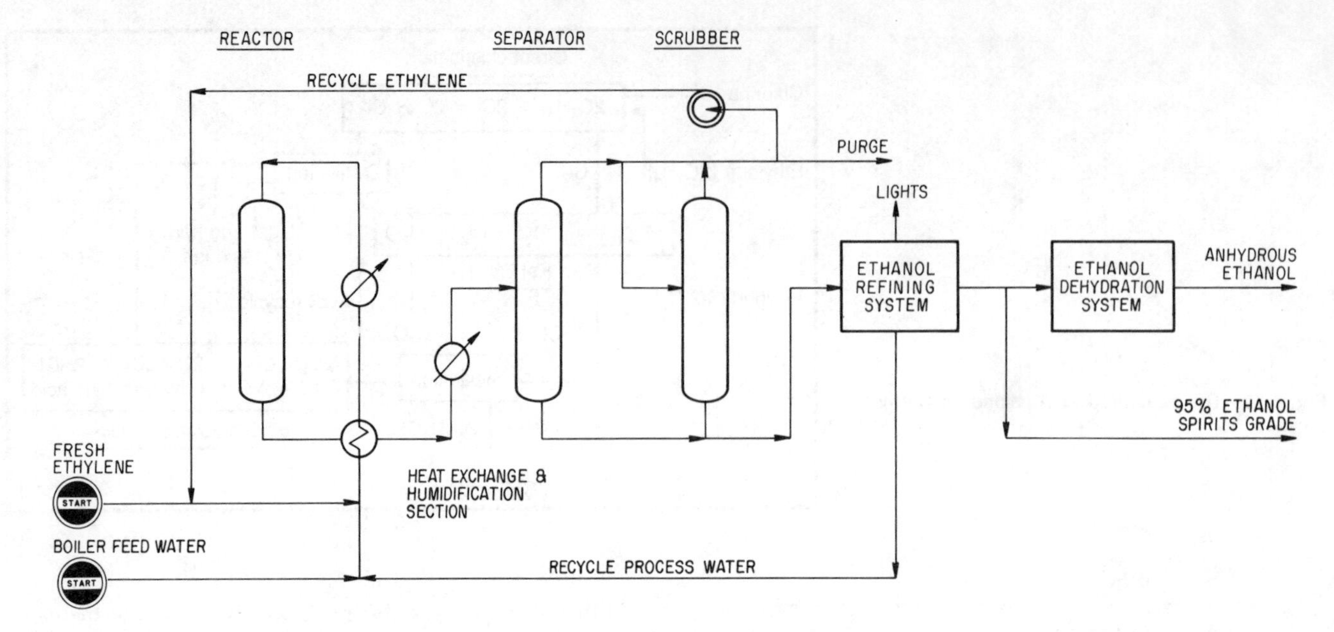

Fig. 8-12—Flow diagram of the Union Carbide process for the hydration of ethylene to ethanol.[29]

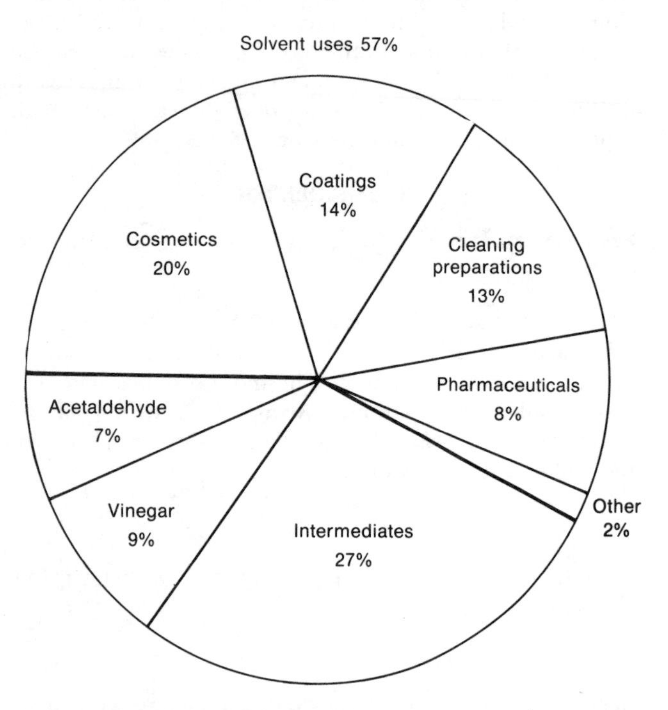

Fig. 8-13—The end uses of the 236 million gallon U.S. production of ethanol in fiscal year 1977.[30]

tion 2. The copper(I) ion is reoxidized to copper (II) ion, by air or oxygen, Equation 3. Equation 4 represents the overall reaction.

$$CH_2 = CH_2 + H_2O + PdCl_2 \rightarrow CH_3 - \overset{\displaystyle O}{\overset{\|}{C}} - H + 2\ HCl + Pd \quad (1)$$

$$Pd + 2\ CuCl_2 \rightarrow PdCl_2 + 2\ CuCl \quad (2)$$

$$2\ CuCl + \tfrac{1}{2}\ O_2 + 2\ HCl \rightarrow 2\ CuCl_2 + H_2O \quad (3)$$

$$2\ CH_2 = CH_2 + O_2 \rightarrow 2\ CH_3 - \overset{\displaystyle O}{\overset{\|}{C}} - H \qquad \Delta H = -58.2\ \text{kcal} \quad (4)$$

The oxidation is carried out as a single-stage process with the oxygen used *in situ* to regenerate the copper(II) ion (Fig. 8-14).[33] In the two-stage process, the catalyst solution, containing copper(I) ion equivalent to the amount of acetaldehyde formed, is transferred into a tube-oxidizer and reoxidized with air at *ca.* 140 psig (Fig. 8-15).[33]

Single-stage process:

Temperature:	130°C
Pressure:	45 psig
Catalyst:	$PdCl_2/CuCl_2$
Yield:	95%

Two-stage process:

Temperature:	130°C
Pressure:	120 psig
Catalyst:	$PdCl_2/CuCl_2$
Yield:	95%
Catalyst regeneration:	140 psig

Jira, Blau and Grimm conclude that "no general answer can be given to the question of which process is preferred. It depends upon the conditions at the plant site as well as on the availability of pure ethylene and oxygen, the price of oxygen and the possibility of nitrogen utilization from the two-stage plant."[33]

Acetaldehyde can be produced by the vapor phase catalytic oxidation of ethylene.

$$CH_2 = CH_2 + O_2 + H_2O \rightarrow CH_3 - \overset{\displaystyle O}{\overset{\|}{C}} - H + CH_3 - \overset{\displaystyle O}{\overset{\|}{C}} - OH$$

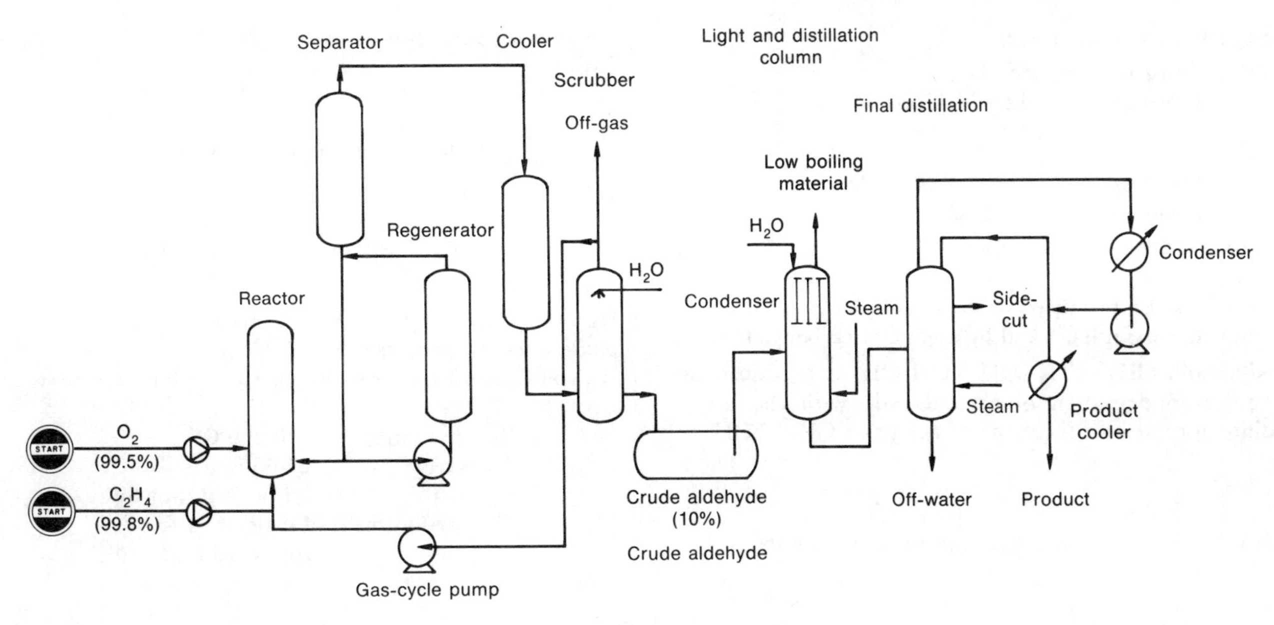

Fig. 8-14—The single-stage acetaldehyde process using pure oxygen and ethylene.[33]

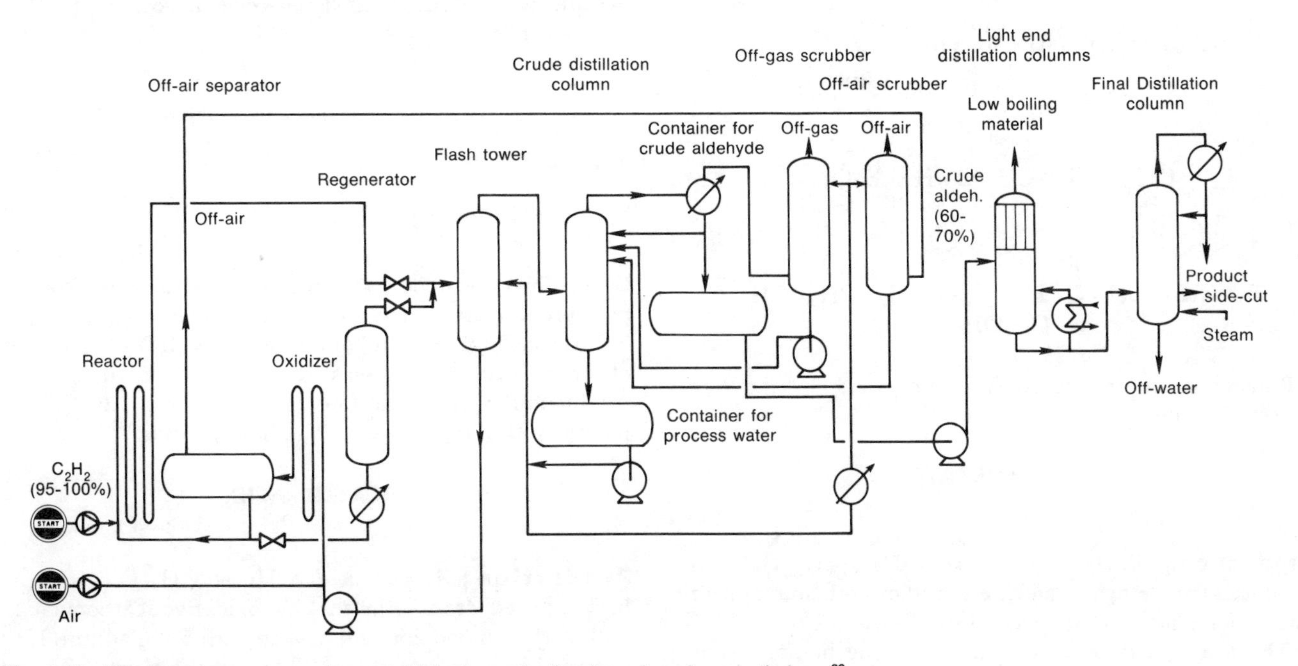

Fig. 8-15—The two-stage process for acetaldehyde production by using air and ethylene.[33]

The reaction conditions are

Temperature:	140° C
Pressure:	nominal
Catalyst:	$Pd/V_2O_5/Ru$ on Al_2O_3
Yields:	Acetaldehyde 64%
	Acetic acid 20%

The byproduct acetic acid is not an economic burden because acetaldehyde is used as one of the precursers of acetic acid.

Acetaldehyde is also produced by the noncatalyzed oxidation of propane, butane or a mixture of the two.[34] The yield is 31% with the products being formaldehyde (33%), methanol (20%), other (16%).

Uses. Acetaldehyde has no important use other than to synthesize other compounds. Among these are acetic acid, peracetic acid, acetic anhydride, chloral, paraldehyde, polyacetaldehyde, *n*-butyraldehyde, *n*-butanol and pentaerythritol.

Acetic acid is obtained by the liquid phase oxidation of acetaldehyde.[35]

$$2\,CH_3\text{-}\overset{\overset{\displaystyle O}{\|}}{C}\text{-}H + O_2 \rightarrow 2\,CH_3\text{-}\overset{\overset{\displaystyle O}{\|}}{C}\text{-}OH$$

The reaction conditions are

Temperature:	66° C
Pressure:	Liquid phase
Catalyst:	$(CH_3-\overset{O}{\overset{\parallel}{C}}-)_2Mn$
Conversion:	*ca* 50%
Yields:	95%

Acetic acid is also produced by the liquid phase oxidation of *n*-butane and naphtha,[34] and by methanol carbonylation.[36]

n-Butanol, $CH_3-CH_2-CH_2-CH_2OH$, is produced by the aldol condensation of acetaldehyde with the intermediate formation of *crotonaldehyde,* $CH_3-CH=CH-\overset{O}{\overset{\parallel}{C}}-H$, which is hydrogenated to *n*-butanol.

$$2\ CH_3-\overset{O}{\overset{\parallel}{C}}-H \xrightarrow{(OH^-)} CH_3-CHOH-CH_2-\overset{O}{\overset{\parallel}{C}}-H$$

$$CH_3-CHOH-CH_2-\overset{O}{\overset{\parallel}{C}}-H \xrightarrow[\text{distill}]{(H^+)} CH_3-CH=CH-\overset{O}{\overset{\parallel}{C}}-H+H_2O$$

$$CH_3-CH=CH-\overset{O}{\overset{\parallel}{C}}-H+H_2 \rightarrow CH_3-CH_2-CH_2-CH_2OH$$

n-Butanol is also produced from propylene by the **oxo** process (Chapter 9).

VINYL ACETATE

Production. *Vinyl acetate,* $CH_2=CHO-\overset{O}{\overset{\parallel}{C}}-CH_3$, is produced from ethylene and acetic acid by both liquid phase and vapor phase catalytic oxidation processes.

The *liquid phase* process is similar to the homogeneous catalytic systems used for the production acetaldehyde from ethylene. The major difference is the presence of acetic acid.

$$CH_2=CH_2 + 2\ CH_3-\overset{O}{\overset{\parallel}{C}}-OH+O_2 \rightarrow CH_2=CHO-\overset{O}{\overset{\parallel}{C}}-CH_3+CH_3-\overset{O}{\overset{\parallel}{C}}-H+H_2O$$

The mole ratio of acetaldehyde to vinyl acetate can be varied from 0.3:1 to 2.5:1. The liquid phase process is not used extensively, if at all, because of corrosion problems and the formation of a fairly wide variety of other by-products.

The *vapor phase* process is currently the most economical.

$$2\ CH_2=CH_2+2\ CH_3-\overset{O}{\overset{\parallel}{C}}-OH+O_2 \rightarrow 2\ CH_2=CHO-\overset{O}{\overset{\parallel}{C}}-CH_3+2\ H_2O$$

Fig. 8-16 represents the U.S. Industrial Chemical Co.'s vapor phase one-step process.[37]

Typical reaction conditions for vapor phase synthesis are

Temperature:	180-200°C
Pressure:	25-60 psig
Catalyst:	Pd on Al_2O_3 or SiO_2/Al_2O_3
Conversion on:	Ethylene 10-15%
	Acetic acid 20-30%
	Oxygen 60%

The now obsolete acetylene process by which acetic acid adds to acetylene at *ca* 100° C in the presence of mercury(II) acetate awaits parity in ethylene-acetylene costs.

$$HC\equiv CH+CH_3-\overset{O}{\overset{\parallel}{C}}-OH \rightarrow CH_2=CHO-\overset{O}{\overset{\parallel}{C}}-CH_3$$

Uses. Vinyl acetate is a versatile monomer used to produce polyvinyl acetate and vinyl acetate copolymers (60%), polyvinyl alcohol (20%), and other polymers (15%).[38] The uses of these polymers are discussed in Chapter 12. In 1978 the production of vinyl acetate monomer was 1.68 billion pounds with a capacity of 2.23 billion pounds.[38]

ACRYLIC ACID

Production. *Acrylic acid,* $CH_2=CH-\overset{O}{\overset{\parallel}{C}}-OH$, can be produced from ethylene by oxidative carbonylation with carbon monoxide and oxygen with a palladium(II)/copper(II) catalyst system. This is the same type of homogeneous liquid phase catalytic reaction used to produce acetaldehyde from ethylene.

The overall reaction is

$$2\ CH_2=CH_2 + 2\ CO+O_2 \rightarrow 2\ CH_2=CH-\overset{O}{\overset{\parallel}{C}}-OH$$

Typical reaction conditions are

Temperature:	140° C
Pressure:	1100 psig
Catalyst:	$PdCl_2/CuCl_2$
Yield on ethylene:	80-85%

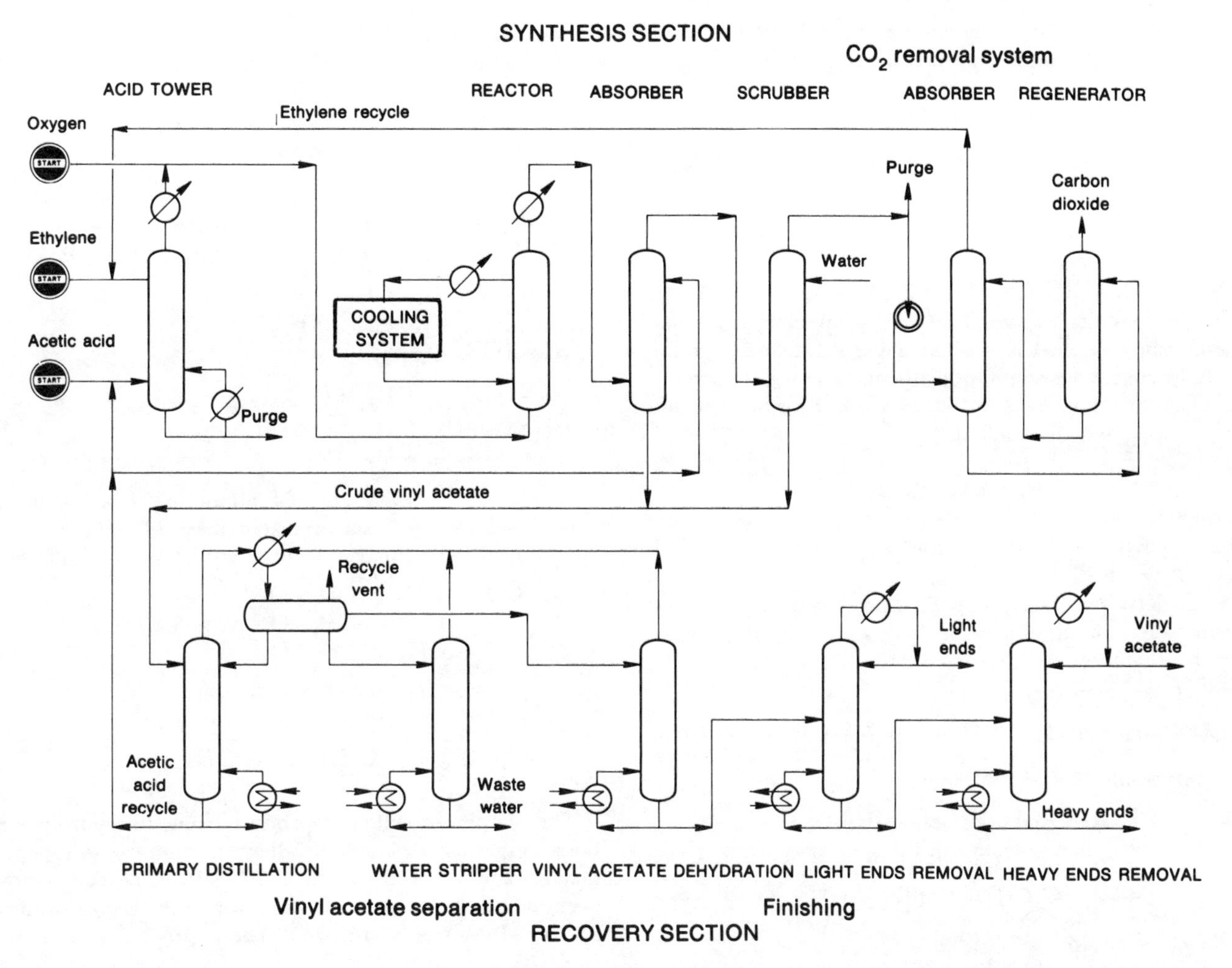

Fig. 8-16—Vapor phase catalytic oxidation process for making vinyl acetate from ethylene, acetic acid, and oxygen.[37]

Currently the process cannot compete with the oxidation of propylene.

Uses. Acrylic acid and its esters are used to make acrylic fibers and plastics.

PROPIONALDEHYDE

Production. *Propionaldehyde,* $CH_3 - CH_2 - \overset{\displaystyle O}{\overset{\|}{C}} - H$, is produced by the hydroformylation of ethylene with carbon monoxide and hydrogen, the **OXO** reaction.

$$CH_2 = CH_2 + CO + H_2 \rightarrow CH_3 - CH_2 - \overset{\displaystyle O}{\overset{\|}{C}} - H \qquad \Delta H = -35 \text{ kcal}$$

Cobalt carbonyl complexes have been the catalyst used and high pressures (3000-6000 psi) and temperatures (145-180° C) are required.

Currently, cobalt is being replaced by a complex of rhodium bonded by labile linkages to an organic ligand. During the reaction, the carbon monoxide and hydrogen also form a complex with the catalyst. The result is a reaction which is effective at lower temperatures and pressures. The conditions for this homogeneous liquid phase reaction are

Temperature:	*ca* 100° C
Pressure:	100-200 pisg
Catalyst:	Rh complex
Selectivity on ethylene:	*ca* 0.5 lb / 1 lb propionaldehyde

The construction costs are said to be less by 30% and operating costs are about 10% lower than for hydroformylation by use of cobalt as a catalyst. Also higher yields and a purer product are obtained.[39]

Uses. As with aldehydes in general, propionaldehyde has essentially no uses other than to make other compounds.

Propionaldehyde is hydrogenated to *propanol,* $CH_3 - CH_2 - CH_2OH$.

$$CH_3 - CH_2 - \overset{\displaystyle O}{\overset{\|}{C}} - H + H_2 \rightarrow CH_3 - CH_2 - CH_2OH$$

Various heribicide syntheses account for about 40% of propanol demand. The large use is in solvents for coatings and for ink used in printing on food containers. Propanol utilizes about 130 million pounds of propionaldehyde per year.

Propionic acid, $CH_3-CH_2-\overset{O}{\overset{||}{C}}-OH$, is obtained from propionaldehyde by oxidation.

$$2\ CH_3-CH_2-\overset{O}{\overset{||}{C}}-H + O_2 \rightarrow 2\ CH_3-CH_2-\overset{O}{\overset{||}{C}}-OH$$

This use accounts for about 50% of propionaldehyde demand. Propionic acid is used as a preservative for grain, especially corn. One pound of propionic acid will preserve a bushel of corn.[40] Various small uses account for the remaining propionaldehyde demand.

LINEAR ALCOHOLS

Production. *Linear alcohols,* $CH_3-(CH_2)_{2\text{-}26}-CH_2OH$, are produced from ethylene by a four-step process—the Alfol process. (Fig. 8-17).[41]

A Ziegler-type "catalyst" is first produced by the reaction between aluminum metal, hydrogen and ethylene to form triethylaluminum, $(CH_3-CH_2-)_3Al$.

Catalyst preparation:

$$6\ CH_2=CH_2 + 3\ H_2 + 2\ Al \rightarrow 2\ (CH_3-CH_2-)_3Al \quad (1)$$

Polymerization:

$$n\ CH_2=CH_2 + (CH_3-CH_2-)_3Al \rightarrow \begin{cases} CH_3-(CH_2-)_x-CH_2\diagdown \\ CH_3-(CH_2-)_y-CH_2-Al \\ CH_3-(CH_2-)_z-CH_2\diagup \end{cases} \quad (2)$$

Typical reactions are

Temperature:	120° C (0-20° C)
Pressure:	2000 psi
"Catalyst":	$(CH_3-CH_2-)_3Al$
Time:	140 min. to give C_{12} ave.
Yield:	Poisson distribution X, Y, Z, 2-26

Oxidation: The trialkylaluminums are oxidized between 20-50° C with "bone-dry" air to aluminum trialkoxides.

$$2\ \begin{matrix} CH_3-(CH_2-)_x-CH_2\diagdown \\ CH_3-(CH_2-)_y-CH_2-Al \\ CH_3-(CH_2-)_z-CH_2\diagup \end{matrix} + 3\ O_2 \rightarrow 2\ \begin{matrix} CH_3-(CH_2)_x-CH_2-O\diagdown \\ CH_3-(CH_2)_y-CH_2-O-Al \\ CH_3-(CH_2)_z-CH_2-O\diagup \end{matrix} \quad (3)$$

Hydrolysis: The mixture of aluminum trialkoxides is hydrolyzed by conc. sulfuric acid to yield a mixture of even numbered primary alcohols and aluminum sulfate.

$$2\ \begin{matrix} CH_3-(CH_2)_x-CH_2-O\diagdown \\ CH_3-(CH_2)_y-CH_2-O-Al \\ CH_3-(CH_2)_z-CH_2-O\diagup \end{matrix} + 3\ H_2SO_4 \rightarrow \begin{cases} 2\ CH_3-(CH_2-)_x CH_2OH \\ 2\ CH_3-(CH_2-)_y CH_2OH + Al_2(SO_4)_3 \\ 2\ CH_3-(CH_2-)_z CH_2OH \end{cases} \quad (4)$$

The aluminum sulfate coproduct from the hydrolysis is not an especially desirable product. Its main use is in paper making and water treatment. The Alfol process has now been modified to permit the use of water instead of sulfuric acid to effect the hydrolysis.[42] The coproduct is alumina,

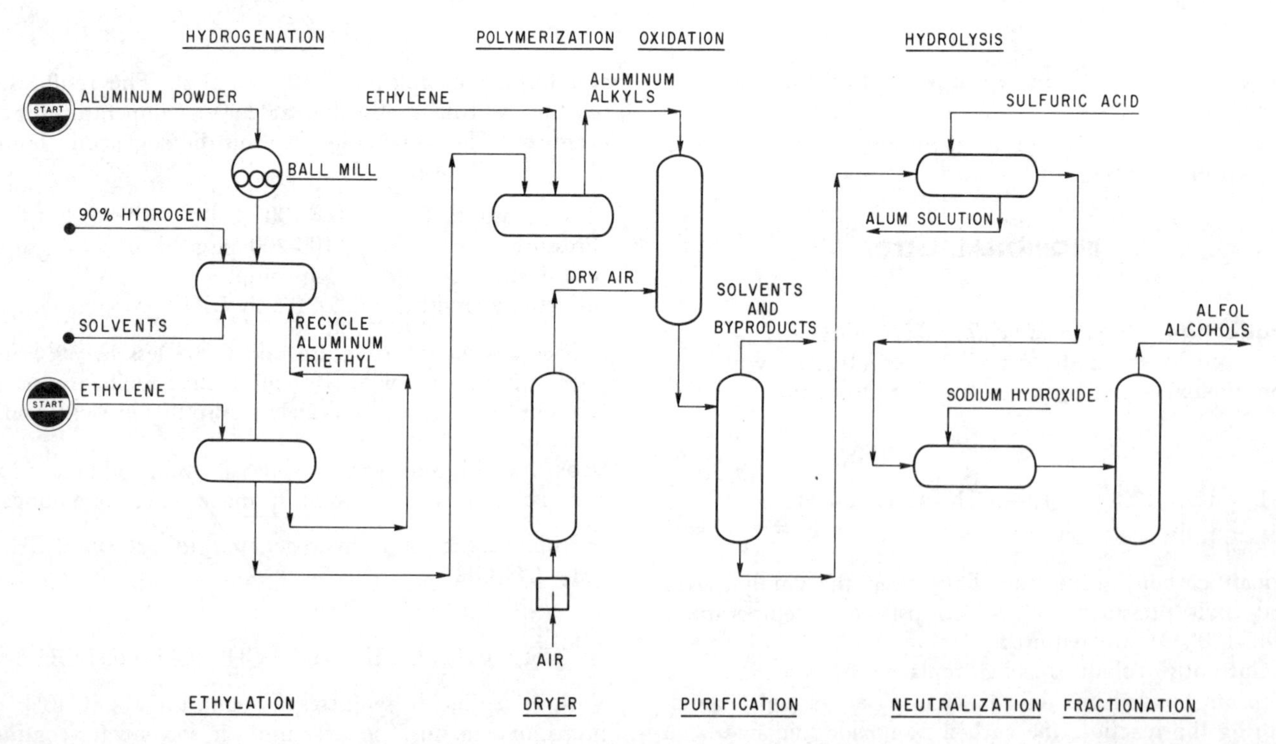

Fig. 8-17—A process for making even numbered straight chain alpha alcohols from aluminum, hydrogen, ethylene, air, and sulfuric acid—the Alfol Process.[41]

Al_2O_3, which is characterized by its activity, surface area and purity. It is used in the production of catalysts.

Uses. These linear alcohols are biodegradable and those in the C_{12-16} range are used to make detergents. Those in the C_{10-12} range are used to make plasticizers and the C_{16-18} alcohols are modifiers for wash-and-wear resins. The higher alcohols, C_{20-26}, are used as lubricants and mold-release agents.

Alpha olefins, R - CH = CH_2, are produced from trialkylaluminum by reaction with 1-butene.

$$[CH_3(-CH_2-)_n]_3Al + 3\ CH_3\text{-}CH_2\text{-}CH{=}CH_2 \rightarrow$$
$$3\ CH_3\text{-}CH_2\text{-}(CH_2\text{-})_{3n-9}\text{-}CH{=}CH_2 +$$
$$[CH_3(-CH_2-)_3]_3Al$$

The triethylaluminum and 1-butene are recovered by reaction between tributylaluminum and ethylene.

$$[CH_3(-CH_2-)_3]_3Al + 3\ CH_2{=}CH_2 \rightarrow$$
$$(CH_3\text{-}CH_2\text{-})_3Al + 3\ CH_3\text{-}CH_2\text{-}CH{=}CH_2$$

The alpha olefins are used in the production of detergents.

Poly-alpha-olefins, PAO, are used as industrial lubricants, hydraulic fluids, turbine and gear lubricants, multigrade greases, and air compressor lubes.[43] It is reported that they conserve fuel by reducing friction, and that their greater thermal stability gives longer service life.

LITERATURE CITED

1. *Chemical and Engineering News,* Nov. 20, 1978, p. 13.
2. *Chemical Week,* Oct. 3, 1979, pp. 32-34.
3. *Chemical Week,* Sept. 20, 1978, pp. 35-37.
4. Gans, M. and B.J. Ozero. *Hydrocarbon Processing,* Vol. 55, No. 3, 1976, pp. 73-77.
5. *Hydrocarbon Processing, 1977 Petrochemical Handbook,* Vol. 56, No. 11, p. 177.
6. Kiguchi, I., T. Kumazawa and T. Nakai. *Hydrocarbon Processing,* Vol. 55, No. 3, (1976) 69-72.
7. DeMaglie, B., *Hydrocarbon Processing,* Vol. 55, No. 3, 1976, pp. 78-80.
8. Johnson, S.C., *Hydrocarbon Processing,* Vol. 55, No. 6, 1976, pp. 109-113.
9. Greek, B.F. and W.F. Fallwell. *Chemical and Engineering News,* April 26, 1976, pp. 8-9.
10. *The Oil and Gas Journal,* Mar. 28, 1977, pp. 32-33.
11. *Chemical Week,* April 21, 1977, p. 37.
12. *The Oil and Gas Journal,* June 13, 1977, p. 23.
13. German Offen. 2,020,770 (Dec. 17, 1970) to Halcon International.
14. Brownstein, A.M., *Trends in Petrochemical Technology,* Tulsa: Petroleum Publishing Company, 1976, pp. 153-154.
15. *Process Development Digest, CHEM SYSTEMS,* Vol. 1, No. 1, ND.
16. *Chemical Week,* Nov. 28, 1979, p. 34.
17. British Patent 1,182,273 (Feb. 25, 1970) to Teijin.
18. *Petroleum Refiner,* November, 1957, pp. 36, 231.
19. *Hydrocarbon Processing, 1975 Petroleum Handbook,* Vol. 54, No. 11, p. 137.
20. Hatch, Lewis F. and Sami Matar. *Hydrocarbon Processing,* Vol. 56, No. 5, 1977, pp. 191-196.
21. *Chemical Week,* Mar. 29, 1978, pp. 27-28.
22. Keane, D.P., R.B. Stobaugh and P.L. Townsend. *Hydrocarbon Processing,* Vol. 52, No. 2, 1973, pp. 99-110.
23. Reich, P., *Hydrocarbon Processing,* Vol. 55, No. 3, 1976, pp. 85-89.
24. *Chlorine—Its Manufacture, Properties and Uses,* ACS Monograph Series, No. 154, pp. 250-260.
25. Wimer, W.E. and R.E. Feathers, *Hydrocarbon Processing,* Vol. 55, No. 3, 1976, pp. 81-88.
26. *The Oil and Gas Journal,* Feb. 19, 1979, pp. 129, 132.
27. Schillmoller, C.M. *Hydrocarbon Processing,* Vol. 58, No. 3, 1979, pp. 89-93.
28. McPherson, R.W., C.M. Starks and G.F. Fryar. *Hydrocarbon Processing,* Vol. 58, No. 3, 1979, pp. 75-88.
29. *Hydrocarbon Processing, Petrochemical Handbook,* November, 1975, p. 134.
30. *Chemical Week,* Jan. 12, 1977, pp. 26-28.
31. Anderson, E.V., *Chemical and Engineering News,* Jan. 10, 1977, pp. 12-13.
32. Winter, O. and M.T. Eng. *Hydrocarbon Processing,* Vol. 55, No. 11, 1976, pp. 125-133.
33. Jira, R., W. Blau and D. Grimm. *Hydrocarbon Processing,* Vol. 55, No. 3, 1976, pp. 97-100.
34. Hatch, Lewis F. and Sami Matar. *Hydrocarbon Processing,* Vol. 56, No. 11, 1977, pp. 349-357.
35. Lowry, R.P. and A. Aquillo. *Hydrocarbon Processing,* Vol. 53, No. 11, 1974, pp. 103-113.
36. Hatch, Lewis F. and Sami Matar. *Hydrocarbon Processing,* Vol. 56, No. 10, 1977, pp. 153-163.
37. *Hydrocarbon Processing, 1977 Petrochemical Handbook,* Vol. 56, No. 11, p. 235; *The Oil and Gas Journal,* March 12, 1973, p. 91.
38. *Chemical and Engineering News,* Oct. 30, 1978, p. 13.
39. *Chemical Week,* Technology Newsletter, Oct. 15, 1975, p. 38.
40. *Chemical and Engineering News,* Oct. 13, 1975, p. 6.
41. *Hydrocarbon Processing, 1975 Petrochemical Handbook,* Vol. 54, No. 11, p. 110.
42. *Chemical Week,* June 1, 1977, pp. 35-36.
43. *Chemical Week,* Mar. 1, 1978, p. 9.
44. *Chemical and Engineering News,* May 28, 1979, pp. 12-13.
45. Kilty, P.A. West German Offen 24,48,449 (Apr. 30, 1975) and Nielson, R.P. U.S. Patent 3,962,136, both to Shell; Winnick, C. U.S. Patent 4,066,575 to Halcon; Bel. Patent 867,045 to BASF.
46. *CHEMTECH,* Vol. 9, No. 8, 1979, p. 465.
47. Kuhn, Wayne. *Hydrocarbon Processing,* Vol. 57, No. 10, 1979, pp. 123-128.
48. Weismantel, G.F. *Chemical Engineering,* Jan. 15, 1979, p. 67.
49. Fryar, G.J. *Hydrocarbon Processing,* Vol. 57, No. 5, 1979, p. 101.
50. *Chemical Week,* Feb. 13, 1980, pp. 26-27.
51. *Chemical and Engineering News,* Oct. 29, 1979, p. 12.
52. *Chemical and Engineering News,* Dec. 3, 1979, pp. 6-7.

9

Petrochemicals from Propylene and Higher Olefins

Propylene, $CH_3 - CH = CH_2$, often referred to as "the crown prince of petrochemicals," is superficially similar to ethylene, but is has many differences both in production and uses. Some of these are obvious and others are subtle. It has been noted that "propylene lies in the shadow world dominated by gasoline and ethylene markets and manufacturing economics, both of which keep changing."[1]

Propylene is always a byproduct, never a main product. About two-thirds of the propylene intended for chemical markets, chemical propylene, comes as a byproduct of refinery operations and one-third from steam cracking of ethane and naphtha for ethylene production. As the production gradually shifts from ethane to steam cracking of naphtha and higher hydrocarbons, nonrefinery byproduct propylene is favored.

Chemical grade propylene is in competition with refinery utilization of propylene as a fuel (**LP**-gas) and to produce alkylate and polymer gasoline components. Currently, about 50% of the total propylene supply is utilized by these divergent uses in the refinery.

Petrochemical demand for propylene is about one-half the demand for ethylene. This is somewhat surprising because the added complexity of the propylene molecule should permit a wider selection of end-products and markets. But this very difference can lead to the production of undesirable byproducts, and it frequently does. An appreciable amount of industrial research effort goes into developing selective catalysts and operating conditions for propylene.

The propylene utilization pattern for the estimated 15.5 billion pound demand for chemical propylene in 1980 is illustrated in Fig. 9-1.[2] Polypropylene will account for only 29 percent of the propylene demand, while the 1980 polyethylene production is estimated to be 45 percent of the propylene demand. Polypropylene production and utilization are discussed in Chapter 12. The overall petrochemical demand for propylene will grow about 6.2 percent per year through 1985.[2] Propylene capacity will be 24-28 billion pounds in 1980.

ACRYLONITRILE

Production. *Acrylonitrile,* AN, $CH_2 = CH - CN$, is produced by the direct ammoxidation, oxidative amination, of propylene.

$$2\,CH_2 = CH - CH_3 + 2\,NH_3 + 3\,O_2 \rightarrow 2\,CH_2 = CH - CN + 6\,H_2O \qquad \Delta H_{25^\circ C} = -123 \text{ kcal}$$

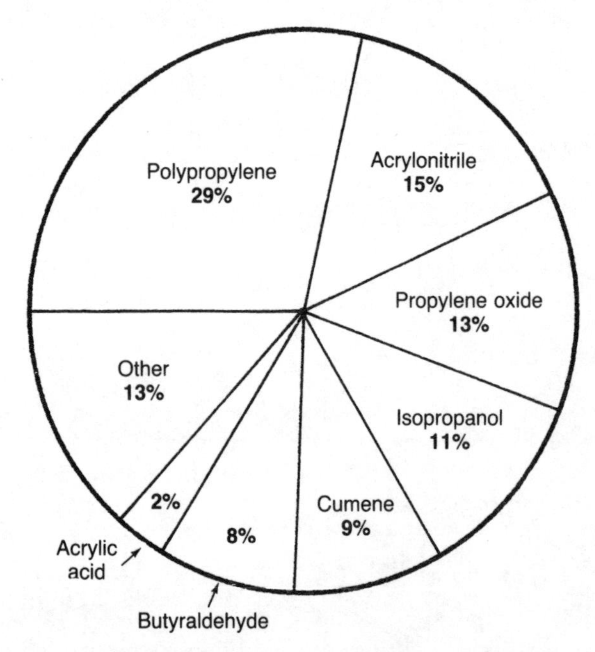

Fig. 9-1—The predicted use pattern of the 15.5 billion pound propylene consumption in 1980.[2]

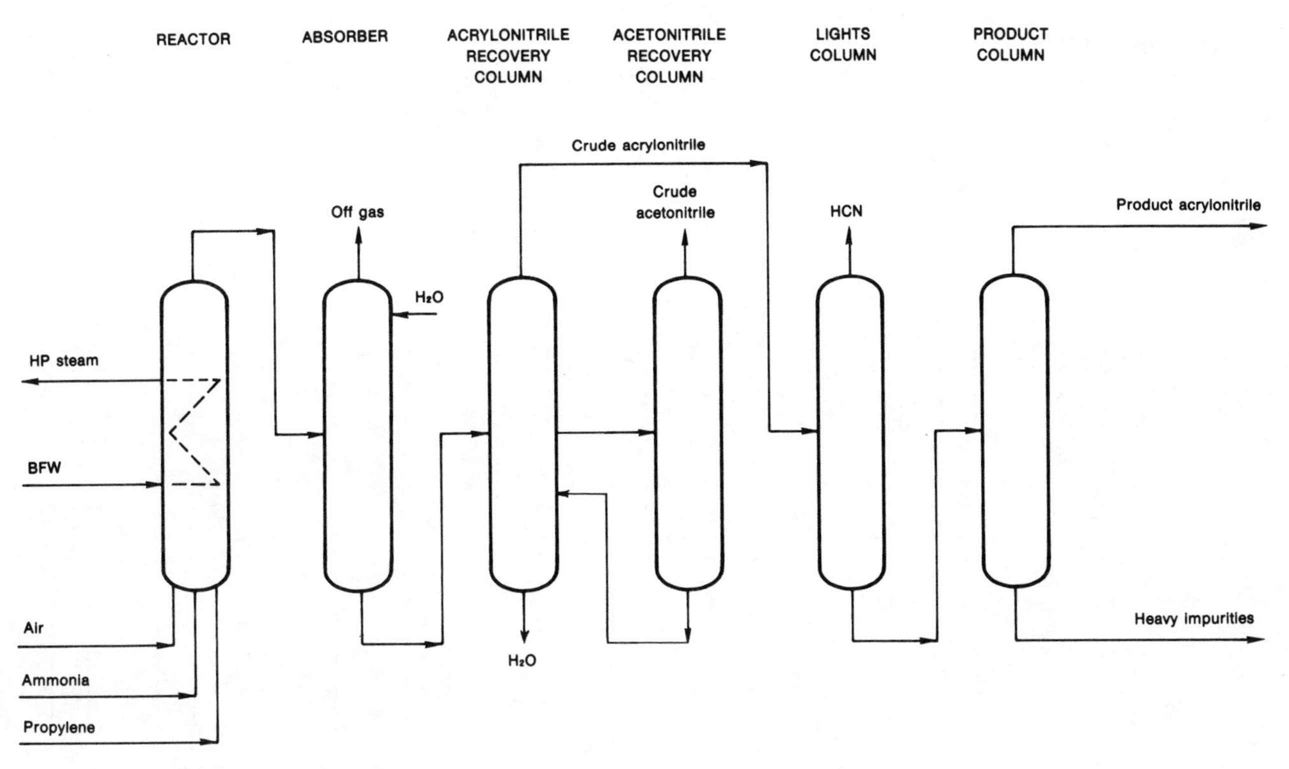

Fig. 9-2—The Sohio process for the production of acrylonitrile by ammoxidation of propylene.[3]

A typical reactor and recovery system are shown in Fig. 9-2.[3] The reaction conditions are:

Temperature	400-500° C	
Pressure	5-30 psig	
Catalyst	"Catalyst 41"	
Residence time	2-25 sec	
Reactor	Fluid bed	
Conversion	"High on a once through basis"	
Yield: on 1 lb. on propylene	Acrylonitrile	0.73 lb.
	Acetonitrile	0.11 lb.
	Hydrogen cyanide	0.13 lb.

Both *acetonitrile,* CH_3 - CN, and *hydrogen cyanide,* HCN, may be recovered for sale.

The heart of the process is the ammoxidation catalyst and the reactor. Several new catalysts have been developed. The Montedison-UOP fluidized bed process uses a "new, extremely active, high-performance" catalyst. This catalyst is reported to give a propylene conversion of over 95% and acrylonitrile selectivity in excess of 80%.[4, 5] Still another new catalyst has been developed by Nitto Chemical Industry Co. (Tokyo).[6] This catalyst may be destined to replace Catalyst 41, at least Sohio has acquired licensing right for this new catalyst. An extensive review of catalysts and the kinetics and mechanism of the ammoxidation of propylene has been made by Hucknall.[7]

Both fluid bed catalyst and fixed-catalyst bed reactors are used. The British Petroleum[8] and the Snamprogetti[9] processes use fixed-catalyst bed reactors. Currently, all modern acrylonitrile processes use fluidized bed reactors.[4]

The ammoxidation of propylene has been reviewed by Dumas and Bulani with special emphasis on reaction variables.[10]

Uses. The major uses of acrylonitrile are in the production of plastics and resins. There are, however, a few interesting uses which depend upon simple chemical reactions.

Acrylonitrile is converted to *acrylamide,* $CH_2 = CH - \overset{\overset{\displaystyle O}{\|}}{C} - NH_2$, which is used primarily in water treatment. *Polyacrylamide* is used in tertiary oil recovery. It has a potential use in drilling for crude oil and natural gas.[11] Acrylonitrile is also used in the synthesis of *lysine,* $H_2N - (CH_2)_4 - CHNH_2 - \overset{\overset{\displaystyle O}{\|}}{C} - OH$, an essential amino acid.

The reaction with the most potential is the production of *adiponitrile,* **ADN,** NC - $(CH_2)_4$ - CN, by direct electrohydrodimerization of acrylonitrile.[12, 13, 14] The following is a representation of the electrochemistry involved and Fig. 9-3 is a flow diagram of a typical process.[12]

Cation exchange membrane

Anode				Cathode
	$H_2O - 2_e^-$	$H^+ \longrightarrow$	$2CH_2 = CHCN$	
	$\downarrow$		$+ 2H^+ + 2_e^-$ $\downarrow$	
	$\frac{1}{2}O_2 + 2H^+$	$H^+ \longrightarrow$	$NC(CH_2)_4CN$	
	Anolyte		Catholyte	

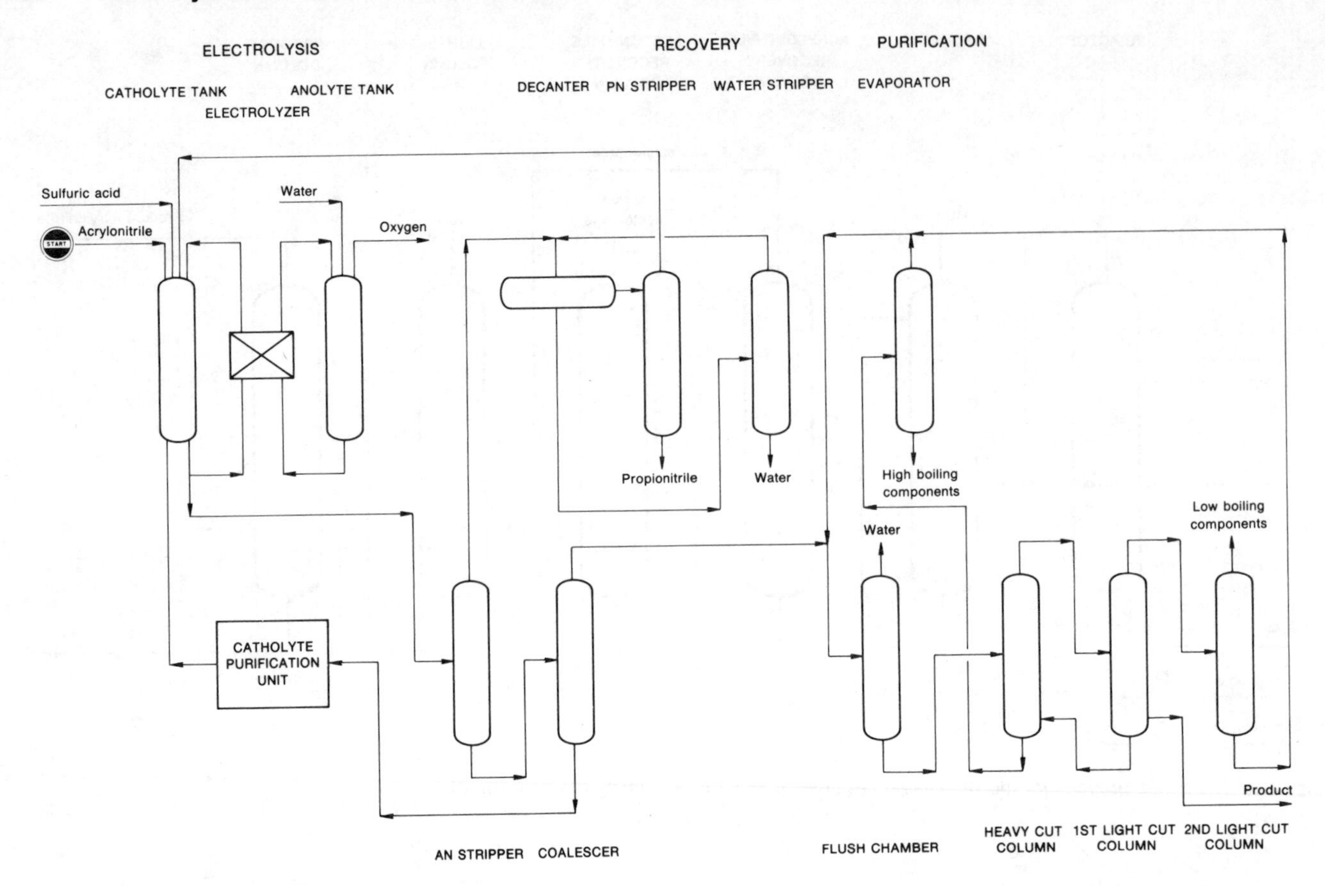

Fig. 9-3—Flow diagram of the Asahi Chemical Industry Co., Ltd., process for producing adiponitrile from acrylonitrile by direct electro-hydrodimerization.[12]

The electro-hydrodimerization takes place on the cathode surface. *Propionitrile,* $CH_3 - CH_2 - CN$, is a byproduct. Typical yields (wt. %) are

Adiponitrile	85 - 90
Propionitrile	4 - 5
Heavy products	11 - 5

A one compartment electrolyzer without a diaphragm is also used.[13]

Monsanto Co. received the Kirkpatrick Chemical Engineering Achievement Award in 1965 for its development of the hydrodimerization of acrylonitrile.

Adiponitrile is the precursor for nylon 6,6. The 1978 production of acrylonitrile was 1.82 billion pounds.

PROPYLENE OXIDE

While *propylene oxide, 1,2-epoxypropane,* $CH_3-\overset{O}{\overbrace{CH - CH_2}}$, is similar in structure to ethylene oxide, it is markedly different in both production and uses.

Production. There are two major processes for the production of propylene oxide. The older and still dominant process is chlorohydrination of propylene followed by epoxidation of the chlorohydrin by calcium hydroxide. The more recent process is oxidation by use of an organic peroxide—the Oxirane process. In 1979 this peroxidation process accounted for 45 percent of the total United States propylene oxide capacity. Both of these processes have been reviewed by Stobaugh, *et al.*[15] More recently, Simmrock has made a definitive review of all current routes to propylene oxide.[16] Landau, Sullivan, and Brown have reviewed the production of propylene oxide, with special emphasis on the economic role played by coproduct production.[78] The most important routes to propylene oxide are shown in Fig. 9-4.[16]

All efforts to effect direct catalytic vapor-phase epoxidation of propylene to propylene oxide have met with economic failure.

Propylene Chlorohydrin Process. The chlorohydrination process consists of the formation of *propylene chlorohydrin,* $CH_3 - CHOH - CH_2Cl$, by the reaction between hypochlorous acid, HOCl, and propylene. The hypochlorous acid is formed by the reaction between chlorine and water.

$$Cl_2 + H_2O \rightarrow HOCl + HCl$$

$$CH_3 - CH = CH_2 + HOCl \rightarrow CH_3 - CHOH - CH_2Cl$$

Typical reaction conditions are

Temperature	35°C	
Pressure	Normal	
Catalyst	None	
Yields	Propylene chlorohydrin	87 - 90%
	Propylene dichloride	9 - 6%.

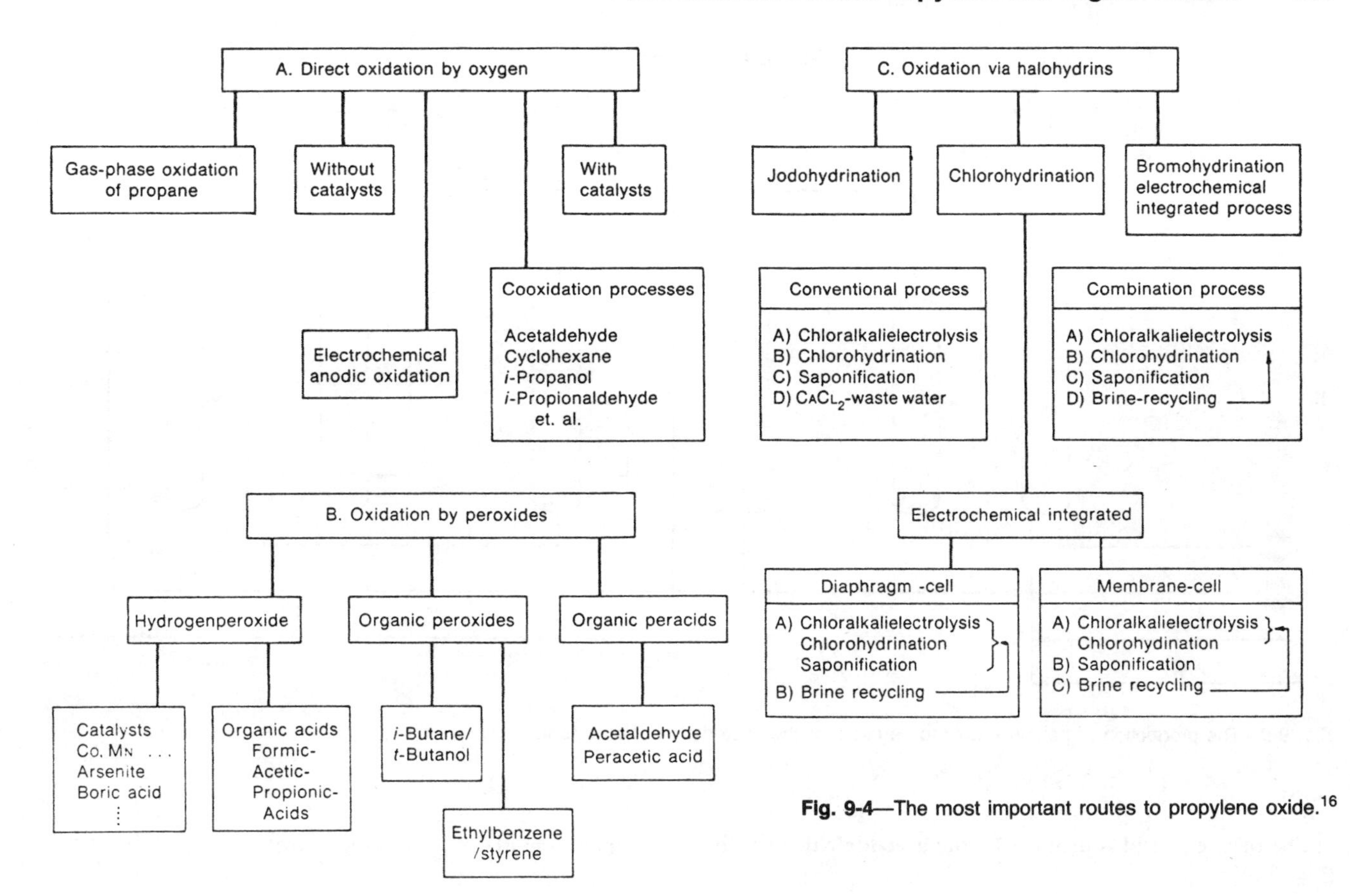

Fig. 9-4—The most important routes to propylene oxide.[16]

Approximately 0.1 pound of dichloride is formed for each pound of chlorohydrin.

The propylene chlorohydrin is epoxidized to propylene oxide by a 10% solution of milk of lime, $Ca(OH)_2$.

$$2\,CH_3\text{-}CHOH\text{-}CH_2Cl + Ca(OH)_2 \rightarrow 2\,CH_3\text{-}\overset{O}{\widehat{CH\text{-}CH_2}} + CaCl_2 + 2\,H_2O$$

The propylene oxide is removed by stripping with live steam. The yield on chlorohydrin is 95%. Fig. 9-5 represents a typical chlorohydrin process for the production of propylene oxide.[15]

There are two disadvantages of the chlorohydrin process: The chlorine is costly and it ends up as very weak calcium chloride solution. It is also necessary to dispose of 0.1 to 0.15 pounds of propylene dichloride per pound of propylene oxide.

Epoxidation by Peroxides. The production of propylene oxide by organic peroxides is also a two-step process but it differs from the chlorohydrination process in several important respects. The main difference is that the coproducts are compounds of appreciable economic value. The coproducts also may be produced in greater quantity, by weight, than the propylene oxide. They may even produce more revenue than propylene oxide.

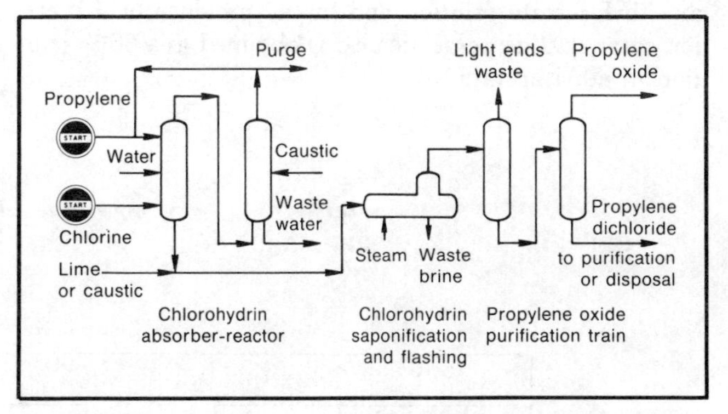

Fig. 9-5—Flow diagram of a typical chlorohydrin process for the production of propylene oxide.[15]

The chemistry involved may be illustrated with the epoxidation of propylene by *peracetic acid*, $CH_3\text{-}\overset{O}{\overset{\|}{C}}\text{-}OOH$, which is a peroxide.

$$CH_3\text{-}CH{=}CH_2 + CH_3\text{-}\overset{O}{\overset{\|}{C}}\text{-}OOH \rightarrow CH_3\text{-}\overset{O}{\widehat{CH\text{-}CH_2}} + CH_3\text{-}\overset{O}{\overset{\|}{C}}\text{-}OH \qquad \Delta H = -60\text{ kcal}$$

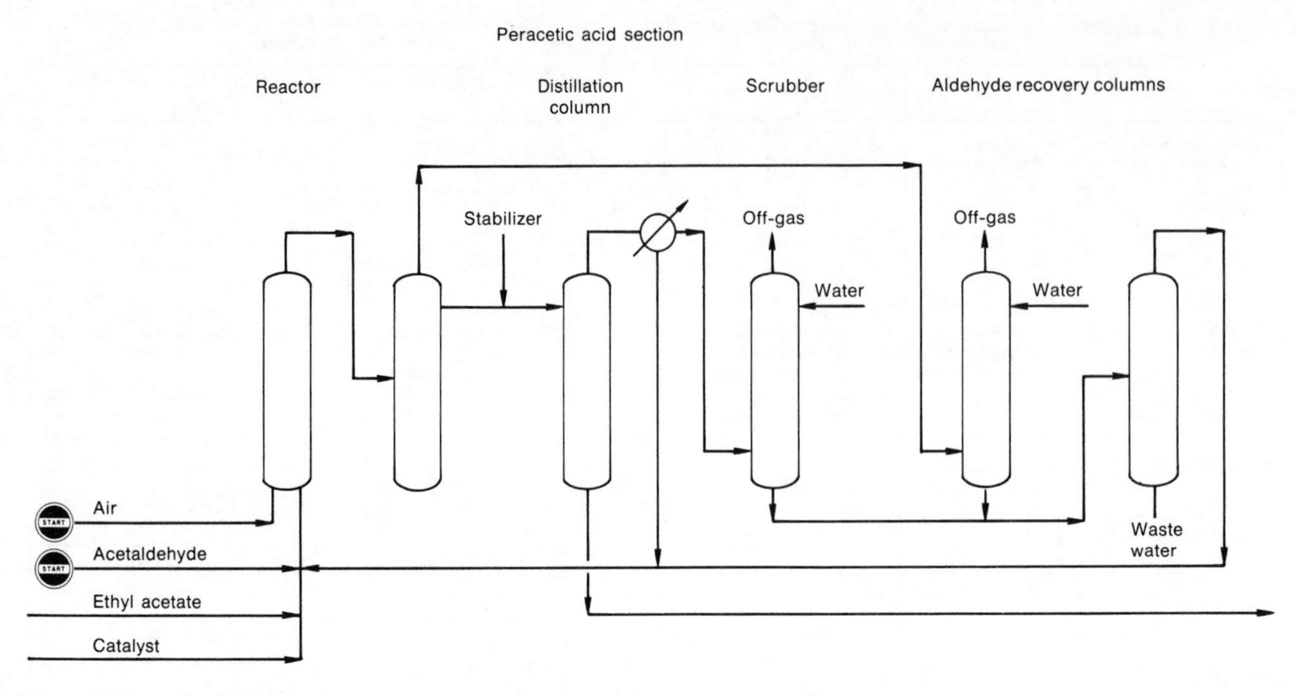

Fig. 9-6—The production of peracetic acid to be used for the epoxidation of propylene.[17]

The peracetic acid is produced from acetaldehyde and air (Fig. 9-6).[17]

$$CH_3 - \overset{O}{\overset{||}{C}} - H + O_2 \rightarrow CH_3 - \overset{O}{\overset{||}{C}} - OOH$$

The oxidation is carried out at 30-50° C and 25-40 atm in an ethyl acetate solution and in the presence of a metal ion catalyst. The peracetic acid is obtained as a 30% solution in ethyl acetate.

The epoxidation conditions are

Temperature	50-80° C
Pressure	130-180 psig
Catalyst	None
Conversion	97-98%
Yield	90-92%.

Fig. 9-7 is a flow diagram of the Daicel process.[18]

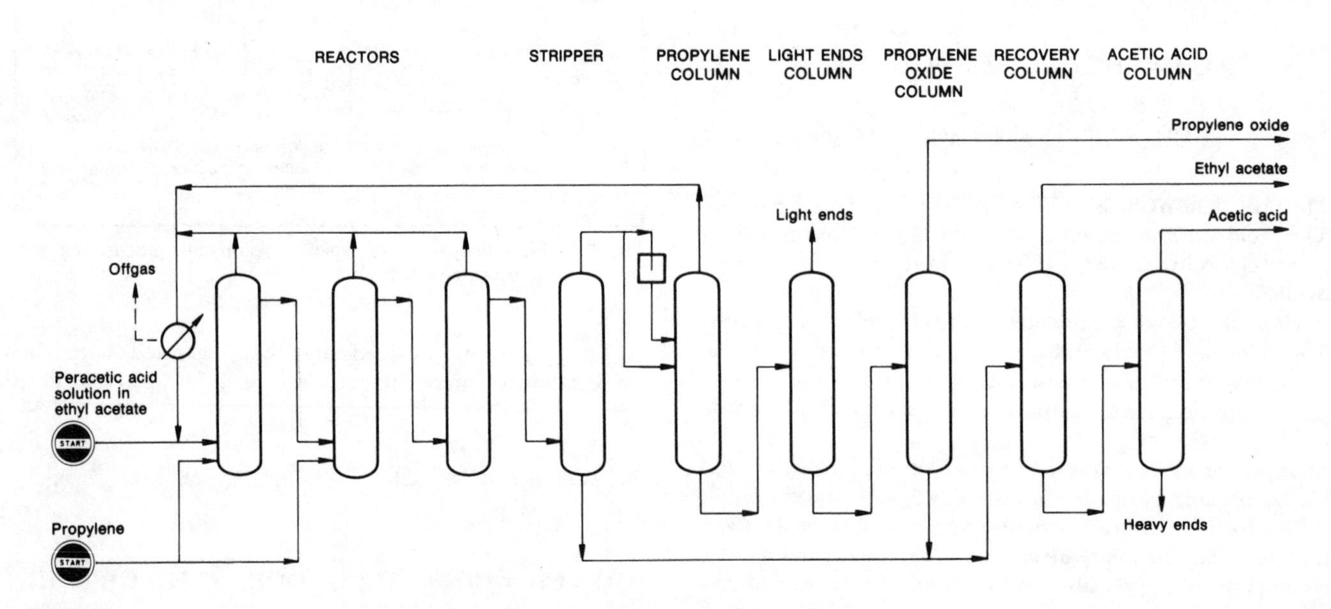

Fig. 9-7—Daicel's direct propylene oxide process by way of peracetic acid epoxidation.[18]

TABLE 9-1—Peroxides Actually or Potentially Used to Epoxidize Propylene[15]

Peroxide Feedstock	Epoxidation Coproduct	Coproduct Derivative
Acetaldehyde	Acetic acid	....
Isobutane	*tert*-Butyl alcohol	Isobutylene
Ethylbenzene	α-Phenylethyl alcohol	Styrene
Isopentane	Isopentanol	Isopentene and isoprene
Isopropanol	Acetone	Isopropanol

Other actual and potential industrial processes use different peroxide feedstocks. These are listed in Table 9-1 along with the epoxidation coproducts and their derivatives.[15] The *Oxirane process* uses either *isobutane hydroperoxide* or *ethylbenzene hydroperoxide*.

$$(CH_3)_3C\text{-}OOH \qquad C_6H_5\text{-}CH(OOH)\text{-}CH_3$$

Which feedstock is used is determined by availability and cost plus the demand for the coproduct derivative—isobutylene or styrene. The hydroperoxides are formed by air oxidation of the feedstock at 125-150° C and 50 psig.[19]

General epoxidation conditions using ethylbenzene hydroperoxide, **EBHP,** are[20]

Temperature	80-130° C
Pressure	550 psig
Catalyst	Mo compounds
Conversion: on propylene	22%
on **EBHP**	98%
Selectivity: on propylene	95+%
on **EBHP**	80-90%

Catalysts used for the epoxidation are coordination compounds of molybdenum hexacarbonyl and molybdenum oxyacetylacetonate.[15, 19] Other epoxidation catalysts include transition metal ions such as Mo, W, Cr, and V.

Other Methods. *Perisobutyric acid,* $CH_3\text{-}CH(CH_3)\text{-}C(=O)\text{-}OOH$, obtained from isobutyraldehyde, has been proposed as an epoxidizing agent.[21] The coproduct isobutyric acid is esterified and dehydrogenated to methyl methacrylate, **MMA.** In another process hydrogen peroxide is used to form a peracid *in situ*. The peracid then epoxidizes the propylene. The coproduct organic acid is recycled to be peroxidized again.[22] The net reaction is

$$CH_3\text{-}CH=CH_2 + H_2O_2 \rightarrow CH_3\text{-}\overset{O}{\overbrace{CH\text{-}CH_2}} + H_2O$$

Hydrogen peroxide in the presence of arsenic compounds in dioxane at 90° C will react directly with propylene to form propylene oxide.[23] Conversion and efficiencies are reported to be 85+%. A similar arsenic catalyst, in the presence of tetracyanoethylene, $(NC)_2C=C(CN)_2$, and ethyl acetate, uses oxygen directly to epoxidize propylene.[24]

$$2\,CH_3\text{-}CH=CH_2 + O_2 \rightarrow 2\,CH_3\text{-}\overset{O}{\overbrace{CH\text{-}CH_2}}$$

The reaction conditions are

Temperature	150° C
Pressure	850 psi
Catalyst	As - cat. - $CH_3\text{-}C(=O)\text{-}OCH_2\text{-}CH_3$
Conversion	48%
Efficiency	52%.

An interesting two-step route to propylene oxide has been described in the patent literature. In the first step, propylene is oxidized in the liquid phase, in acetic acid, to the diol monoacetates.[25]

$$2CH_3\text{-}CH=CH_2 + 2CH_3\text{-}C(=O)\text{-}OH + O_2 \rightarrow$$

$$CH_3\text{-}CH(OH)\text{-}CH_2\text{-}OC(=O)\text{-}CH_3 + CH_3\text{-}CH(OC(=O)\text{-}CH_3)\text{-}CH_2OH$$

The reaction conditions are

Temperature	65° C
Pressure	5 atm
Catalyst	$PdCl_2/LiNO_3$
Conversion	72%
Selectivity	74%.

The mixed acetates are then pyrolyzed to propylene oxide and acetic acid.[26]

$$CH_3\text{-}CH(OH)\text{-}CH_2\text{-}OC(=O)\text{-}CH_3 + CH_3\text{-}CH(OC(=O)\text{-}CH_3)\text{-}CH_2OH \rightarrow$$

$$2CH_3\text{-}\overset{O}{\overbrace{CH\text{-}CH_2}} + 2CH_3\text{-}C(=O)\text{-}OH$$

The acetic acid is recycled.

The pyrolysis conditions are

Temperature	400 °C	
Pressure	1 atm	
Catalyst	$CH_3\text{-}C(=O)\text{-}OK$/alundum	
Conversion	31%	
Selectivity:	propylene oxide	77%
	propionaldehyde	15%
	acetone	8%.

The byproduct propionaldehyde and acetone places a minor economic burden on the process.

Uses. The major uses of propylene oxide are in the production of flexible foams (48%) and propylene glycol (25%). Propylene oxide had a 2 billion pound production in 1978 (Fig. 9-8).[79] The remaining propylene oxide went to rigid foams, non-foams, dipropylene glycol, polypropylene glycol and isopropylamines.

When producing unsaturated polyester resins, substitution by propylene oxide for 75 percent, by weight, of the propy-

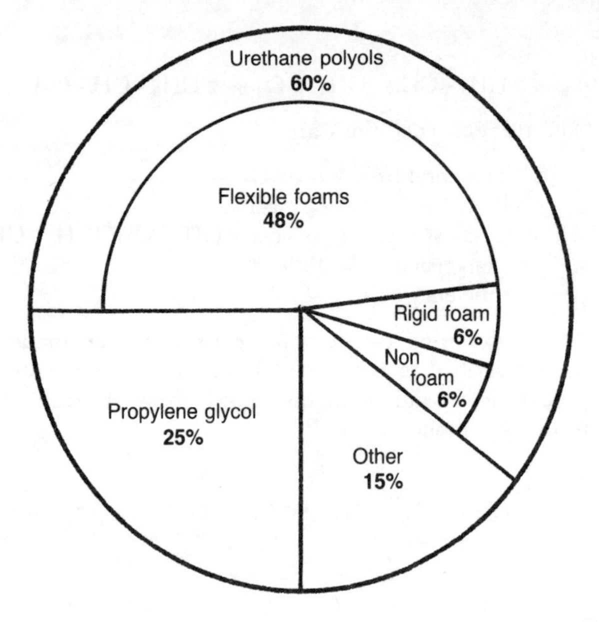

Fig. 9-8—1978 U.S. propylene oxide demand of 2 billion pounds.[79]

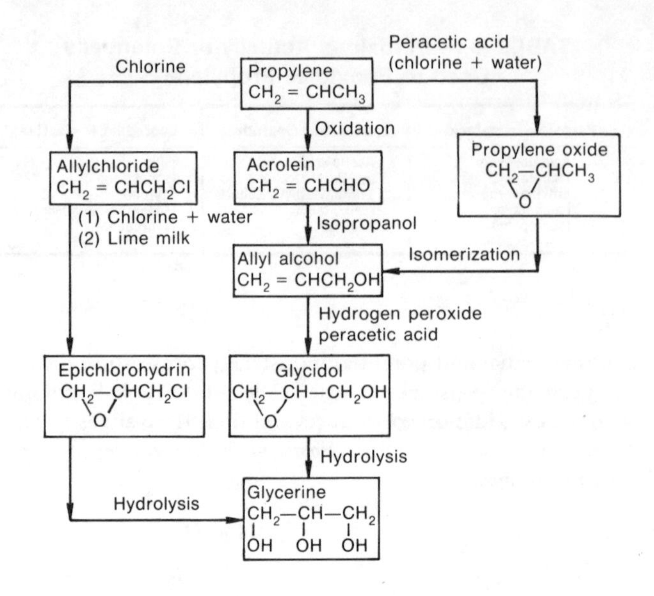

Fig. 9-9—The various commercial routes from propylene to glycerol (glyerine).[28]

lene glycol normally used results in savings in raw material, utilities and investment costs.[27]

Propylene glycol, CH_3 - CHOH - CH_2OH, is produced by the hydration of propylene oxide in a manner similar to the production of ethylene glycol from ethylene oxide.[27] The glycol is used as an intermediate for unsaturated polyester resins and for the softening of cellophane. Other uses include tobacco humectants, in cosmetics, brake fluids, as plasticizers, and as additives in such varied things as pet food, inks and soft drink syrups. *Dipropylene glycol* is consumed in the production of unsaturated polyester resins and plasticizers.

The *isopropylamines* are produced and utilized in much the same way as the ethanolamines.[27]

Propylene carbonate, $CH_3 - \underset{|}{CH} - \underset{|}{CH_2}$ (ring: O - C(=O) - O bridging the CH and CH_2), is produced by the reaction between propylene oxide and carbon dioxide at 200° C and 80 atm in a 95% yield. It is used as a specialty solvent. For example, it is used in the Fluor process for removing acid constituents from natural gas.

Allyl alcohol, $CH_2=CH$ - CH_2OH, is produced by the isomerization of propylene oxide. This reaction is one of the steps in a process for producing *glycerol,* CH_2OH - CHOH - CH_2OH, from propylene (Fig. 9-9).[28]

The reaction is

$$CH_3\overset{O}{\widehat{CH - CH_2}} \rightarrow CH_2 = CH - CH_2OH$$

The reaction conditions are

Temperature	280°C
Catalyst	Li_3PO_4
Vapor phase	
Conversion	20 - 30%
Selectivity	94-98%.

The allyl alcohol is epoxidized to *glycidol,* $\overset{O}{\widehat{CH_2 - CH}}$ - CH_2OH, which is then hydrolyzed to glycerol (glycerine).[28]

Fig. 9-10 shows the relationships between propylene oxide supply and demand from 1978 through 1983.[79]

ISOPROPANOL

Isopropanol, CH_3 - CHOH - CH_3, is a compound with several names. Isopropanol is the name used in industry, *2-Propanol* is the official name (*IUPAC* name). *Isopropyl alcohol,* **IPA,** is the common as well as the historical name. Isopropanol, under the name "isopropyl alcohol," was the first industrial chemical synthesized from a petroleum derived olefin (1920).

Production. There are two industrial processes for the production of isopropanol. The older method, and one still used,

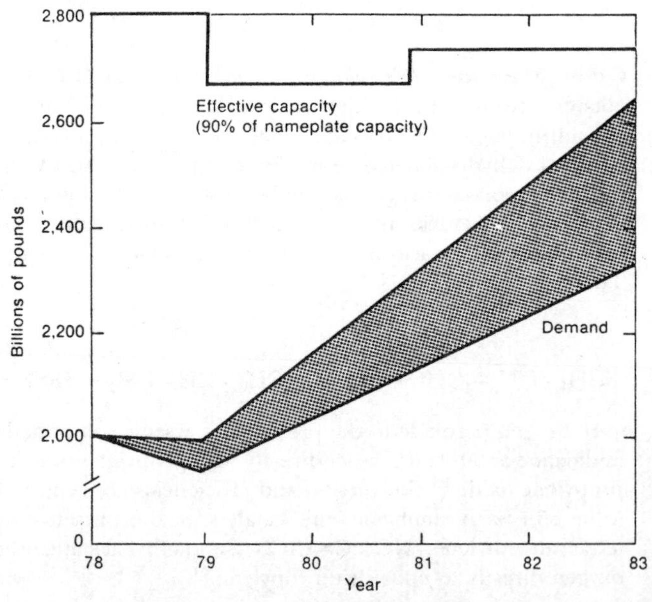

Fig. 9-10—U.S. propylene oxide supply/demand.[79]

involves the sulfation of propylene followed by hydrolysis of the propylene sulfates. The current process of choice is direct hydration of propylene to isopropanol.

Sulfation-Hydrolysis Process. The reactions are

$$3\ CH_3\text{-}CH{=}CH_2 + 2\ H_2SO_4 \rightarrow$$

$$CH_3\text{-}\underset{}{\overset{CH_3}{\overset{|}{C}}}H\text{-}OSO_3H + CH_3\text{-}\overset{CH_3}{\overset{|}{C}}H\text{-}OSO_2O\text{-}\overset{CH_3}{\overset{|}{C}}H\text{-}CH_3$$

Both isopropyl hydrogen sulfate and diisopropyl sulfate are formed.

Typical reaction conditions are (Fig. 9-11)[29]

Temperature	60° C
Pressure	300-400 psi
Catalyst	None (75% H_2SO_4)
Yield (after hydrolysis)	93-95%

Diisopropyl ether, $CH_3\text{-}\overset{CH_3}{\overset{|}{C}}H\text{-}O\text{-}\overset{CH_3}{\overset{|}{C}}H\text{-}CH_3$, is the only byproduct (*ca* 5%).

The primary variables are temperature and acid concentration. The interrelationship between these two is shown in Table 9-2.[30] The higher the temperature, the lower the acid concentration. The relationship between olefin structure and these variables is also indicated in Table 9-2.

Hydrolysis is effected by diluting the acid of the reaction mixture to under 40%

TABLE 9-2—Acid Concentrations and Temperatures for the Sulfation of Various Olefins[30]

Olefins	Formula	Acid Conc. Range, %	Temperature Range °C
Ethylene	$CH_2{=}CH_2$	90—98	60—80
Propylene	$CH_3—CH{=}CH_2$	75—85	25—40
Butylenes	$CH_3—CH_2—CH{=}CH_2$	75—85	15—30
	$CH_3—CH{=}CH—CH_3$	75—85	15—30
Isobutylene	$CH_3—\overset{CH_3}{\overset{|}{C}}{=}CH_2$	50—65	0—25

$$CH_3\text{-}\overset{CH_3}{\overset{|}{C}}H\text{-}OSO_3H + CH_3\text{-}\overset{CH_3}{\overset{|}{C}}H\text{-}OSO_2\text{-}\overset{CH_3}{\overset{|}{C}}H\text{-}CH_3 +$$

$$3\ H_2O \rightarrow 3\ CH_3\text{-}CHOH\text{-}CH_3 + 2\ H_2SO_4$$

The main disadvantage of the sulfation process is the necessary reconcentration of the sulfuric acid after the hydrolysis step. An advantage is that a propylene concentration as low as 65% can be used as feedstock.

Direct Hydration. The newer isopropanol plants utilize direct catalytic hydration.

$$CH_3\text{-}CH{=}CH_2 + H_2O \rightarrow CH_3\text{-}CHOH\text{-}CH_3$$
$$\Delta H = -12.3\ kcal$$

Several processes are used to effect this reaction. A liquid phase process is illustrated in Fig. 9-12.[31] The reaction conditions are

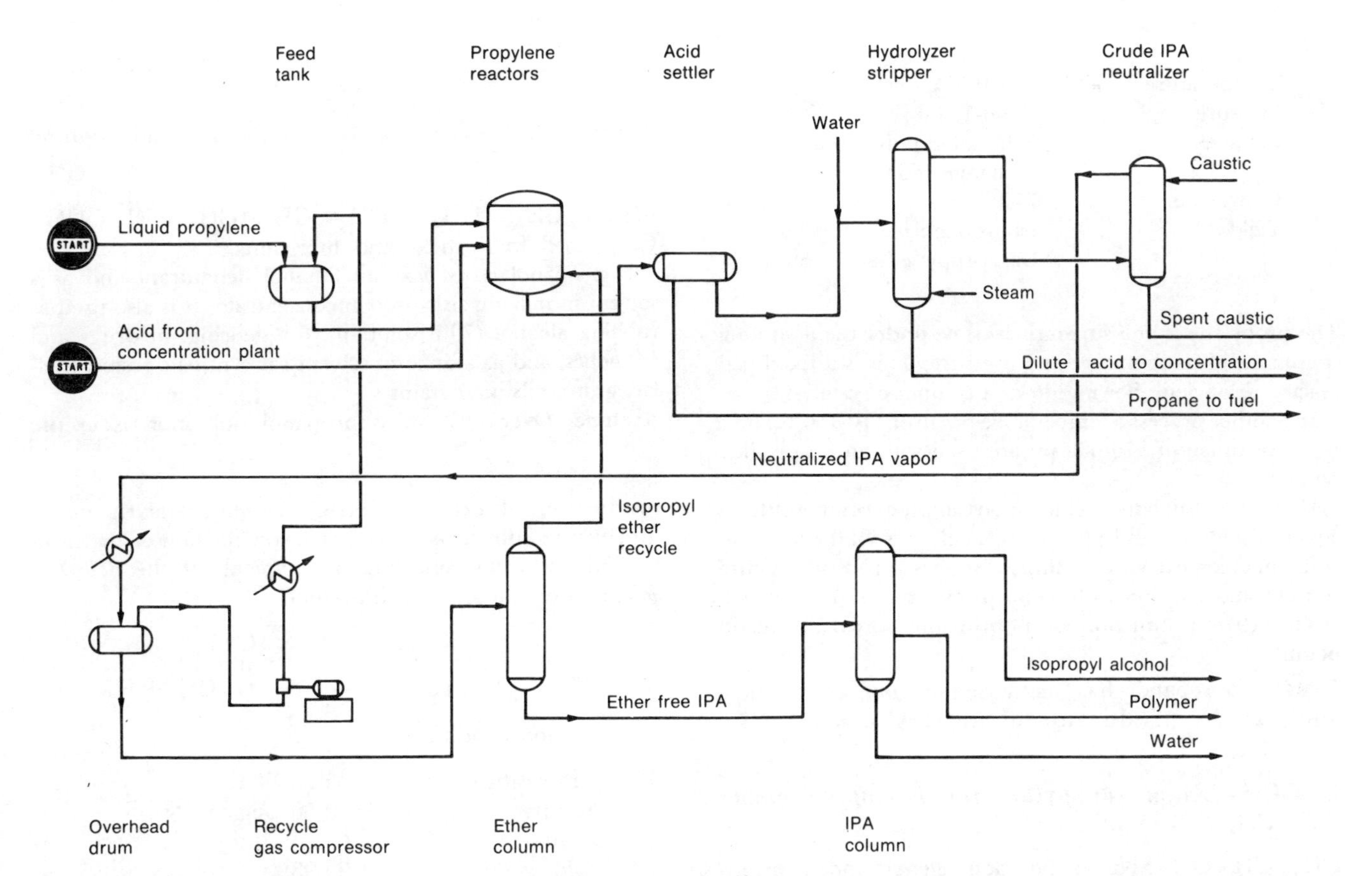

Fig. 9-11—The BP Chemicals Ltd. process for the production of isopropanol.[29]

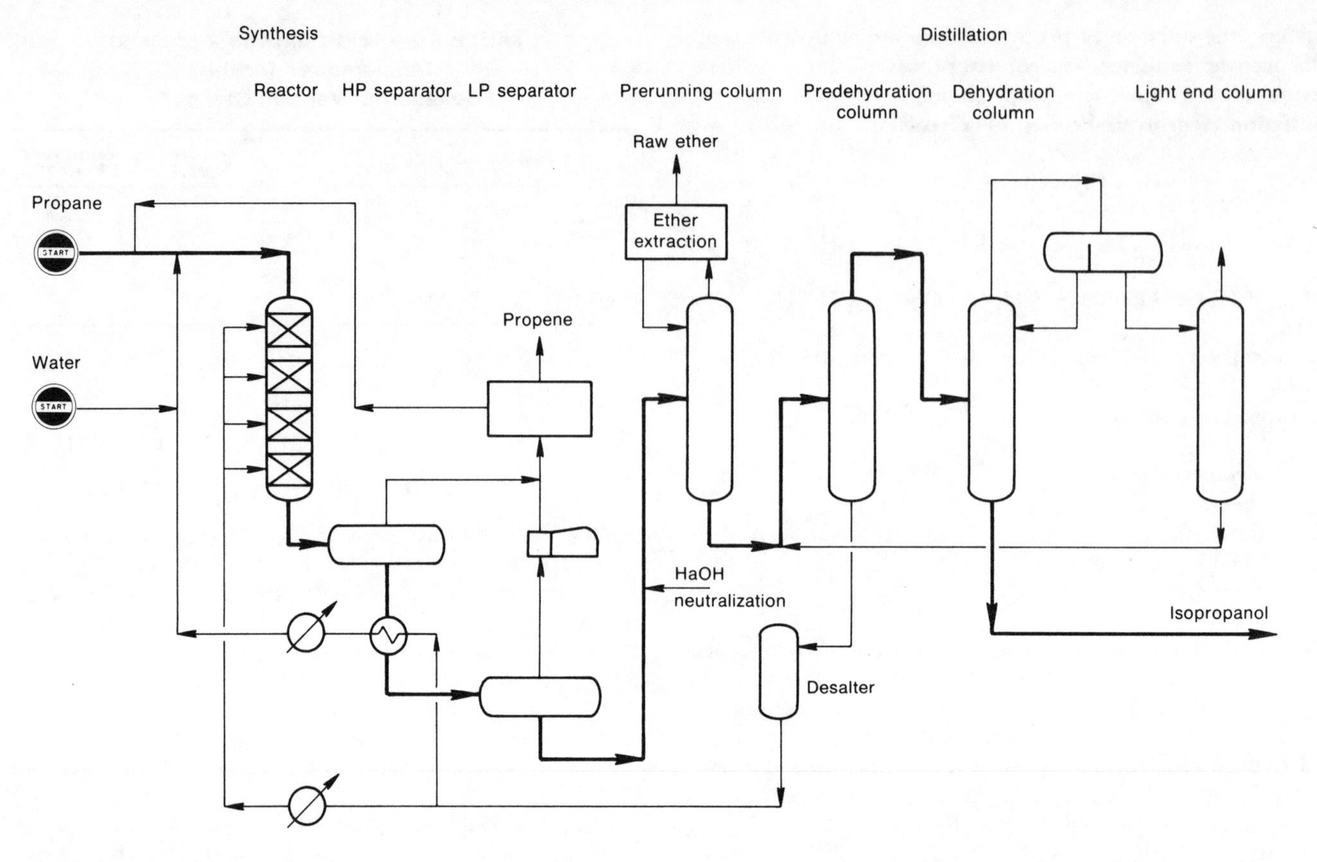

Fig. 9-12—Liquid phase hydration of propylene to isopropanol over an ion exchange catalyst.[31]

Temperature	130-160° C	
Pressure	1,200-1,500 psi	
Catalyst	Sulfonated polystyrene cation exchange resin	
Conversion	75%	
Yield	Isopropanol	93.5%
	Diisopropyl ether	5.0%.

The propylene is in a supercritical state under the operating conditions of temperature and pressure. This is a fixed bed reactor, propylene flowing down a column of catalyst.

In another process a temperature range of 170-190°C and a pressure range of 370-660 psi are used with an unspecified catalyst.

Hydration catalysts include polytungsten compounds in aqueous solutions and phosphoric acid on a solid carrier, in addition to ion exchange resins. All of these processes can use concentrations as low as 65% propylene and 35% propane.

The 1978 production of isopropanol was 1.71 billion pounds.

Uses. Isopropanol has many minor uses in synthesis such as the production of *isopropyl acetate,* $CH_3-\overset{O}{\overset{||}{C}}-O\overset{CH_3}{\overset{|}{C}}H-CH_3$, *isopropylamines, isopropyl xanthate,* $CH_3-\overset{CH_3}{\overset{|}{C}}H-O\overset{S}{\overset{||}{C}}-SNa$, (a flotation agent) and *isopropyl myristate,* $CH_3-(CH_2)_{10}-\overset{O}{\overset{||}{C}}-O\overset{CH_3}{\overset{|}{C}}H-CH_3$, and *isopropyl oleate,* $CH_3-(CH_2)_7-CH=CH-(CH_2)_7-\overset{O}{\overset{||}{C}}-O\overset{CH_3}{\overset{|}{C}}H-CH_3$, used in lipsticks and lubricants.

Isopropanol is used as an ethanol denaturant and as a solvent in making fish protein concentrate. It is also used as rubbing alcohol (70% solution), for de-icing, in drugs and cosmetics, and as a general solvent for synthetic resins, shellac, gums, oils, and stains.

Acetone. Over 50% of isopropanol utilization is in the production of *acetone,* $CH_3-\overset{O}{\overset{||}{C}}-CH_3$ by catalytic dehydrogenation, direct oxidation, or a combination of the two.

The following conditions are typical for the *dehydrogenation* of isopropanol to acetone.

$$CH_3-CHOH-CH_3 \rightarrow CH_3-\overset{O}{\overset{||}{C}}-CH_3 + H_2$$

The reaction conditions are

Temperature	450-550° C
Pressure	40-50 psig
Catalyst	Cu
Yield	95-98%.

Both brass and zinc oxide, ZnO, are also used as catalysts for this reaction. A temperature of 380°C is used with zinc oxide.

Direct oxidation with oxygen yields hydrogen peroxide with acetone as a coproduct.

$$CH_3-CHOH-CH_3 + O_2 \rightarrow H_2O_2 + CH_3-\overset{\overset{\large O}{\|}}{C}-CH_3$$

The reaction conditions are

Temperature	90-140° C
Pressure	200-300 psi
Catalyst	None
Conversion	15%
Yield	Hydrogen peroxide 87%
	Acetone 93%.

Oxidation-dehydration process utilizes air as the oxidant.

$$2\,CH_3-CHOH-CH_3 + O_2 \rightarrow 2\,CH_3-\overset{\overset{\large O}{\|}}{C}-CH_3 + 2\,H_2O$$

The reaction conditions are

Temperature	400-600° C
Pressure	40-50 psig
Catalyst	Ag or Cu
Yield	85-90%.

Acetone is a *coproduct* in the reaction between isopropanol and *acrolein*, $CH_2=CH-\overset{\overset{\large O}{\|}}{C}-H$, for the production of *allyl alcohol*, $CH_2=CH-CH_2OH$, (Fig. 9-9).[28]

$$CH_3-CHOH-CH_3 + CH_2=CH-\overset{\overset{\large O}{\|}}{C}-H \rightarrow$$

$$CH_2=CH-CH_2OH + CH_3-\overset{\overset{\large O}{\|}}{C}-CH_3$$

The reaction conditions are

Temperature	400° C
Pressure	15+ psig
Catalyst	MgO + ZnO
Yield	Allyl alcohol 77%.

Acetone is also produced as a coproduct in the production of phenol from cumene, a reaction which is noted in Chapter 10. The production of acetone as a *byproduct* of the oxidation of propane and butane is noted in Chapter 6.

It might appear that acetone is always a coproduct or a byproduct, never a main product directly from an olefin. The development of the Wacker process has rectified this situation. Acetone is now produced *directly* from *propylene*.[32]

$$2\,CH_3-CH=CH_2 + O_2 \rightarrow 2\,CH_3-\overset{\overset{\large O}{\|}}{C}-CH_3$$

The reaction conditions are similar to those used to produce acetaldehyde by catalytic oxidation of ethylene by use of the Wacker catalyst system ($PdCl_2/CuCl_2$).

Acetone is primarily used as a solvent but it has many synthesis applications. For example, it is used in the synthesis of *methyl isobutyl ketone*[33] with lesser quantities going into the production of *methyl methacrylate*, $CH_2=\overset{\overset{\large CH_3}{|}}{C}-\overset{\overset{\large O}{\|}}{C}-OCH_3$, *diacetone alcohol*, $CH_3-\underset{\underset{\large OH}{|}}{\overset{\overset{\large CH_3}{|}}{C}}-CH_2-\overset{\overset{\large O}{\|}}{C}-CH_3$, *mesityl oxide*, $CH_3-\overset{\overset{\large CH_3}{|}}{C}=CH-\overset{\overset{\large O}{\|}}{C}-CH_3$, *phoron*, $CH_3-\overset{\overset{\large CH_3}{|}}{C}=CH-\overset{\overset{\large O}{\|}}{C}-CH=\overset{\overset{\large CH_3}{|}}{C}-CH_3$, *ketene*, $CH_2=C=O$, and *bisphenol-A*, $HO-C_6H_4-\underset{\underset{\large CH_3}{|}}{\overset{\overset{\large CH_3}{|}}{C}}-C_6H_4-OH$.

Acetone in combination with sulfur dioxide is used in the production of *dicalcium phosphate*, $CaHPO_4$, from phosphate rock.[34]

$$CH_3-\overset{\overset{\large O}{\|}}{C}-CH_3 + SO_2 + 2H_2O \rightarrow CH_3-\underset{\underset{\large OH}{|}}{\overset{\overset{\large CH_3}{|}}{C}}-SO_3^- + H_3O^+$$

$$6\,CH_3-\underset{\underset{\large OH}{|}}{\overset{\overset{\large CH_3}{|}}{C}}-SO_3H + Ca_{10}(PO_4)_6F_2 + 3\,H_2O \rightarrow$$

$$6\,CaHPO_4 + CaF_2 + 3\,Ca\left[CH_3-\underset{\underset{\large OH}{|}}{\overset{\overset{\large CH_3}{|}}{C}}-SO_3^-\right]_2 \cdot H_2O$$

The dicalcium phosphate remains in solution and the impurities are separated by filtration. The yield of dicalcium phosphate is 95%+. The acetone and sulfur dioxide are recovered from the filter cake by heating. Dicalcium phosphate is an effective fertilizer, especially for direct application.

The 1979 production of acetone was 2.48 billion pounds.

ACROLEIN

Production. *Acrolein*, $CH_2=CH-\overset{\overset{\large O}{\|}}{C}-H$, is produced by the catalytic oxidation of propylene with either air or oxygen.

$$CH_3-CH=CH_2 + O_2 \rightarrow CH_2=CH-\overset{\overset{\large O}{\|}}{C}-H + H_2O$$

$$\Delta H = -81.4\ kcal$$

Two processes, Shell and Sohio, are used with the fundamental differences being the source of oxygen (oxygen or air) and the catalyst. Operating conditions for the *Shell process* are

Temperature	350° C
Pressure	30 psig
Catalyst	CuO (supported)
Conversion	20%
Yield	85%.

A flow diagram for the *Sohio process* is given in Fig. 9-13.[35] The reaction conditions are

Temperature	300-360° C
Pressure	15-30 psig
Catalyst	BiO_3/MoO_3 (fixed bed)
Yield	"Excellent".

This process uses air as the source of oxygen. The by-products are *acetaldehyde*, $CH_3-\overset{O}{\overset{\|}{C}}-H$, and *acrylic acid*, $CH_2=CH-\overset{O}{\overset{\|}{C}}-OH$.

Acrylic acid is made the main product by the addition of a second catalytic reactor that oxidizes the acrolein to the acid.

$$2\,CH_2=CH-\overset{O}{\overset{\|}{C}}-H+O_2 \rightarrow 2\,CH_2=CH-\overset{O}{\overset{\|}{C}}-OH$$

$$\Delta H=-60.7 \text{ kcal}$$

A temperature of 250° C is used for the acrolein oxidation. An over-all yield from propylene to acrylic acid is 85%.

Acrylic acid may be produced by oxidative carbonylation of ethylene with carbon monoxide and oxygen with a Pd^{2+}/Cu^{2+} catalyst system. The original method for the production of acrylic acid was the carbonylation of acetylene with nickel carbonyl, $Ni(CO)_4$ in hydrochloric acid.

$$4\,HC\equiv CH+Ni(CO)_4+2\,HCl+4\,H_2O \rightarrow$$

$$4\,CH_2=CH-\overset{O}{\overset{\|}{C}}-OH+NiCl_2+H_2$$

Acrylic acid is esterified to *acrylic esters* by addition of an esterification reactor at the end of the propylene → acrolein → acrylic acid oxidation system (Fig. 9-14).[36] The esterification takes place in the liquid phase over an ion-exchange resin catalyst.

$$CH_2=CH-\overset{O}{\overset{\|}{C}}-OH+ROH \rightarrow$$

$$CH_2=CH-\overset{O}{\overset{\|}{C}}-OR+H_2O$$

The 1978 production of acrylic acid was 31 million pounds.

Uses. Acrolein's primary use is in the production of acrylic acid which, in turn, is used to produce acrylic esters, acrylates. These esters can be produced not only from propylene by an in-line process (Fig. 9-14) but also from formaldehyde and *ketene*, $CH_2=C=O$, by way of *β-propiolactone*,

$CH_2-CH_2-C=O$ (ring closed through –O–).

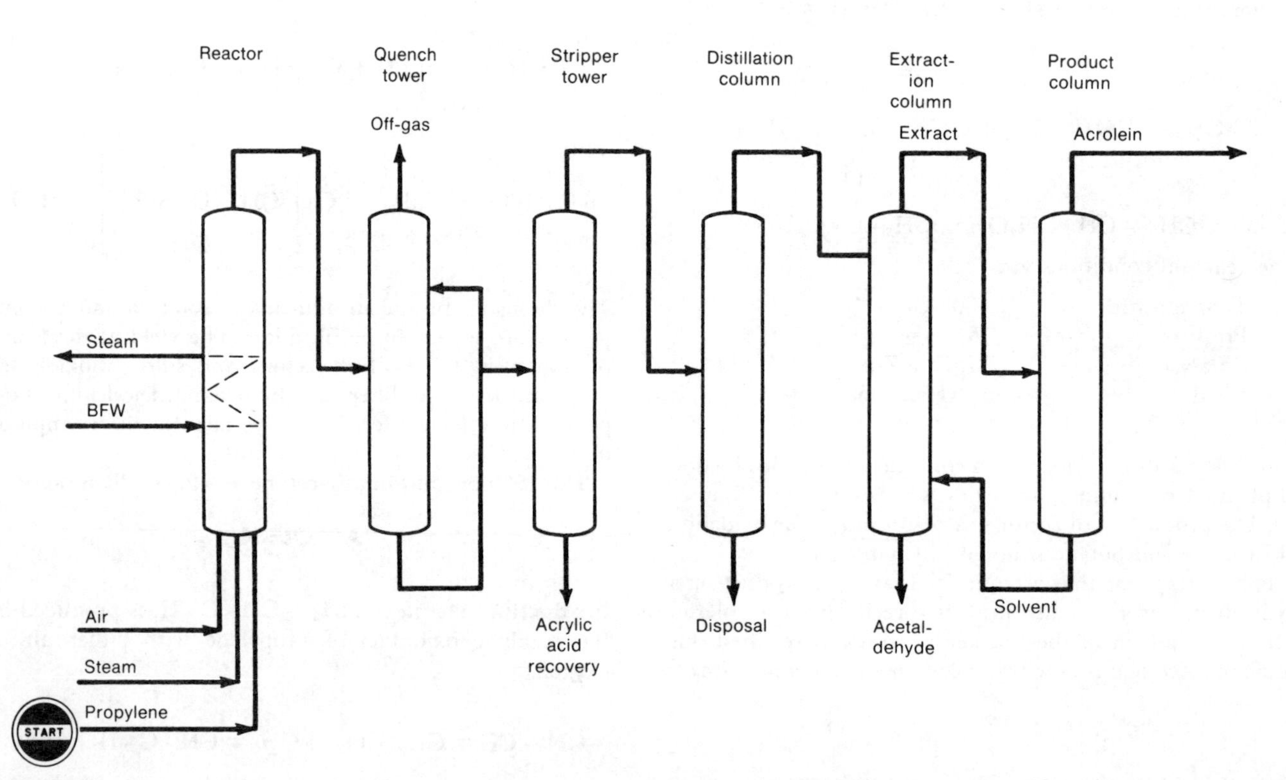

Fig. 9-13—The Sohio process for the production of acrolein by catalytic oxidation of propylene.[35]

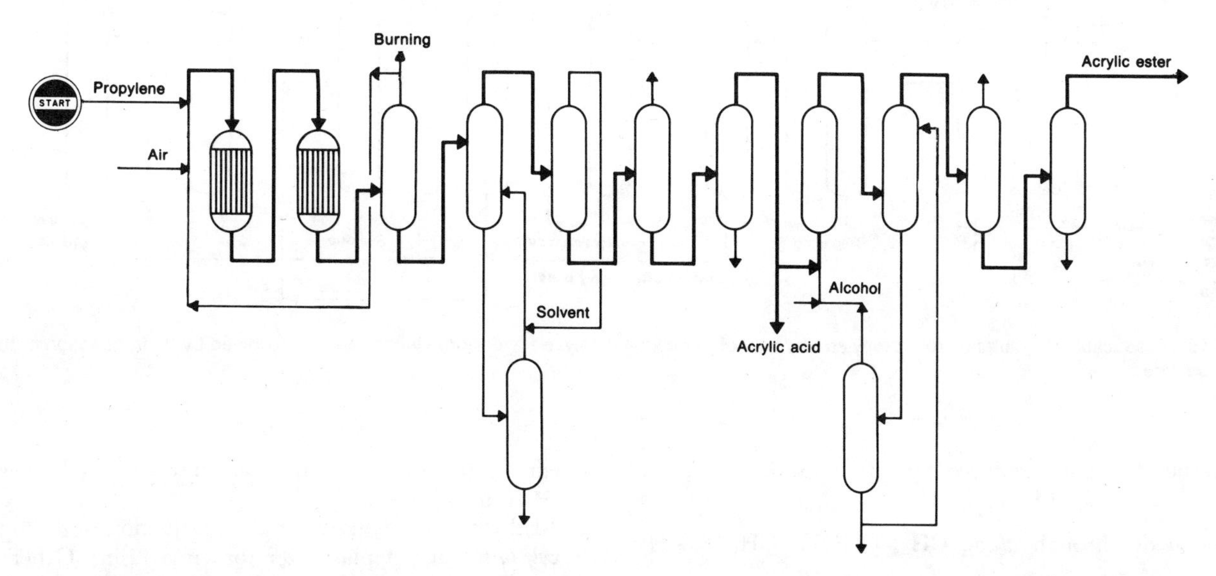

Fig. 9-14—The Nippon Shokubai Kagaku Kogyo Co., Ltd. process for the continuous production of acrylic esters from propylene.[36]

$$CH_2 = C = O + H\text{-}\overset{\overset{\displaystyle O}{\|}}{C}\text{-}H \rightarrow \underbrace{CH_2\text{-}CH_2\text{-}C}_{\text{ring with O}} = O$$

$$CH_2\text{-}CH_2\text{-}C = O + ROH \xrightarrow{\text{acid}}$$

$$CH_2 = CH\text{-}\overset{\overset{\displaystyle O}{\|}}{C}\text{-}OR + H_2O$$

The ketene is produced by the high temperature pyrolysis of either acetic acid or acetone.

The acrylates, especially *ethyl acrylate,* $CH_2{=}CH\text{-}\overset{\overset{\displaystyle O}{\|}}{C}\text{-}OCH_2\text{-}CH_3$, are used in latex coatings, textile finishes, thermosetting finishes, leather finishes and many other general polymer and copolymer uses.

The use of acrolein in the production of glycerol has been noted previously (Fig. 9-9).

BUTYRALDEHYDES

Production. The butyraldehydes, *n-butyraldehyde,* $CH_3\text{-}CH_2\text{-}CH_2\text{-}\overset{\overset{\displaystyle O}{\|}}{C}\text{-}H$, and *isobutyraldehyde,* $CH_3\text{-}\overset{\overset{\displaystyle CH_3}{|}}{CH}\text{-}\overset{\overset{\displaystyle O}{\|}}{C}\text{-}H$, are produced by the catalytic hydroformylation of propylene with carbon monoxide and hydrogen—the oxo reaction.

$$2\,CH_3\text{-}CH = CH_2 + 2\,CO + 2\,H_2 \rightarrow$$

$$CH_3\text{-}CH_2\text{-}CH_2\text{-}\overset{\overset{\displaystyle O}{\|}}{C}\text{-}H + CH_3\text{-}\overset{\overset{\displaystyle CH_3}{|}}{CH}\text{-}\overset{\overset{\displaystyle O}{\|}}{C}\text{-}H$$

Two processes are commercial, the older one uses cobalt compounds and the other process uses rhodium compounds as the catalyst. A comparison of the variables in the cobalt and rhodium processes are given in Table 9-3.[37]

There are three advantages to the rhodium catalyst over the cobalt catalyst: lower reaction temperature, lower pressure, and better *n*- to iso- ratio.[38] These and other aspects of the hydroformylation reaction have been discussed in depth by Cornils, Payer and Traenckner.[39] Further comments on the economics of the modified rhodium catalyst vs. the third generation cobalt catalyst processes have been made by Cornils and Mullen.[40]

Uses. *n-Butanol,* $CH_3\text{-}CH_2\text{-}CH_2\text{-}CH_2OH$, production is the major utilization of ***n*-butyraldehyde.**

$$CH_3\text{-}CH_2\text{-}CH_2\text{-}\overset{\overset{\displaystyle O}{\|}}{C}\text{-}H + H_2 \rightarrow$$

$$CH_3\text{-}CH_2\text{-}CH_2\text{-}CH_2OH$$

This hydrogenation can be integrated into the hydroformylation production line as illustrated in Fig. 9-15.[41] The hydrogenation reaction conditions are

Temperature	130-160° C
Pressure	440-740 psi
Catalyst	A variety may be used
Yield	95%.

TABLE 9-3—Comparison of Variables in the Cobalt and Rhodium Processes[37]

Variable	Cobalt processes	Rhodium process
Pressure	1,500-6,000 psig	100-350 psig
Temperature	145-180° C	ca. 100° C
Normal/iso ratio	3:1 or 4:1	Controllable over wide range normally 8:1-16:1

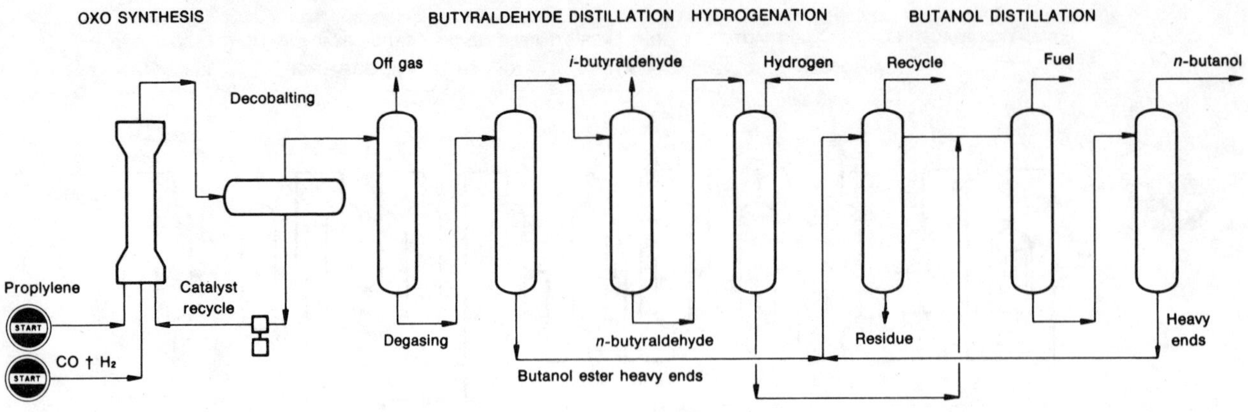

OXO SYNTHESIS
BUTYRALDEHYDE DISTILLATION
HYDROGENATION
BUTANOL DISTILLATION
Off gas
i-butyraldehyde
Hydrogen
Recycle
Fuel
n-butanol
Decobalting
Propylene
START
Catalyst recycle
CO + H_2
START
Degasing
n-butyraldehyde
Residue
Heavy ends
Butanol ester heavy ends

Fig. 9-15—Production of *n*-butanol from propylene by way of the cobalt catalyzed hydroformylation reaction followed by hydrogenation of the *n*-butyraldehyde.[41]

n-Butanol is also produced by fermentation and from acetaldehyde through *aldol,* $CH_3-CHOH-CH_2-\overset{\overset{O}{\|}}{C}-H$, to *crotonaldehyde,* $CH_3-CH=CH-\overset{\overset{O}{\|}}{C}-H$, which is hydrogenated to *n*-butanol.

n-Butanol is used as a solvent and extractant and to produce solvents, such as *butyl acetate,* $CH_3-\overset{\overset{O}{\|}}{C}-OCH_2-CH_2-CH_2-CH_3$, and other butyl esters. *n-Butyl acrylate,* $CH_2=CH-\overset{\overset{O}{\|}}{C}-OCH_2-CH_2-CH_2-CH_3$, is used in the production of plastics. It adds tackiness, softness, plasticity, elongation and low water absorption.

n-Butyraldehyde is also used to produce *2-ethylhexanol,* **2-EH**, $CH_3-(CH_2)_3-\overset{\overset{C_2H_5}{|}}{CH}-CH_2OH$, through an aldol condensation, dehydration and hydrogenation.

$$2\,CH_3-CH_2-CH_2-\overset{\overset{O}{\|}}{C}-H \rightarrow$$

$$CH_3-(CH_2)_2-\overset{\overset{OH}{|}}{CH}-\overset{\overset{C_2H_5}{|}}{CH}-\overset{\overset{O}{\|}}{C}-H$$

$$CH_3-(CH_2)_2-\overset{\overset{OH}{|}}{CH}-\overset{\overset{C_2H_5}{|}}{CH}-\overset{\overset{O}{\|}}{C}-H \rightarrow$$

$$CH_3-(CH_2)_2-CH=\overset{\overset{C_2H_5}{|}}{C}-\overset{\overset{O}{\|}}{C}-H+H_2O$$

$$CH_3-(CH_2)_2-CH=\overset{\overset{C_2H_5}{|}}{C}-\overset{\overset{O}{\|}}{C}-H+2\,H_2 \rightarrow$$

$$CH_3-(CH_2)_2-CH_2-\overset{\overset{C_2H_5}{|}}{CH}-CH_2OH$$

The temperatures are 80-130° C for the aldolization step and 100-150° C for the hydrogenation. Fig. 9-16 shows an integrated process for the production of either *n*-butanol or **2-EH** from propylene.[42, 43]

2-Ethylhexanol is used for the production of *di-2-ethylhexyl phthalate,* a plasticizer for vinyl resins. Other uses include the production of its succinate and acrylates, of antioxidants, antifoams for water solutions and as a special solvent. The U.S. 2-ethylhexanol production is about 500 million lb./yr.

n-Butyric acid, $CH_3-CH_2-CH_2-\overset{\overset{O}{\|}}{C}-OH$, is obtained by the liquid phase oxidation of *n*-butyraldehyde.

$$2\,CH_3-CH_2-CH_2-\overset{\overset{O}{\|}}{C}-H+O_2 \rightarrow$$

$$2\,CH_3-CH_2-CH_2-\overset{\overset{O}{\|}}{C}-OH$$

Its major use is in the production of cellulose acetobutyrate. Other esters of *n*-butyric acid are used as solvents.

Isobutyraldehyde is hydrogenated to *isobutanol, isobutyl alcohol,* $CH_3-\overset{\overset{CH_3}{|}}{CH}-CH_2OH$, by conventional hydrogenation processes.

$$CH_3-\overset{\overset{CH_3}{|}}{CH}-\overset{\overset{O}{\|}}{C}-H+H_2 \rightarrow CH_3-\overset{\overset{CH_3}{|}}{CH}-CH_2OH$$

From 1 to 10% water is added to minimize ether formation.

Isobutanol is used as a solvent, an additive in lubricating oils and for the production of amide resins. Isobutanol's main solvent use is in the form of its acetate ester, a lacquer solvent.

Isobutyraldehyde is oxidized to *isobutyric acid,* $CH_3-\overset{\overset{CH_3}{|}}{CH}-\overset{\overset{O}{\|}}{C}-OH$, which is used to produce esters such as *isobutyl isobutyrate,* $CH_3-\overset{\overset{CH_3}{|}}{CH}-\overset{\overset{O}{\|}}{C}-OCH_2-\overset{\overset{CH_3}{|}}{CH}-CH_3$.

The catalytic oxidation of isobutyraldehyde to acetone and isopropanol has been reported.[44]

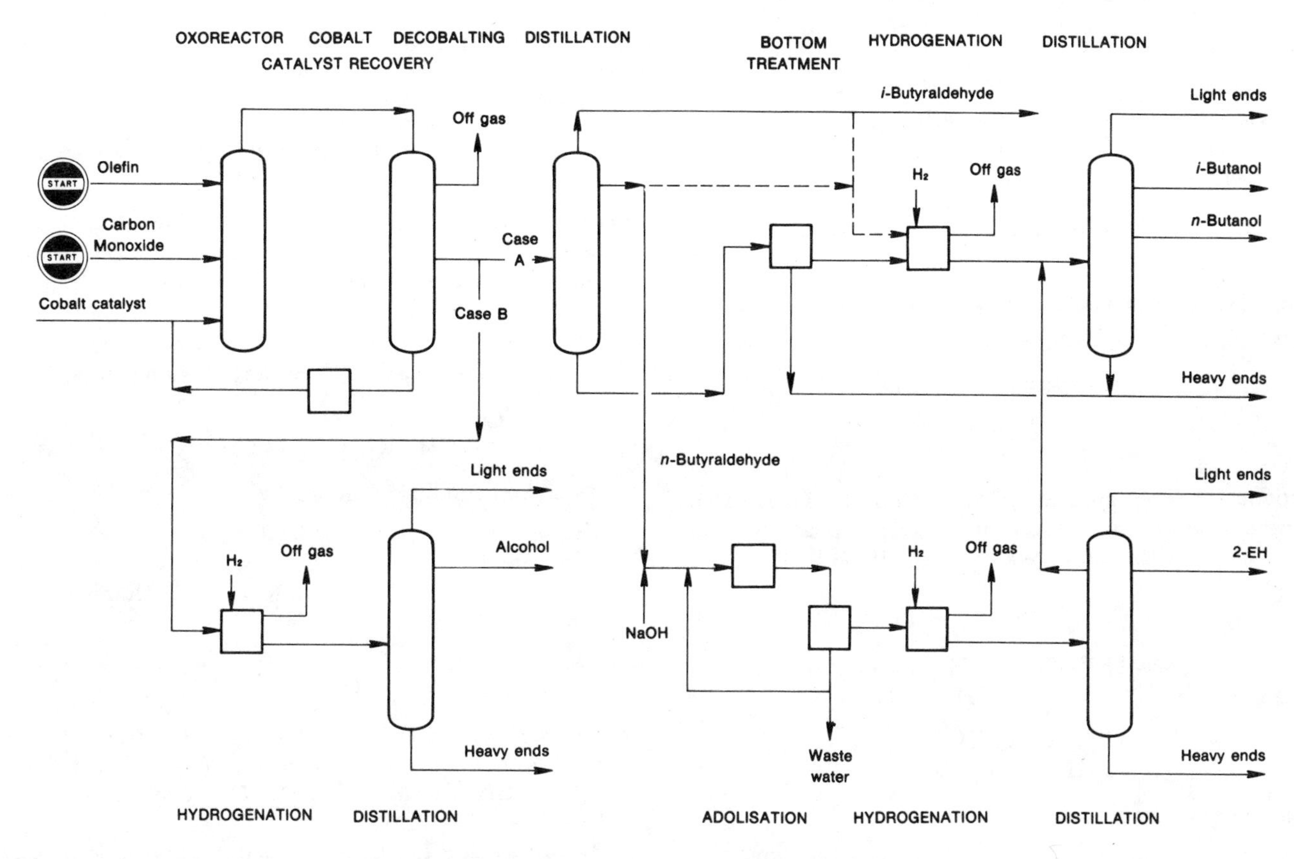

Fig. 9-16—The oxo/2-ethylhexanol process of Hoechst AG/Ruhrchemie/Rhone Poulenc SA.[42,43]

$$2\,CH_3-\overset{CH_3}{\overset{|}{C}H}-\overset{O}{\overset{||}{C}}-H + 2\tfrac{1}{2}\,O_2 \rightarrow CH_3-\overset{O}{\overset{||}{C}}-CH_3 +$$
$$CH_3-CHOH-CH_3 + H_2O + 2\,CO_2$$

Transition metal halides such as $CoBr_2$ and $NiBr_2$ are used as catalysts. This would appear to be finding a use for isobutyraldehyde the hard way.

Neopentyl glycol, $HOCH_2-\underset{CH_3}{\underset{|}{\overset{CH_3}{\overset{|}{C}}}}-CH_2OH$, is produced by the aldol condensation of isobutyraldehyde with formaldehyde, followed by hydrogenation.

$$CH_3-\overset{CH_3}{\overset{|}{C}H}-\overset{O}{\overset{||}{C}}-H + H-\overset{O}{\overset{||}{C}}-H \rightarrow$$
$$HOCH_2-\underset{CH_3}{\underset{|}{\overset{CH_3}{\overset{|}{C}}}}-\overset{O}{\overset{||}{C}}-H$$

$$HOCH_2-\underset{CH_3}{\underset{|}{\overset{CH_2}{\overset{|}{C}}}}-\overset{O}{\overset{||}{C}}-H + H_2 \rightarrow HOCH_2-\underset{CH_3}{\underset{|}{\overset{CH_3}{\overset{|}{C}}}}-CH_2OH$$

Neopentyl glycol is used mainly in the production of saturated and unsaturated polyesters, alkyl and polyurethane resins, as well as plasticizers and synthetic lubricants. These are exceptionally stable because of neopentyl glycol's primary alcohol structure.

ALLYL CHLORIDE

Production. *Allyl chloride,* $CH_2=CH-CH_2Cl$, is produced by the high temperature chlorination of propylene.

$$CH_2=CH-CH_3 + Cl_2 \rightarrow CH_2=CH-CH_2Cl + HCl$$

The reaction conditions are

Temperature	500° C
Pressure	15 psig
Catalyst	None
Yield	85%.

The major byproducts are *cis-* and *trans-1,3-dichloropropene,* $CHCl=CH-CH_2Cl$, and *1,2-dichloropropane,* $CH_2Cl-CHCl-CH_3$.

Uses. Allyl chloride is used in the production of glycerol (Fig. 9-9).[28] The 1,3 dichloropropenes are used as a pesticide, Telone II. They are reported to cause various problems in humans: irritability, breathing difficulties, and personality changes.[45]

ISOPROPYL ACRYLATE

Production. *Isopropyl acrylate,* $CH_2=CH-\overset{O}{\overset{||}{C}}-O\overset{CH_3}{\overset{|}{C}H}-CH_3$ can be produced directly from propylene by reaction with acrylic acid.[46]

$$CH_2=CH-CH_3 + CH_2=CH-\overset{O}{\overset{||}{C}}-OH \rightarrow$$
$$CH_2=CH-\overset{O}{\overset{||}{C}}-O\overset{CH_3}{\overset{|}{C}H}-CH_3$$

The reaction conditions are

Temperature	95-100° C
Catalyst	Amberlyst 15, H^+
Pressure	Liquid phase
Yield	99%.

Amberlyst 15 is a macroporous sulfonated polystyrene resin.

Uses. Isopropyl acrylate may be used as a plasticizing copolymer.

ISOPROPYL ACETATE

Production. *Isopropyl acetate,* **IPAC,** $CH_3-\overset{O}{\overset{\|}{C}}-OCH(CH_3)-CH_3$ can be produced by a direct catalytic vapor phase reaction between propylene and acetic acid (Fig. 9-17).[47]

$$CH_2=CH-CH_3+CH_3-\overset{O}{\overset{\|}{C}}-OH \rightarrow CH_3-\overset{O}{\overset{\|}{C}}-O\overset{CH_3}{\overset{|}{C}H}-CH_3$$

$$\Delta H = -16.4 \text{ kcal}$$

The reaction conditions are

Temperature	120-160° C
Pressure	100-180 psi
Catalyst	"Fixed bed"
Yield: on reacted propylene	99%.

Dilute refinery grade propylene and technical grade acetic acid are used.

Uses. IPAC is used as a solvent for coatings and printing inks. In general, it is interchangeable with methyl ethyl ketone and ethyl acetate.

ALLYL ACETATE

Production. *Allyl acetate,* $CH_3-\overset{O}{\overset{\|}{C}}-OCH_2-CH=CH_2$, is produced by the vapor phase reaction between propylene and acetic acid in the presence of oxygen.[48]

$$CH_2=CH-CH_3+CH_3-\overset{O}{\overset{\|}{C}}-OH \xrightarrow{O_2} CH_3-\overset{O}{\overset{\|}{C}}-OCH_2-CH=CH_2$$

The reaction conditions are

Temperature	180° C
Pressure	25-60 psig
Catalyst	Pd/KOAc (on alumina)
Yield	High.

Uses. Allyl acetate is hydroformulated to *4-acetoxybutyraldehyde,* $CH_3-\overset{O}{\overset{\|}{C}}-OCH_2-CH_2-CH_2-\overset{O}{\overset{\|}{C}}-H$, which is hydrogenated to *1,4-butanediol,* 1,4-**BDO,** $CH_2OH-CH_2-CH_2-CH_2OH$. The over-all yield of 1,4-**BDO** is 77%.

Brownstein and List have reviewed this and various other methods for 1,4-**BDO** production.[49] These include another propylene-based process involving acrolein. 1,4-Butanediol is also produced from butadiene, and from maleic anhydride.

The utilization of 1,4-**BDO** is indicated in Table 9-4.[49] The current demand is about 200 million pounds per year.

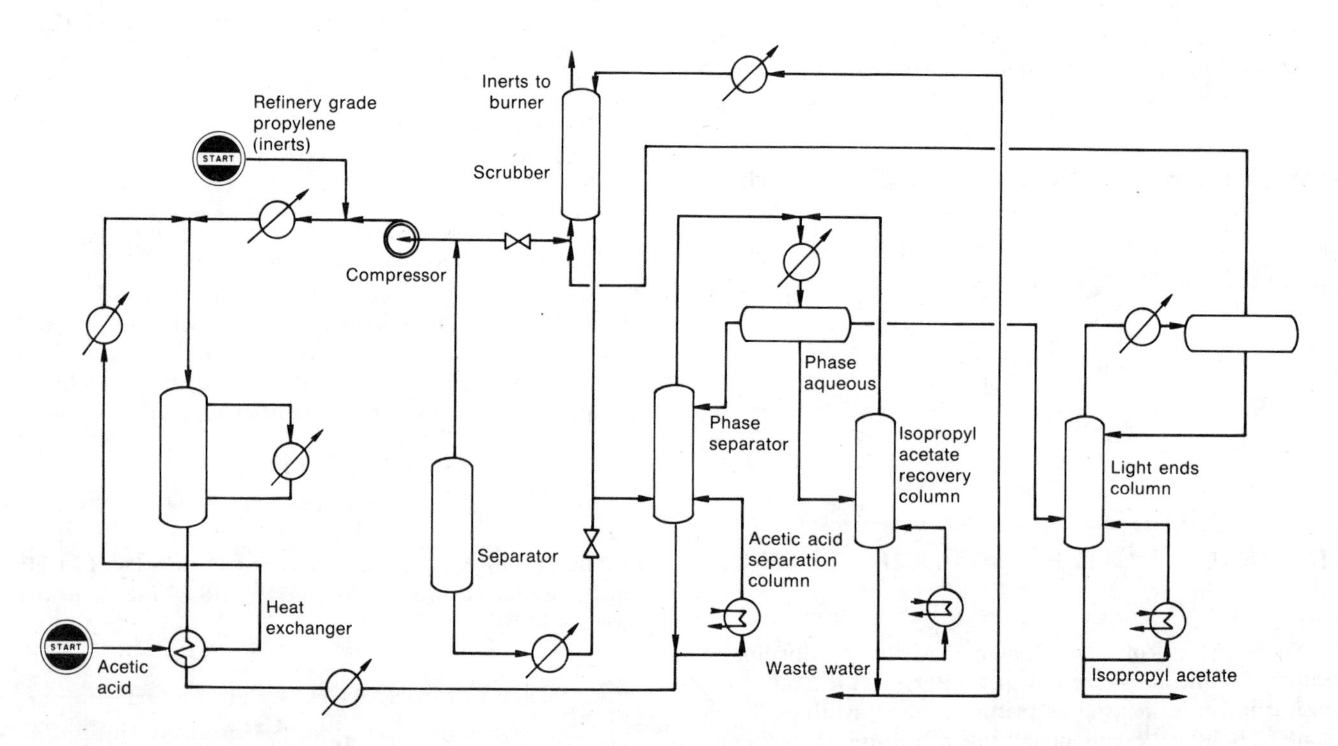

Fig. 9-17—The production of isopropyl acetate from propylene and acetic acid.[47]

TABLE 9-4—United States Demand for 1,4-Butanediol[49]

	Percent
Tetrahydrofuran	57
Acetylenic chemicals	26
Polyurethanes	9
PBT	7
Other	1
Total	100

TABLE 9-5—Representative Disproportionation Catalysts[50]

Transition metal compound Heterogeneous	Support
M $(CO)_6$*	Al_2O_3
MoO_3	Al_2O_3
$CoO.MoO_3$	Al_2O_3
Re_2O_7	Al_2O_3
WO_3	SiO_2
Homogeneous	**Cocatalyst**
WCl_6 (EtOH)	$EtALCl_2$
MX_2 $(NO)_2L_2$*	$R_3Al_2Cl_3$
R_4N [M $(CO)_5X$]*	$RAlX_2$
$ReCl_5/O_2$	$RAlCl_2$

* M = Mo or W; X = halogen (Cl, Br, I); L = Lewis base (e.g., triphenyl-phosphine, pyridien, etc.); R = Alkyl groups (butyl)

DISPROPORTIONATION

Olefin disproportionation is a catalytic process by which an olefin is converted into shorter and longer-chain olefins. With propylene the reaction is

$$2\ CH_2 = CH - CH_3 \rightleftharpoons CH_2 = CH_2 + CH_3 - CH = CH - CH_3$$

The reaction conditions are

Temperature	360-450° C
Pressure	100 psig
Catalyst	See Table 9.5

Table 9-5 indicates the wide variety of catalysts that will effect this type of disproportionation reaction and Fig. 9-18 represents a single-step propylene disproportionation process.[50] Anderson and Brown have discussed extensively this type of reaction and its general utilization.[50] The utility in respect to propylene is to convert excess propylene to olefins of greater economic value.

CUMENE

The production of *cumene, isopropylbenzene,* C_6H_5-CH_3-CH-CH_3 (with the CH branch bonded below the first CH_3), from benzene to propylene alkylation is discussed in Chapter 10.

THE BUTYLENES

In common with propylene, the butylenes and butadiene (C_4's) are byproducts of refinery processes and of the production of ethylene. The butylenes and butadiene have, in general, similar chemical and physical properties. They differ, however, in their utilization with the butylenes being used more for chemical syntheses and less for polymer formation than butadiene.

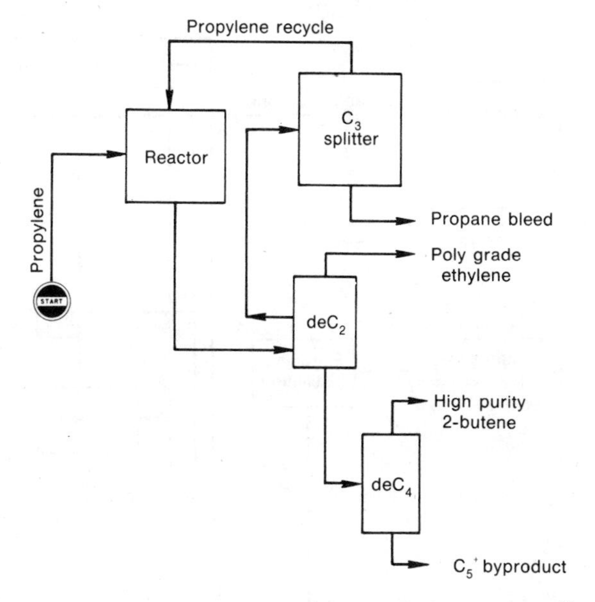

Fig. 9-18—A single-step propylene disproportionation process.[50]

The structures, names and boiling points of the C_4 mono- and diolefins are given in Table 9-6.

The *n*-butenes are characterized by having an unbranched, straight-chain, carbon structure, C-C-C-C while isobutylene has a branched-chain structure, C-C-C with a C attached to the middle carbon. This makes an appreciable difference in the type of reaction, rate of reaction, and general chemical utilization of the two types of C_4 olefins. The relationship between the sources of the butylenes and the interrelationship of the butylenes in respect to their utilization have been illustrated by Guerico (Fig. 9-19).[51]

The United States' refineries produce more than 350,000 b/d of butylenes; nearly all of it used for alkylation. This

TABLE 9-6—The Structure and Names of the C_4 Olefins and Diolefins

Structure	Names[2]	Boiling point °C[3]
	Olefins[1]	
$CH_3-CH_2-CH{=}CH_2$	1-Butene α-Butylene	− 6.3
(H)(CH₃)C=C(H)(CH₃), both H on the same side	*cis*-2-Butene β-Butylene 2-Butene	+ 3.7
(CH₃)(H)C=C(H)(CH₃), H on opposite sides	*trans*-2-Butene β-Butylene 2-Butene	+ 0.9
$CH_3-C(CH_3){=}CH_2$	2-Methylpropene Isobutylene Isobutene	− 6.6
	Diolefins[4]	
$CH_3-CH{=}C{=}CH_2$	1,2-Butadiene Methylallene	+10.8
$CH_2{=}CH-CH{=}CH_2$	1,3-Butadiene Butadiene	− 4.4

[1] Alkenes
[2] The official (IUPAC) name is given first followed by the common name or names
[3] 760 torr
[4] Dienes

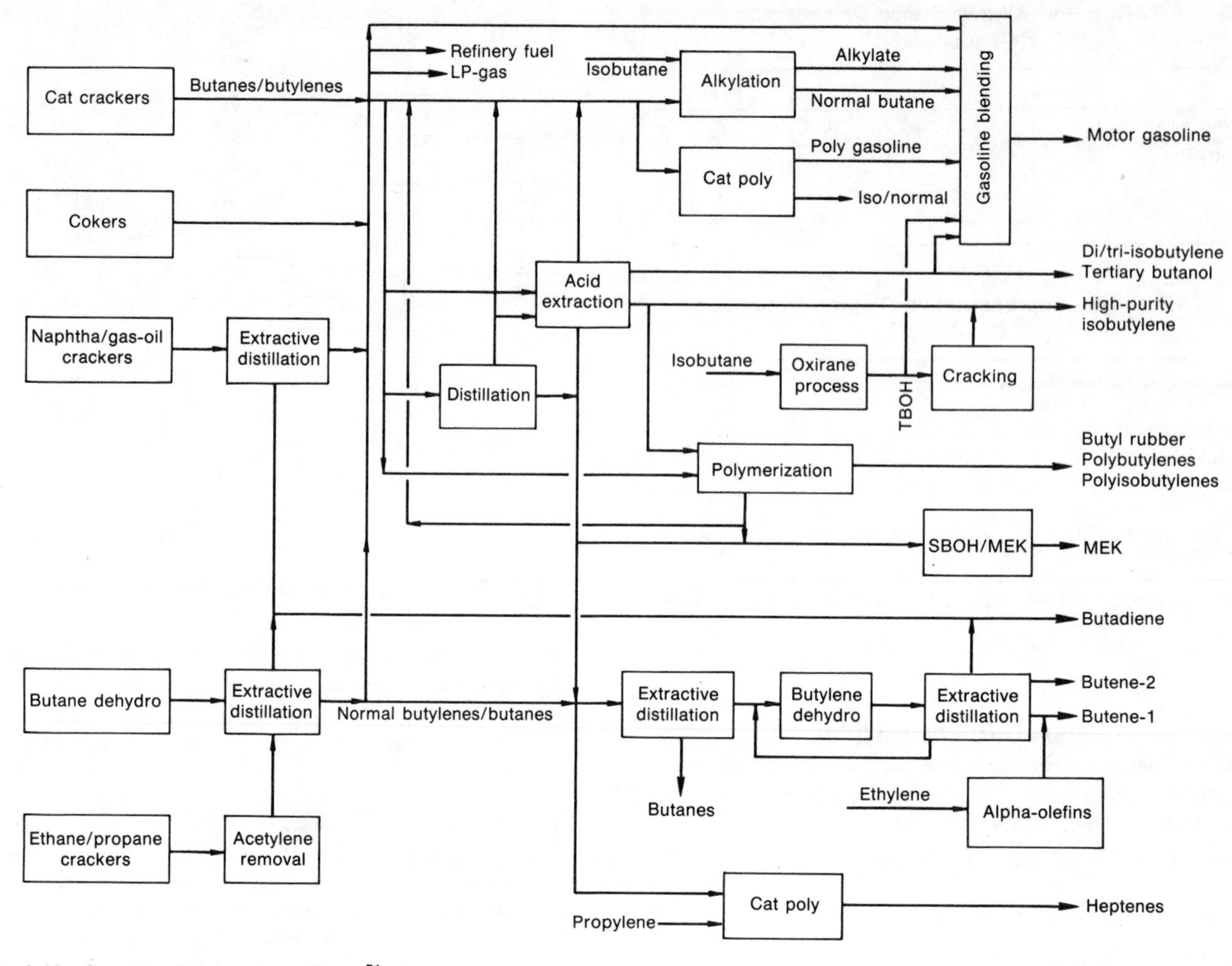

Fig. 9-19—Overall U.S. butylenes industry.[51]

source represents almost 90 percent of the available butylenes. Ethylene plants contribute 7.9 percent of the total; miscellaneous sources contribute 2.4 percent. Increasing amounts of the butylenes will be available as a coproduct of ethylene production as ethylene production changes to liquid feedstocks. This byproduct butylene is predicted to have a growth average of 13 percent to 1985. The total available butylene in 1983 is predicted to be 33 billion pounds.[60]

n-Butenes

There are three *n*-butenes: 1-butene, *cis*-2-butene, and *trans*-2-butene (Table 9-6). The industrial reactions involving *cis*- and *trans*-2-butene are the same and produce same products. Because of this, they will be treated as an entity and referred to collectively as 2-butene. There are also addition reactions where both 1-butene and 2-butene give the same products, such as hydration to produce *sec*-butanol.

This similarity in the reaction products of the *n*-butenes makes it economically feasible to separate a mixture containing 1-butene (bp −6.3°C), 2-butene (bps +0.9 and +3.7°C), and isobutylene (bp −6.6°C) by isomerizing the 1-butene to 2-butene, followed by fractionation (Fig. 9-20).[52] The isomerization process yields two streams, one of 2-butene and the other of isobutylene, each with a purity of 80-90%. The Olex process is reported to separate a feed stream of 53% isobutylene, 27% 1-butene, and 16% 2-butene, and butadiene, *n*-butane and isobutane, to streams of 95% isobutylene and 90% 2-butene.[52]

The standard method for the separation of C_4 olefins is to remove the butadiene by extraction and the isobutylene by absorption in cold sulfuric acid (Fig. 9-19). The isobutylene polymerizes to di- and triisobutylene which go to the gasoline pool.

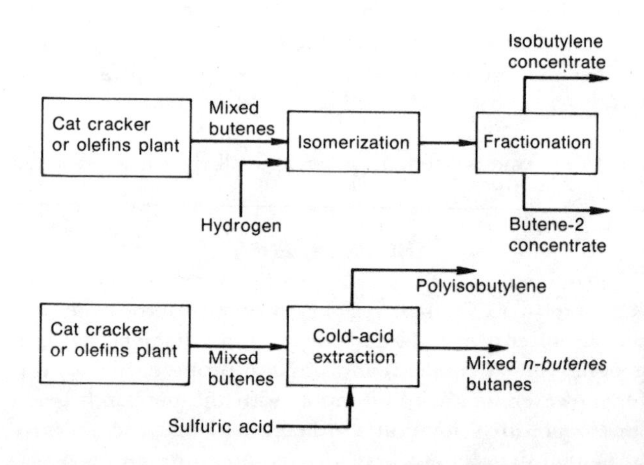

Fig. 9-20—The two processes for the separation of the *n*-butenes and isobutylene.[52]

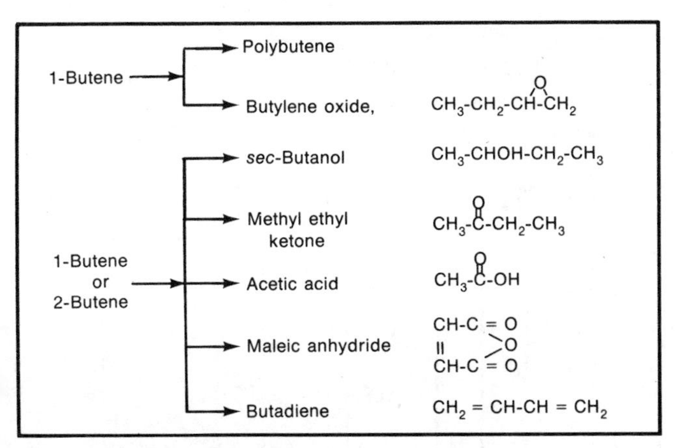

Fig. 9-21—The chemical utilization of the *n*-butenes.

Relative to the potential supply of the *n*-butenes, their chemical utilization is small (Fig. 9-21). In addition, the chemical derivatives are frequently in competition with similar compounds of ethylene and propylene, and even of benzene.

sec-Butanol

Production. *sec*-Butanol (2-butanol, *sec*-butyl alcohol, **SBA**), CH_3-CHOH-CH_2-CH_3, is produced by sulfuric acid esterification of the *n*-butenes followed by hydrolysis of the resulting mixture of *sec*-butyl hydrogen sulfate and di-*sec*-butyl sulfate. The reactions with 1-butene are

Sulfation:

$$3\ CH_3\text{-}CH_2\text{-}CH{=}CH_2 + 2\ H_2SO_4 \rightarrow CH_3\text{-}CH_2\text{-}CH(CH_3)\text{-}OSO_3H + CH_3\text{-}CH_2\text{-}CH(CH_3)\text{-}OSO_2O\text{-}CH(CH_3)\text{-}CH_2\text{-}CH_3$$

Hydrolysis:

$$CH_3\text{-}CH_2\text{-}CH(CH_3)\text{-}OSO_3H + CH_3\text{-}CH_2\text{-}CH(CH_3)\text{-}OSO_2O\text{-}CH(CH_3)\text{-}CH_2\text{-}CH_3 + 3\ H_2O \rightarrow 3\ CH_3\text{-}CH_2\text{-}CHOH\text{-}CH_3 + 2\ H_2SO_4$$

The reaction conditions are (Fig. 9-22)[53]

Temperature	*ca* 35° C
Pressure	Liquid phase
Catalyst	None (75% H_2SO_4)
Yield:	*sec*-Butanol 85 wt.%
	Di-*sec*-butyl ether + polymers 10 wt. %

One ton of *n*-butenes yields 1.12 tons of **SBA.**

The reaction conditions are similar to those used for the production of isopropanol from propylene by the sulfuric acid esterification process. The temperature used will vary with the acid concentration.

Uses. About 90% of the **SBA** production is converted to *methyl ethyl ketone,* **MEK,** CH_3-C(=O)-CH_2-CH_3, by dehydrogenation.

Methyl Ethyl Ketone

Production. Methyl ethyl ketone (2-butanone, **MEK**), CH_3-C(=O)-CH_2-CH_3, is produced directly from the *n*-butenes by a liquid phase Wacker-type process similar to the process used to produce acetaldehyde from ethylene, Chapter 8.

$$2\ CH_2{=}CH\text{-}CH_2\text{-}CH_3 + O_2 \rightarrow 2\ CH_3\text{-}C(=O)\text{-}CH_2\text{-}CH_3$$

The reaction conditions are:

Temperature	120°C
Pressure	150-300 psi
Catalyst	$PdCl_2/CuCl_2$
Yield	88%

MEK can also be produced by the dehydrogenation of *sec*-butyl alcohol.

$$CH_3\text{-}CHOH\text{-}CH_2\text{-}CH_3 \rightarrow CH_3\text{-}C(=O)\text{-}CH_2\text{-}CH_3 + H_2$$

The reaction conditions are:

Temperature	400-550°C
Pressure	Atmospheric
Catalyst	ZnO or brass (Zn-Cu)
Yield	95%

The process is similar to that used to produce acetone from propylene.

In Europe a liquid phase process uses Raney nickel or copper chromite as the dehydrogenation catalyst at 150°C. **MEK** is also a byproduct of the oxidation of butane, Chapter 6.

The 1979 demand for MEK was 569 million pounds.

Uses. Essentially all **MEK** is used as a solvent with about 7% used for lubricating-oil refining (Fig. 9-23).[54] The **MEK** selectivity dissolves the oil from the wax. **MEK** is used as a reaction solvent in one of the terephthalic acid processes.

MEK is also used in the synthesis of various compounds including *methyl ethyl ketoxime,*

$CH_3\text{-}CH_2\text{-}C(CH_3){=}N\text{-}OH$, an antiskinning agent, *methyl ethyl ketone peroxide,*

$CH_3\text{-}CH_2\text{-}C(CH_3)(OH)\text{-}O\text{-}O\text{-}C(CH_3)(OH)\text{-}CH_2\text{-}CH_3$, a polymerization catalyst, especially for acrylic and polyester polymers, and

methyl pentynol, $CH_3\text{-}CH_2\text{-}C(CH_3)(OH){-}C{\equiv}CH$, a corrosion inhibitor.

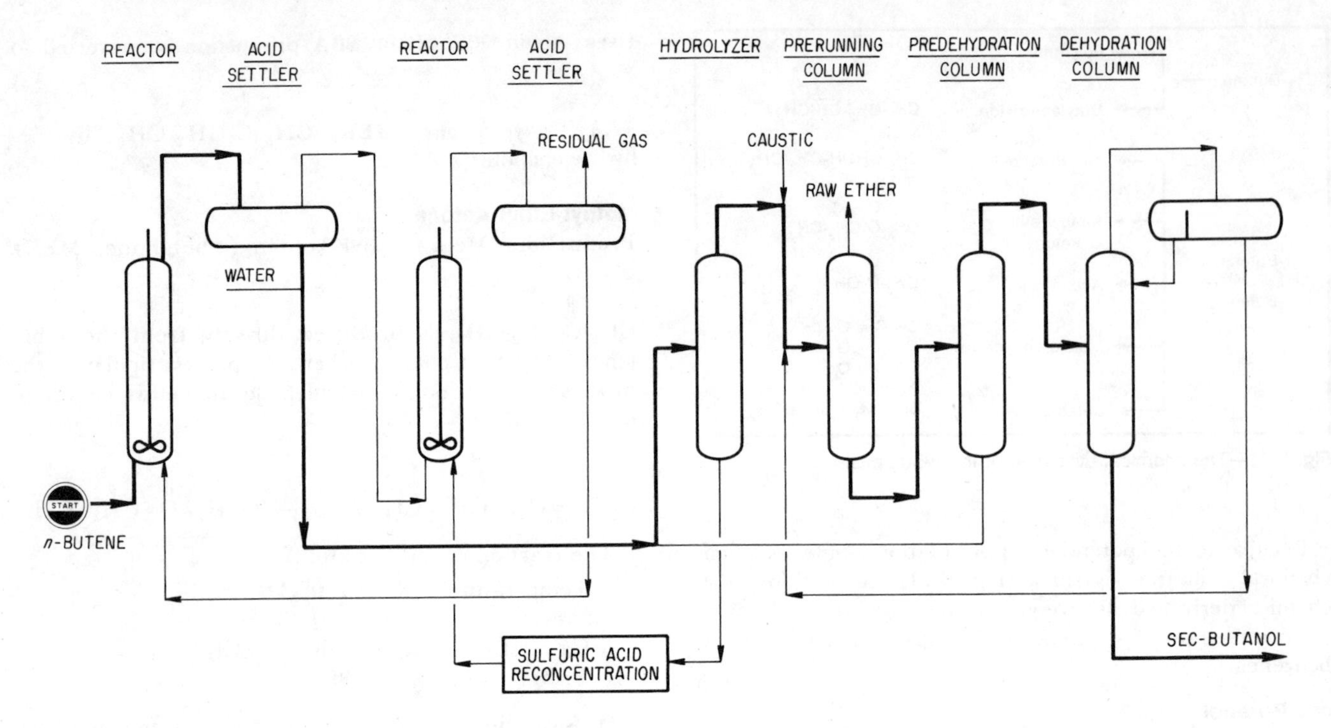

Fig. 9-22—A two-stage process for the production of *sec*-butanol from *n*-butenes.[53]

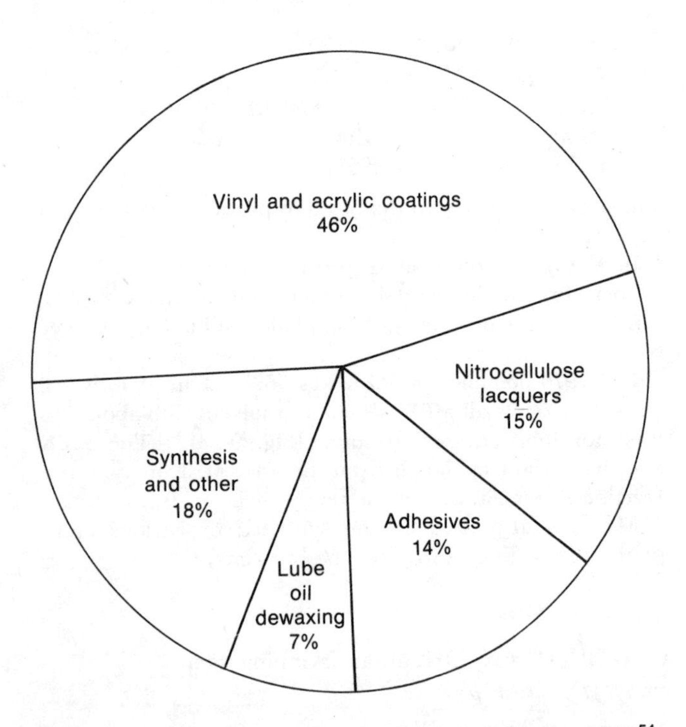

Fig. 9-23—The solvent and other uses of methyl ethyl ketone, MEK.[54]

Acetic Acid

Production. *Acetic acid,* (ethanoic acid) $CH_3\text{-}\overset{\displaystyle O}{\overset{\|}{C}}\text{-}OH$, is produced by several commercial processes including the oxidation of acetaldehyde, the carbonylation of methanol and the oxidation of butane and other paraffin hydrocarbons.

There is little information on the direct catalytic oxidation of *n*-butene to acetic acid outside the patent literature. The catalysts used consist of vanadates of Ti, Al, Sn, Sb and Zn. The temperatures are in the range of 240-275°C. The ideal reaction is:

$$CH_3\text{-}CH{=}CH\text{-}CH_3 + 2\ O_2 \rightarrow 2\ CH_3\text{-}\overset{\displaystyle O}{\overset{\|}{C}}\text{-}OH$$

The following reaction conditions have been reported.[57]

Temperature	270°C	
Pressure	Vapor phase	
Catalyst	Titanium vanadate (Ti:V = 1.0:0:98)	
Conversion	73%	
Yield	Acetic acid	70%
	Maleic acid	3%
	Oxides of carbon	25%

A two-step process for the oxidation of *n*-butenes to acetic acid has been developed.[55, 56] The *n*-butenes are esterified with acetic acid to *sec*-butyl acetate, $CH_3\text{-}\overset{\displaystyle O}{\overset{\|}{C}}\text{-}O\underset{\displaystyle CH_3}{\overset{|}{C}}H\text{-}CH_2\text{-}CH_3$, which is then oxidized to three moles of acetic acid (Fig. 9-24).[56]

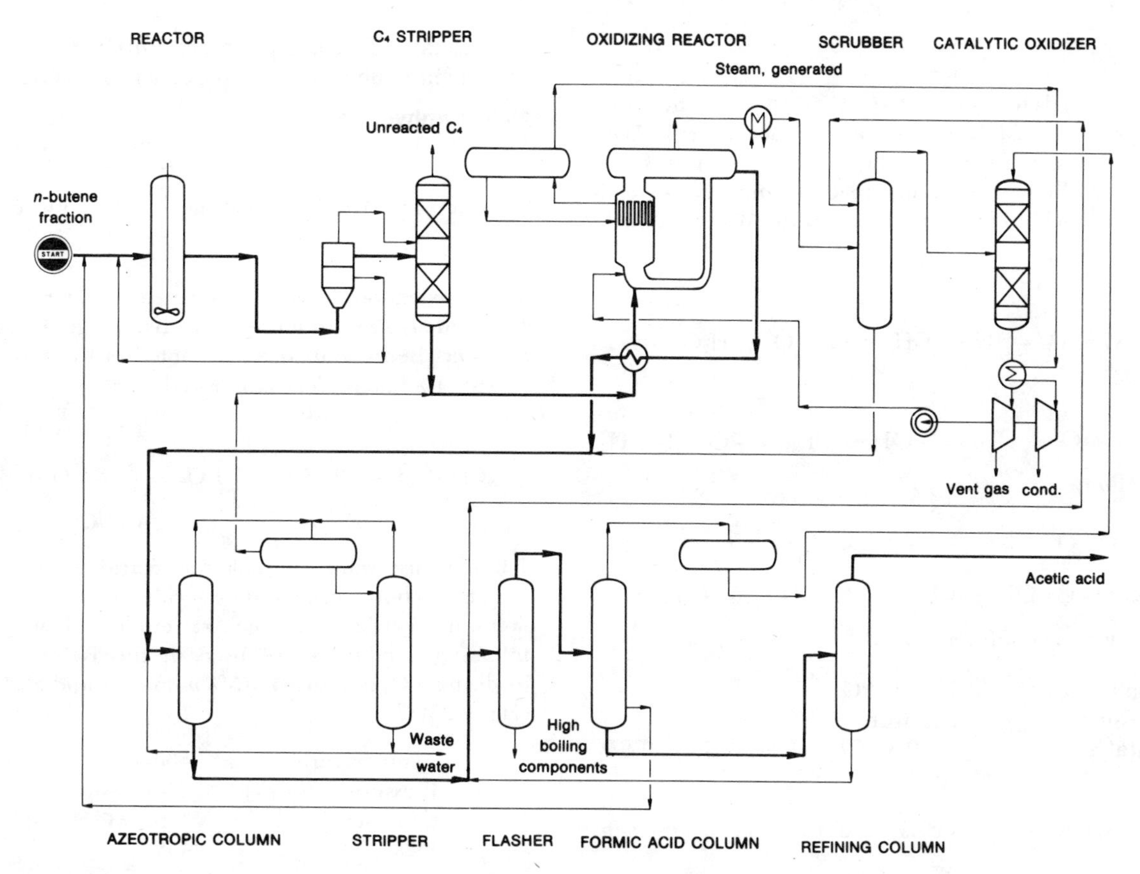

Fig. 9-24—The Bayer AG two-step process for producing acetic acid from *n*-butenes.[56]

Esterification:

$$CH_3-CH_2-CH{=}CH_2 + CH_3-\overset{\overset{\displaystyle O}{\|}}{C}-OH \rightarrow$$

$$CH_3-\overset{\overset{\displaystyle O}{\|}}{C}-O\overset{\overset{\displaystyle CH_3}{|}}{C}H-CH_2-CH_3$$

Oxidation:

$$CH_3-\overset{\overset{\displaystyle O}{\|}}{C}-O\overset{\overset{\displaystyle CH_3}{|}}{C}H-CH_2-CH_3 + 2\,O_2 \rightarrow 3\,CH_3-\overset{\overset{\displaystyle O}{\|}}{C}-OH$$

Net reaction:

$$CH_3\text{-}CH_2\text{-}CH{=}CH_2 + 2O_2 \rightarrow 2\,CH_3\text{-}\overset{\overset{\displaystyle O}{\|}}{C}\text{-}OH$$

The reaction conditions are

Esterification:

Temperature	100-120°C
Pressure	220-370 psi
Catalyst	Acid exchange resin

Oxidation:

Temperature	*ca* 200°C	
Pressure	880 psi	
Catalyst	None	
Net yield	Acetic acid	58%
	Formic acid	6%
	Higher boiling	3%
	Carbon oxides	28%

This process for acetic acid production will be hard pressed, economically, by the methanol carbonylation process.

Uses. The 2.8 billion pound per year production of synthetic acid is utilized primarily in the form of esters. These include *vinyl acetate,* $CH_3-\overset{\overset{\displaystyle O}{\|}}{C}-OCH=CH_2$, *ethyl acetate,* $CH_3-\overset{\overset{\displaystyle O}{\|}}{C}-OCH_2-CH_3$, *butyl acetate,* $CH_3-\overset{\overset{\displaystyle O}{\|}}{C}-OCH_2-CH_2-CH_2-CH_3$, and *amyl acetate,* $CH_3-\overset{\overset{\displaystyle O}{\|}}{C}-OCH_2-(CH_2)_3-CH_3$. Acetic acid is also used to produce acetic anhydride.

Acetic Anhydride. $CH_3\text{-}\overset{O}{\overset{\|}{C}}\text{-}O\text{-}\overset{O}{\overset{\|}{C}}\text{-}CH_3$ may be produced from acetaldehyde, acetone or acetic acid. With both acetone and acetic acid the initial product is *ketene*, $CH_2{=}C{=}O$. Ketene is highly reactive and reacts readily with acetic acid to form acetic anhydride. With acetic acid the reactions are

$$CH_3\text{-}\overset{O}{\overset{\|}{C}}\text{-}OH \rightarrow CH_2 = C = O + H_2O$$

$$CH_2 = C = O + CH_3\text{-}\overset{O}{\overset{\|}{C}}\text{-}OH \rightarrow CH_3\text{-}\overset{O}{\overset{\|}{C}}\text{-}O\text{-}\overset{O}{\overset{\|}{C}}\text{-}CH_3$$

Net reaction:

$$2\ CH_3\text{-}\overset{O}{\overset{\|}{C}}\text{-}OH \rightarrow CH_3\text{-}\overset{O}{\overset{\|}{C}}\text{-}O\text{-}\overset{O}{\overset{\|}{C}}\text{-}CH_3 + H_2O$$

The reaction conditions are

Temperature	700-800°C
Pressure	200 torr
Catalyst	0.2-0.3 $(CH_3\text{-}CH_2\text{-})_3PO_4$ .TEP
Yield	85-89%

Uses. Acetic anhydride is used to make acetic acid esters.

$$CH_3\text{-}\overset{O}{\overset{\|}{C}}\text{-}O\text{-}\overset{O}{\overset{\|}{C}}\text{-}CH_3 + ROH \rightarrow CH_3\text{-}\overset{O}{\overset{\|}{C}}\text{-}OR + CH_3\text{-}\overset{O}{\overset{\|}{C}}\text{-}OH$$

It is especially effective where acetylations by acetic acid are difficult, such as the production of aspirin and cellulose acetate. The latter use accounts for *ca* 85% of the 1.7-billion-pound annual production of acetic anhydride.

Maleic Anhydride

Production. *Maleic anhydride,*

$$\begin{array}{l} CH - C = O \\ \| \qquad\quad \diagdown \\ \| \qquad\qquad O \\ \| \qquad\quad \diagup \\ CH - C = O \end{array}$$

Production by the oxidation of butane is noted in Chapter 6. Maleic anhydride is also produced by the oxidation of benzene and *n*-butene. Benzene oxidation is noted in Chapter 10.

With a *n*-butene feed, the reaction is

$$CH_3\text{-}CH = CH\text{-}CH_3 + 3\ O_2 \rightarrow \begin{array}{l} CH\text{-}C = O \\ \| \qquad \diagdown \\ \| \qquad\quad O \\ \| \qquad \diagup \\ CH\text{-}C = O \end{array} + 3\ H_2O$$

1-Butene also will form maleic anhydride on oxidation.

This reaction is catalyzed by a wide variety of catalysts, the literature of which has been reviewed by Hucknall.[58] The following conditions are used in the Bayer process for oxidizing a C_4 mixture of 75% *n*-butenes and 25% *n*-butane (Fig. 9-25).[59]

Temperature	400-440°C
Pressure	25-50 psig
Catalyst	"Special catalyst"
Yield	45%

The special catalyst may be oxides of Mo-V-P on silica gel in a ratio of 9-3-1.

Uses. Maleic anhydride is used to modify plastic properties because it readily copolymerizes with various other substances but does not polymerize with itself. It is used to modify alkyd resins and drying oils such as linseed, soy and safflower oils. Malathion, an important insecticide, and maleic hydrazide, a plant growth regulator, are also made from maleic anhydride.

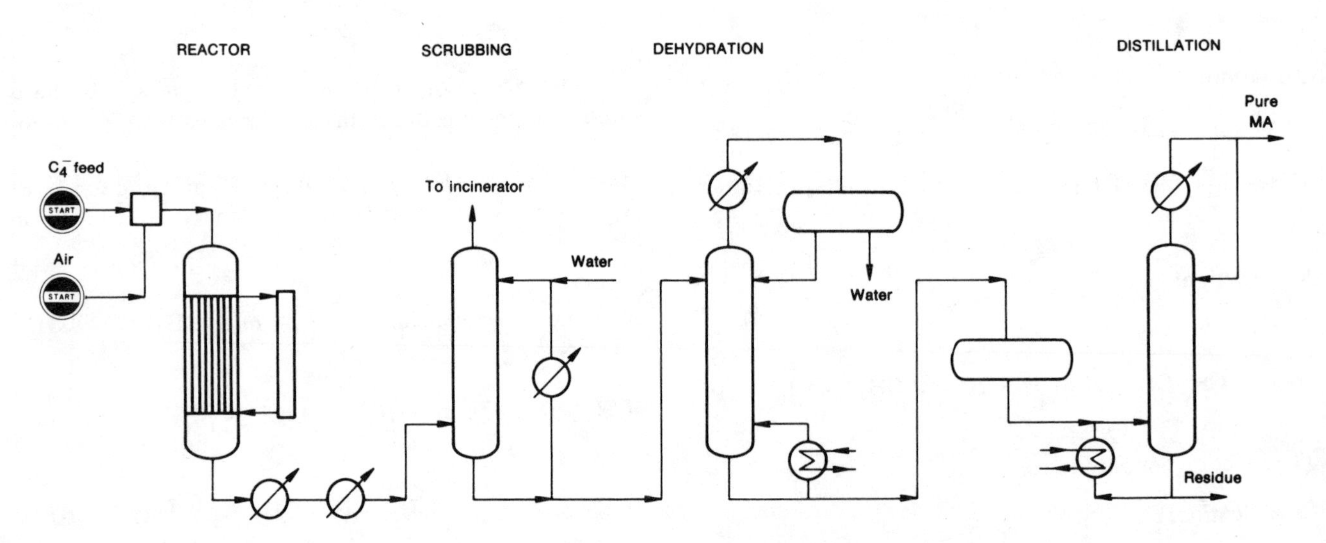

Fig. 9-25—The Bayer process for the catalytic air oxidation of the *n*-butenes to maleic anhydride.[59]

$$(CH_3O)_2-\overset{S}{\overset{\|}{P}}S-CH(-CH_2-\underset{\underset{O}{\|}}{C}-OCH_2-CH_3)-\overset{O}{\overset{\|}{C}}-OCH_2-CH_3$$

Malathion

$$\begin{array}{l} CH-C=O \\ \| \quad\quad\quad \backslash \\ \| \quad\quad\quad N-NH_2 \\ \| \quad\quad\quad / \\ CH-C=O \end{array}$$

Maleic hydrazide

Butylene Oxide

Production. *Butylene oxide,* $CH_3-CH_2-\overbrace{CH-CH_2}^{O}$, is produced from 1-butene by chlorohydrination with hypochlorous acid followed by epoxidation.

Chlorohydrination—

$$CH_3-CH_2-CH=CH_2+HOCl \rightarrow$$
$$CH_3-CH_2-CHOH-CH_2Cl$$

Epoxidation—

$$2\ CH_3-CH_2-CHOH-CH_2Cl+Ca(OH)_2 \rightarrow$$
$$2\ CH_3-CH_2-\overbrace{CH-CH_2}^{O}+CaCl_2+2\ H_2O$$

The reaction conditions are similar to those used in the chlorohydrination process for the production of propylene oxide from propylene.

Uses. Butylene oxide is hydrolyzed to *butylene glycol,* $CH_3-CH_2-CHOH-CH_2OH$,

$$CH_3-CH_2-\overbrace{CH-CH_2}^{O}+H_2O \xrightarrow{H^+}$$
$$CH_3-CH_2-CHOH-CH_2OH$$

The glycol is used in the production polymeric plasticizers.

1,2-Butylene oxide is a stabilizer for *1,1,1,-trichloroethane* (methyl chloroform), CH_3-CCl_3, and other chlorinated solvents. Other uses include applications such as pharmaceuticals, surfactants and agro chemicals.

ISOBUTYLENE

Isobutylene (isobutene), $CH_3-\overset{CH_3}{\overset{|}{C}}=CH_2$, is not used extensively as a chemical precursor because many of its derivatives have the reactive tertiary structure which has

$$CH_3-\overset{CH_3}{\overset{|}{\underset{CH_3}{\underset{|}{C}}}}-$$

a tendency to revert to isobutylene. *tert*-Butyl alcohol, and its derivatives, are examples of this tendency. However, alkylation reactions such as the alkylation of *p*-cresol to 2,6-di-*tert*-butyl-*p*-cresol produce stable compounds. Reactions which preserve the carbon-carbon double bond also produce stable compounds. The oxidation of isobutylene to methacrylic acid is an example.

Isobutylene dimerizes readily with itself and with other olefins to form higher molecular weight olefins such as diisobutylene and "heptene." It also readily alkylates benzene and its derivatives.

Fig. 9-26 indicates some of the compounds synthesized from isobutylene. These are produced in relatively small amounts. Butyl rubber and polybutenes account for about 75% of the chemical utilization of isobutylene.

tert-Butyl Alcohol

Production. *tert-Butyl alcohol* (**TBA**), $CH_3-\overset{CH_3}{\overset{|}{C}}OH-CH_3$ is produced by the sulfation-hydrolysis process used to produce *sec*-butanol from the *n*-butenes. The over-all reaction is

$$CH_3-\overset{CH_3}{\overset{|}{C}}=CH_2+H_2O \xrightarrow{H^+} CH_3-\overset{CH_3}{\overset{|}{C}}OH-CH_3$$

The reaction conditions are

Temperature	10-30° C
Pressure	Liquid phase
Catalyst	None (50-65% H_2SO_4)
Yield (after hydrolysis)	*ca* 95%

tert-Butyl alcohol is also produced as a coproduct in the *tert*-butyl epoxidation of propylene to propylene oxide. This alcohol is usually dehydrated to obtain pure isobutylene.

Uses. *tert*-Butyl alcohol is used to a small extent as a solvent. It is also used as a raw material in the production of *p-tert*-butyl phenol

$$HO-C_6H_4-\overset{CH_3}{\overset{|}{\underset{CH_3}{\underset{|}{C}}}}-CH_3$$

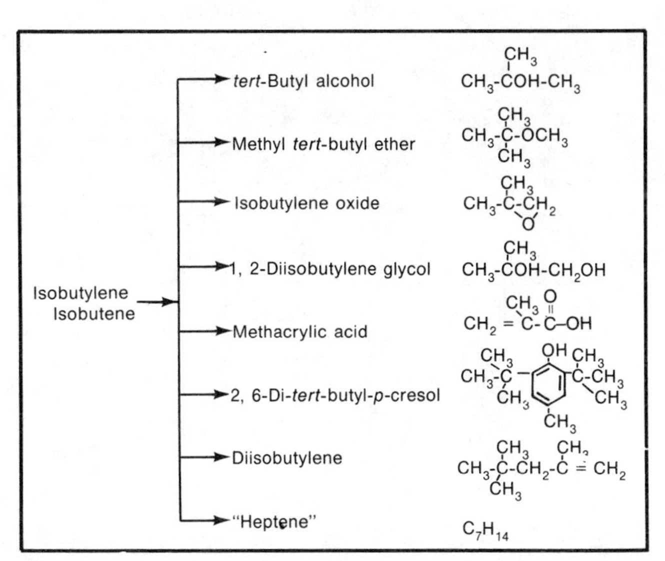

Fig. 9-26—The chemical utilization of isobutylene, exclusive of polymer formation.

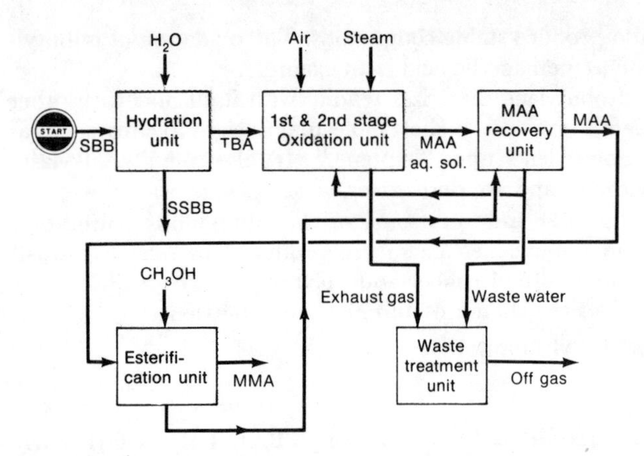

Fig. 9-27—Block flow diagram of Mitsubishi Rayon's new methyl methacrylate monomer (MMA) process. SBB: Spent-BB, the raffinate of C_4 fraction from naphtha cracker after butadiene has been extracted. SSBB: Spent-spent-BB, the raffinate of spent-BB after isobutylene has been extracted in the form of TBA. TBA: Tertiary butyl alcohol. MAA: Methacrylic acid. MMA: Methyl methacrylate monomer.

which is an intermediate for oil-soluble phenol-formaldehyde resins. **TBA** has been proposed as feedstock for methyl methacrylate, **MMA.** The alcohol is oxidized to methacrolein, then the methacrolein is converted to methacrylic acid which is esterified with methanol to yield **MMA.** Fig. 9-27 illustrates this process.[61] **TBA** has a research octane number, **RON,** of 108 and has been proposed as a gasoline additive.

Methyl *tert*-Butyl Ether

Production. *Methyl tert-butyl ether* (**MTBE**), production is potentially the single largest use

$$CH_3-\overset{\displaystyle CH_3}{\underset{\displaystyle CH_3}{\overset{|}{\underset{|}{C}}}}-OCH_3$$

TABLE 9-7—Potential MTBE Production (10^6 Metric TPY)[81]

	1982			1987		
	U.S.A.	West Europe	Japan	U.S.A.	West Europe	Japan
From cat cracking	8.8	1.8	0.5	8.6	2.5	0.5
From ethylene plants	2.5	3.0	1.0	4.0	4.0	1.5
Total	11.3	4.8	1.5	12.6	6.5	2.0

for isobutylene, perhaps to the extent of utilizing all of the isobutylene available from ethylene production. Table 9-7 summarizes the world **MTBE** potential production estimate for the 1980-1985 period.[62] The estimate is based only on the isobutylene produced as a byproduct of ethylene plants.

MTBE is produced by the reaction between isobutylene and methanol.

$$CH_3-\overset{\displaystyle CH_3}{\overset{|}{C}}=CH_2+CH_3OH\rightarrow CH_3-\overset{\displaystyle CH_3}{\underset{\displaystyle CH_3}{\overset{|}{\underset{|}{C}}}}-OCH_3$$

The general reaction conditions are[62]

Temperature	Mild
Pressure	Liquid phase
Catalyst	Sulfonated polystyrene resin
Yield	95%

Fig. 9-28 shows a flow diagram for this process.[63] The feed is a mixed steam cracker product with the butadiene removed. The reaction conditions are mild enough to permit the *n*-butenes to pass through without ether formation. Case 1 contains 60 wt.% **MTBE** and may be blended directly with gasoline. Case 2 **MTBE** has a purity of 95 wt.% and Case 3 has a purity of 99.8 wt.%.

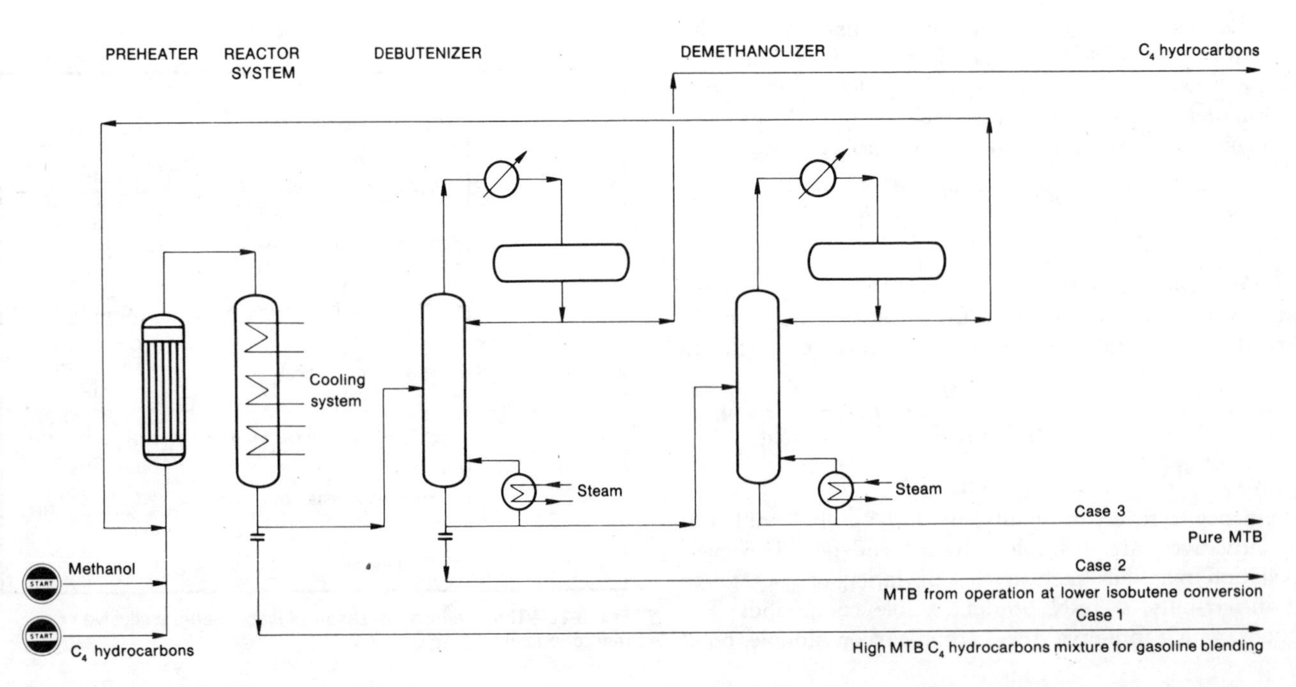

Fig. 9-28—A process for the production of *tert*-butyl ether, MTBE, from a mixed C_4 stream and methanol.[63]

Uses. Methyl *tert*-butyl ether is an excellent octane booster for both low lead and unleaded fuels. Its use in this respect has been reviewed by Pecci and Floris.[62] The use of **MTBE**-*sec*-butanol mixtures has been reported by Csikos, *et al.*[64] Fuels blended with a **MTBE** and *sec*-butanol mixture are said to have good anti-knock, increased power, reduced fuel consumption and CO emission without fuel system modification.[64] An extensive review of all aspects of the use of **MTBE** in fuels has been made by Taniguchi and Johnson.[82]

tert-Amylmethyl ether, **TAME,** has been proposed as comparable to **MTBE** as a fuel additive.[65] With typical unleaded gasoline the blend octane numbers for **MTBE** are **RON**-118 and **MON**-101. Those for **TAME** are **RON**-112 and **MON**-99. **TAME** is produced from isoamylenes (2-methyl-1-butene and 2-methyl-2-butene). The production of **MTBE** in 1980 is expected to reach 1-3 billion pounds.

Isobutylene Oxide

Production. *Isobutylene oxide,* $CH_3-\overset{CH_3}{\overset{|}{C}}-CH_2$ (with O bridging C and CH_2), is produced by chlorohydrination of isobutylene followed by epoxidation of the chlorohydrin by reaction with a base. The process is similar to that used to produce butylene oxide and propylene oxide.

Direct noncatalytic liquid phase oxidation of isobutylene to isobutylene oxide has been reported.[66]

$$2\ CH_3-\overset{CH_3}{\overset{|}{C}}=CH_2 + O_2 \rightarrow 2\ CH_3-\overset{CH_3}{\overset{|}{C}}-CH_2 \quad (\text{epoxide, O bridging C and } CH_2)$$

The reaction conditions are

Temperature	120° C	
Pressure	735 psi	
Catalyst	None	
Yield	Isobutylene oxide	28.7%
	Isobutylene glycol	9.6%
	Isobutylene glycol esters	6.7%
	Acetone	23.0%
	tert-Butyl alcohol	5.8%
	Other	26.2%

Other compounds include formic acid, acetic acid, β-methallyl alcohol, methyl alcohol, *tert*-butyl formate, dimethyldioxane, peroxide compounds, water, carbon dioxide and "other products."

A direct catalytic liquid phase oxidation process in an acetic acid-water-tetrahydrofuran solution has been proposed. The reported reaction conditions are[67]

Temperature	70° C	
Pressure	Liquid phase	
Catalyst	$Tl(O-\overset{O}{\overset{\|}{C}}-CH_3)_3$ (Stoic. amts.)	
Yield	Isobutylene oxide	82%
	Glycol monoacetate	15%

At a slightly higher temperature the epoxide is hydrolyzed to the glycol.

Uses. Isobutylene oxide is hydrolyzed to isobutylene glycol, $CH_3-\overset{CH_3}{\overset{|}{C}}OH-CH_2OH$, in an acid solution. The glycol, in turn, can be oxidized to α-hydroxyisobutyric acid.[68]

$$CH_3-\overset{CH_3}{\overset{|}{C}}OH-CH_2OH + O_2 \rightarrow$$

$$CH_3-\overset{CH_3}{\overset{|}{C}}OH-\overset{O}{\overset{\|}{C}}-OH + H_2O$$

The reaction conditions are

Temperature	70-80° C
Pressure	Liquid phase
Catalyst	5% Pt/C (pH 2-7)
Yield	High

The hydroxy acid is readily dehydrated to give a 95% yield of methacrylic acid, $CH_2=\overset{CH_3}{\overset{|}{C}}-\overset{O}{\overset{\|}{C}}-OH$.

Isobutylene Glycol

Production. Isobutylene glycol synthesis by direct catalytic liquid phase oxidation of isobutylene has been reported.[69]

$$2\ CH_3-\overset{CH_3}{\overset{|}{C}}=CH_2 + O_2 + 2\ H_2O \rightarrow$$

$$2\ CH_3-\overset{CH_3}{\overset{|}{C}}OH-CH_2OH$$

The isobutylene is oxidized by Tl^{3+} ions to isobutylene glycol. The Tl^{3+} ions are regenerated from Tl^{+} ions by a $CuCl_2/O_2$ couple. This is a Wacker-type process.

$$TlCl + 2\ CuCl_2 \rightarrow TlCl_3 + 2\ CuCl$$

$$4\ CuCl + 4\ HCl + O_2 \rightarrow 4\ CuCl_2 + 2\ H_2O$$

Coupled with the glycol oxidation to α-hydroxyisobutyric acid, this may present a viable route from isobutylene to methacrylic acid.

Methacrolein-Methacrylic Acid

The methyl ester of *methacrylic acid* (**MAA**), $CH_2=\overset{CH_3}{\overset{|}{C}}-\overset{O}{\overset{\|}{C}}-OH$, is a useful vinyl monomer produced by the acetone cyanohydrin process. This process has toxicity problems and a large ammonium sulfate waste stream. The reactions are

$$CH_3-\overset{O}{\overset{\|}{C}}-CH_3 + HCN \rightarrow CH_3-\overset{CH_3}{\overset{|}{C}}OH-CN$$

$$CH_3-\overset{CH_3}{\overset{|}{C}}OH-CN + H_2SO_4 \rightarrow$$

$$CH_2=\overset{CH_3}{\overset{|}{C}}-\overset{O}{\overset{\|}{C}}-NH_2 \cdot H_2SO_4$$

$$CH_2 = C(CH_3) - C(=O) - NH_2 \cdot H_2SO_4 + CH_3OH \rightarrow CH_2 = C(CH_3) - C(=O) - OCH_3 + NH_4HSO_4$$

Production. Extensive research has gone into processes which will oxidize isobutylene directly to *methacrylic acid*, or indirectly to *methacrolein*, $CH_2 = C(CH_3) - C(=O) - H$, as an intermediate.

Ammoxidation of isobutylene to *methacrylonitrile*, $CH_2 = C(CH_3) - CN$, in a process similar to that used to produce acrylonitrile from propylene has been under study. When *nitrogen dioxide*, NO_2, is used as the oxidant, both methacrolein and methacrylonitrile are produced in low yields.[70]

The literature on the catalytic oxidation of isobutylene to methacrolein and methacrylonitrile has been reviewed by Hucknall.[71]

A viable process for the air oxidation of isobutylene has been reported by Oda, *et al.*[72] The reactions are

$$CH_2 = C(CH_3) - CH_3 + O_2 \rightarrow CH_2 = C(CH_3) - C(=O) - H + H_2O$$

$$2\ CH_2 = C(CH_3) - C(=O) - H + O_2 \rightarrow 2\ CH_2 = C(CH_3) - C(=O) - OH$$

Because of the different oxidation characteristics of isobutylene and methacrolein, this is a two-step process (Fig. 9-29).[72]

Isobutylene oxidation to methacrolein reaction conditions are

Temperature	350 450° C
Pressure	15-25 psig
Catalyst	*
Yield	80-90%

* "Complex molybdenum oxide promoted with several selected oxides, supported on a grain carrier."

Methacrolein oxidation to methacrylic acid reaction conditions are

Temperature	250-350° C
Pressure	15-25 psig
Catalyst	*
Yield	70-80%

* "A molybdenum compound with some specific promoters, supported."

Uses. Methacrylic acid is esterified with methanol to methyl methacrylate (**MMA**), $CH_2 = C(CH_3) - C(=O) - OCH_3$. Methyl methacrylate is used to produce polymers: cast sheet, molding and extrusion powders and coatings. It polymerizes readily to a homopolymer or various copolymers.

HEPTENES

Production. Isobutylene and propylene can be dimerized in the presence of phosphoric acid or aluminum chloride to a mixture of *heptenes*, C_7H_{14}.

Uses. The heptene mixture is hydroformulated, OXO reaction, then hydrogenated to "isooctanol." This alcohol mix-

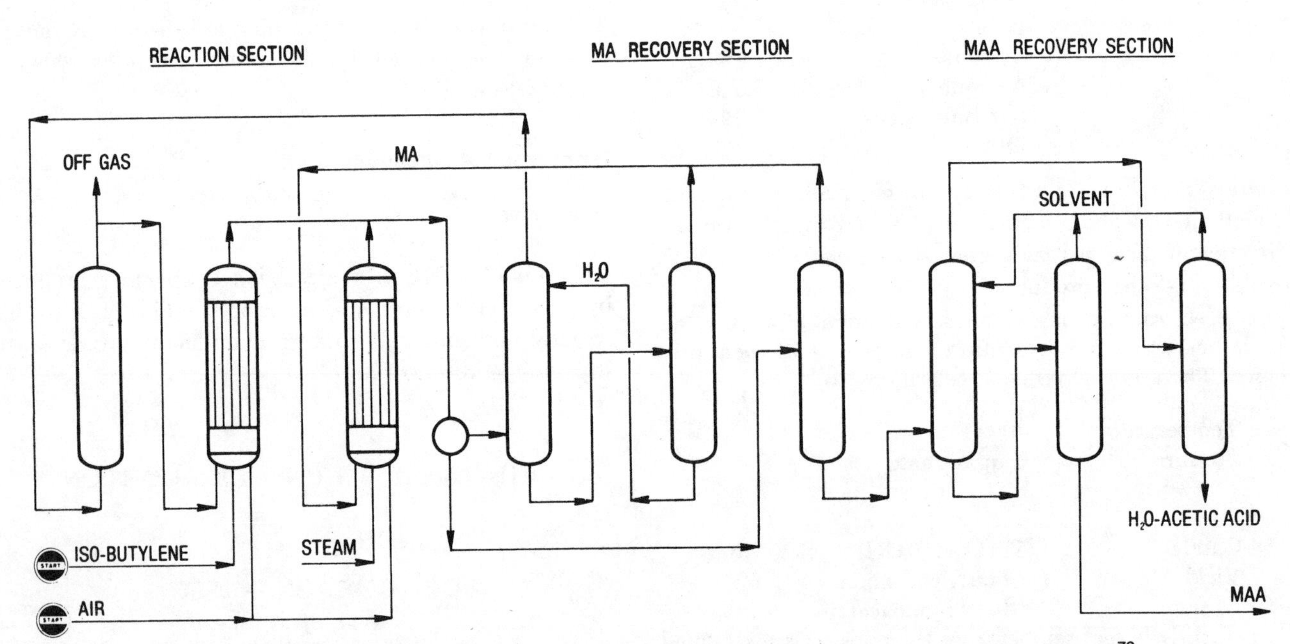

Fig. 9-29—A proposed process for the commercial production of methacrylic acid from isobutylene by two-stage oxidation.[72]

ture is used to make phthalate plasticizers. This and other uses are similar to those of 2-ethylhexanol.

DIISOBUTYLENE

Production. *Diisobutylene,* C_8H_{16}, is a byproduct of isobutylene extraction with sulfuric acid.

$$2CH_2=C(CH_3)-CH_3 \xrightarrow{H^+} CH_3-C(CH_3)_2-CH_2-C(CH_3)=CH_2 + \quad (80\%)$$

$$CH_3-C(CH_3)_2-CH=C(CH_3)-CH_3 \quad (20\%)$$

Diisobutylene, as well as the heptenes, are produced by the Dimersol process. This is a selective, liquid phase, codimerization or dimerization of propylene and/or butylene cuts from various refinery streams.[73] The process is characterized by low pressure and ambient temperature in the presence of a soluble catalyst system.

Uses. Diisobutylene is used to make *octylphenol* for the production of nonionic detergents. *Nonyl alcohols* are produced by the OXO reaction. The nonyl alcohols are used to make plasticizers, etc. The use of both the heptenes and the diisobutylenes is for octane improvement of gasoline.

Neo-Pentanoic Acid

Production. *Neo-pentanoic acid,*

$$CH_3-C(CH_3)_2-C(=O)-OH,$$

is produced by the high pressure addition of carbon monoxide to isobutylene in the presence of an acid catalyst to produce a CO-catalyst-olefin complex—an acyl carbonium ion.

$$CH_2=C(CH_3)-CH_3 + H^+ + CO \rightarrow \left[CH_3-C(CH_3)_2-CO\right]^+$$

This is followed by low pressure hydrolysis.[74]

$$\left[CH_3-C(CH_3)_2-CO\right]^+ + H_2O \rightarrow CH_3-C(CH_3)_2-C(=O)-OH + H^+$$

Uses. Neo-pentanoic acid has the very stable *neo* structure. It is used where a very stable acid and esters are required.

BUTADIENE

In 1955 it was noted that "the assured future of *butadiene,* $CH_2=CH-CH=CH_2$, lies with synthetic rubber, the potential of butadiene is in its chemical versatility. Its low cost, ready availability, and great reactivity tempt researchers."[75] Today only about 7% of the 3.3-billion-pound butadiene production is utilized in other than direct polymer formation. Butadiene has not lived up to its potential as a chemical intermediate (Fig. 9-30). The polymer distribution is: styrene-butadiene rubber (50%), polymers (20%), and other rubbers (10%).

Hexamethylenediamine

Production. *Hexamethylenediamine* (**HMDA**), $H_2N-(CH_2)_6-NH_2$, was initially produced from butadiene by the addition of chlorine at 150°C followed by cyanation and hydrogenation.

$$2\,CH_2=CH-CH=CH_2 + 2\,Cl_2 \rightarrow$$
$$CH_2Cl-CH=CH-CH_2Cl +$$
$$CH_2Cl-CHCl-CH=CH_2$$

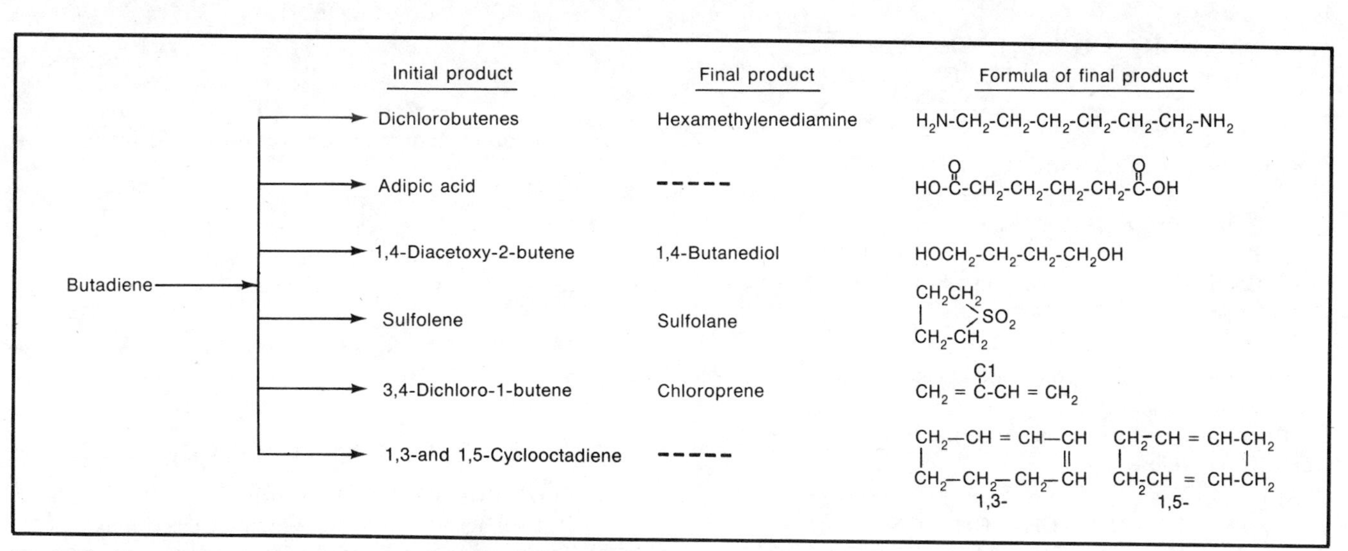

Fig. 9-30—The principle industrial compounds derived from butadiene.

The 1,4-addition product predominates but the 1,2-addition product gives the same 1,4-dinitrile with NaCN (or HCN).

$$CH_2Cl - CH = CH - CH_2Cl + 2\ NaCN \rightarrow NC - CH_2 - CH = CH - CH_2 - CN$$

The reaction conditions are

Temperature	80-95° C
Pressure	Nominal
Catalyst	Cu_2Cl_2
Yield	95%

The *1,4-dicyano-2-butene* is hydrogenated to adiponitrile, $NC - (CH_2)_4 - CN$.

$$NC - CH_2 - CH = CH - CH_2 - CN + H_2 \rightarrow NC - CH_2 - CH_2 - CH_2 - CH_2 - CN$$

The reaction conditions are

Temperature	250°C
Pressure	Nominal
Catalyst	Pd on C
Yield	95-97%

Adiponitrile can be produced by addition of hydrogen cyanide, HCN, to butadiene. This is a two-step process.

$$2\ CH_2 = CH - CH = CH_2 + 2\ HCN \rightarrow CH_3 - CH = CH - CH_2 - CN + CH_2 = CH - CH_2 - CH_2 - CN$$

The reaction conditions are

Temperature	215°C
Pressure	15 psig
Catalyst	CuMgCrO (+HCl, N_2)
Yield	88%

In the second step, HCN reacts with the mononitriles to form to 1,4-dinitrile

$$CH_3 - CH = CH - CH_2 - CN + CH_2 = CH - CH_2 - CH_2 - CN + 2\ HCN \rightarrow 2\ NC - CH_2 - CH_2 - CH_2 - CH_2 - CN$$

The reaction conditions are

Temperature	90°C
Pressure	Liquid phase in THF
Catalyst	Ni (tolylphosphites)$_3$ + $SnCl_2$
Yield	90%

Adiponitrile is also produced by electrodimerization of acrylonitrile, Chapter 8.

Adiponitrile is hydrogenated in the liquid phase to hexamethylenediamine.

$$NC - CH_2 - CH_2 - CH_2 - CH_2 - CN + 4\ H_2 \rightarrow H_2N - CH_2 - CH_2 - CH_2 - CH_2 - CH_2 - CH_2 - NH_2$$

Typical reaction conditions are

Temperature	200°C
Pressure	440 psi
Catalyst	Co
Yield	98%

Hexamethylenediamine is also produced by the reaction between adipic acid, $HO - \overset{O}{\overset{\|}{C}} - (CH_2)_4 - \overset{O}{\overset{\|}{C}} - OH$ and ammonia followed by dehydration.

Uses. Hexamethylenediamine plus adipic acid polymerize forms nylon-66.

Adipic Acid

Production. It has been proposed to produce adipic acid, **AA,** by liquid phase catalytic carbonylation of butadiene.[76]

$$CH_2 = CH - CH = CH_2 + 2\ CO + 2\ H_2O \rightarrow HO - \overset{O}{\overset{\|}{C}} - CH_2 - CH_2 - CH_2 - CH_2 - \overset{O}{\overset{\|}{C}} - OH$$

The reaction conditions are

Temperature	220°C	
Pressure	1100 psi	
Catalyst	$RhCl_2 + CH_3I$ promoter	
Yield	Adipic acid	49%
	α-Methyl glutaric acid	25%
	Valeric acid	26%

Uses. Adipic acid is used to produce nylon-66. Adipic acid also can be used to produce sebasic acid, **SA,**

$$HO - \overset{O}{\overset{\|}{C}} - (CH_2)_8 - \overset{O}{\overset{\|}{C}} - OH$$

by a three-step electrooxidation process.[80]

1. Esterification

$$HO - \overset{O}{\overset{\|}{C}} - (CH_2)_4 - \overset{O}{\overset{\|}{C}} - OH + CH_3OH \xrightarrow{\text{Catalyst}} CH_3 - O - \overset{O}{\overset{\|}{C}} - (CH_2)_4 - \overset{O}{\overset{\|}{C}} - OH$$

Cation exchange resin of sulfonic acid form

2. Electrolysis

$$2\ CH_3 - O - \overset{O}{\overset{\|}{C}} - (CH_2)_4 - \overset{O}{\overset{\|}{C}} - O^-$$

$$2H^+ \xrightarrow{+e} H_2$$

$$\xrightarrow{-e} CH_3 - O - \overset{O}{\overset{\|}{C}} - (CH_2)_8 - \overset{O}{\overset{\|}{C}} - O - CH_3 + 2\ CO_2$$

Current density	10-30 A/dm^2
Cell voltage	5-12 V/pair of electrodes
Temperature	50-60°C

3. Hydrolysis

$$CH_3-O-\overset{O}{\overset{\|}{C}}-(CH_2)_8-\overset{O}{\overset{\|}{C}}-O-CH_3 + 2\,H_2O$$

$$\rightarrow HO-\overset{O}{\overset{\|}{C}}-(CH_2)_8-\overset{O}{\overset{\|}{C}}-OH + 2\,CH_3OH$$

Fig. 9-31 is a flow diagram of this process.[80]

1,4-Butanediol

Production. The production of *1,4-butanediol* (1,4-**BDO**), $HOCH_2-CH_2-CH_2-CH_2OH$, from propylene by way of allyl acetate was noted previously in this Chapter. The over-all yield is 77%. Butadiene also can serve as a starting material for production of 1,4-**BDO** by a three step process to give an over-all yield of about 84%.[77]

The first step is the liquid phase acetoxylation of *butadiene* to *1,4-diacetoxy-2-butene.*

$$2\,CH_2=CH-CH=CH_2 + 4\,CH_3-\overset{O}{\overset{\|}{C}}-OH \rightarrow$$

$$CH_3-\overset{O}{\overset{\|}{C}}-OCH_2-CH=CH-CH_2O-\overset{O}{\overset{\|}{C}}-CH_3 +$$

$$CH_2=CH-\underset{}{CH}(O-\overset{O}{\overset{\|}{C}}-CH_3)-CH_2O-\overset{O}{\overset{\|}{C}}-CH_3$$

The reaction conditions are

Temperature	80°C	
Pressure	27 kg/cm^2	
Catalyst	Pd-Te on carbon	
Yield	1,4-diacetoxy-2 butene	91%
	3,4-Diacetoxy-1-butene	8%

The second step consists of hydrogenation of the *1,4-diacetoxy-2-butene* to *1,4-diacetoxybutane.*

$$CH_3-\overset{O}{\overset{\|}{C}}-OCH_2-CH=CH-CH_2O-\overset{O}{\overset{\|}{C}}-CH_3 + H_2 \rightarrow$$

$$CH_3-\overset{O}{\overset{\|}{C}}-OCH_2-CH_2-CH_2-CH_2O-\overset{O}{\overset{\|}{C}}-CH_3$$

The reaction conditions are

Temperature	80°C
Pressure	880 psi

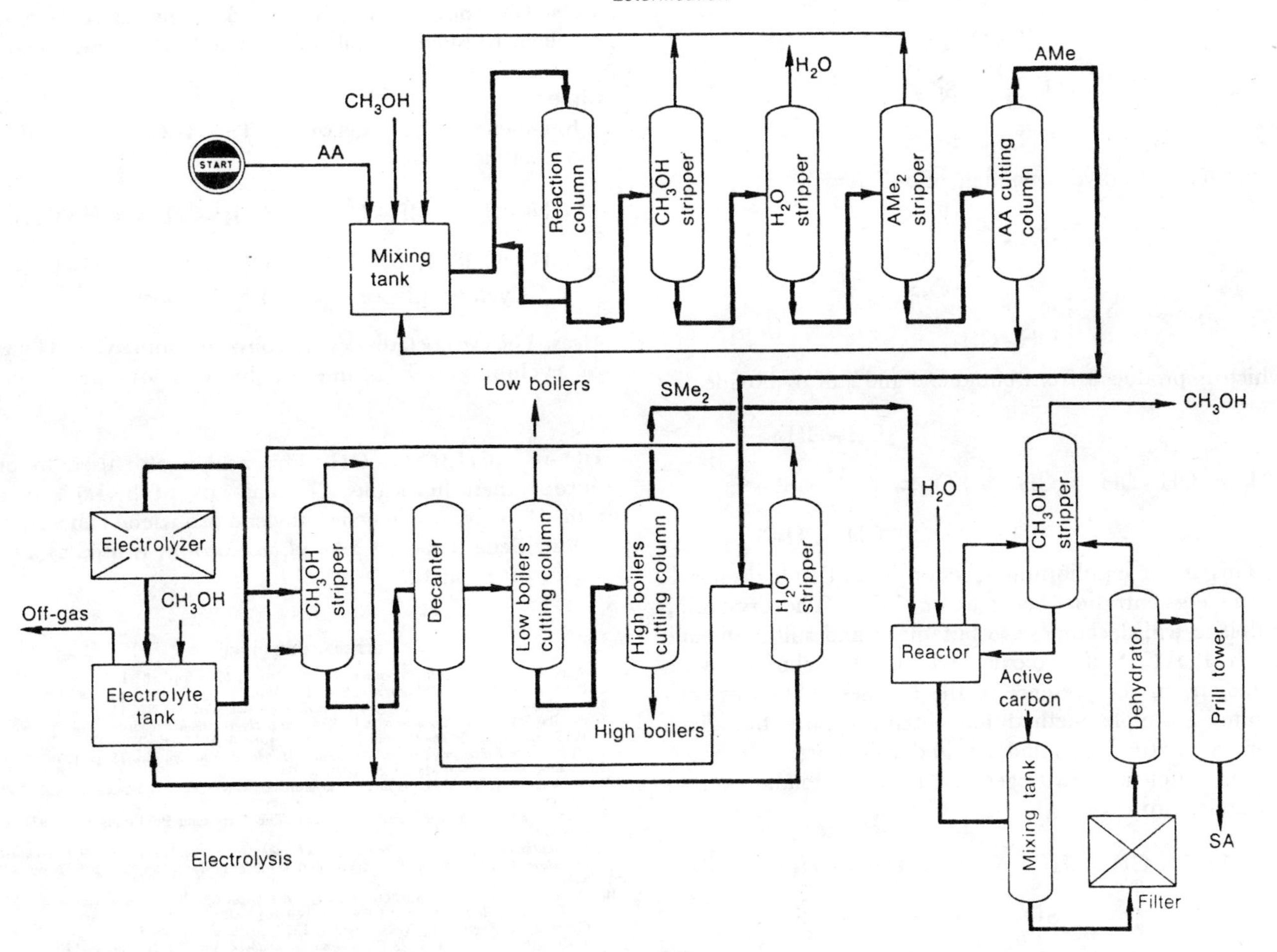

Fig. 9-31—Flow diagram for manufacture of sebasic acid from adipic acid.[80]

Catalyst	Ni-Zn on diatomaceous earth
Yield	98%

The third step is conventional hydrolysis to 1,4-**BDO.**

$$CH_3-\overset{O}{\overset{\|}{C}}-OCH_2-CH_2-CH_2-CH_2O-\overset{O}{\overset{\|}{C}}-CH_3 + 2\,H_2O \rightarrow$$

$$HOCH_2-CH_2-CH_2-CH_2OH + 2\,CH_3-\overset{O}{\overset{\|}{C}}-OH$$

The over-all reaction is

$$CH_2 = CH-CH = CH_2 + 2\,H_2O \rightarrow$$
$$HOCH_2-CH_2-CH_2-CH_2OH$$

The production of 1,4-**BDO** from maleic anhydride is noted in Chapter 10.

Uses. The two main uses of 1,4-butanediol are the production of *tetrahydrofuran*,

$$\begin{array}{l} CH_2-CH_2 \\ \;|\qquad\;\; \diagdown \\ \;|\qquad\quad O, \\ \;|\qquad\;\; \diagup \\ CH-CH_2 \end{array}$$

(57%) and acetylene chemicals (26%). It also goes into the polyurethane, synthetic rubber, thermoplastic polyester and plasticizer industries.

Sulfolane

Production. *Sulfolane* (tetramethylene sulfone),

$$\begin{array}{l} CH_2-CH_2 \\ \;|\qquad\;\; \diagdown \\ \;|\qquad\quad SO_2, \\ \;|\qquad\;\; \diagup \\ CH_2-CH_2 \end{array}$$

is produced by hydrogenation of *sulfolene,*

$$\begin{array}{l} CH-CH_2 \\ \;\|\qquad\;\; \diagdown \\ \;\|\qquad\quad SO_2, \\ \;\|\qquad\;\; \diagup \\ CH-CH_2 \end{array}$$

which is produced from butadiene and sulfur dioxide.

$$CH_2 = CH-CH = CH_2 + SO_2 \leftrightarrows \begin{array}{l} CH-CH_2 \\ \;\|\qquad\;\; \diagdown \\ \;\|\qquad\quad SO_2 \\ \;\|\qquad\;\; \diagup \\ CH-CH_2 \end{array}$$

This is an equilibrium reaction with the highest sulfolene concentration being at *ca* 75°C. The crystalline sulfolene will decompose to butadiene and sulfur dioxide at *ca* 125°C. Sulfur dioxide will react exclusively with butadiene in the presence of the butenes. Thus this constitutes a simple method for obtaining pure butadiene from a mixture of butadiene and *n*-butenes.

The sulfolene is hydrogenated to the sulfolane by conventional processes.

$$\begin{array}{l} CH-CH_2 \\ \;\|\qquad\;\; \diagdown \\ \;\|\qquad\quad SO_2 \\ \;\|\qquad\;\; \diagup \\ CH-CH_2 \end{array} + H_2 \rightarrow \begin{array}{l} CH_2-CH_2 \\ \;|\qquad\;\; \diagdown \\ \;|\qquad\quad SO_2 \\ \;|\qquad\;\; \diagup \\ CH_2-CH_2 \end{array}$$

Uses. A mixture of sulfolane and diisopropanolamines is used for acid gas removal, especially carbon dioxide, the Sulfinol process. Sulfolane is also used for the extraction of aromatics from petroleum or coke-oven sources. High purity aromatics are produced. High octane number aromatic concentrates for gasoline blending also can be produced by this process.

Chloroprene

Production. *Chloroprene* (2 - Chloro - 1,3 - butadiene), $CH_2 = \overset{Cl}{\overset{|}{C}}-CH = CH_2$, is produced from butadiene by high temperature chlorination followed by isomerization to *3,4-dichloro-1-butene*, $CH_2 = CH-CHCl-CH_2Cl$, which is dehydrochlorinated to chloroprene.

$$CH_2 = CH-CHCl-CH_2Cl \rightarrow$$
$$CH_2 = CH-\overset{Cl}{\overset{|}{C}} = CH_2 + HCl$$

The conventional synthesis of chloroprene is the addition of hydrogen chloride to *vinyl acetylene*, $CH_2 = CH - C \equiv CH$.

$$CH_2 = CH-C \equiv CH + HCl \rightarrow CH_2 = CH-\overset{Cl}{\overset{|}{C}} = CH_2$$

Uses. Chloroprene is polymerized to give a rubber with excellent resistance to oil, solvents and ozone-cracking.

Dimers

Butadiene can be dimerized by $TiCl_3/AlCl_3$ to 1.3- and 1,5-*cyclooctadienes*.

$$\begin{array}{l} CH_2-CH = CH-CH \\ | \qquad\qquad\qquad\quad \| \\ CH_2-CH_2-CH_2-CH \end{array} \qquad \begin{array}{l} CH_2-CH = CH-CH_2 \\ | \qquad\qquad\qquad\quad | \\ CH_2-CH = CH-CH_2 \end{array}$$

1,3-Cyclooctadiene   1,5-Cyclooctadiene

Uses. The cyclooctadienes are converted into nylon-8 by way of cyclooctanone oxime or by way of *suberic acid*, $HO-\overset{O}{\overset{\|}{C}}-(CH_2)_6-\overset{O}{\overset{\|}{C}}-OH$. This acid is also used to produce synthetic lubricants. The major use of the 1,5-isomer is a the third comonomer in ethylene-propylene rubber.

Butadiene trimer, *1,5,9-cyclododecatriene,* is used as a precursor of nylon-12.

LITERATURE CITED

1. Spitz, P.H. *Hydrocarbon Processing*, Vol. 55, No. 7, 1976, pp. 131-135.
2. Ponder, Thomas C. *Hydrocarbon Processing*, Vol. 57, No. 7, 1978, pp. 187-189.
3. *Hydrocarbon Processing, 1979 Petrochemical Handbook*, Vol. 58, No. 11, p. 124.
4. Pujada, P.R., B.V. Vora and A.P. Krueding. *Hydrocarbon Processing*, Vol. 56, No. 5, 1977, pp. 169-172; *The Oil and Gas Journal*, June 6, 1977, pp. 171-172.
5. *Hydrocarbon Processing, 1977 Petrochemical Handbook*, Vol. 56, No. 11, p. 125.
6. *Chemical Engineering*, March 29, 1976, p. 64.
7. Hucknall, D.J. *Selective Oxidation of Hydrocarbons*, New York: Academic Press Inc., 1974, pp. 55-69.
8. *International Petroleum Encyclopedia*, Tulsa: The Petroleum Publishing Co., 1977, pp. 416-417.
9. *Hydrocarbon Processing, 1975 Petrochemical Handbook*, Vol. 54, No. 11, p. 109; *International Petrochemical Encyclopedia*, Tulsa: The Petroleum Publishing Company, 1977, pp. 417-418.
10. Dumas, T. and W. Bulani, *Oxidation of Petrochemicals; Chemistry and Technology*, New York: John Wiley and Sons, 1974, pp. 140-161.
11. *Chemical and Engineering News*, April 4, 1977, p. 12.
12. *Hydrocarbon Processing, 1977 Petrochemical Handbook*, Vol. 56, No. 11, p. 126.
13. *Hydrocarbon Processing, 1977 Petrochemical Handbook*, Vol. 56, No. 11, p. 127.
14. Davis, J.C. *Chemical Engineering*, Vol. 82, No. 14, 1975, pp. 44-48.

15. Stobaugh, R.B., V.A. Caloro, R.A. Morris and L.W. Stroud. *Hydrocarbon Processing,* Vol. 52, No. 1, 1973, pp. 99-108.
16. Simmrock, K.H. *Hydrocarbon Processing,* Vol. 57, No. 11, 1978, pp. 107-113.
17. Yamagishi, K. O. Kageyama, H. Haruki, and Y. Numa. *Hydrocarbon Processing,* Vol. 55, No. 7, 1976, pp. 102-104.
18. *Hydrocarbon Processing, 1979 Petrochemical Handbook,* Vol. 58, No. 11, p. 240.
19. Brownstein, A.M. *Trends in Petrochemical Technology,* Tulsa: Petroleum Publishing Co., 1976, p. 163.
20. Landau, R., D. Brown, J.R. Russell and J. Kollar. "Proceedings of the Seventh World Petroleum Congress," Vol. 5, *Petrochemicals,* 1967, pp. 67-72.
21. Imacura, J., R. Wakasa and K. Kataoka. Japan Patent 43,926, 1976, to Ashi Co.; *CHEMTECH,* Vol. 5, No. 10, 1975, p. 579.
22. *Chemical Week,* June 8, 1977, p. 39.
23. McMullen, C.H., U.S. Patent 3,993,673 to Union Carbide Corp.; *CHEMTECH,* Vol. 7, No. 9, 1977, p. 536.
24. *CHEMTECH,* Vol. 7, No. 9, 1977, p. 536; Japan Patent 76,040,050 to Mitsubishi Chem. Ind.
25. British Patent 1,124,862; *CHEMTECH,* Vol. 6, No. 4, 1976, p. 219.
26. Brownstein, A., J. Jung and R. Hansen. West German Offen 2,412,136, 1974, to ChemSystems; *CHEMTECH,* Vol. 6, No. 4, 1976, p. 219.
27. Zimmer, J.C. *Hydrocarbon Processing,* Vol. 56, No. 12, 1977, pp. 115-116.
28. Yamagishi, K. and O. Kageyama. *Hydrocarbon Processing,* Vol. 55, No. 11, 1976, pp. 139-144.
29. *Hydrocarbon Processing, 1973 Petrochemical Handbook,* Vol. 52, No. 11, p. 142.
30. Hatch, Lewis F. *The Chemistry of Petrochemical Reactions,* Houston: Gulf Publishing Company, 1955, p. 76.
31. *Hydrocarbon Processing, 1979 Petrochemical Handbook,* Vol. 58, No. 11, p. 181.
32. *Hydrocarbon Processing, 1979 Petrochemical Handbook,* Vol. 58, No. 11, p. 122.
33. Onoue, V., Y. Mizutani, S. Akiyama, Y. Izumi and Y. Watanabe. *CHEMTECH,* Vol. 7, No. 1, 1977, pp. 36-39.
34. *Chemical and Engineering News,* Sept. 6, 1977, p. 32.
35. *Hydrocarbon Processing, 1975 Petrochemical Handbook,* Vol. 54, No. 11, p. 105.
36. *Hydrocarbon Processing, 1979 Petrochemical Handbook,* Vol. 58, No. 11, p. 123.
37. Fowler, R., H. Connor, and R.A. Bael. *Hydrocarbon Processing,* Vol. 55, No. 9, 1976, pp. 247-249; *CHEMTECH,* Vol. 6, No. 12, 1976, pp. 772-775; *The Oil and Gas Journal,* Sept. 13, 1976, pp. 92, 94.
38. Brewester, E.A.V. *Chemical Engineering,* Nov. 8, 1976, pp. 90-91.
39. Cornils, B., R. Payer and K.C. Traencker. *Hydrocarbon Processing,* Vol. 54, No. 6, 1975, pp. 83-91.
40. Cornils, B. and A. Mullen. *Hydrocarbon Processing,* Vol. 56, No. 4, 1977, p. 88.
41. *Hydrocarbon Processing, 1979 Petrochemical Handbook,* Vol. 58, No. 11, p. 143.
42. *Hydrocarbon Processing, 1979 Petrochemical Handbook,* Vol. 58, No. 11, p. 172.
43. Weber, H., W. Dimmling and A.M. Desai. *Hydrocarbon Processing,* Vol. 55, No. 4, 1976, pp.129-130.
44. *CHEMTECH,* Vol. 7, No. 6, 1977, p. 365; U.S. Patent 3,987,103 to Rhone-Progil.
45. *Chemical and Engineering News,* Mar. 6, 1978, p. 12.
46. *CHEMTECH,* Vol. 5, No. 9, 1975, p. 517.
47. Rona, P., L. Mednick and Y. Ehrenreich. *Hydrocarbon Processing,* Vol. 54, No. 4, 1975, pp. 185-186.
48. *CHEMTECH,* Vol. 6, No. 2, 1976, p. 133.
49. Brownstein, A.M. and H.L. List. *Hydrocarbon Processing,* Vol. 56, No. 10, 1977, pp. 159-162.
50. Anderson, K.L. and T.D. Brown. *Hydrocarbon Processing,* Vol. 55, No. 8, 1976, pp. 119-122.
51. Guerico, V.J. *The Oil and Gas Journal,* Feb. 21, 1977, pp. 68-71.
52. *Chemical Week,* Nov. 16, 1977, p. 49.
53. *Hydrocarbon Processing, 1975 Petrochemical Handbook,* Vol. 54, No. 11, p. 118.
54. Austin, G.T. *Chemical Engineering,* June 24, 1974, p. 153.
55. Krönig, W. "The Proceedings of the Seventh World Petroleum Congress," Vol. 5, *Petrochemicals,* 1967, pp. 59-65.
56. *Hydrocarbon Processing, 1979 Petrochemical Handbook,* Vol. 58, No. 11, p. 120.
57. Brockhaus, R. German Patent 1,279,011, 1968.
58. Hucknall, D.J. *Selective Oxidation of Hydrocarbons,* London, New York: Academic Press, 1974, pp. 97-101.
59. *Hydrocarbon Processing, 1979 Petrochemical Handbook,* Vol. 58, No. 11, p. 188.
60. *The Oil and Gas Journal,* Apr. 23, 1979, p. 42.
61. Hasuike, Tooru and H. Matsuzawa, *Hydrocarbon Processing,* Vol. 58, No. 2, 1979, pp. 105-107.
62. Pecci, G. and T. Floris. *Hydrocarbon Processing,* Vol. 56, No. 12, 1977, pp. 98-101.
63. *Hydrocarbon Processing, 1979 Petrochemical Handbook,* Vol. 58, No. 11, p. 197.
64. Csikos, R., I. Pallay, J. Laky, E.D. Radcsenko, B.A. Englin and J.A. Robert. *Hydrocarbon Processing,* Vol. 55, No. 7, 1976, pp. 121-125.
65. Chase, J.D. and H.J. Woods. *The Oil and Gas Journal,* Apr. 9, 1979, pp. 149-152.
66. Boboleva, S.P., M.G. Bulgin and E.A. Blyumberg. *Petroleum Chemistry,* Vol. 14, No. 3, 1974, pp. 193-200.
67. *CHEMTECH,* Vol. 5, No. 3, 1975, p. 189.
68. West German Offen. 2,354,331, to Atlantic Richfield.
69. British Patent 1,182,273 to Teijin.
70. Abu-Elgheit, M. *Preprints,* Vol. 20, No. 1, Division of Petroleum Chemistry, Inc., 1975, pp. 77-81.
71. Hucknall, D.J., *Selective Oxidation of Hydrocarbons,* London, New York: Academic Press, 1974, pp. 104-111.
72. Oda, Y., I Gotoh, K. Uchida, T. Morimoto, J. Endoh, and T. Ueno. *Hydrocarbon Processing,* Vol. 54, No. 10, 1975, pp. 115-117.
73. *Hydrocarbon Processing, 1977 Petrochemical Handbook,* Vol. 56, No. 11, p. 170.
74. *Hydrocarbon Processing, 1977 Petrochemical Handbook,* Vol. 56, No. 11, p. 186.
75. Hatch, Lewis F. *The Chemistry of Petrochemical Reactions,* Houston: Gulf Publishing Company, 1955, p. 149.
76. Belgian Patent 770,615 to BASF, 1971.
77. Brownstein, A.M. and H.L. List. *Hydrocarbon Processing,* Vol. 56, No. 9, 1977, pp. 159-162.
78. Landau, Ralph, G.A. Sullivan and D. Brown. *CHEMTECH,* Vol. 9, No. 10, 1979, pp. 602-607.
79. Kuhn, Wayne. *Hydrocarbon Processing,* Vol. 58, No. 9, 1979, pp. 123-128.
80. Seko, S., A. Yomiyama and T. Isoya. *Hydrocarbon Processing,* Vol. 58, No. 12, 1979, pp. 117-118.
81. Clementi, A., G. Oriani, F. Ancillotti and G. Pecci. *Hydrocarbon Processing,* Vol. 58, No. 12, 1979, pp. 109-113.
82. Taniguchi, Brian and R.T. Johnson, *CHEMTECH,* Vol. 9, No. 8, 1979, pp. 502-510.

10

Petrochemicals from Benzene, Toluene, the Xylenes

Benzene, toluene and the xylenes, BTX, are aromatic hydrocarbons with a wide use as petrochemicals. They are important precursors for plastics such as nylon, polyurethane, polyesters, alkyd resins, and phthalate plasticizers. These represent the large scale applications. On the lesser scale, they are precursors for insecticides, weed killers, medicinals, and dyes.

BENZENE

Benzene, C_6H_6, is the simplest aromatic hydrocarbon and by far the most widely utilized one. Several methods are employed to represent the structure of benzene.

C_6H_6

The dotted circle and the alternating single and double bonds represent delocalized pi, π, electrons. The simple hexagon representation of benzene is understood to have a carbon atom and a hydrogen atom at each angle. When an atom or group of atoms replace one or more of the hydrogen atoms of benzene, a substituted benzene is formed and the groups or atoms are shown:

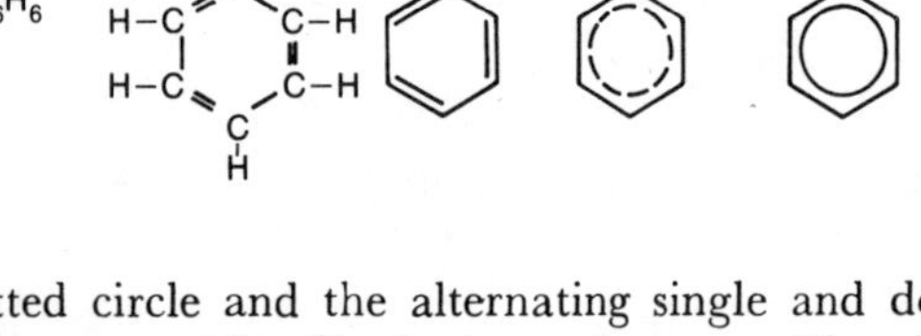

OH — Phenol; CH_2-CH_3 — Ethylbenzene; CH_3 — Toluene, Methylbenzene

The delocalized electrons associated with the benzene ring impart very special properties to aromatic hydrocarbons. They have chemical properties of neither single-bond compounds such as paraffin hydrocarbons nor double-bond compounds such as the olefins. They have some properties of both and many of their own. The aromatic hydrocarbons, like the paraffin hydrocarbons, react by substitution but by a different reaction mechanism and under milder conditions. They react by addition only under severe conditions.

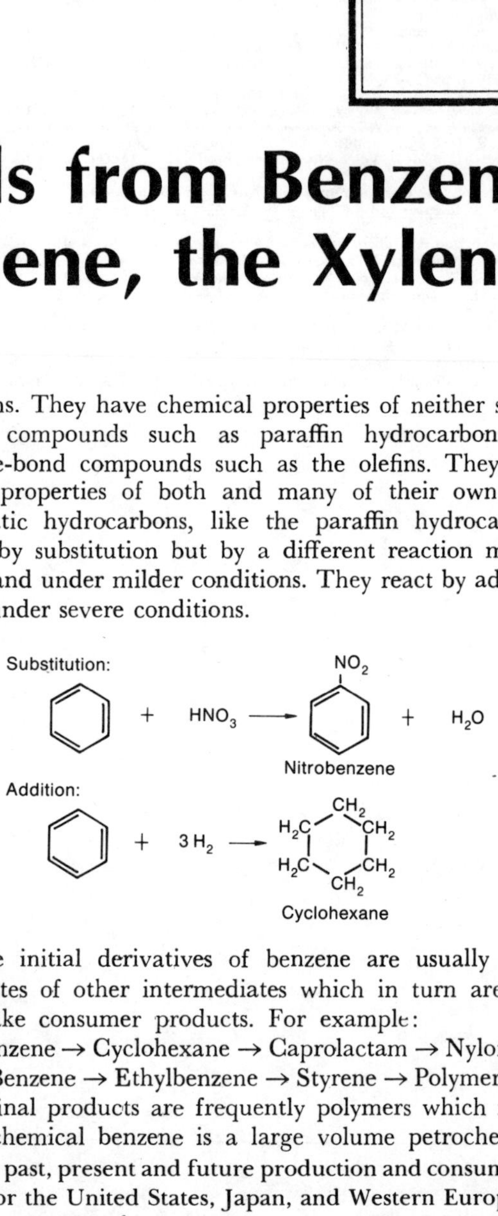

Substitution:

$$C_6H_6 + HNO_3 \longrightarrow C_6H_5NO_2 + H_2O$$

Nitrobenzene

Addition:

$$C_6H_6 + 3H_2 \longrightarrow C_6H_{12}$$

Cyclohexane

The initial derivatives of benzene are usually intermediates of other intermediates which in turn are used to make consumer products. For example:

Benzene → Cyclohexane → Caprolactam → Nylon 6

Benzene → Ethylbenzene → Styrene → Polymers

The final products are frequently polymers which means that chemical benzene is a large volume petrochemical.

The past, present and future production and consumption data for the United States, Japan, and Western Europe are given in Fig. 10-1.[1] Fig. 10-2 shows the various United States sources of benzene in 1978 and the predicted sources in 1985.[2] The main difference is in the estimated quantity of benzene associated with ethylene production. This, in turn, is dependent on the feedstock. As the U.S. ethylene production moves to heavier feeds, more benzene will come from this source.

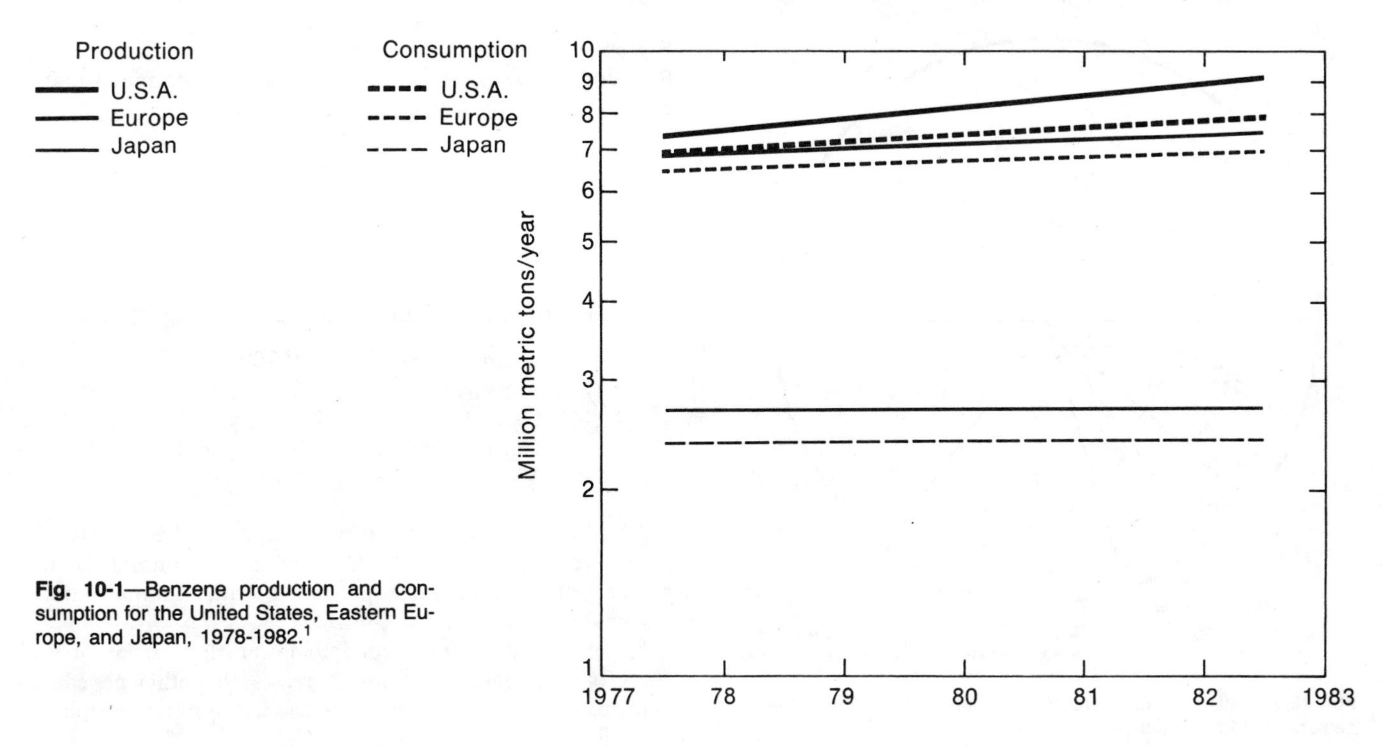

Fig. 10-1—Benzene production and consumption for the United States, Eastern Europe, and Japan, 1978-1982.[1]

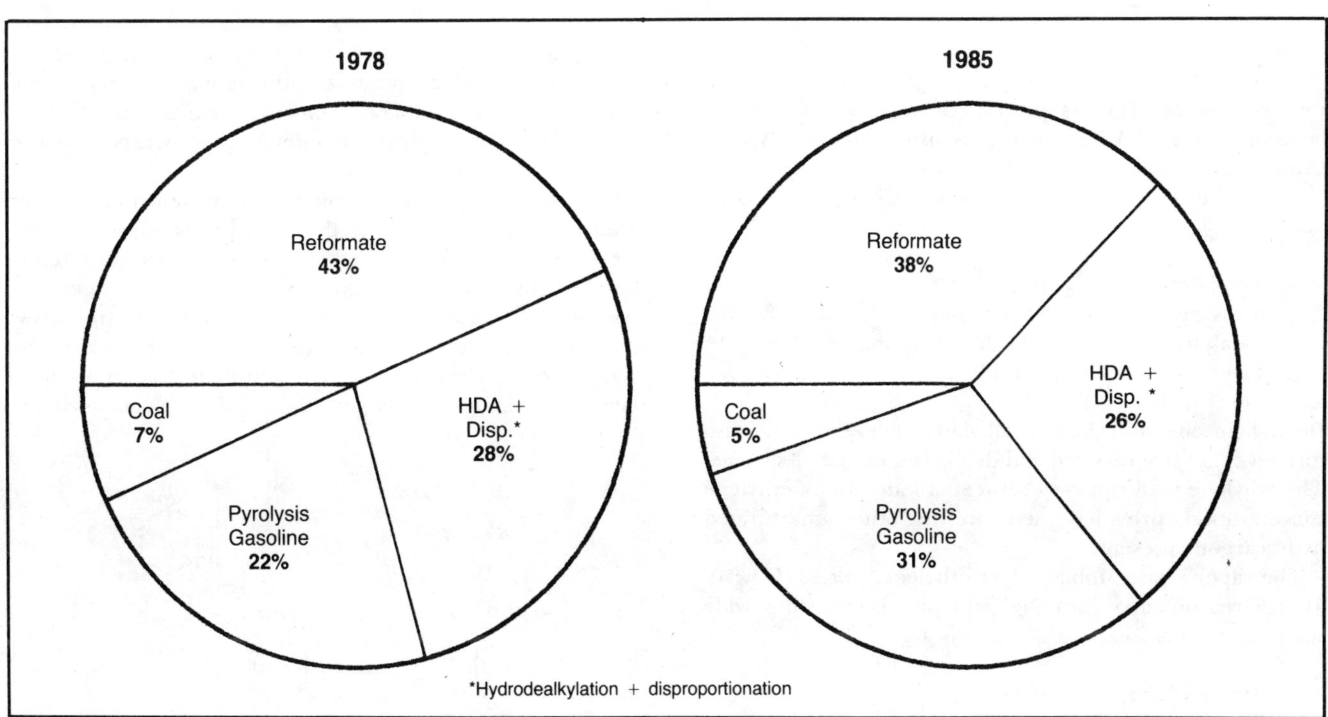

Fig. 10-2—The sources of benzene in 1978 (5,037 million metric tons) and predicted sources in 1985 (7,650 million metric tons).[2]

The chemical utilization of benzene is illustrated in Fig. 10-3.

Ethylbenzene

Production. A small quantity of *ethylbenzene*, C_6H_5-CH_2-CH_3, is obtained by intensive fractionation of reformate and other refinery streams. Over 90% is synthesized by the alkylation reaction between benzene and ethylene.

$$+ \quad CH_2 = CH_2 \longrightarrow \quad (CH_2-CH_3)$$

Many different catalysts are available for this reaction with the very flexible and reliable $AlCl_3$•HCl-hydrocarbon complex being the most common. *Ethyl chloride,* CH_3CH_2Cl, may be substituted for hydrogen chloride on a mole-for-mole basis. The aluminum chloride alkylation process has been

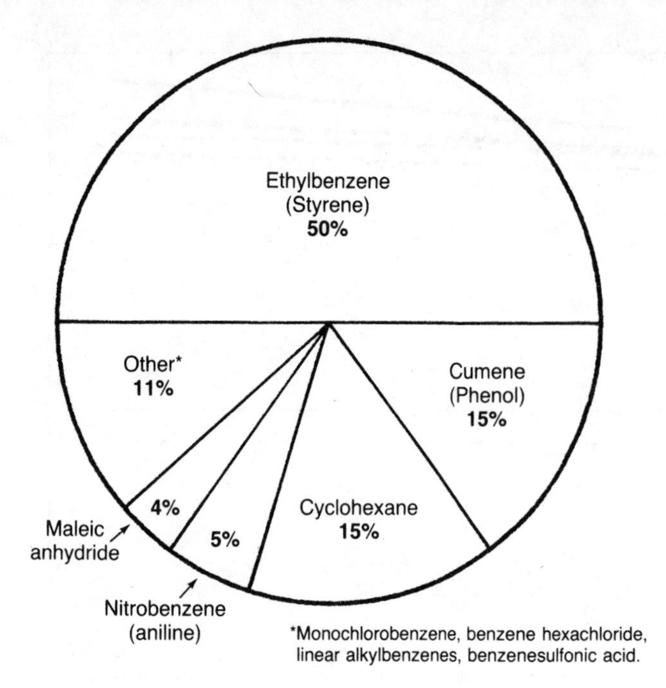

Fig. 10-3—Initial use (and end use) of benzene in 1978 based on demand of 11.13 billion pounds.

reviewed by MacFarlane in respect to Monsanto's ethylbenzene process. [3] A BF_3 complex is also a viable catalyst for benzene alkylation.

Typical reaction conditions for the $AlCl_3$ catalyzed reaction are:

Temperature:	40-100°C
Pressure:	30-100 psig
Catalyst:	$AlCl_3$/hydrocarbon complex
Yield:	*ca* 100%

Diethylbenzene and higher alkylated benzenes are also formed. They are recycled and dealkylated to ethylbenzene. The ethylene feed can vary between 10 and 100% ethylene concentration, provided there are no other unsaturated hydrocarbons present.

The vapor-phase Mobil-Badger ethylene process (Fig. 10-4) is a recent entry into the field and is receiving wide acceptance.[4] The reaction conditions are:[4, 5, 6, 7]

Temperature:	420°C
Pressure:	200-300 psig
Catalyst:	Crystalline aluminosilicate zeolite
Conversion:	85-90%
Yield:	98-100%

The mole ratio of benzene to ethylene is *ca* 7.5 to 1. Two reactors are used—one onstream while the other is being regenerated.

Another high temperature (260-285°C) process utilizes a solid phosphoric acid catalyst and a 98% ethylene feed.[8] The pressure is 900-950 psig.

Uses. Ethylbenzene has a limited use as a solvent and for the production of dyes. Essentially its only reason for being is to produce *styrene,* $C_6H_5\text{-}CH{=}CH_2$.

Styrene

Production. *Styrene* is produced by the endothermic, vapor phase, catalytic dehydrogenation of ethylbenzene.

$$C_6H_5\text{-}CH_2\text{-}CH_3 \longrightarrow C_6H_5\text{-}CH{=}CH_2 + H_2$$

Typical reaction conditions are:

Temperature:	600-700° C
Pressure:	Atmospheric
Catalyst:	Fe-Cr oxides*
Conversion:	30-40 percent
Yield:	90 percent

*A wide variety of catalysts appears in the literature including Fe_2O_3, $SiO_2-Al_2O_3$, solid phosphoric acid, cobalt oxides, and ZnO promoted with alumina and chromates.

Super-heated steam is used to lower the partial pressures of the reactants. This increases the ethylene conversion. The steam also decokes the catalyst and supplies sensible heat.

Optimization of a styrene process is very complex. Portes and Escourrou, however, have produced a mathematical model which permits optimization of a new plant or improves operating conditions of an existing unit.[9]

Frequently the ethylbenzene unit and the styrene unit are combined as illustrated by Fig. 10-5.[10]

Because the dehydrogenation of ethylbenzene to styrene is relatively energy intensive, there have been quite a few alternative methods proposed for the production of styrene. They are, in general, characterized more by their ingenuity than by their feasibility. Styrene can be produced from raw materials and processes as divergent as direct oxidative coupling of ethylene and benzene and dimerization of butadiene to 4-vinyl-1-cyclohexene followed by catalytic dehydrogenation to styrene.[11]

$$2\,CH_2{=}CH\text{-}CH{=}CH_2 \longrightarrow \text{4-vinyl-1-cyclohexene}$$

$$\text{4-vinyl-1-cyclohexene} \longrightarrow C_6H_5\text{-}CH{=}CH_2 + 2\,H_2$$

Styrene has also been produced by the oxidative coupling of toluene to form stilbene followed by disproportionation to styrene and benzene.[11]

$$2\,C_6H_5\text{-}CH_3 + O_2 \longrightarrow C_6H_5\text{-}CH{=}CH\text{-}C_6H_5 + 2\,H_2O$$

$$C_6H_5\text{-}CH{=}CH\text{-}C_6H_5 + H_2 \longrightarrow C_6H_5\text{-}CH{=}CH_2 + C_6H_6$$

The temperature is high (600°C) and yields are low.

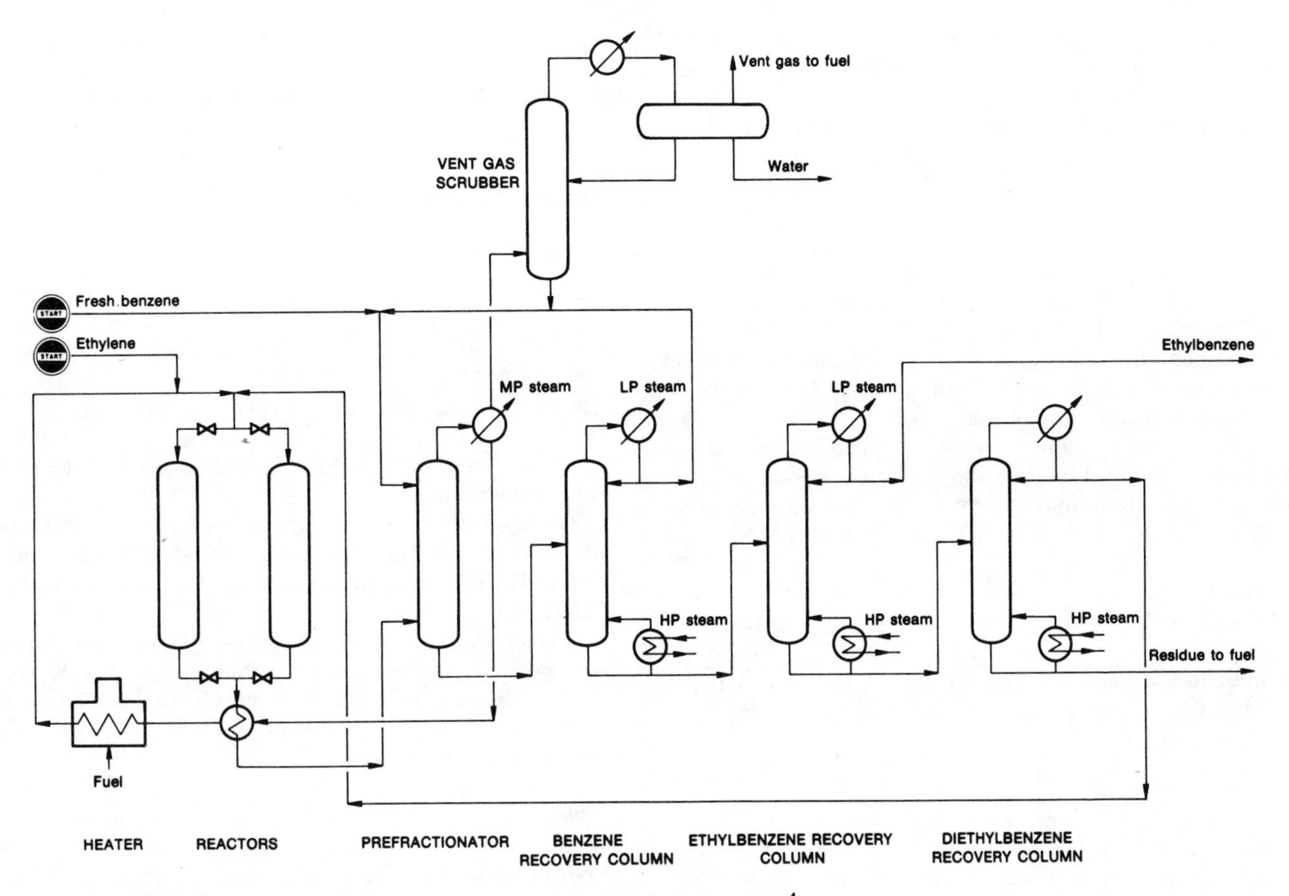

Fig. 10-4—The Mobil-Badger process for the vapor-phase production of ethylbenzene.[4]

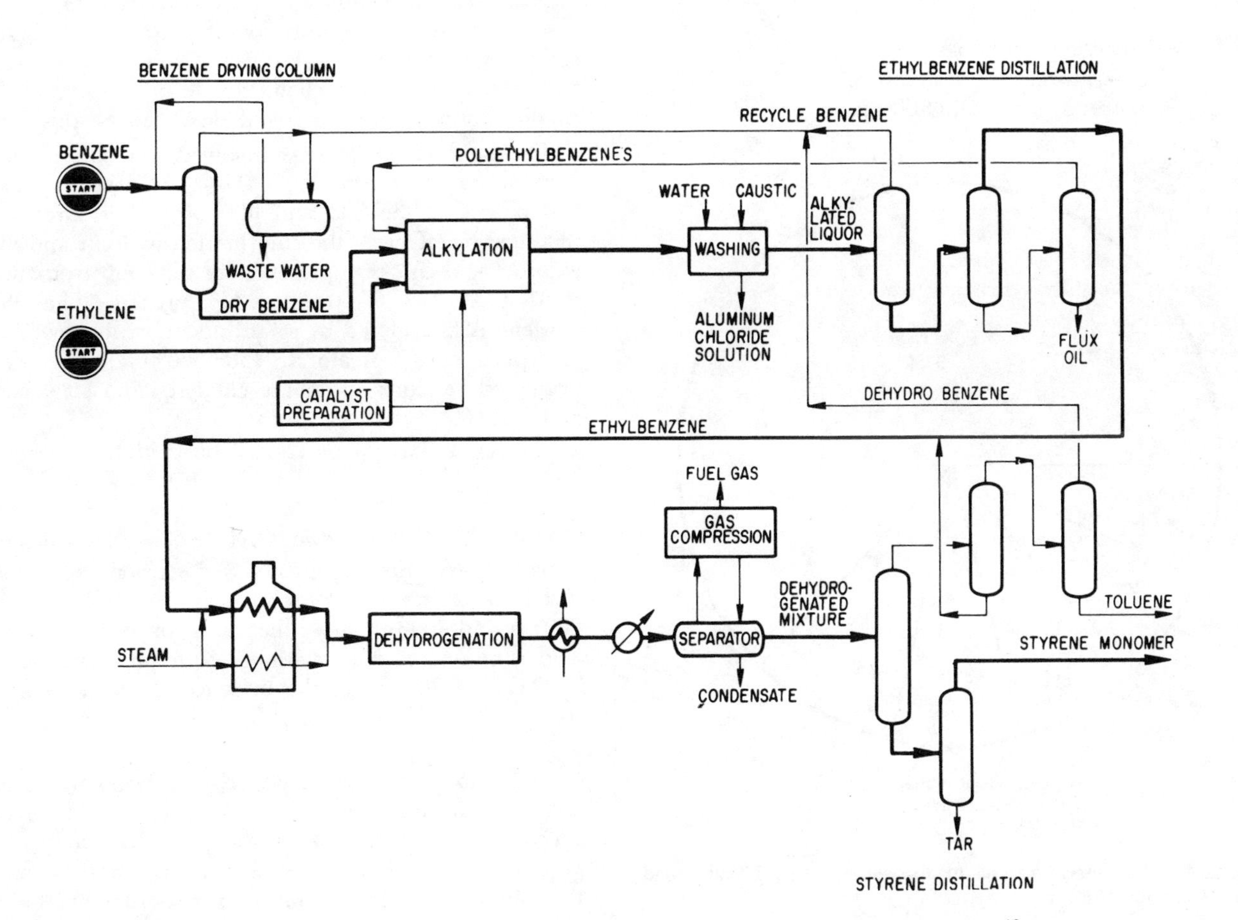

Fig. 10-5—Monsanto-C-E Lummus continuous process for the production of styrene from ethylene and benzene.[10]

These and other processes are reviewed by Brownstein.[12] Ethylbenzene and styrene production has also been reviewed extensively by Landau, *et al.*[13]

The production of styrene by the dehydration of *α-phenylethyl alcohol,* C_6H_5 - CHOH - CH_3, has been noted in Chapter 9. The alcohol is a coproduct of propylene oxide production by ethylbenzene hydroperoxide epoxidation of propylene.

It has been predicted that during the 1977-1982 period styrene demand in the U.S. will grow 5.8% per year, but export markets will cut net growth in U.S. production requirements to 4-4.5% growth per year.[14] The 1982 styrene operating rate will be only 69% of capacity. The worldwide annual growth rate will average 7%.[15]

The 1978 production of styrene was 6.88 billion pounds. Capacity at the end of 1978 was 8.43 billion pounds.

Uses. Styrene is an important monomer used to produce a variety of elastomers, plastics, and resins (Fig. 10-6).[16]

Cumene

Production. *Cumene* (isopropylbenzene), $C_6H_5\text{-}\overset{\overset{CH_3}{|}}{CH}\text{-}CH_3$, is produced by catalytic vapor phase, propylene alkylation of benzene (Fig. 10-7).[17]

$$C_6H_6 + CH_3\text{-}CH=CH_2 \longrightarrow C_6H_5\text{-}\overset{\overset{H}{|}}{C}(CH_3)\text{-}CH_3 \qquad \Delta H_{25°} = -23.8 \text{ kcal}$$

Typical reaction conditions are:

Temperature:	200-250° C
Pressure:	400-600 psi

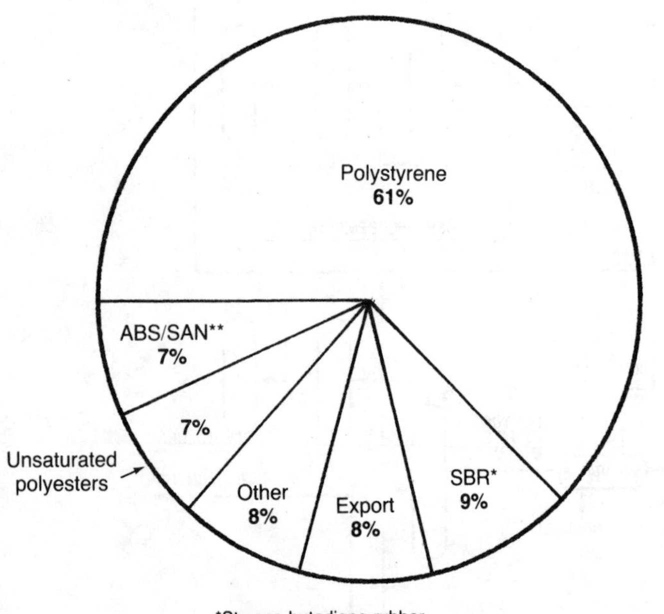

*Styrene-butadiene rubber
**Acrylonitrile-butadiene-styrene/styrene-acrylonitrile

Fig. 10-6—The styrene markets for the estimated 8.2 billion pound U.S. demand in 1982.[16]

Catalyst:	H_3PO_4 on kieselguhr or pumice
Yields:	On benzene: 96-97%
	On propylene: 91-92%

Sulfuric acid has been used as a catalyst in liquid phase alkylation.

The propylene feed can be as low as 40% propylene with propane as the other component. Small amounts of ethylene can be tolerated because under propylene alkylation conditions ethylene is quite unreactive. However, ethylene is more readily removed during the feed gas preparation than ethylbenzene is from the cumene product stream. Butylenes are relatively unimportant because butylbenzene can be removed as bottoms from the cumene column.

Uses. Like ethylbenzene, cumene has only one utilization that justifies its synthesis. Essentially all of its 3.2 billion pound 1978 production was used in the synthesis of *phenol,* C_6H_5OH.

Unlike styrene from ethylbenzene, phenol from cumene does not have new processes breathing down its neck. There are no challengers in the foreseeable future. The other existing processes are old ones still operating because of unique economic situations.

Phenol

Production. The first commercial synthesis of phenol was from benzene by sulfonation to *benzenesulfonic acid,* $C_6H_5SO_3H$. The sulfonic acid was converted to its sodium salt and then by caustic fusion at 300° to 340° C to the sodium salt of phenol, C_6H_5ONa. The phenol was sprung from its solution by carbon dioxide or sulfur dioxide. A small amount of phenol is still produced by this process.

The next viable process involved the chlorination of benzene to *chlorobenzene,* C_6H_5Cl, followed by caustic fusion at 320°-350° C and 4000 psi. This process was modified by effecting the chlorination with air and hydrochloric acid in the presence of a copper-on-alumina catalyst at 275° C—the Raschig process. The chlorobenzene is hydrolyzed by steam under various conditions. A typical one is steam at 450°-500° C over a copper promoted calcium phosphate catalyst. The conversion is about 12%.

Another industrial process is the oxidation of *toluene,* $C_6H_5\text{-}CH_3$, to *benzoic acid,* $C_6H_5 - \overset{\overset{O}{\|}}{C} - OH$, followed by a second oxidation to phenol and carbon dioxide. The first step utilizes cobalt bromide as the catalyst at 140°-200° C and 400 psi with either air or oxygen. The benzoic acid is caused to react with air or oxygen and steam at 230° C and 20 psig in the presence of cupric ion, Cu^{2+}, to give phenol and carbon dioxide. The overall yield is about 80%.

The cumene process, however, can be regarded as the only one currently with industrial significance. It probably accounts for about 90% of the phenol capacity in the Western world.[17] The process not only accounts for most of the phenol production but it is also one of the most

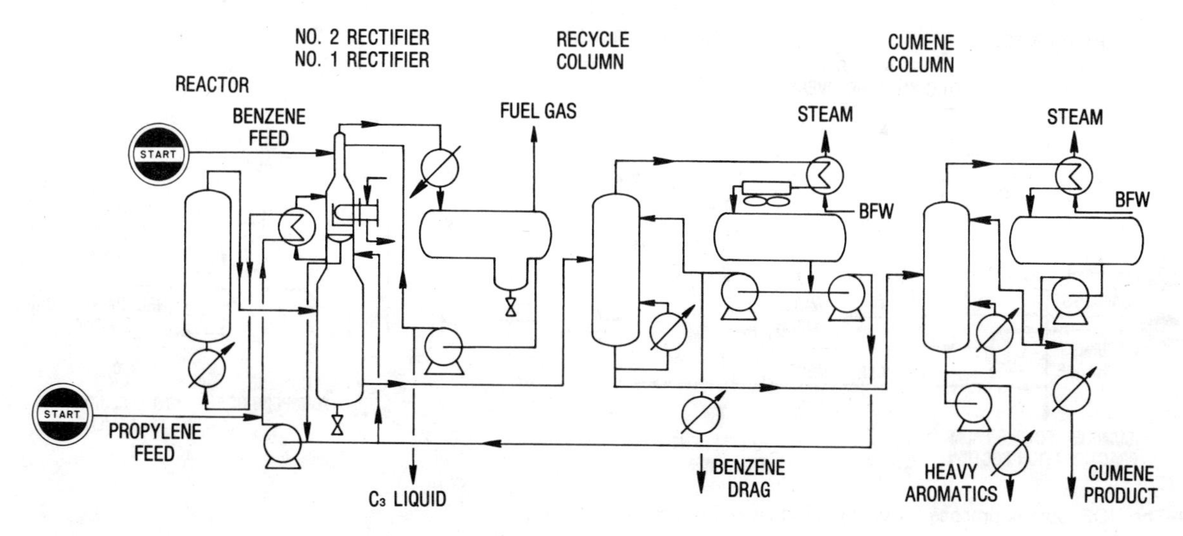

Fig. 10-7—The UOP catalytic alkylation process for cumene synthesis.[17]

competitive processes for *acetone,* $CH_3-\overset{\overset{\displaystyle O}{\|}}{C}-CH_3$, production—its coproduct.

Phenol from cumene is a two step process with the first step being the production of *cumene hydroperoxide,* **CHP,** $C_6H_5-\underset{\underset{\displaystyle CH_3}{|}}{\overset{\overset{\displaystyle CH_3}{|}}{C}}-OOH$. The second step is the decomposition of the hydroperoxide to phenol and acetone.

The cumene hydroperoxidation reaction is:

$$C_6H_5C(H)(CH_3)_2 + O_2 \longrightarrow C_6H_5C(O\text{-}OH)(CH_3)_2 \qquad \Delta H_{25^\circ C} = -27.7 \text{ kcal}$$

General reaction conditions for the liquid phase reaction are:

Temperature:	100-130° C (130-140° C)
Pressure:	25-50 psig (20-250 psi)
Catalyst:	*
Conversion:	30-50%
Yields:	85-95%

* May be metal salts.

Usually mild bases (Na_2CO_3) are added to inhibit acid catalyzed decomposition of the hydroperoxide.

The hydroperoxide decomposition reaction is:

$$C_6H_5C(O\text{-}OH)(CH_3)_2 \xrightarrow{H^+} C_6H_5OH + CH_3-\overset{O}{\overset{\|}{C}}-CH_3 \qquad \Delta H_{25^\circ C} = -60.4 \text{ kcal}^*$$

* Also reported to be 53 kcal/mole.[18]

The general decomposition conditions are:

Temperature:	60-80°C
Pressure:	5 psig
Catalyst:	H^+
Yields:	90-95%

The hydrogen ion catalyst may be any strong mineral acid (10% H_2SO_4), ion-exchange resins or even sulfur dioxide.

Both the oxidation and cleavage reactions are exothermic and care must be taken to prevent runaway conditions which can result in spontaneous thermal decomposition of the **CHP.** Each step involves many safety considerations. These have been reviewed by Fleming, Lambrix, and Nixon.[18]

Fig. 10-8 represents the UOP Cumox process for phenol production from cumene.

The cumene to phenol synthesis gives an overall 92-98% yield of phenol and 91-93% yield of acetone. The primary byproducts are *dimethylphenylcarbinol,* **DMPC,** (methylbenzyl alcohol), $C_6H_5-\underset{\underset{\displaystyle CH_3}{|}}{\overset{\overset{\displaystyle CH_3}{|}}{C}}-OH$, and *acetophenone,* **ACP,** $C_6H_5-\overset{\overset{\displaystyle O}{\|}}{C}-CH_3$. The **DMPC** dehydrates to *α-methylstyrene,* $C_6H_5-\overset{\overset{\displaystyle CH_3}{|}}{C}=CH_2$, during the decomposition of the hydroperoxide. The weighted average of the heats of reaction for the formation of **CHP** as well as **DMPC** and **ACP** is approximately 40 kcal/mole.[18]

Uses. About 50 percent of the 2.8 billion pound production of phenol is used to produce phenolic resins which are noted in Chapter 13. As an example, an appreciable amount of phenol (20%) goes into the production of *bisphenol* A.

$$2\,C_6H_5OH + CH_3-\overset{O}{\overset{\|}{C}}-CH_3 \xrightarrow{H^+} HO-C_6H_4-C(CH_3)_2-C_6H_4-OH + H_2O$$

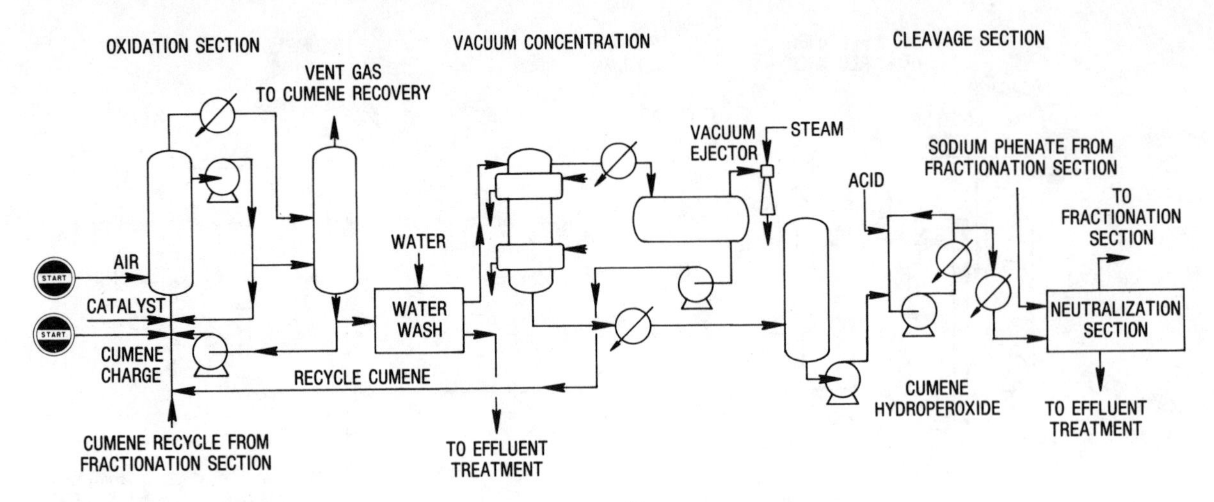

Fig. 10-8—The UOP Cumox process for phenol production from cumene.[17]

Bisphenol A, **BPA,** is used to produce epoxy resins and polycarbonates. These utilize about 240-250 million pounds of the United States 565 million pound production of **BPA.** Total capacity is 910 million pounds.[19]

Phenol's use in synthesis is not restricted to polymers or polymer intermediates. Phenol is used to produce a wide range of compounds, each of which does not account for a large amount of the total phenol demand. Both directly and indirectly, however, it contributes markedly to our life and comfort—a truly versatile compound.

Synthesis from phenol includes the synthesis of salicylic acid and acetylsalicyclic acid (aspirin). 2,4-Dichlorophenoxy acetic acid, 2,4-D, and 2,4,5-trichlorophenoxy acetic acid 2,4,5-T, are selective herbicides; pentachlorophenol is a wood preservative.

Salicylic acid | Acetyl-salicylic acid | **2, 4–D** | Pentachlorophenol

Other halophenols are miticides, leather preservatives and bactericides. Halophenols account for about 5% of phenol utilization. The importance of these compounds, however, cannot be measured by pounds alone.

Alkylphenols serve as monomers in resin polymer applications, antioxidants, and non-ionic surfactants. Their phosphate esters are used as plasticizers and for lube oil and gasoline additives. Alkylphenols account for about 4% of phenol demand in the U.S. *Caprolactam,*

$$\underbrace{CH_2 - CH_2 - CH_2 - CH_2 - CH_2 - C = O}_{NH},$$

is produced from phenol rather extensively in Europe and is reported to account for 16% of phenol demand in the U.S. Caprolactam is used to make nylon 6.

Both phenol and nitrobenzene are used in the production of *aniline,* $C_6H_5NH_2$. Fig. 10-9 shows these two commercial routes to aniline.[20]

Fig. 10-10 represents the predicted use pattern for phenol in 1990.[21] This is based on a projected demand of about 5.6 billion pounds.

Nitrobenzene

Production. The production of *nitrobenzene,* $C_6H_5NO_2$, is as old as the coal tar industry and the process has changed little with time.

$$C_6H_6 + HNO_3 \xrightarrow{H_2SO_4} C_6H_5NO_2 + H_2O \qquad \Delta H = -27.0 \text{ kcal}$$

Typical reaction conditions are:

Temperature:	50-55°C
Pressure:	Liquid phase

MOLAR YIELDS
95-98% BENZENE TO NITROBENZENE
98% NITROBENZENE TO ANILINE
95% BENZENE TO ANILINE

Benzene $\xrightarrow{HNO_3,\ H_2SO_4}$ Nitrobenzene (NO_2) + H_2O $\xrightarrow{H_2,\ CAT}$ Aniline (NH_2) + H_2O

Benzene $\xrightarrow{CH_3\text{-}CH=CH_2}$ Cumene ($CH_3\text{-}CH\text{-}CH_3$) $\xrightarrow{O_2}$ Phenol (OH) + $CH_3\text{-}C(=O)\text{-}CH_3$ $\xrightarrow{NH_3}$ Aniline (NH_2) + H_2O

MOLAR YIELDS
95% BENZENE TO CUMENE
96% CUMENE TO PHENOL
99% PHENOL TO ANILINE
90% BENZENE TO ANILINE

Fig. 10-9—Commercial routes to aniline from benzene with approximate yields.[20]

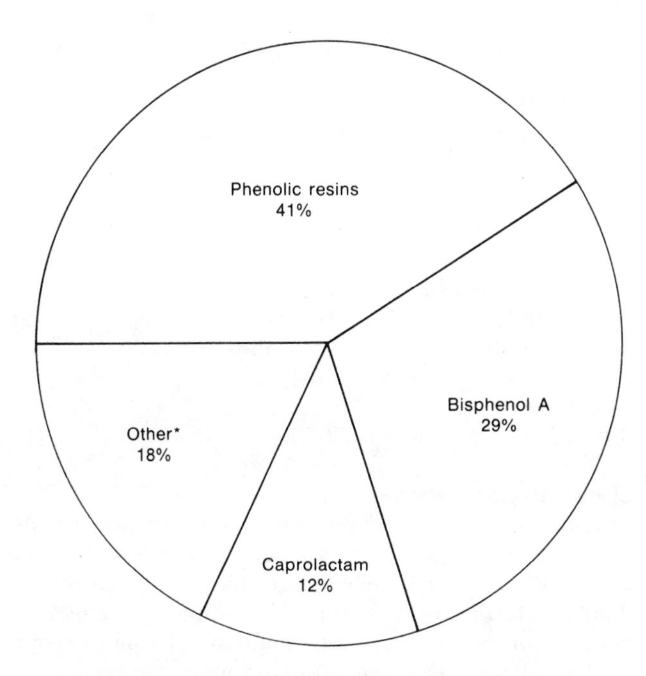

Fig. 10-10—The predicted 1990 use pattern for phenol, based on a 5.6 MMM pound demand.[21]

Catalyst:	*
Yield:	95-98%

*A mixed acid is used: 60-53% H_2SO_4, 32-39% HNO_3, 8% water.

The concentrated sulfuric acid has two functions. It reacts with the nitric acid to form a *nitronium ion,* NO^+_2, which is the nitrating entity.

$$HONO_2 + 2\,H_2SO_4 \rightleftarrows 2\,HSO^-_4 + H_3O^+ + NO^+_2$$

The sulfuric acid also absorbs the water formed during the nitration, thus preventing the nitric acid from ionizing to hydrogen ions, H^+, and nitrate ions, NO^-_3, instead of forming nitronium ions.

Uses. About 97% of nitrobenzene is used to produce aniline.

Aniline

Production. The traditional process for aniline production is hydrogenation of nitrobenzene (Fig. 10-11).[20]

$$C_6H_5NO_2 + 3\,H_2 \longrightarrow C_6H_5NH_2 + 2\,H_2O \qquad \Delta H = -130 \text{ kcal}$$

Typical reaction conditions are:

Temperature:	250-270°C
Pressure:	20 psig
Catalyst:	Cu on silica
Yield:	95%

The hydrogenation of nitrobenzene to aniline and the Halcon catalytic phenol-ammonolysis process (Fig. 10-9) has been carefully evaluated to select the process best for each local situation.[20]

The third industrial process for the production of aniline from benzene is by the chlorination of benzene to *chlorobenzene,* C_6H_5Cl, followed by ammonolysis.

Uses. The use patterns of aniline for 1961-1974-1980 are given in Table 10-1.[20] The projected demand for aniline in 1980 is 800 million pounds.[21] About 50 percent of aniline demand is for isocyanates, but this may change in the 1980s. New technology is being developed which permits the production of isocyanates directly from nitrobenzene and carbon monoxide.[22]

Chlorobenzene

Production. *Chlorobenzene* (monochlorobenzene), C_6H_5Cl, is produced by batch and continuous processes. Monochlorobenzene and both ortho-dichlorobenzene and para-dichlorobenzene are formed with their ratios determined by temperature, mole ratios of benzene and chlorine, residence time, catalyst and conversion.

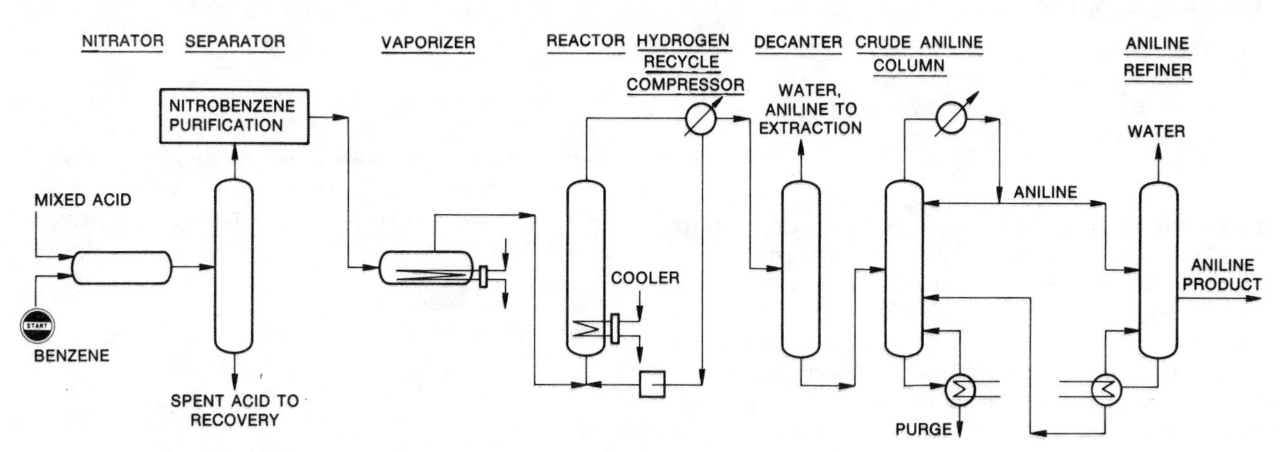

Fig. 10-11—Production of aniline from benzene by way of nitrobenzene.[20]

TABLE 10-1—The Past and Future Use Pattern for Aniline[20]

Application	1961	1974	1980 (Est.)*
Isocyanates	(in other)	40	52
Rubber chemicals	66	35	28
Dyes and intermediates	16	6	4
Drugs and pharmaceuticals	6	4	3
Photographic chemicals	6	6	6
Other	6	9	7
Total	100	100	100

*Demand projected to be 800 million pounds.[21]

$$C_6H_6 + Cl_2 \rightarrow C_6H_5Cl + o\text{-}C_6H_4Cl_2 + p\text{-}C_6H_4Cl_2 + HCl$$

A typical liquid phase batch process has the following general operating conditions:

Temperature:	80-100° C
Pressure:	Atmospheric
Catalyst:	Iron ($FeCl_3$)*
Yields:	Monochlorobenzene: 80%
	p-Dichlorobenzene: 15%
	o-Dichlorobenzene: 5%

*Aluminum, tin, and Fuller's earth have been used.

**Including polychlorobenzenes, mainly 1,2,4-trichlorobenzene.

Continuous chlorination processes permit removal of the monochlorobenzene as formed with a resulting 5% or lower yield of higher chlorinated benzenes.

Monochlorobenzene is also produced in a vapor phase process utilizing a regenerative oxychlorination process which utilizes hydrogen chloride as the chlorinating agent through oxidation to chlorine—the Raschig-Hooker process.

$$4\,HCl + O_2 \rightarrow 2\,Cl_2 + 2\,H_2O$$
$$2\,Cl_2 + 2\,C_6H_6 \rightarrow 2\,C_6H_5Cl + 2\,HCl$$

$$\text{Net: } 2\,C_6H_6 + 2\,HCl + O_2 \rightarrow 2\,C_6H_5Cl + 2\,H_2O$$

The reaction conditions are:

Temperature:	220-260° C
Pressure:	Atmospheric
Catalyst:	CuO promoted, on silica
Conversion:	10%
Yield:	95+%

Higher conversions have been reported when temperatures of 234-315°C and pressures of 40-80 psi are used.[23]

Uses. Monochlorobenzene is the starting material for a wide variety of compounds including phenol and aniline. Others are **DDT,** chloronitrobenzenes, *o*-aminophenol-*p*-sulfonic acid, biphenyl, and polychlorobenzenes. Except for phenol and aniline, the demand is small, but collectively they, plus phenol and aniline, add up to an annual chlorobenzene demand of *ca* 350 million pounds.

Aniline is produced from chlorobenzene by ammonolysis.

$$C_6H_5Cl + 2\,NH_3 \xrightarrow{H_2O} C_6H_5NH_2 + NH_4Cl$$

The reaction conditions are:

Temperature:	210-220° C
Pressure:	880-1030 psi
Catalyst:	Cu^+ salts
Yield:	96%

The aqueous ammonia is 28-30%.

Linear Alkylbenzenes

Production. The production of *linear alkylbenzenes* (straight-chain alkylbenzenes) starts with the production of linear mono-olefins either by dehydrogenation of the corresponding *n*-paraffins extracted from kerosine fractions, wax cracking, or from polymerization of ethylene using a Ziegler catalyst. Benzene is then alkylated with this mixture of olefins to produce linear alkylbenzene, also called *detergent alkylate.*

$$C_{12}H_{26} \rightarrow C_{12}H_{24} + H_2$$

$$C_6H_6 + C_{12}H_{24} \rightarrow C_6H_5C_{12}H_{25}$$

Typical vapor phase dehydrogenation conditions over a fixed bed catalyst are:

Temperature:	870° C
Pressure:	30 psig
Catalyst:	Pt or Pd on alumina
Conversion:	"Less than complete"
Yield:	Mono-olefins: 91 wt%
	Diolefins: 2 wt%

Typical alkylation conditions are:

Temperature:	40-70° C
Pressure:	Liquid phase
Catalyst:	HF
Yield:	95%

The dehydrogenation and alkylation can be combined into a single integrated process as shown in Fig. 10-12.[24]

Uses. The *ca* 600 million pound annual production of detergent alkylate is sulfonated to produce *linear alkylbenzene-sulfonate,* **LAS,** $R\text{-}C_6H_4\text{-}SO^-{}_3Na^+$. These detergents are biodegradable.

Maleic Anhydride

Production. Maleic anhydride,

$$\begin{matrix} CH - C{=}O \\ \| \quad\quad \diagdown \\ \quad\quad\quad O \\ \| \quad\quad \diagup \\ CH - C{=}O \end{matrix}$$

, is produced by

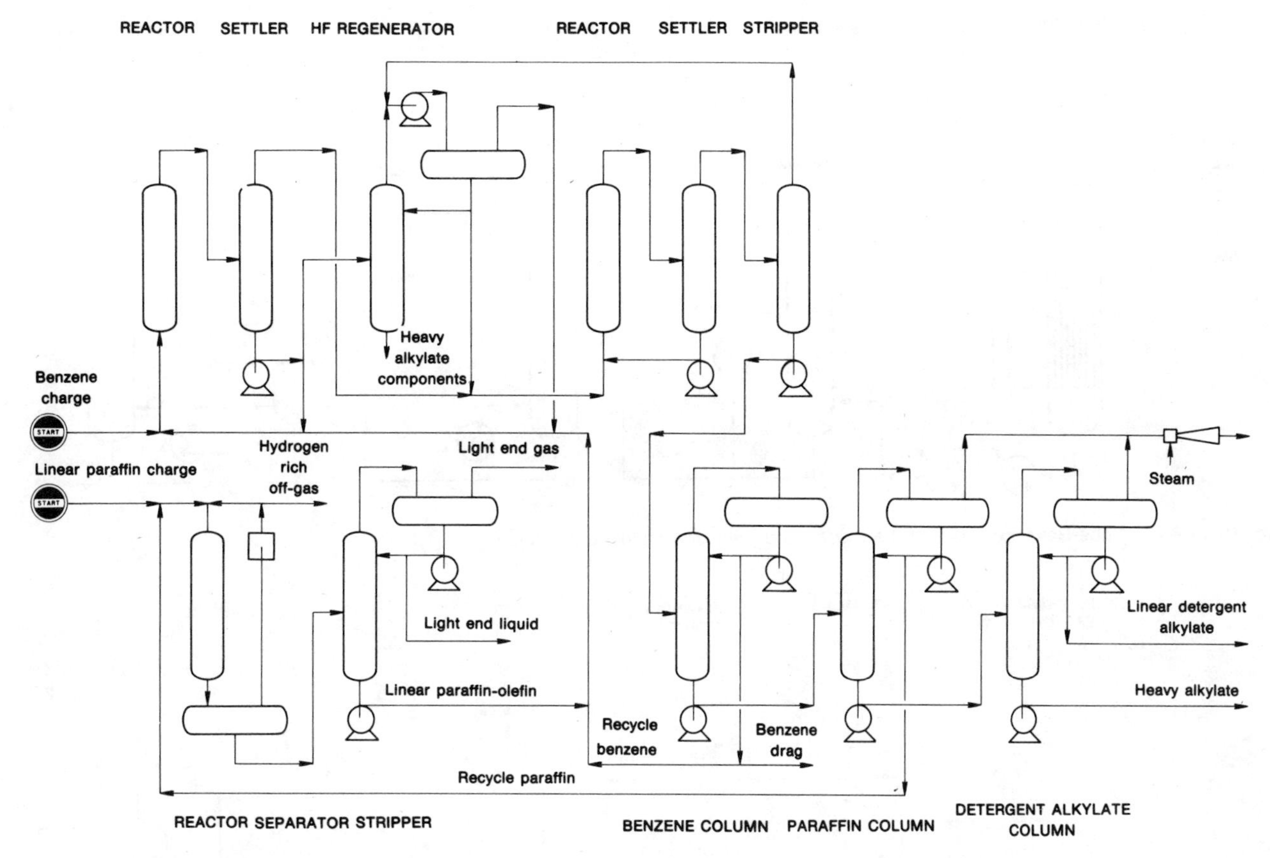

Fig. 10-12—Pacol-detergent alkylation process for the production of linear alkylbenzene from *n*-paraffins.[24]

the oxidation of butane, the butenes, and benzene. Benzene oxidation is the oldest process and is still viable.

$$2\,C_6H_6 + 9\,O_2 \longrightarrow 2\,(CH-CO)_2O + 4\,CO_2 + 4\,H_2O$$

Typical operating conditions for a fixed bed reactor using air are:

Temperature:	380° C (350-450° C)
Pressure	Atmospheric
Catalyst:	V_2O_5/MoO_3
Yield:	65-70%

Fig. 10-13 represents the Veba-Bayer process for benzene oxidation to maleic anhydride.[25]

The feasibility of using oxygen instead of air has been evaluated by Maux.[26] His conclusion is that it is obvious that the use of pure oxygen will not be of any technical advantage. The competitive economics of the three feedstocks—butane, butenes and benzene—have been reviewed by Brownstein[27] and others.[28] Their conclusions are that butane capacity is on the increase and new benzene capacity will be adaptable when a switch to butane is desired. Total capacity will be over 500 million pounds by 1981 with supply and demand in balance.[29]

Uses. The major uses of maleic anhydride are given in Chapter 9.

Cyclohexane

Production. *Cyclohexane*, C_6H_{12}, is produced by the hydrogenation of benzene (Fig. 10-14).[30]

$$C_6H_6 + 3\,H_2 \longrightarrow C_6H_{12}$$

General reaction conditions for the liquid phase hydrogenation are:

Temperature:	160-220° C
Pressure:	400 psi
Catalysf:	Ni on alumina
Yield:	99.6%

Nickel-palladium catalysts also are used. The temperatures and pressures are dependent on catalyst activity. When sulfided nickel or palladium catalysts are used, a temperature of *ca* 450° C and a pressure of 4,500 psi are required.

About 15% of the total cyclohexane capacity (*ca* 3 billion pounds) is from natural gasoline. The 1978 United States production of cyclohexane was 2.34 billion pounds.

Uses. Cyclohexane is used almost exclusively in the production of intermediates for nylon fibers and resins, with 60 percent to nylon 66 and 30 percent to nylon 6.

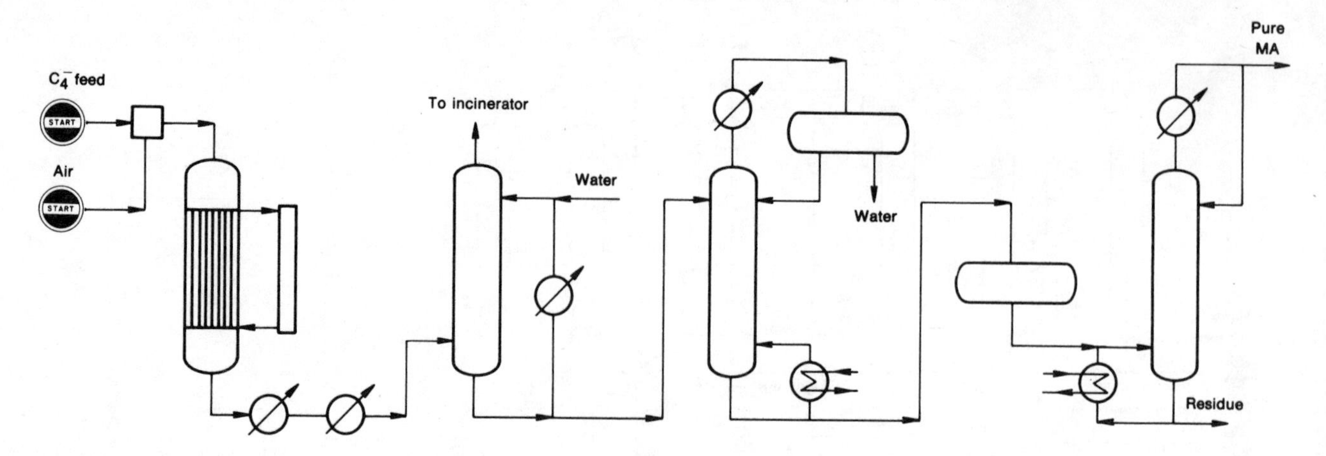

Fig. 10-13—The Veba-Bayer process for the oxidation of benzene maleic anhydride.[25]

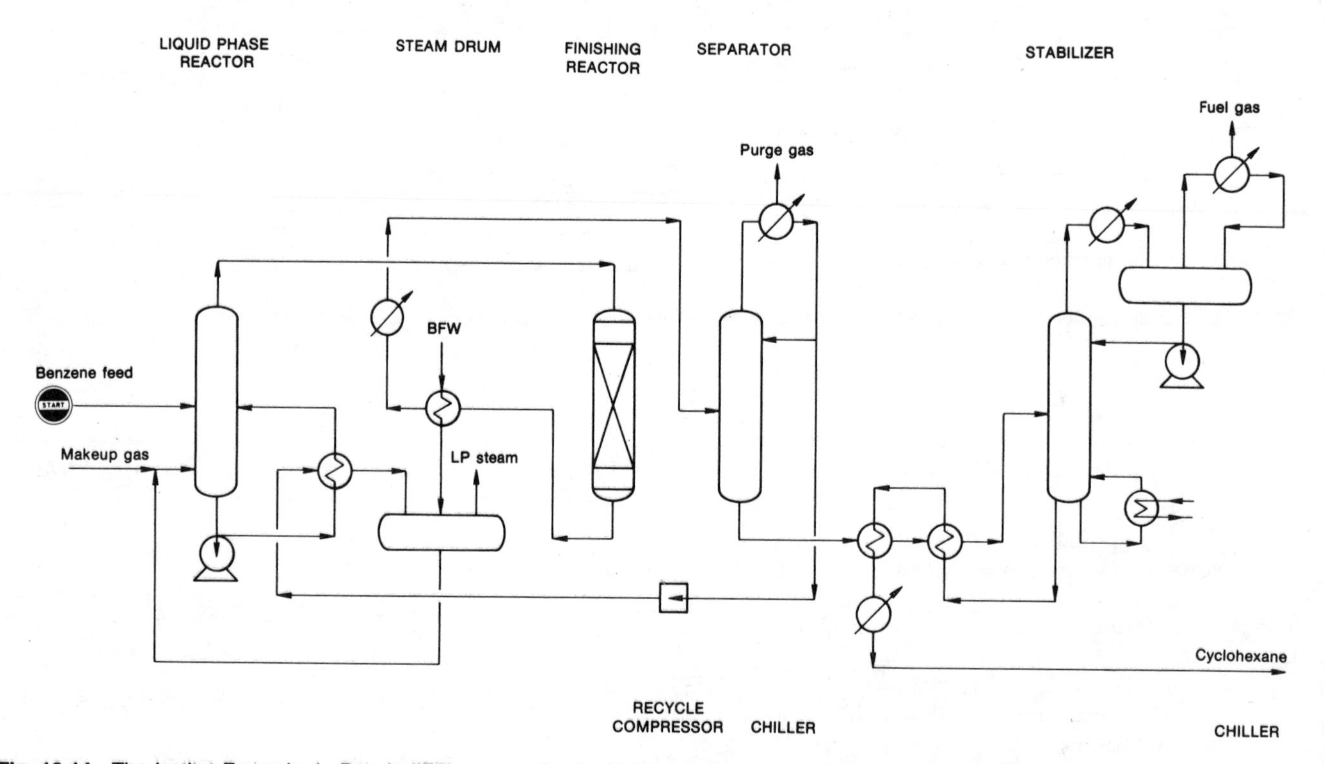

Fig. 10-14—The Institut Francais du Petrole (IFP) process for the hydrogenation of benzene to cyclohexane.[30]

Cyclohexanol-Cyclohexanone

Production. Cyclohexane is oxidized to a mixture of cyclohexanol and cyclohexanone, **KA**-oil (Fig. 10-15).[31]

$$4\,(CH_2)_6 + 3O_2 \rightarrow 2\,(CH_2)_5CH{-}OH + 2\,(CH_2)_5C{=}O + 2\,H_2O$$

Typical liquid phase oxidation conditions are:

Temperature:	95-120° C
Pressure:	150 psi
Catalyst:	$(CH_3-\overset{O}{\overset{\|}{C}}-O)_2\,Co + H_3BO_3$
Yield:	90-95%

Uses. **KA**-oil is used for the production of *caprolactam, adipic acid* and *phenol.* Caprolactam is used to produce nylon 6. Adipic acid is used in the production of nylon 66. The distribution of **KA**-oil utilization for nylon production is about one-third to nylon 6 and two-thirds to nylon 66.[32]

Adipic Acid

Production. *Adipic acid,* $HO-\overset{O}{\overset{\|}{C}}-(CH_2)_4-\overset{O}{\overset{\|}{C}}-OH$, can be prepared by a direct liquid phase air oxidation of cyclohexane in acetic acid.[33]

$$2\,(CH_2)_6 + 5\,O_2 \rightarrow 2\,HO-\overset{O}{\overset{\|}{C}}-(CH_2)_4-\overset{O}{\overset{\|}{C}}-OH + 2\,H_2O$$

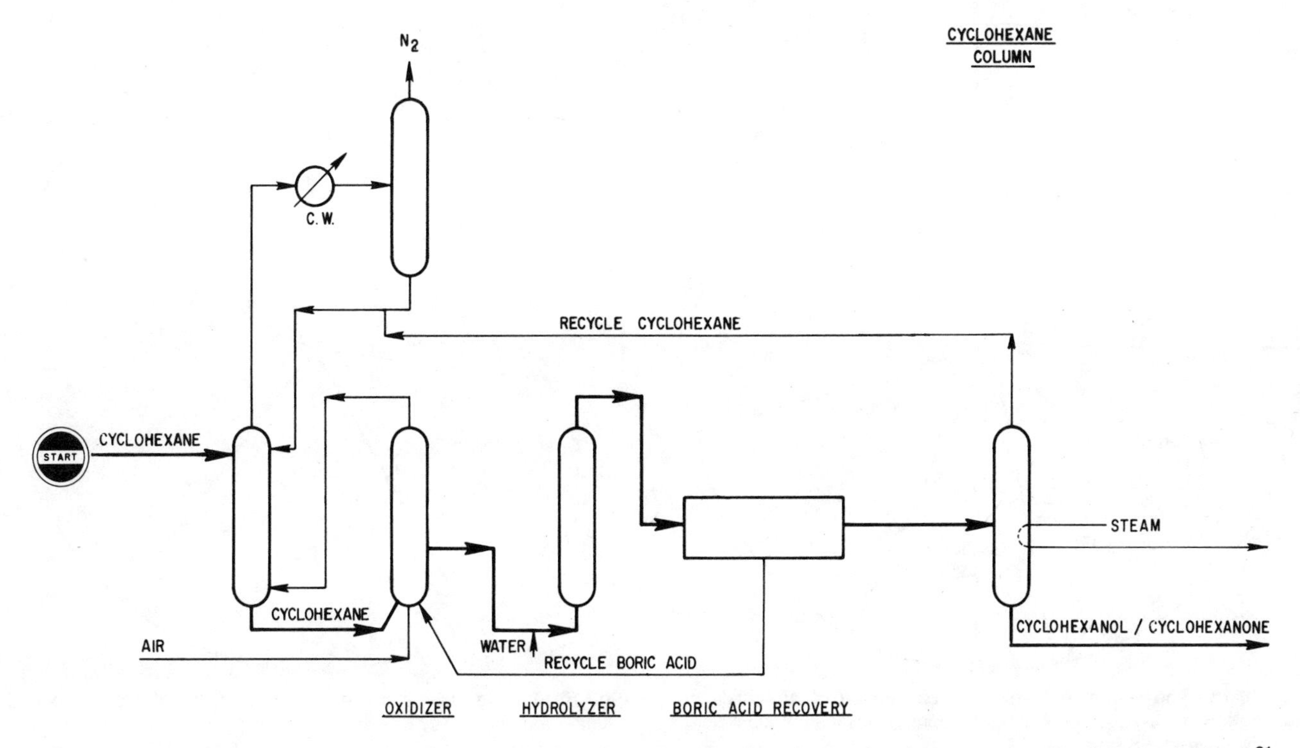

Fig. 10-15—Scientific Design Co., Inc. process for the oxidation of cyclohexane to a mixture of cyclohexanone and cyclohexanol, KA-oil.[31]

The reaction conditions are:

Temperature:	90° C
Pressure:	Liquid phase
Catalyst:	$(CH_3-\overset{O}{\overset{\|\|}{C}}-O)_2Co$
Conversion (9 hrs.):	88%
Yield:	73%

The adipic acid is used both as the acid and also converted to adiponitrile; then to hexamethylenediamine for the other half of nylon 66.

THE METHYLBENZENES

Methylbenzenes occur in naphtha and higher boiling fractions of petroleum. The ones of present commercial importance are **toluene,** *ortho*-**xylene** and *para*-**xylene,** and to a much lesser extent, *meta*-**xylene.** The structure of these aromatic hydrocarbons and their melting points and boiling points are given in Table 10-2. There are also eight polymethylbenzenes, some of which have potential as petrochemicals. These potentials have been reviewed extensively by Earhart[34] and Ockerbloom.[35]

The primary sources of toluene and the xylenes are refinery streams, especially from catalytic reforming, and pyrolysis gasoline from the production of ethylene by cracking naphtha. The yields of aromatics in the reformate of light Arabian naphtha (62-136°C) are given in Fig. 10-16.[36] These aromatics are separated by solvent extraction.[37] Only a relatively small amount of the total toluene and xylenes available from these sources is separated and used to produce petrochemicals.

The methylbenzenes have chemical characteristics similar to benzene and methane. Each group's chemical reactions are modified by the presence of the other group. For example, substitution on the benzene ring goes more readily with toluene than with benzene and the methyl group is oxidized more readily than methane. But in spite of these favorable reaction conditions, toluene and especially the xylenes have relatively few chemical derivatives.

TABLE 10-2—The Structure, Name, Boiling Point and Melting Point of Toluene and the C_8 Aromatic Hydrocarbons

Structure	Name	Freezing Point °C	Melting Point °C	Boiling Point °C*
CH_3 (benzene ring)	Toluene	−94.97	−95.0	110.6
CH_3, CH_3 (benzene ring)	*o*-Xylene	−25.17	−25.2	144.4
CH_3, CH_3 (benzene ring)	*m*-Xylene	−46.84	−47.9	139.1
CH_3, CH_3 (benzene ring)	*p*-Xylene	+13.25	+13.3	138.4
CH_2-CH_3 (benzene ring)	Ethylbenzene	−94.95	−95.0	136.2

*760 torr

Toluene

The total *toluene* (methylbenzene, toluol), $C_6H_5-CH_3$, available from the gasoline pool in the United States in 1977 was 24.4 million metric tons. Of this amount only 17 percent

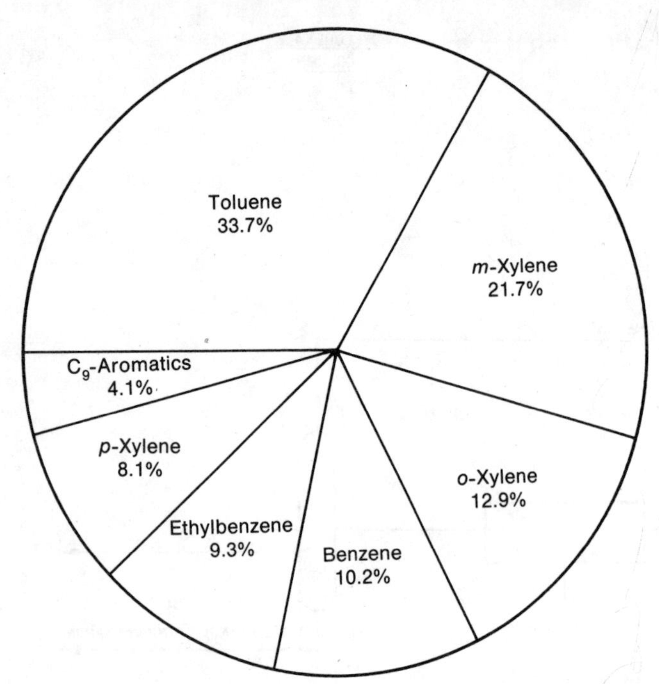

Fig. 10-16—The weight percent yield distribution of aromatics in reformate from 62-136°C light Arabian naphtha. Total aromatic weight percent on feed: 58.9 percent.[36]

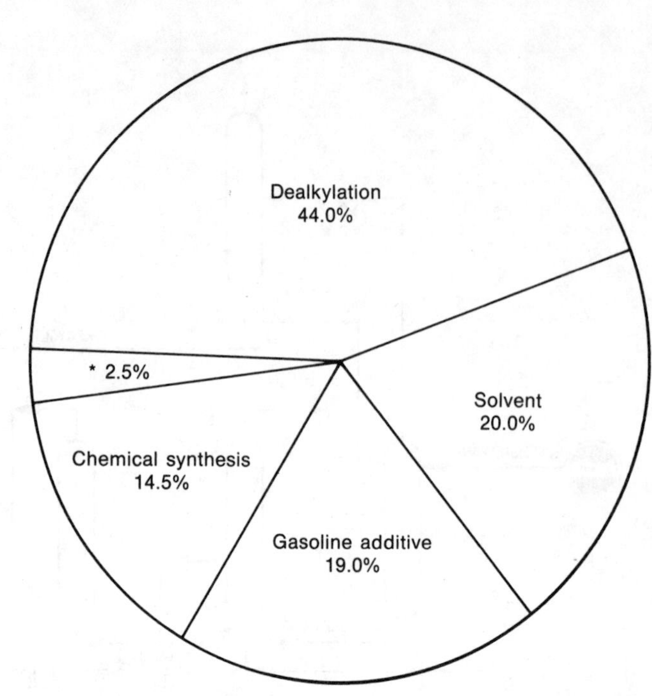

Fig. 10-17—The U.S. end-use pattern from the 4.2 billion pound toluene demand in 1977.[38,39]

was isolated and used, more or less, as a petrochemical (Fig. 10-17).[38, 39] The single largest use of this toluene is hydrodealkylation, **HDA,** to benzene. It is predicted that this source of benzene will supply 29 percent of the 6.26-million-metric-ton demand for benzene in 1980 and 17 percent of the 9.80-million-metric-ton benzene demand in 1990.[40] Currently about 20 percent of Europe's benzene supply comes from toluene hydrodealkylation. This represents 63 percent of the isolated toluene. The non-HDA toluene capacities, production and consumption, for the United States, Western Europe and Japan, 1978-1982, are given in Fig. 10-18.[1]

Only 14.5 percent of the isolated toluene in 1977 was used in chemical syntheses. But even a small percentage of a very large number is still a large number. In general, the first product is further processed to produce consumer products, primarily polymers, Fig. 10-19. The chemical use of the predicted 290 million gallon chemical demand for toluene in 1980 is shown in Fig. 10-20.[41] Benzene production is excluded.

Benzene Production. *Benzene,* C_6H_6, is produced by both catalytic and noncatalytic *dehydroalkylation* of toluene.

$$C_6H_5CH_3 + H_2 \rightarrow C_6H_6 + CH_4$$

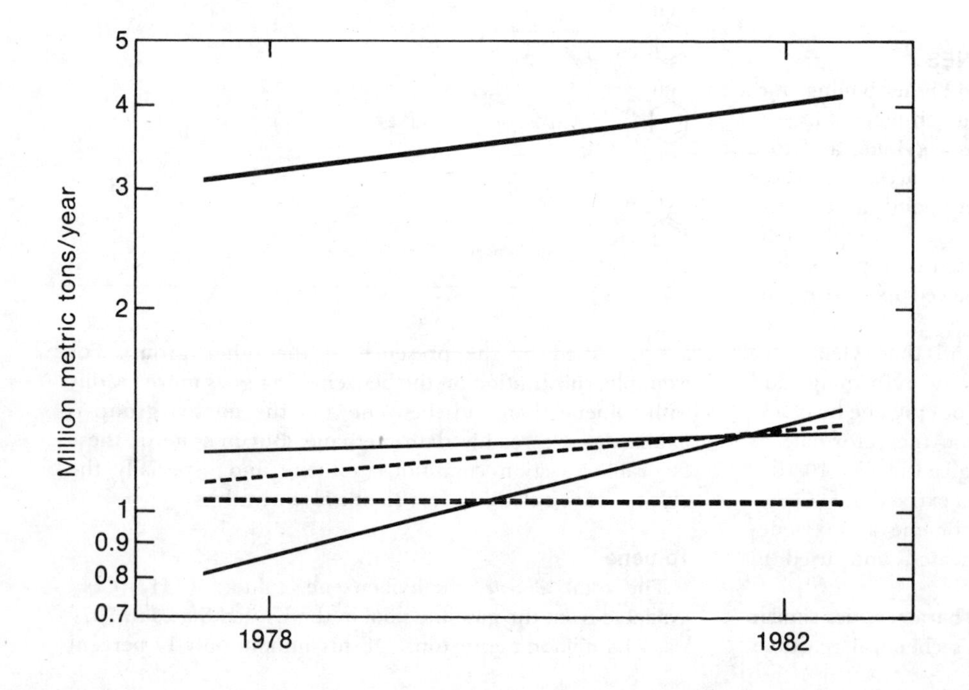

Fig. 10-18—Non-HDA toluene capacities, production, and consumption for the United States, Western Europe, and Japan, 1978-1982.[1]

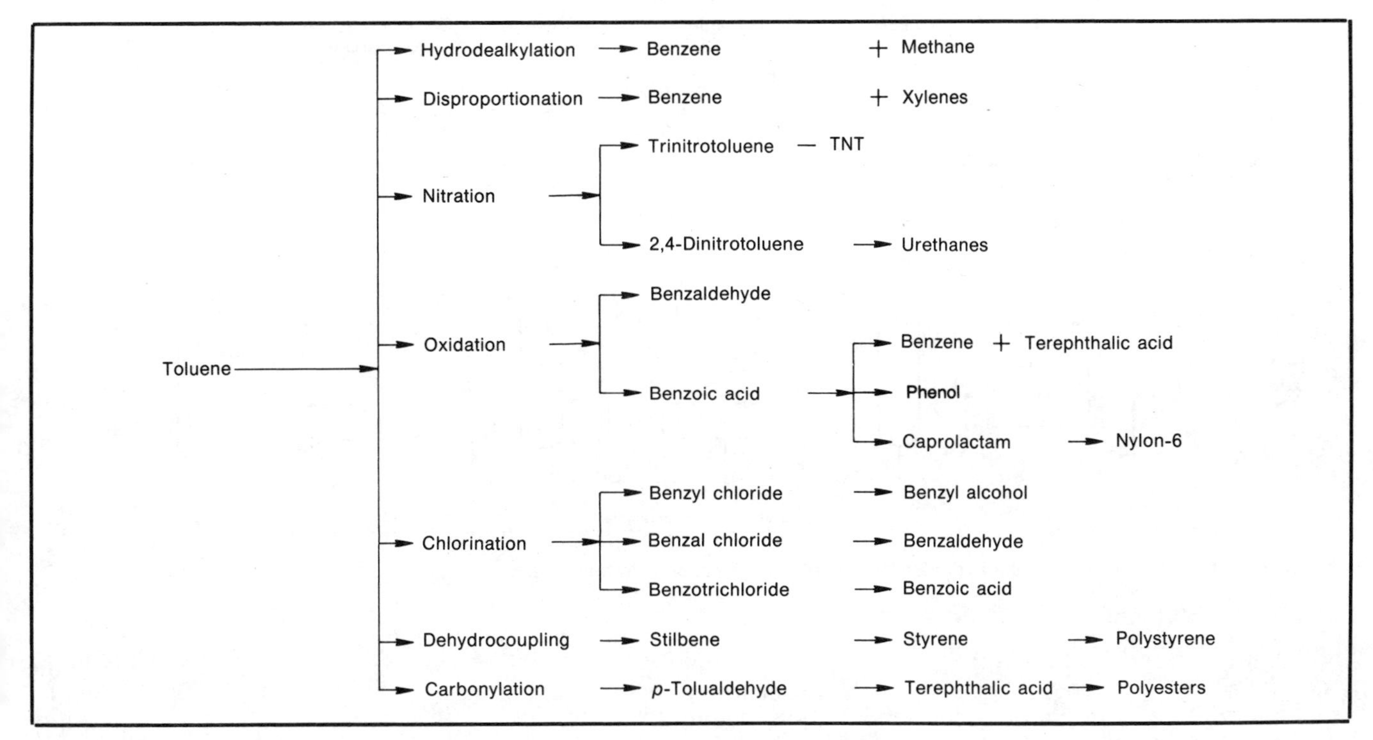

Fig. 10-19—The major and end products from the chemical use of toluene.

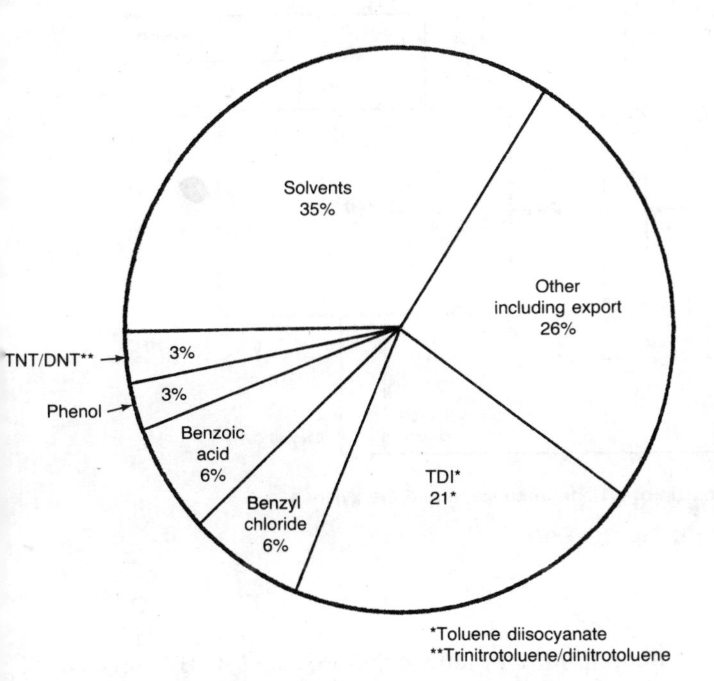

Fig. 10-20—The nongasoline 290 million gallon utilization of toluene in 1980, less toluene dehydroalkylated to benzene.[41]

Catalytic hydrodealkylation is the favored process (Fig. 10-21).[42]

The following reaction conditions are representative of current practice.[43]

Temperature	680-720° C
Pressure	570 psi
Catalyst	"a dealkylation catalyst"*
Yield	96 + %

*Such as nickel.

Dealkylation also can be effected by the use of steam in place of hydrogen.

$$C_6H_5CH_3 + 2\,H_2O \rightarrow C_6H_6 + CO_2 + 3\,H_2$$

The reaction takes place at 600-800°C over Y, La, Ce, Pr, Nd, Sm or Th compounds[44] and over Ni-Cr_2O_3 catalysts, and over Ni-Al_2O_3 catalysts at temperatures between 320-630°C. Yields of about 90 percent are obtained. This process has the advantage of producing hydrogen rather than using hydrogen.

Benzene and Xylenes Production. *Benzene* plus the *xylenes*, $C_6H_4(CH_3)_2$, are obtained by the catalytic disproportionation of toluene in the presence of hydrogen (Fig. 10-22).[45] *Disproportionation* is the conversion of 2 moles of a single aromatic compound to one mole each of two different aromatic compounds.

$$2\,C_6H_5CH_3 \rightleftharpoons C_6H_6 + C_6H_4(CH_3)_2$$

This is an equilibrium reaction with a 58 percent conversion per pass theoretically attainable. Attempting to obtain a conversion higher than about 40 percent results in more side reactions and a greater catalyst deactivation rate than is desirable.[46]

Typical reaction conditions are:

Temperature	450-530° C
Pressure	300 psig
Catalyst	CoO-MoO_3 on aluminosilicate/Alumina

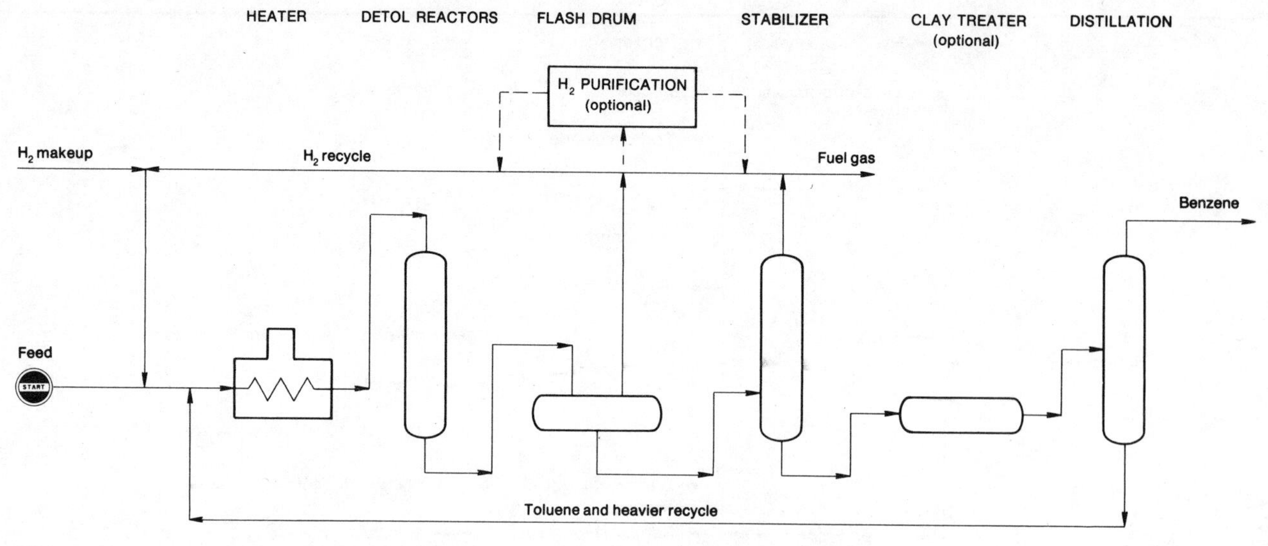

Fig. 10-21—The Detol process for the hydrodealkylation of methylbenzenes to benzene.[42]

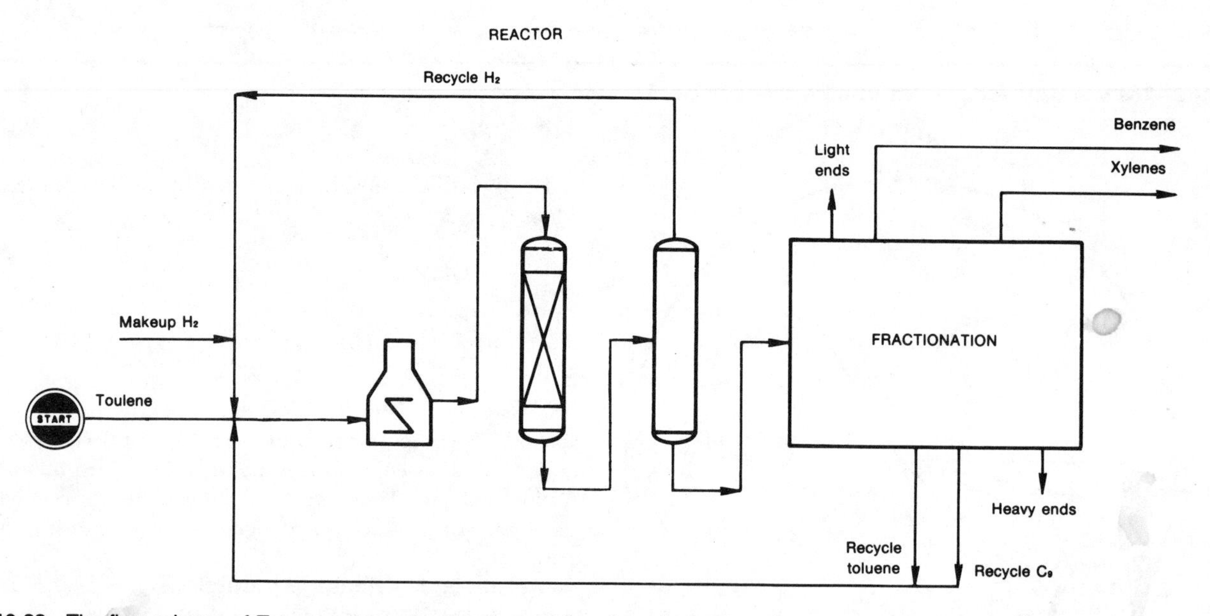

Fig. 10-22—The flow scheme of Tatoray process for the disproportionation of toluene to benzene and the xylenes.[45]

Conversion	*ca* 40%
Yield	97%

Essentially all the metals of Group VIB and Group VIII of the Periodic Table have been used as primary catalysts. Metals of Groups I, II, IV and rare earth metals have been suggested as activity enhancers. A typical ratio of xylenes is:

o-xylene/*m*-xylene/*p*-xylene: 22/55/23

The ubiquitous ZSM-5 of the Mobil Catalyst System is the most recent, and perhaps the most promising, catalyst for the disproportionation of toluene to the xylenes. It is effective at 480°C. The ZSM-5s are crystalline aluminosilicates with high SiO_2/Al_2O_3 ratios. They are among the most thermally stable zeolites.

Benzoic Acid Production. *Benzoic acid,* $C_6H_5-\overset{\overset{\Large O}{\|}}{C}-OH$, is produced by the liquid phase catalytic oxidation of toluene (Fig. 10-23).[47]

$$2\,C_6H_5CH_3 + 3\,O_2 \rightarrow 2\,C_6H_5COOH + 2\,H_2O$$

Typical reaction conditions are:

Temperature	165°C
Pressure	150 psig
Catalyst	Cobalt acetate

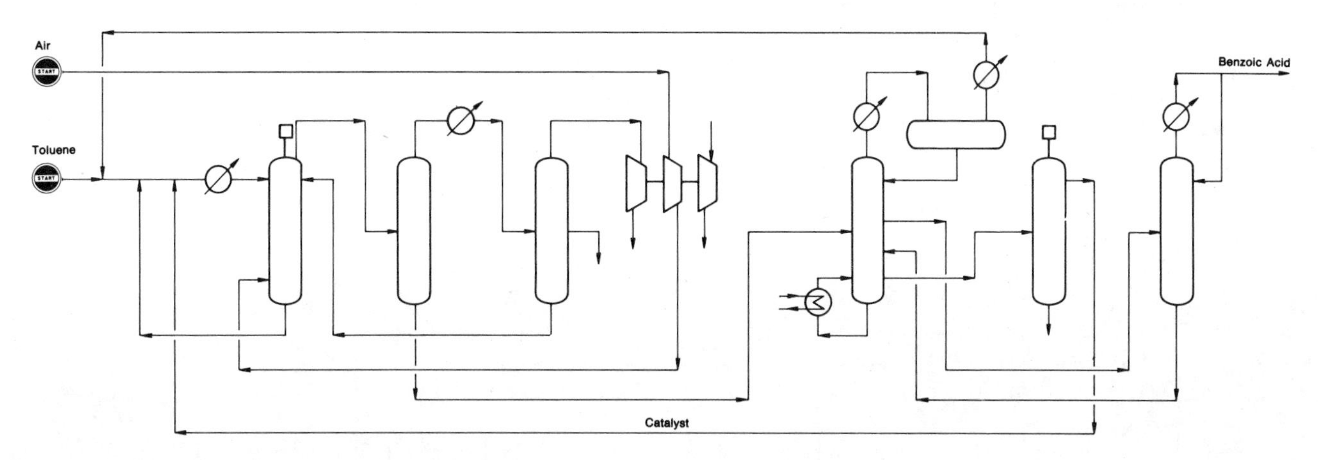

Fig. 10-23—The flow diagram for SNIA Viscosa catalytic liquid phase oxidation of toluene to benzoic acid.[47]

Conversion	30+%
Yield	90%

Toluene is also oxidized to benzoic acid at 200°C and 440 psig with a bromide compound as an initiator and a cobalt or manganese catalyst. The yield is about 90 percent.

Uses. Benzoic acid is used primarily as a mordant in calico printing, seasoning tobacco, in dentifrices and medicines as a germicide, in textiles, and as a plasticizer and resin intermediate. It is also used for the synthesis of *caprolactam, phenol* and *terephthalic acid.* The principal use for benzoic acid is in the form of its sodium salt used as a preservative in canned goods and fruit beverages.

Caprolactam. *Caprolactam,* $CH_2-(CH_2)_4-C=O$ (ring closed through NH), the precursor of nylon 6, is synthesized from benzoic acid by hydrogenation to *cyclohexane carboxylic acid* (Equation 1).

$$C_6H_5\text{-}COOH + 3\,H_2 \rightarrow \underset{H_2C-CH_2-CH_2-CH_2-CH_2}{CH}\text{-}C(=O)\text{-}OH \qquad (1)$$

This acid is converted to caprolactam by reaction with *nitrosyl-sulfuric acid*, $NOHSO_4$, (Equation 2).

$$C_6H_{11}\text{-}C(=O)\text{-}OH + NOHSO_4 \rightarrow CH_2-(CH_2)_4-C=O\ (\text{ring closed through } NH) + CO_2 + H_2SO_4 \qquad (2)$$

A flow diagram of this process is shown in Fig. 10-24.[48] The over-all yield from toluene is 85 percent.

The advantage of the SNIA-Viscosa process over the conventional process is its freedom from ammonium sulfate byproduct.

Phenol. *Phenol,* C_6H_5 - OH, can be produced from benzoic acid by several different routes. Fig. 10-25 shows the chemistry involved in two of these routes.[49] The overall reaction is

$$2\,C_6H_5\text{-}C(=O)\text{-}OH + O_2 \rightarrow 2\,C_6H_5\text{-}OH + 2\,CO_2$$

The reaction conditions for the Dow process are:

Temperature	220-240°C
Pressure	Liquid phase
Catalyst	Mg^{2+} + Cu^{2+} benzoates
Yield	80%

The magnesium benzoate is an initiator, with the copper (II) ion, Cu^{2+}, being reduced to copper (I), Cu^{1+}. The copper (I) ion is reoxidized to copper (II) ion by air.

The C-E Lummus process is a recently proposed vapor phase oxidation process over a copper-containing catalyst and operates at very high velocities.[49] Conversion is 50 percent and the yield of phenol approaches 90 percent. The byproducts are benzene and diphenyl ether. Fig. 10-26 illustrates the C-E Lummus process.

Uses. The uses of phenol are noted in Chapter 9.

Terephthalic Acid. *Terephthalic acid,* HO - C(=O) - C_6H_4 - C(=O) - OH, can be produced by solid phase disproportionation of potassium benzoate in the presence of carbon dioxide.

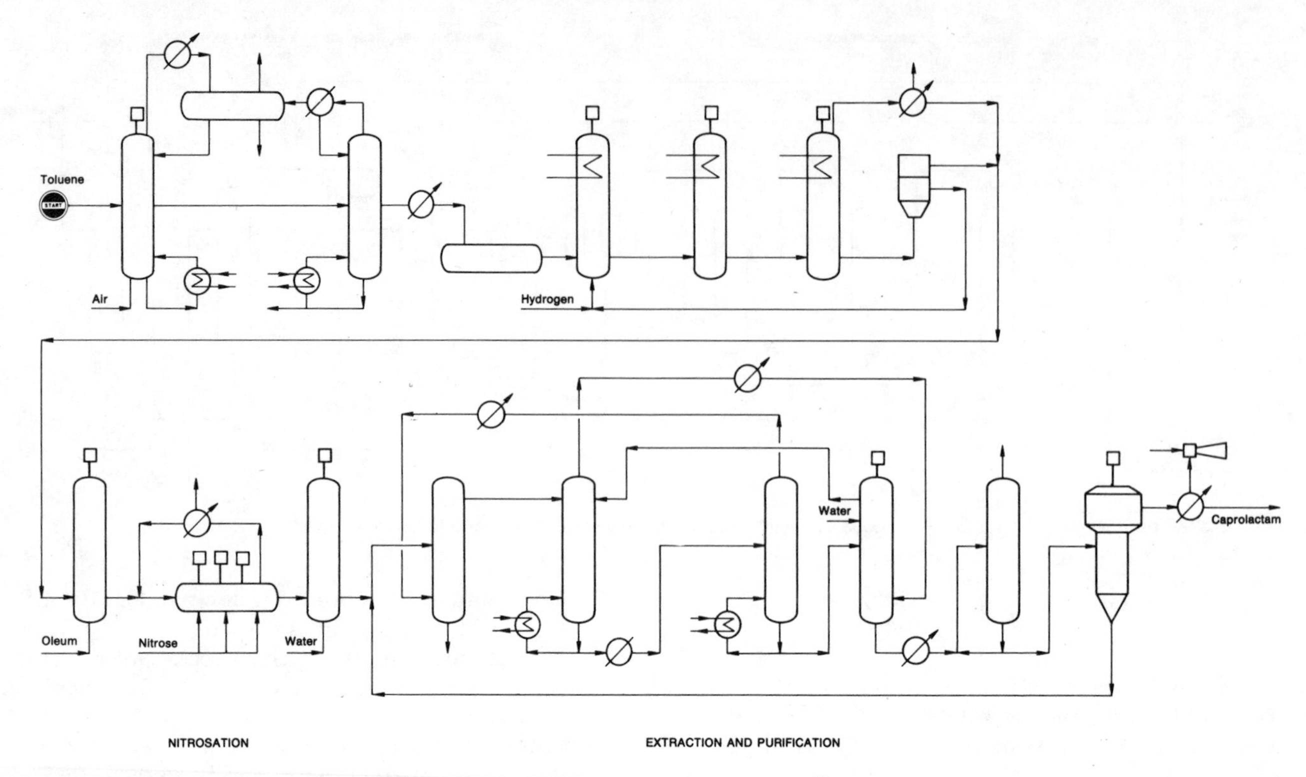

Fig. 10-24—The SNIA Viscosa process for the production of caprolactam from toluene.[48]

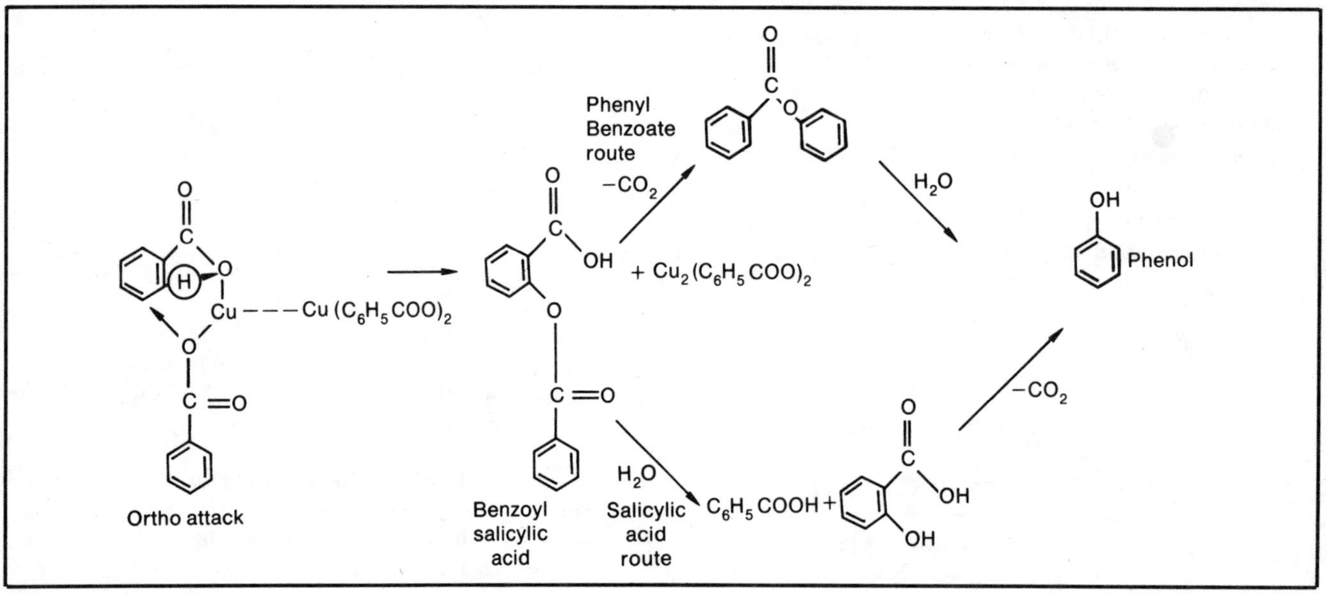

Ortho attack

$Cu---Cu(C_6H_5COO)_2$

Benzoyl salicylic acid

$+ Cu_2(C_6H_5COO)_2$

Phenyl Benzoate route

$-CO_2$

H_2O

Salicylic acid route

H_2O

$C_6H_5COOH +$ (salicylic acid)

$-CO_2$

Phenol

Fig. 10-25—The chemistry of the oxidation of benzoic acid to phenol.[49]

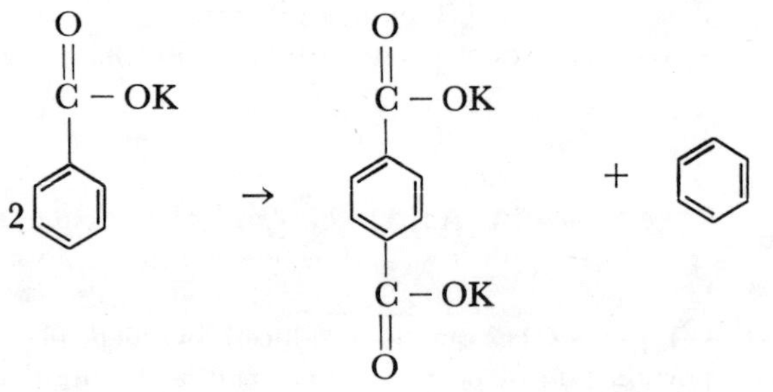

$$2\,C_6H_5COOK \rightarrow KOOC{-}C_6H_4{-}COOK + C_6H_6$$

Typical reaction conditions are:[50, 51]

Temperature	400°C
Pressure	Liquid phase*
Catalyst	Zinc compounds**
Yield	Not given

*Under carbon dioxide at 740 psi.
**Also cadmium compounds such as CdO.

The terephthalic acid is obtained by mineral acid treatment and the byproduct potassium salt is recycled.

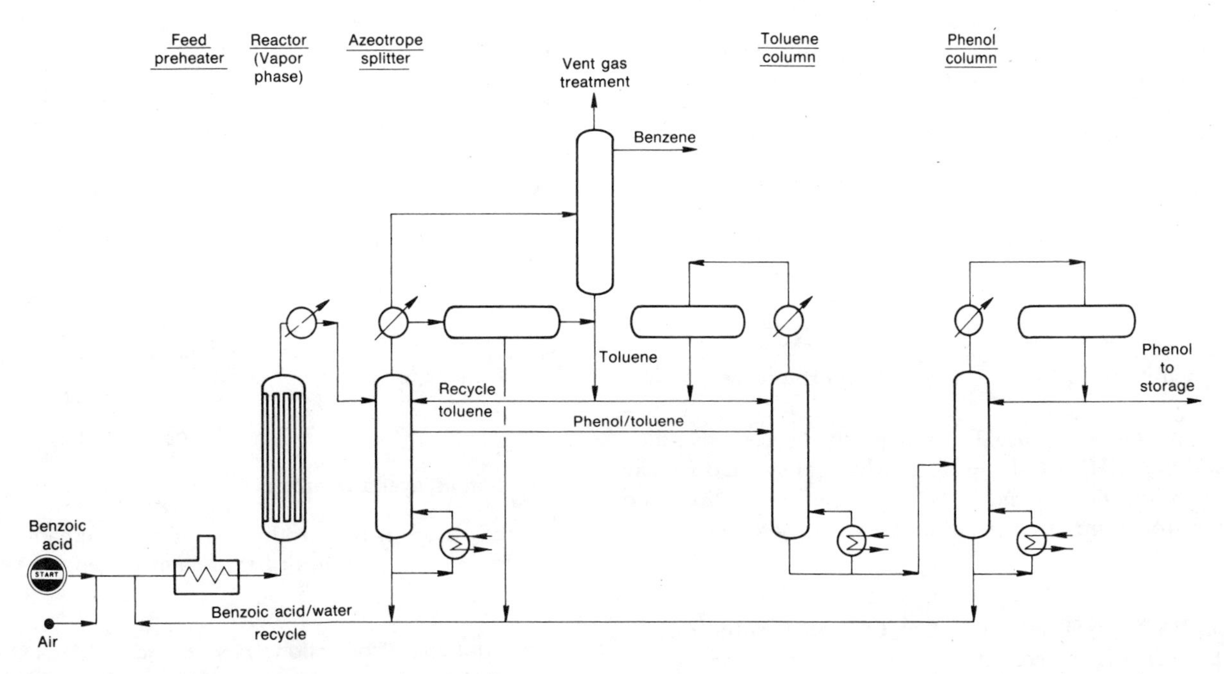

Fig. 10-26—The Lummus benzoic acid to phenol process.[49]

The Phillips Petroleum and Rhone Poulenc processes for the conversion into terephthalic acid are shown in Fig. 10-27.[52] The toluene is oxidized to benzoic acid over a cobalt catalyst under mild conditions to give a 90-92 percent yield. A zinc oxide catalyst is used to disproportionate the potassium benzoate to dipotassium terephthalate and benzene.

Uses. Essentially all of the 1978 production of TPA, 5.97 billion pounds, was used in the production of polyesters.

Nitrotoluenes Production. The art of nitrating aromatic hydrocarbons is very old. The reaction is so simple that there has been very little change since benzene was first nitrated to produce nitrobenzene. Nitration is essentially the only reaction of toluene that involves the aromatic ring rather than the aliphatic methyl group, —CH_3. The reaction with toluene takes place under milder conditions than those used for benzene. This is made possible because of the activating

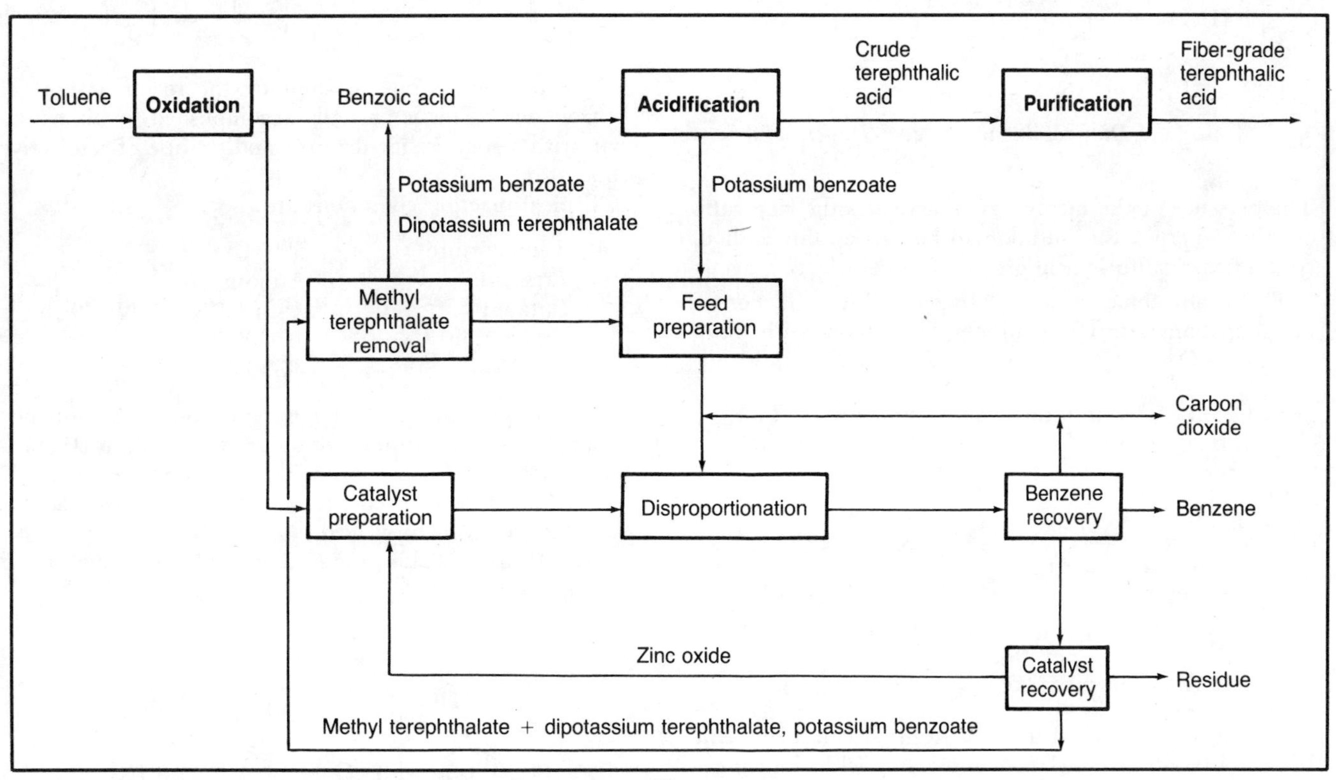

Fig. 10-27—The PRP process to convert toluene to terephthalic acid.[52]

influence of the methyl group. Barona and Prengle have tabulated the characteristics of industrial aromatic nitration processes.[53]

Uses. The mononitrotoluenes produced by direct nitra-

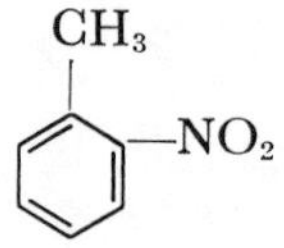

CH_3 $-NO_2$ CH_3 NO_2

o-Nitrotoluene 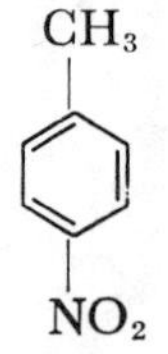 *p*-Nitrotoluene

tion of toluene are usually reduced to the corresponding *toluidines,* $CH_3 - C_6H_5 - NH_2$, which are used to make dyes and rubber chemicals. *p*-Nitrotoluene is oxidized and the nitro group is then reduced to give *p-aminobenzoic acid,* **PABA,** $NH_2 - C_6H_4 - \overset{O}{\overset{\|}{C}} - OH$, an essential metabolite. **PABA** is also converted to *4-carboxyphenylhydrazine,* $H_2N - NH - C_6H_4 - \overset{O}{\overset{\|}{C}} - OH$, an intermediate for several optical bleaches used as brighteners in soap. *Amino-stilbene,* $H_2N - C_6H_4 - CH = CH - C_6H_4 - NH_2$, is another important derivative of *p*-nitrotoluene. It is a dye intermediate for optical bleaches.

2, 4-Dinitrotoluene is the important dinitrotoluene formed by the dinitration of toluene.

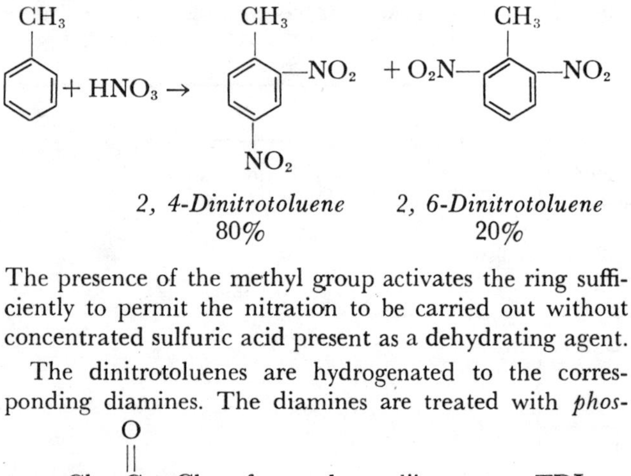

CH_3 $+ HNO_3 \rightarrow$ CH_3 $-NO_2$ NO_2 $+ O_2N-$ CH_3 $-NO_2$

2, 4-Dinitrotoluene 80% *2, 6-Dinitrotoluene* 20%

The presence of the methyl group activates the ring sufficiently to permit the nitration to be carried out without concentrated sulfuric acid present as a dehydrating agent.

The dinitrotoluenes are hydrogenated to the corresponding diamines. The diamines are treated with *phosgene,* $Cl - \overset{O}{\overset{\|}{C}} - Cl$, to form *toluene diisocyanate,* **TDI.**

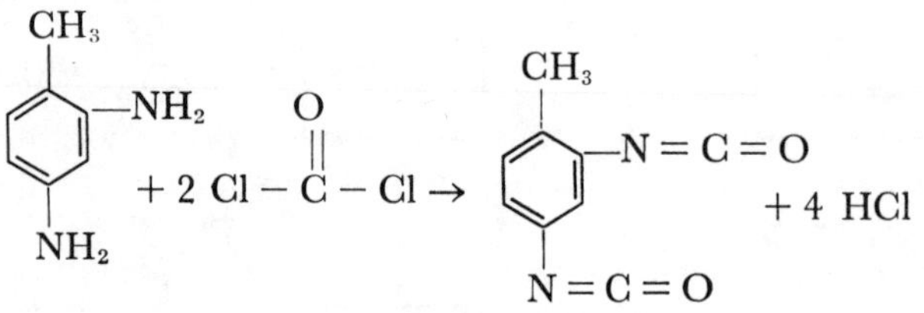

CH_3 $-NH_2$ NH_2 $+ 2\ Cl - \overset{O}{\overset{\|}{C}} - Cl \rightarrow$ CH_3 $-N=C=O$ $N=C=O$ $+ 4\ HCl$

The yield from toluene is 85%.

TDI is also produced directly from the dinitrotoluene by liquid phase carbonylation in a slurry with *o*-dichlorobenzene.

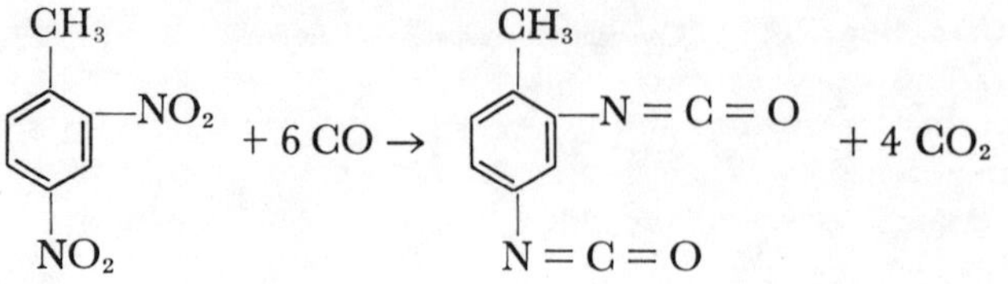

CH_3 $-NO_2$ NO_2 $+ 6\ CO \rightarrow$ CH_3 $-N=C=O$ $N=C=O$ $+ 4\ CO_2$

The reaction conditions are:

Temperature	250° C	
Pressure	3000 psi	
Catalyst	$PdCl_2$	
Conversion	63%	
Yield	**TDI**	22%
	Monoisocyanates	56%

The monoisocyanates are recycled.

TDI is used in the production of polyurethanes with a worldwide consumption of 618 million pounds in 1979.[54]

Benzaldehyde Production. *Benzaldehyde,* $C_6H_5 - \overset{O}{\overset{\|}{C}} - H$, is produced by the oxidation of toluene. This a low yield process because the first step is the oxidation of toluene to *benzyl alcohol,* $C_6H_5 - CH_2OH$, which in turn is oxidized to benzaldehyde which can be oxidized to *benzoic* acid, $C_6H_5 - \overset{O}{\overset{\|}{C}} - OH$. Each successive oxidation goes more readily than the preceding one.

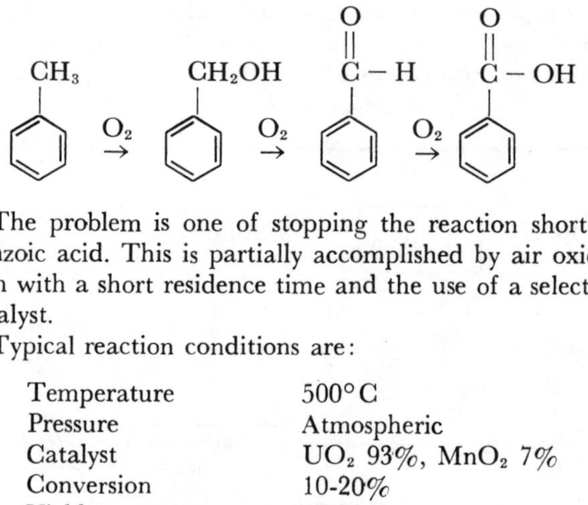

CH_3 $\xrightarrow{O_2}$ CH_2OH $\xrightarrow{O_2}$ $\overset{O}{\overset{\|}{C}} - H$ $\xrightarrow{O_2}$ $\overset{O}{\overset{\|}{C}} - OH$

The problem is one of stopping the reaction short of benzoic acid. This is partially accomplished by air oxidation with a short residence time and the use of a selective catalyst.

Typical reaction conditions are:

Temperature	500° C
Pressure	Atmospheric
Catalyst	UO_2 93%, MnO_2 7%
Conversion	10-20%
Yield	30-50%

In another process the reaction is carried out in the presence of methanol at 100-140°C with a mixture of iron (II) bromide, $FeBr_2$, and cobalt bromide, $CoBr_2$, as the catalyst.[55] The methanol is reported to retard the further oxidation of benzaldehyde to benzoic acid.

Benzaldehyde is also produced by the hydrolysis of *benzal chloride,* $C_6H_5 - CHCl_2$.

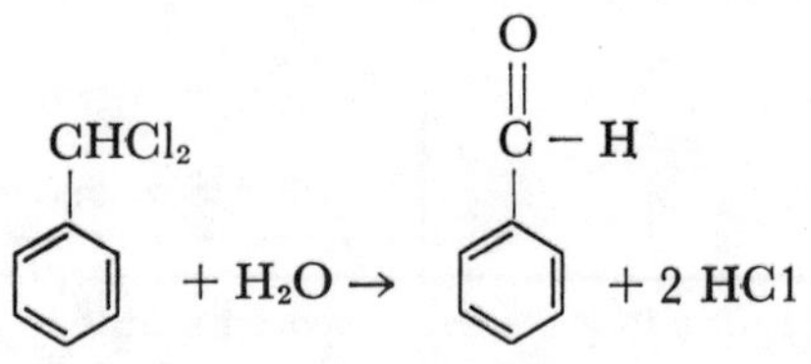

$CHCl_2$ $+ H_2O \rightarrow$ $\overset{O}{\overset{\|}{C}} - H$ $+ 2\ HCl$

Uses. Benzaldehyde is used as a solvent for oils, resins and various cellulose esters and ethers. It is also used in flavoring compounds and in synthetic perfumes. It has some minor uses as an intermediate for perfumery chemicals, certain triarylmethane dyes, *benzyl benzoate,* and *cinnamic acid,* $C_6H_5\text{-}CH{=}CH\text{-}C({=}O)\text{-}OH$.

Chlorination. The methyl group of toluene is chlorinated to give a mixture of chlorides; *benzyl chloride,* $C_6H_5\text{-}CH_2Cl$, *benzal chloride,* $C_6H_5\text{-}CHCl_2$, and *benzotrichloride,* $C_6H_5\text{-}CCl_3$. A preponderance of either benzyl chloride or benzotrichloride can be obtained by adjusting the chlorine-toluene ratio. Benzyl chloride is produced by passing dry chlorine into boiling toluene, 110°C, until a density of 1.283 is obtained. At this density the concentration of benzyl chloride is at its maximum. Light is sometimes used as a catalyst.

Uses. Benzyl chloride is hydrolyzed to *benzyl alcohol,* $C_6H_5\text{-}CH_2OH$, which is used to produce *butylbenzyl phthalate,* $C_4H_9O\text{-}C({=}O)\text{-}C_6H_4\text{-}C({=}O)\text{-}OCH_2\text{-}C_6H_5$, a vinyl chloride plasticizer, and other esters. These esters are used in perfumes. *Benzyl acetate,* $C_6H_5\text{-}CH_2O\text{-}C({=}O)\text{-}CH_3$, is a low-cost perfume used in soap.

Benzyl chloride reacts with sodium cyanide to form *benzyl cyanide,* $C_6H_5-CH_2CN$, which is hydrolyzed to *phenylacetic acid,* $C_6H_5-CH_2-C({=}O)-OH$. Phenylacetic acid is the precursor of *penicillin G* and other pharmaceuticals such as *amphetamine* (benzedrene). Benzyl cyanide is also the precursor for "Phenobarbital."

Benzal chloride is hydrolyzed to *benzaldehyde* and *benzotrichloride* is hydrolyzed to *benzoic acid.*

The chlorinated toluenes are not strong in tonnage but they are precursors for many of the compounds that help make life worth while, or, at least, more bearable.

Reactions With a Future. Because of the abundance of toluene and its low cost, there is continuing research directed toward its chemical utilization to produce established end products. The production of stilbene and its reaction with ethylene to produce styrene is an example.[56] The carbonylation of toluene to *p*-tolualdehyde and subsequent oxidation of the *p*-tolualdehyde is another example.

Stilbene, $C_6H_5-CH=CH-C_6H_5$, is produced by the oxidative dehydrocoupling of two molecules of toluene in the presence of steam.

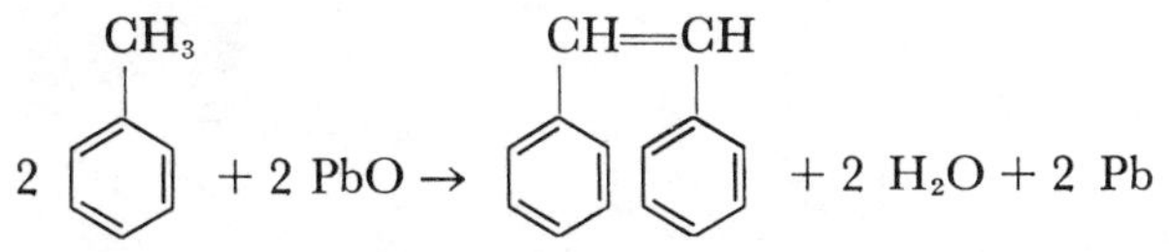

$$2\,C_6H_5CH_3 + 2\,PbO \rightarrow C_6H_5CH{=}CHC_6H_5 + 2\,H_2O + 2\,Pb$$

The reaction conditions are:

Temperature	600° C
Pressure	Atmospheric
Catalyst	*
Conversion	41%
Yield	Stilbene 67%
	Benzene 27%
	Other 6%

*The PbO is supported on $MgAl_2O_4$ particles. The PbO is regenerated by air oxidation.

Styrene, $C_6H_5-CH=CH_2$, is formed by the reaction between stilbene and ethylene at a mole ratio of 1 to 5.

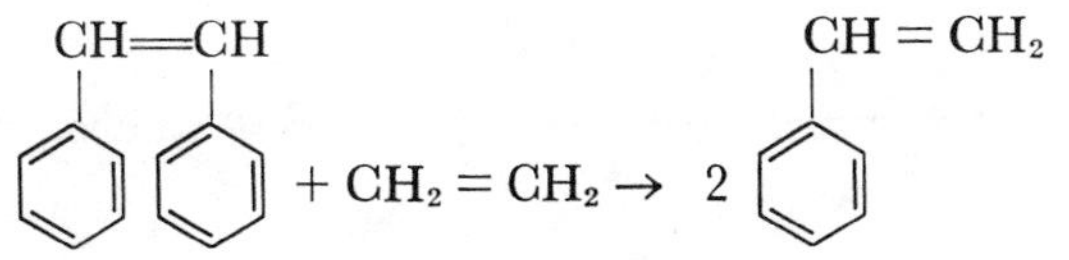

$$C_6H_5CH{=}CHC_6H_5 + CH_2=CH_2 \rightarrow 2\,C_6H_5CH=CH_2$$

The reaction conditions are:

Temperature	500° C
Pressure	1 kg/cm²
Catalyst	$CaO/WO_2/SiO_2$
Conversion (on stilbene)	74%
Yield (on stilbene)	99%

The over-all reaction is

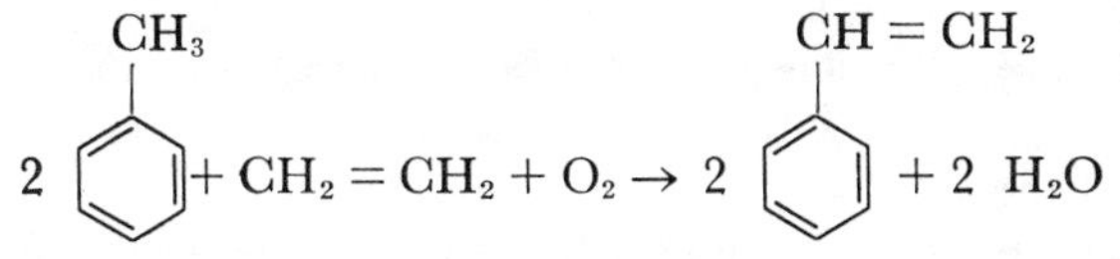

$$2\,C_6H_5CH_3 + CH_2=CH_2 + O_2 \rightarrow 2\,C_6H_5CH=CH_2 + 2\,H_2O$$

This process requires only one-half the amount of ethylene that the ethylbenzene process requires and toluene cost is less than benzene.

***p*-Tolualdehyde, PTAL,** $CH_3\text{-}C_6H_4\text{-}C({=}O)\text{-}H$, is formed by the reaction between toluene and carbon monoxide—a carbonylation reaction.

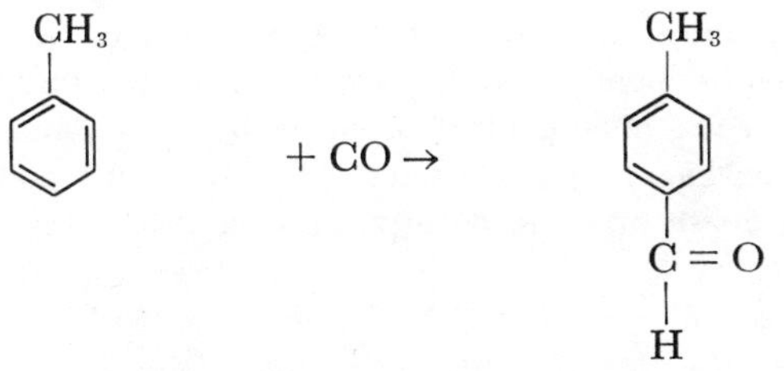

$$C_6H_5CH_3 + CO \rightarrow CH_3C_6H_4CHO$$

This reaction is catalyzed by HF/BF_3. The yield is 96 percent based on toluene and 98 percent based on carbon monoxide. The **PTAL** is subsequently oxidized to *terephthalic acid* which is a principal polyester fiber precursor.

The Xylenes

The xylenes are obtained from refinery reformate streams where the yield of mixed xylenes plus ethylbenzene (the C_8 aromatics) is greater than that of the combined yield of benzene and toluene. The weight percent of the C_8 aromatics is given in Fig. 10-28.[58] Minor amounts for the xylenes are obtained from pyrolysis gasoline, coal tar, and the disproportionation of toluene.

About 16 percent of the isolated xylenes are blended with gasolines, 17 percent are used as solvents, and 67 percent are used for the production of the individual isomers.[38] The

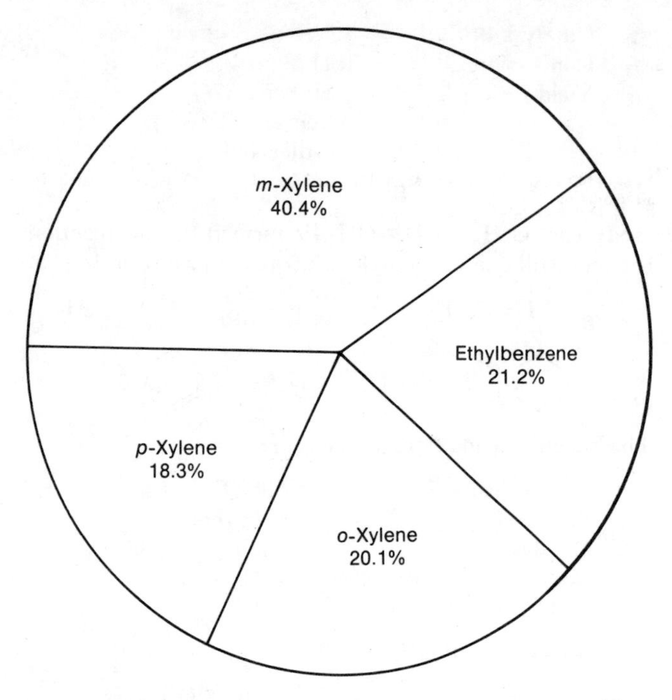

Fig. 10-28—Weight percent composition of a typical reformate.[58]

separation of *o-*, *m-*, and *p*-xylene is the key to their chemical use and this separation is difficult. It is further complicated by the co-present ethylbenzene. Ethylbenzene has a boiling point only 2.2°C lower than the boiling point of *p*-xylene and 2.9°C below *m*-xylene (Table 10-2). However, the separation of ethylbenzene from the xylenes has been carried out commercially using a column with over 300 plates and a high reflux ratio. The closeness of the boiling points of *m*-xylene and *p*-xylene (0.7°C) precludes their separation by fractional distillation.

Because *p*-xylene is the most important xylene for chemical utilization, most separation techniques involve the production of chemical grade *p*-xylene. Until 1970 high-purity *p*-xylene was obtained by low-temperature fractional crystallization which gives a 60 percent yield. Currently, continuous liquid-phase *p*-xylene adsorption separation processes are used.[46, 59, 60, 61, 67] These processes give a one pass yield of over 90 percent. The over-all yield of *p*-xylene is increased by incorporating an isomerization unit in the process to isomerize the *o*-xylene and *m*-xylene.[60] A flow diagram of a combined process for the production of *p*-xylene is shown in Fig. 10-29.[62] Table 10-3 gives operating data for the Aromax portion of this process.[60] High conversions of ethylbenzene can also be obtained by these isomerization processes and in some cases the *m*-xylene in the mixture is separated and isomerized.[63, 64]

Combined separation-isomerization processes can be operated to produce either *p*-xylene or *p*-xylene plus *o*-xylene. When *p*-xylene alone is desired, an adsorption separation process is used.[61] When both *o*-xylene and *p*-xylene are wanted, the *o*-xylene is separated by efficient fractionation.[65] Product distribution from these two processes is given in Table 10-4.[65]

The favored isomerization catalyst is ZSM-5, which at 370° C gives a *p*-xylene content of 24 percent. The advantage of this catalyst is its failure to cause intermolecular methyl migration. The ratio of isomerization to disproportionation activity is 1,000 to 1. Other isomerization catalysts have ratios of about 70 to 1.[66]

Fig. 10-30 gives the production and consumption of the xylenes for the United States and Western Europe.[1] The 1978 production of the xylenes, all grades, was 6.16 billion pounds. *p*-Xylene was 3.49 billion pounds.

o-Xylene, Phthalic Anhydride. Phthalic anhydride,

C=O

O, was first produced commercially by the oxi-

C=O

dation of *naphthalene*, , in a process similar to that now used for the oxidation of *o*-xylene. In 1980 approximately 75 percent of the 1.55-billion-pound phthalic anhydride capacity used *o*-xylene as the feedstock.[68]

Essentially, the only use for *o*-xylene is in the production of phthalic anhydride by vapor phase oxidation over a solid catalyst (Fig. 10-31).[69]

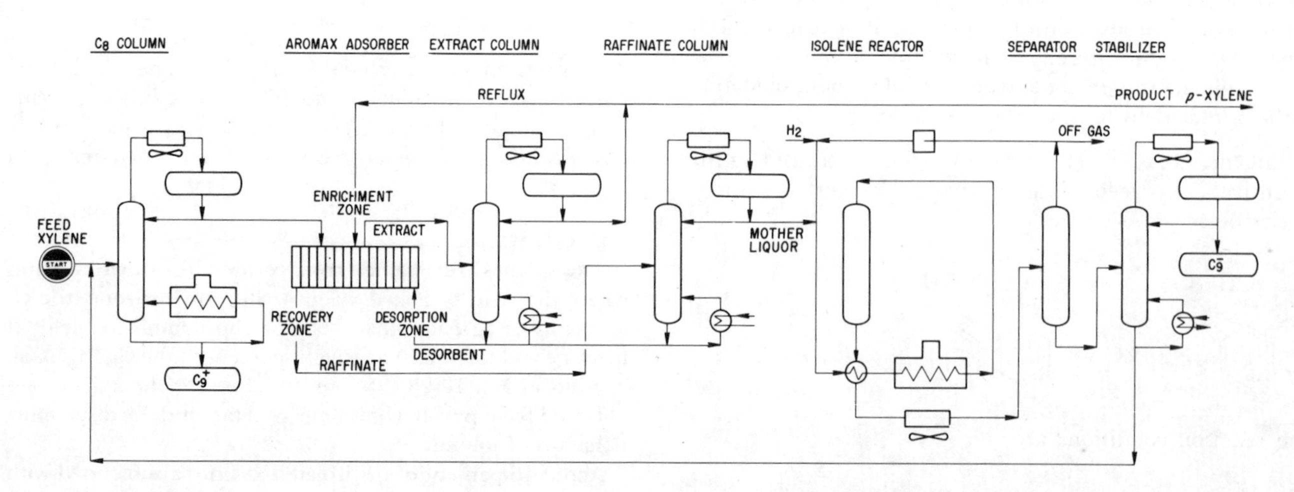

Fig. 10-29—Flow diagram for the production of *p*-xylene by a combined adsorption-isomerization process.[62]

TABLE 10-3—Product and Raffinate Data for the Aromax Process[60]

Compound	Feed*	Product*	Raffinate*
o-Xylene	13.3	....	17.3
m-Xylene	45.3	0.1	51.8
p-Xylene	18.3	99.5	2.0
Ethylbenzene	14.9	0.4	18.3
Light ends	8.2	....	10.6

* Weight percent

TABLE 10-4—Weight Percent Yields from the Isomer Isomerization Process[65]

	Feed composition*	Product composition*	Feed composition†	Product composition†
o-Xylene	21.8	37.2	21.7	0.0
m-Xylene	43.3	....	40.9	....
p-Xylene	17.6	54.6	21.7	83.2
Ethylbenzene	17.3	....	17.1	....

* Operated to recovery of o-xylene and p-xylene
† Operated for recovery of p-xylene only

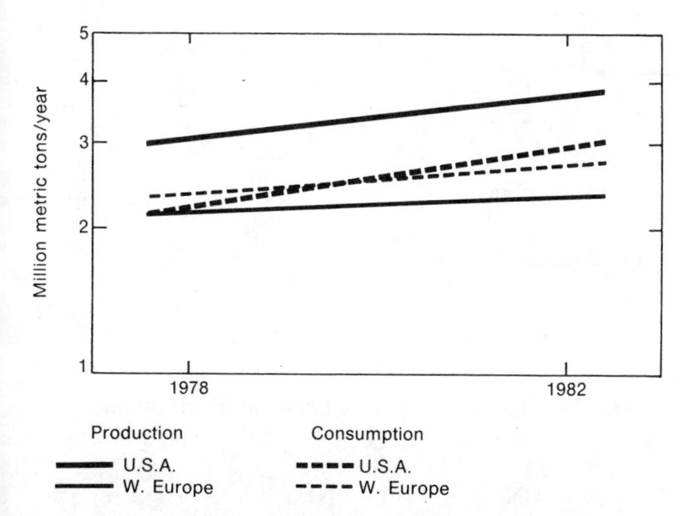

Fig. 10-30—Xylene isomer capacities, production, and consumption for the United States and Western Europe, 1978-1982.[1]

$$\text{o-xylene } (C_6H_4(CH_3)_2) + 3\,O_2 \rightarrow \text{phthalic anhydride } (C_6H_4(CO)_2O) + 3\,H_2O$$

Reaction conditions will vary with equipment and catalyst. The following are typical.

Temperature	375-435° C
Pressure	10 psig
Catalyst	$V_2O_5 + TiO_2/Sb_2O_3$*
Yield	85%

*Also $V_2O_5 + MoO_3/CaO$ or Mn_3O_6, and other similar catalysts. "The technology of catalyst preparation is an intricate art."[70]

Along with the phthalic anhydride, varying amounts of *maleic anhydride, citraconic anhydride* and *benzoic acid* are formed.

Maleic anhydride (H–C–C=O / H–C–C=O, ring-closed through O) — Citraconic anhydride (CH_3–C–C=O / H–C–C=O, ring-closed through O) — Benzoic acid (C_6H_5–C(=O)–OH)

Maleic anhydride Citraconic anhydride Benzoic acid

The maleic anhydride can be recovered economically.[71, 72]

The various processes differ mainly in the equipment used to control the highly exothermic oxidation reaction. Much of this energy can be used to meet energy requirements of the phthalic anhydride plant.[73]

Uses. The single largest use (48%) of phthalic anhydride is for the production of plasticizers for polyvinyl chloride, **PVC.** The second most important use (24%) is in the production of polyester resins; 20 percent is used for alkyd resins.[74] The 1978 United States production of phthalic anhydride was just over 1 billion pounds. Other, relatively minor uses are in dyes, fire retardants, cross-linking agent for polyester resins, polyester polyols, phenolphthalein and drying-oil modifiers.

Because the major application of phthalic anhydride is to produce the corresponding diesters, a continuous phthalate production process has been proposed.[75]

$$C_6H_4(C{=}O)_2O + ROH \rightarrow C_6H_4(COOR)(COOH)$$

$$C_6H_4(COOR)(COOH) + ROH \rightarrow C_6H_4(COOR)_2 + H_2O$$

The usual phthalic anhydride condensers are replaced with a packed absorption tower fed with the desired alcohol. The reaction to produce the monoester is fast while the formation of the desired diester is relatively slow and requires a high temperature or a catalyst.

Phthalonitrile. *Phthalonitrile* can be produced by an oxidative ammonolysis process (Fig. 10-32).[76] The reaction is carried out in the absence of molecular oxygen using lattice oxygen from a metal oxide in a fluid bed catalytic system. The catalyst is reoxidized for reuse. The over-all reaction system is similar to a fluid bed catalytic cracking, **FCC,** system.

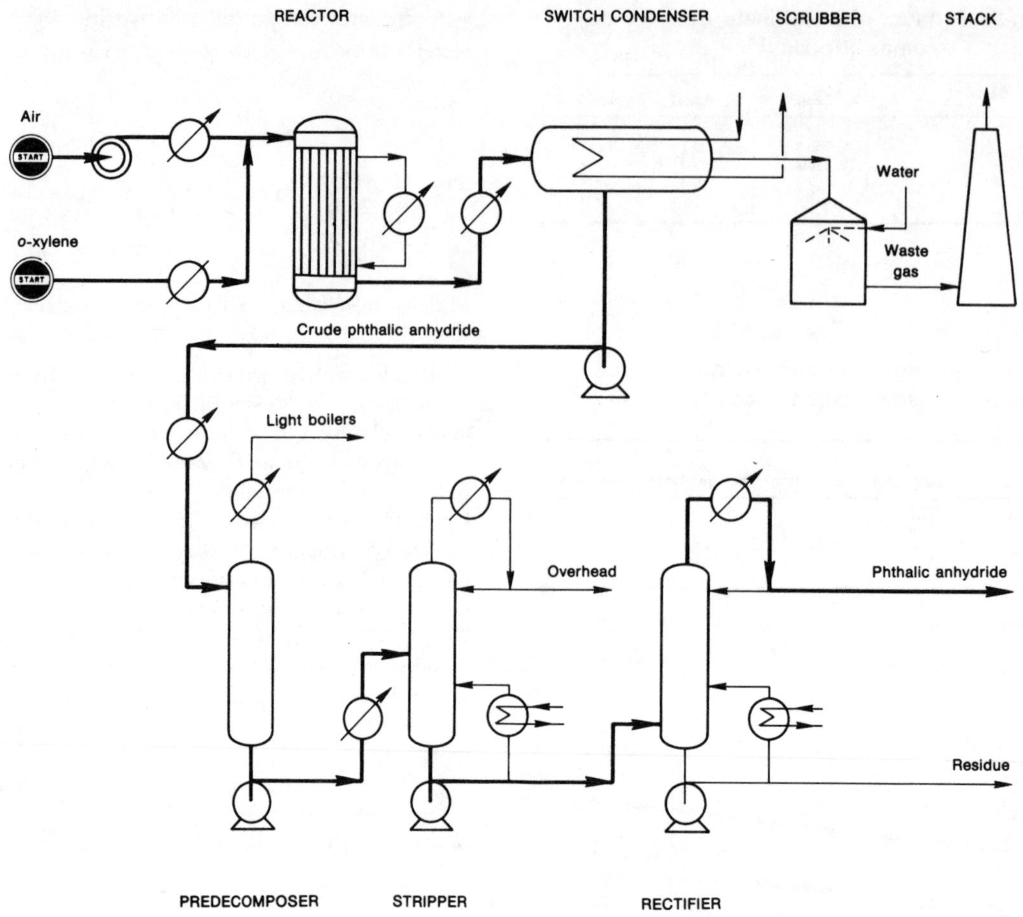

REACTOR SWITCH CONDENSER SCRUBBER STACK
Air
START
o-xylene
START
Water
Waste gas
Crude phthalic anhydride
Light boilers
Overhead
Phthalic anhydride
Residue
PREDECOMPOSER STRIPPER RECTIFIER

Fig. 10-31—The BASF AG process for the air oxidation of *o*-xylene to phthalic anhydride.[69]

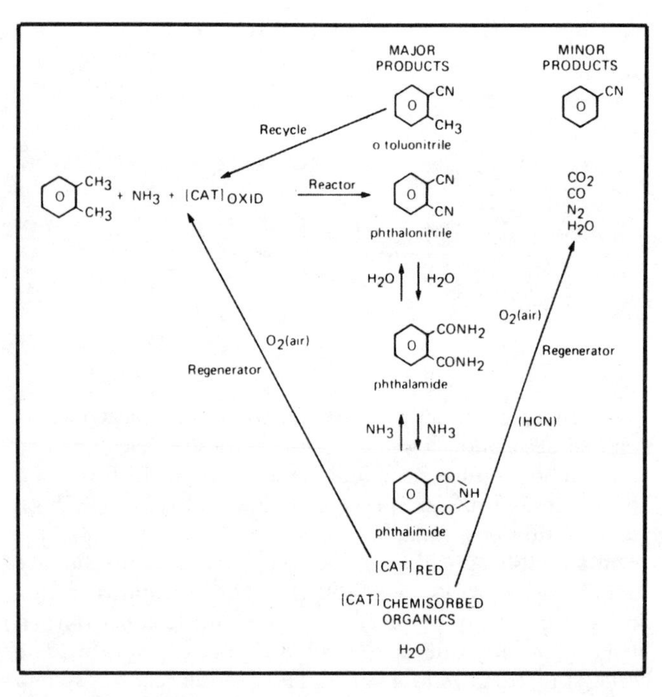

Fig. 10-32—The reaction scheme for *o*-xylene to phthalonitrile.[76]

The byproducts are *phthalamide* and *phthalimide.*

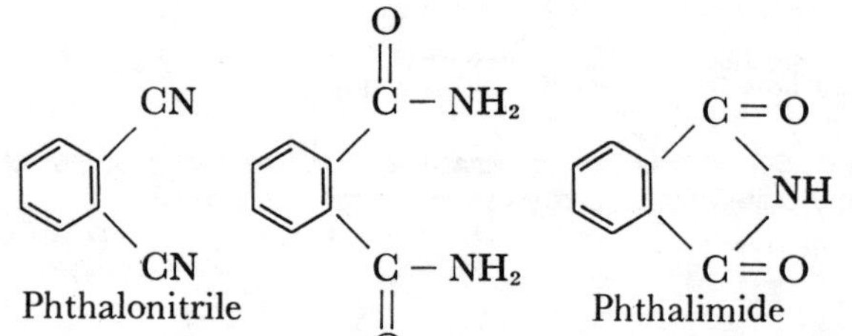

Phthalonitrile Phthalamide Phthalimide

Operating conditions dictate which of these three compounds will be the major product. Because of the equilibria, phthalonitrile can be the only final product obtained.

Uses. *Phthalonitrile* is used to produce *copper phthalocyanine* pigments. Phthalic anhydride plus urea are also used for the production of this type of pigment.

***m*-Xylene.** *m-Xylene* is the least utilized of the three isomeric xylenes. World consumption was only about 150 million pounds in 1976. New capacity, now onstream, may encourage wider chemical use of *m*-xylene.

Uses. At present the major use of *m*-xylene is isomerization to the more desirable *o*- and *p*-xylenes. The most important petrochemical use is its oxidation to *isophthalic acid.* An interesting new synthesis is oxidative ammonolysis to *isophthalonitrile.*

Isophthalic Acid. *Isophthalic* acid is produced by the liquid phase oxidation of *m*-xylene by use of *ammonium sulfite,* $(NH_4)_2SO_3$.

$$C_6H_4(CH_3)_2 + 2\ (NH_4)_2SO_2 \rightarrow C_6H_4(COOH)_2 + 2\ H_2S + 4\ NH_3 + 2\ H_2O$$

This is a two-step process because the second methyl group is much more difficult to oxidize than the first one.

Isophthalic acid is also produced by the liquid phase oxidation of mixed xylenes in the presence of a solvent, usually acetic acid. The reaction conditions are:

Temperature	200° C
Pressure	400 psi
Catalyst	Heavy-metal salts + a bromine compound.
Conversion	95%
Yield	80%

Uses. The uses of isophthalic acid are similar to those of the other phthalic acids. Its polyesters have better toughness, resistance to abrasion, resiliency and freedom from cracking and crazing. Because of these characteristics polyesters are used for pressure molding applications. Alkyd resins of isophthalic acid have better resistance to abrasion and corrosion than phthalic acid alkyd resins but they are more difficult to process.

The growth area for isophthalic acid polymers is in fibers, films and glass-reinforced plastics. They can be used at temperatures up to 225°C for film and fiber and even higher for the reinforced resins. Isophthalic acid has joined with terephthalic acid to form a copolyester with 1,4-cyclohexylenedimethanol. The product is a high molecular weight thermoplastic, Kodar, used in blister packaging for individual servings.

Isophthalonitrile. *Isophthalonitrile* can be produced by oxidative ammonolysis by a reaction similar to that used for the production of phthalonitrile. The initial products are isophthalonitrile and *m-toluonitrile.* The *m*-toluonitrile is recycled. The minor organic product is *benzonitrile.* (Fig. 10-33).[76, 77]

Uses. *Isophthalonitrile* is used as a precursor for agricultural chemicals. It is readily hydrogenated to the corresponding diamine which can be used to form polyamides or converted to isocyanates for polyurethane production.

p-Xylene. Virtually the only use of *p*-xylene is for the production of *terephthalic acid,* **TPA,** and *dimethyl terephthalate,* **DMT.**

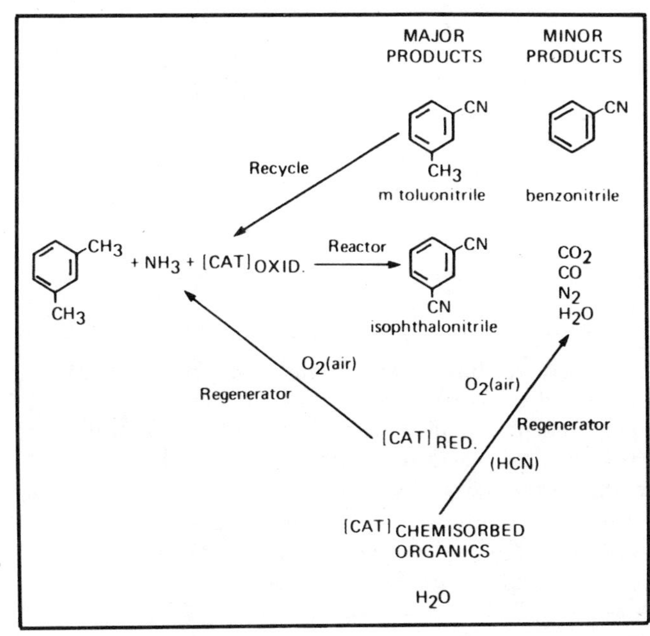

Fig. 10-33—The reaction scheme for *m*-xylene to isophthalonitrile.[76]

$$p\text{-}C_6H_4(COOH)_2 \qquad\qquad p\text{-}C_6H_4(COOCH_3)_2$$

Terephthalic acid **TPA** — Dimethyl terephthalate **DMT**

Pure, fiber grade terephthalic acid is called **PTA.**

Current production capacity is about evenly split between **PTA** and **DMT.**[78] The trend is toward **PTA** expansion rather than **DMT.** This is the result of technological advances which permit both better purification of **TPA** and the use of the acid for polymer formation. The 1980 terephthalate production is 6 billion pounds with a capacity of 7.8 billion pounds. A demand growth of 5 to 8 percent is predicted for the next several years. It has also been predicted that demand by 1982 will exceed 100 percent of capacity.[79]

Production. There are many good processes for **PTA** and **DMT** which give a high yield. The initial commercial process used aqueous nitric acid (30-40%) oxidation at 165°C and 140 psig. This process has now been superseded by various direct air or oxygen catalyzed oxidation.

It is not within the scope of this book to delineate the various current processes with their different modifications of the basic cobalt-catalyzed, liquid-phase oxidation. Temperatures and pressures may vary, either air or oxygen is used, and acetic acid is used as a reaction solvent. **TPA** is recovered and purified by crystallization or solvent extraction. The recent patent literature is summarized by M. Sittig.[80]

p-Xylene may be partially oxidized to *p-toluic acid* or *p-tolualdehyde.*

$CH_3C_6H_4COOH$ — *p*-Toluic acid

$CH_3C_6H_4CHO$ — *p*-Tolualdehyde

Because of the presence of a carboxyl group, $-\overset{\overset{O}{\|}}{C}-OH$, the second oxidation is more difficult than the first. The toluic acid is esterified with methanol to deactivate the carbonyl group and the methyl group is then oxidized to the carboxylic acid.

The Mid-Century process is representative of current practice. *p*-Xylene is oxidized to *terephthalic acid* in acetic acid as the solvent.

$$CH_3C_6H_4CH_3 + 3\,O_2 \rightarrow HOOC{-}C_6H_4{-}COOH + 2H_2O$$

The reaction conditions are:

Temperature	200° C
Pressure	200 psig
Catalyst	Co acetate
Yield	95%

The cobalt catalyst concentration can be as high as 5 percent of the total weight of the reaction mixture. The cobalt reacts with the *p*-xylene by an electron transfer to produce a specie which will combine with oxygen to produce the terephthalic acid. The reduced cobalt must be reoxidized. This is accomplished by oxygen with the aid of a promoter.[81]

Bromine compounds, usually NaBr or HBr, are used as promoters. In other processes, methyl ethyl ketone, MEK, or acetaldehyde are used. They both oxidize to acetic acid under the reaction conditions. The acetic acid constitutes a substantial byproduct.

The major impurity is 4-*carboxybenzaldehyde,* **CBA,** which is hydogenated to *p-toluic acid.*

$$HOOC{-}C_6H_4{-}CHO + 2\,H_2 \rightarrow HOOC{-}C_6H_4{-}CH_3 + H_2O$$

The toluic acid is oxidized to **TPA**. The current trend is to oxidize the *p*-xylene directly to **TPA**.

A flow diagram of a typical *p*-xylene oxidation process to produce either **PTA** or **DMT** is shown in Fig. 10-34.[82]

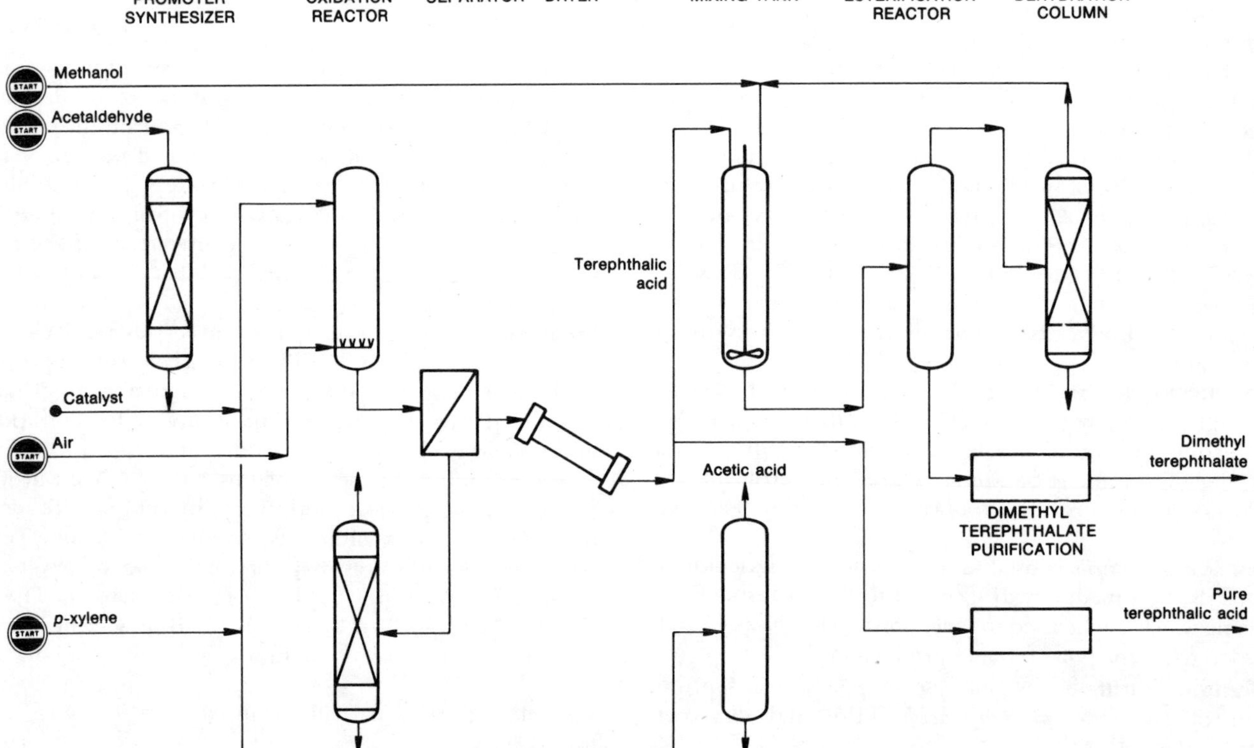

Fig. 10-34—Toray process for the production of dimethyl terephthlate, DMT, and pure terephthalic acid, PTA.[82]

Uses. The two *p*-xylene derivatives, **PTA** and **DMT,** have only one major use, the synthesis of *polyethylene terephthalate fiber* (85 percent) and film.

However, at least two other types of terephthalate esters are growing in importance. These are unsaturated polyesters and *polybutylene terephthalate,* **PBT,** a thermoset engineering plastic. The 1978 production of **PBT** used more than 100 million pounds of **PTA/DMT,** [83] and is one of the fastest growing injection molding plastics. To a lesser extent **TPA** is used as an intermediate for herbicides, in adhesives, in printing inks and coatings, in paints and in animal feed supplements.

LITERATURE CITED

1. Ponder, Thomas C. *Hydrocarbon Processing,* Vol. 58, No. 1, 1979, pp. 149-151.
2. *The Oil and Gas Journal,* Nov. 26, 1979, pp. 26-27.
3. MacFarlane, A.C. *The Oil and Gas Journal,* Feb. 9, 1976, pp. 99-102, See also "Letters," *OGJ,* April 26, 1976, p. 56.
4. *Hydrocarbon Processing, 1979 Petrochemical Handbook,* Vol. 58, No. 11, p. 157.
5. *International Petroleum Encyclopedia,* Vol. 10, 1977, pp. 421-422.
6. Lewis, P.J. and F.G. Dwyer. *The Oil and Gas Journal,* Sept. 26, 1977, pp. 55-58.
7. Dwyer, F.G., P.J. Lewis and F.H. Schneider. *Chemical Engineering,* Jan. 5, 1976, pp. 90-91.
8. Maier, W.H. and D.J. Ward. U.S. Patent 3,478,119, Nov. 11, 1969, to Universal Oil Products Co.
9. Portes, A. and J. Escourrou. *Hydrocarbon Processing,* Vol. 56, No. 9, 1977, pp. 154-155.
10. *Hydrocarbon Processing, 1979 Petrochemical Handbook,* Vol. 58, No. 11, p. 245.
11. *CHEMTECH,* Vol. 7, No. 6, 1977, pp. 334, 351.
12. Brownstein, A.M. *Trends in Petrochemical Technology,* Tulsa: Petroleum Publishing Co., 1976, pp. 220-228.
13. Landau, L., R.E. Lidov, and J. Habeshaw. *Benzene and Its Industrial Derivatives,* E.G. Hancock, editor, London, Tonbridge: Ernest Benn, Ltd., 1975, pp. 285-316.
14. *Chemical Week,* March 8, 1978, p. 33.
15. *Chemical Age,* May 20, 1977, pp. 10-11.
16. *Chemical Week,* Sept. 6, 1978, pp. 42, 45, 47.
17. Pujado, P.R., J.R. Salazar and C.V. Berger. *Hydrocarbon Processing,* Vol. 55, No. 3, 1976, pp. 91-96.
18. Fleming, J.B., J.R. Lambrix, and J.R. Nixon. *Hydrocarbon Processing,* Vol. 55, No. 1, 1976, pp. 185-186, 188-190, 193, 195-196.
19. *Chemical Week,* Dec. 5, 1979, p. 50.
20. Gans, M. *Hydrocarbon Processing,* Vol. 55, No. 11, 1976, pp. 145-150.
21. Starr, H.G. *Chemical Week,* July 20, 1977, pp. 11-12.
22. *CHEMTECH,* Vol. 8, No. 3, 1978, p. 132.
23. Frontier Chemical Co., U.S. Patent 3,148,222 (1964).
24. *Hydrocarbon Processing, 1979 Petrochemical Handbook,* Vol. 58, No. 11, p. 184.
25. *Hydrocarbon Processing, 1975 Petrochemical Handbook,* Vol. 54, No. 11, p. 159.
26. Maux, R. *Hydrocarbon Processing,* Vol. 55, No. 3, 1976, p. 90.
27. Brownstein, A.M. *Trends in Petrochemical Technology,* Tulsa: Petroleum Publishing Co., 1976, pp. 253-259.
28. *Chemical Week,* Oct. 13, 1976, pp. 79.
29. *Chemical Week,* Feb. 2, 1977, pp. 37-38.
30. *Hydrocarbon Processing, 1979 Petrochemical Handbook,* Vol. 58, No. 11, p. 149.
31. *Hydrocarbon Processing, 1979 Petrochemical Handbook,* Vol. 58, No. 11, p. 151.
32. *Chemical and Engineering News,* Dec. 5, 1977, p. 13.
33. Tanaka, K. *Hydrocarbon Processing,* Vol. 53, No. 11, 1974, pp. 114-120.
34. Earhart, H.W. *Polymethylbenzenes,* Park Ridge, N.J.: Noyes Development Corp., 1969, p. 158.
35. Ockerbloom, N.E. *Hydrocarbon Processing,* Vol. 51, No. 4, 1972, pp. 114-118.
36. Nahas, R.S. and M.R. Nahas. *Paper 8 (P-2),* Second Arab Conference on Petrochemicals, Abu Dhabi, March 15-22, 1976.
37. Cinelli, E., S. Noe, and G. Paret. *Hydrocarbon Processing,* Vol. 51, No. 5, 1972, pp. 141-144.
38. Vervalin, C.H. *Hydrocarbon Processing,* Vol. 57, No. 4, 1978, p. 27.
39. *The Oil and Gas Journal,* Dec. 19, 1977, p. 36.
40. Ponder, Thomas C. *Hydrocarbon Processing,* Vol. 56, No. 12, 1977, pp. 111-112.
41. *Chemical and Engineering News,* Dec. 11, 1978, pp. 12-13.
42. *Hydrocarbon Processing, 1979 Petrochemical Handbook,* Vol. 58, No. 11, p. 139.
43. DeGraff, R.R. U.S. Patent 3,558,729, Jan. 26, 1971 to Universal Oil Products Co.
44. Ohsumi, Y. and Y. Komatsuzaki. U.S. Patent 3,903,186, Sept. 2, 1975 to Mitsubishi Chemical Industries, Ltd. and Asis Oil Co., Ltd., Japan.
45. *Hydrocarbon Processing, 1979 Petrochemical Handbook,* Vol. 58, No. 11, p. 140.
46. Vora, B.V., R.H. Jensen and K.W. Rockett. *Paper No. 20 (p-1),* Second Arab Conference on Petrochemicals, Abu Dhabi, March 15-22, 1976.
47. *Hydrocarbon Processing, 1979 Petrochemical Handbook,* Vol. 58, No. 11, p. 141.
48. *Hydrocarbon Processing, 1977 Petrochemical Handbook,* Vol. 56, No. 11, p. 137.
49. Gelbein, A.P. and A.S. Nislick. *Hydrocarbon Processing,* Vol. 57, No. 11, 1978, pp. 125-128.
50. Cines, M.R. U.S. Patent 3,746,754, July 17, 1973 to Phillips Petroleum Co.; U.S. Patent 2,905,709 and 2,794,830.
51. Sittig, M. *Aromatic Hydrocarbons, Manufacture and Technology,* Park Ridge, N.J.: Noyes Data Corp., 1976, pp. 303-306.
52. *Chemical and Engineering News,* June 12, 1978, pp. 34, 36.
53. Barona, N. and H.W. Prengle, Jr. *Hydrocarbon Processing,* Vol. 52, No. 12, 1973, p. 77.
54. *Modern Plastics International,* Vol. 9, No. 4, 1979, p. 25.
55. Massie, S.N. U.S. Patent 3,732,314, May 8, 1973 to Universal Oil Products Co.
56. *CHEMTECH,* Vol. 7, No. 3, 1977, p. 140.
57. Fujiyama, Susumu and T. Kasahara. *Hydrocarbon Processing,* Vol. 57, No. 11, 1978.
58. Sittig, M. *Aromatic Hydrocarbons, Manufacture and Technology,* Park Ridge, N.J.: Noyes Data Corp., 1976, p. 8
59. Biesser, H.J. and G.R. Winter. *The Oil and Gas Journal,* Aug. 11, 1975, p. 74-75.
60. *The Oil and Gas Journal,* Jan. 2, 1978, pp. 101-102.
61. *Hydrocarbon Processing, 1977 Petrochemical Handbook,* Vol. 56, No. 11, p. 240.
62. *Hydrocarbon Processing, 1973 Petrochemical Handbook,* Vol. 52, No. 11, p. 196.
63. Masseling, J.J.H. *CHEMTECH,* Vol. 6, No. 11, 1976, pp. 714-716.
64. *Hydrocarbon Processing, 1977 Petrochemical Handbook,* Vol. 56, No. 11, p. 238.
65. *Hydrocarbon Processing, 1977 Petrochemical Handbook,* Vol. 56, No. 11, p. 239.
66. *CHEMTECH,* Vol. 7, No. 10, 1977, p. 588.
67. Seko, Maomi. *The Oil and Gas Journal,* July 2, 1979, pp. 81-86.
68. Anderson, E.V. *Chemical and Engineering News,* Oct. 31, 1977, pp. 8-9.
69. *Hydrocarbon Processing, 1979 Petrochemical Handbook,* Vol. 58, No. 11, p. 207.
70. Dumas, T. and W. Bulani. *Oxidation of Petrochemicals: Chemistry and Technology,* New York: John Wiley & Sons, 1974, p. 57.
71. Weyens, E. *Hydrocarbon Processing,* Vol. 53, No. 11, 1974, pp. 132-134.
72. *Hydrocarbon Processing, 1977 Petrochemical Handbook,* Vol. 56, No. 11, p. 179.
73. Dow, R.M. and C.D. Miserlis. *Hydrocarbon Processing,* Vol. 56, No. 4, 1977, pp. 167-170.
74. *Chemical and Engineering News,* Dec. 11, 1978, p. 13.
75. Wu, W.H. and J.R. Maa. *Hydrocarbon Processing,* Vol. 53, No. 4, 1974, pp. 117-118.
76. Sze, M.C. and A.P. Gelbein, *Hydrocarbon Processing,* Vol. 55, No. 2, 1976, pp. 103-106.
77. *Hydrocarbon Processing, 1977 Petrochemical Handbook,* Vol. 56, No. 11, p. 174.
78. *Chemical Week,* April 5, 1978, pp. 26-27.
79. *Chemical Week,* March 8, 1978, p. 43.
80. Sittig, M. *Aromatic Hydrocarbon, Manufacture and Technology,* Park Ridge, N.J.: Noyes Data Corp., 1976, pp. 269-297.
81. Saunby, J.B. and B.W. Kiff. *Hydrocarbon Processing,* Vol. 55, No. 11, 1976, pp. 247-252.
82. *Hydrocarbon Processing, 1977 Petrochemical Handbook,* Vol. 56, No. 11, p. 230.
83. *Chemical and Engineering News,* Dec. 5, 1977, p. 11.

11

Introduction to Polymer Chemistry

Naturally occurring polymers, macromolecules, are the substance of life; synthetic polymers add to comfort and enjoyment of life. These synthetic polymers represent the end product of a very great portion of the petrochemicals noted in previous chapters.

In general, polymers are used in three broad fields: plastics, fibers, and elastomers. Many other specific fields of application for polymers include coatings, adhesives, insulation, etc. The polymer industry is also integrated with many other industries such as the automobile, appliance, electrical, and electronic industries. They are also used to produce plasticizers, additives, and catalysts. Polymeric materials are increasingly replacing natural materials and metals in many applications.

A **polymer** is a macromolecule which contains large numbers of building blocks, or molecules, joined together. The building units are monomers. When the same monomer is used, the polymer is called a **homopolymer.** Polymers formed from more than one type of monomer but joined in a regular sequence may also be called homopolymers. For example, condensation polymers such as polyamides and polyesters are homopolymers.

A **copolymer** is a polymer formed from more than one type of monomer by addition polymerization. Several different structures are possible for copolymers depending on the method of polymerization.

Block copolymers have blocks of one homopolymer attached to blocks of another as in styrene-isoprene.

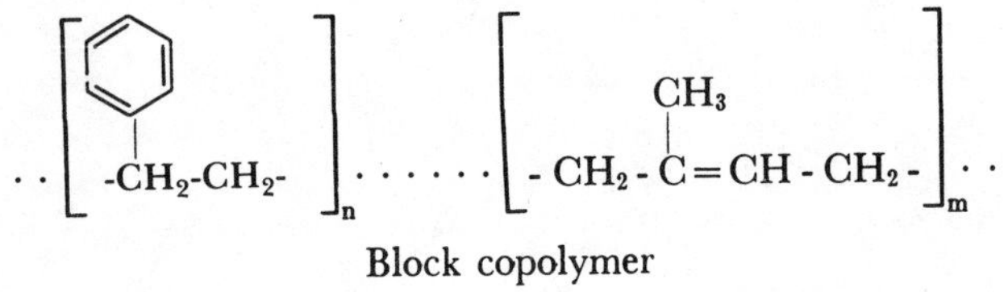

$$\cdot\cdot\left[\begin{matrix}C_6H_5\\ |\\ -CH_2\text{-}CH_2\text{-}\end{matrix}\right]_n\cdots\cdots\left[\begin{matrix}\quad CH_3\\ \quad |\\ -CH_2-C=CH-CH_2-\end{matrix}\right]_m\cdot\cdot$$

Block copolymer

Alternating copolymers have the monomers alternating in a regular manner as in the polymer formed between styrene and butadiene.

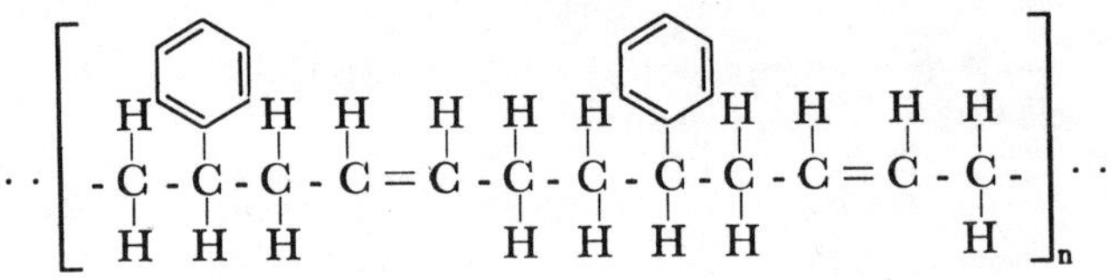

$$\cdot\cdot\left[\begin{matrix}H & C_6H_5 & H & H & H & H & H\,C_6H_5 & H & H & H & H\\ | & | & | & | & | & | & | & | & | & | & |\\ -C- & C- & C- & C= & C- & C- & C- & C- & C- & C= & C- & C-\\ | & | & | & & & | & | & | & | & & & |\\ H & H & H & & & H & H & H & H & & & H\end{matrix}\right]_n\cdot\cdot$$

Alternating copolymer

Random copolymers have the different monomer molecules distributed in a random manner along the macromolecule.

Polymers may be described as being linear or branched. *Branched polymers* have a side chain branching from the polymer backbone. Branching affects the physical properties of the polymer. For example, the degree of crystallinity decreases with the increase in branching. The methyl groups of polypropylene and the phenyl groups of polystyrene are not considered a branch, since a branch should be formed from complete monomer (or monomers) units as in branched polypropylene.

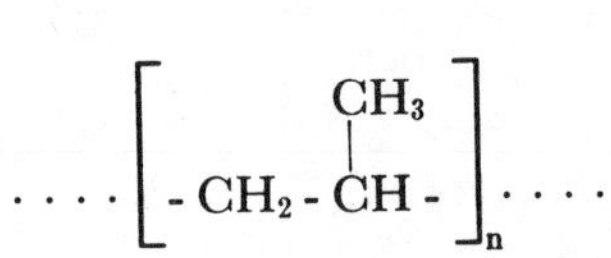

$$\cdots\left[\begin{matrix}\quad CH_3\\ \quad |\\ -CH_2-CH-\end{matrix}\right]_n\cdots\ \text{Linear polypropylene}$$

$$\cdots\left[\begin{matrix}\quad CH_3\\ \quad |\\ -CH_2-C--\\ \quad |\\ \quad CH_2\\ \quad |\\ \cdots\cdots-CH-CH_3\end{matrix}\right]_n\cdots\ \text{Branched polypropylene}$$

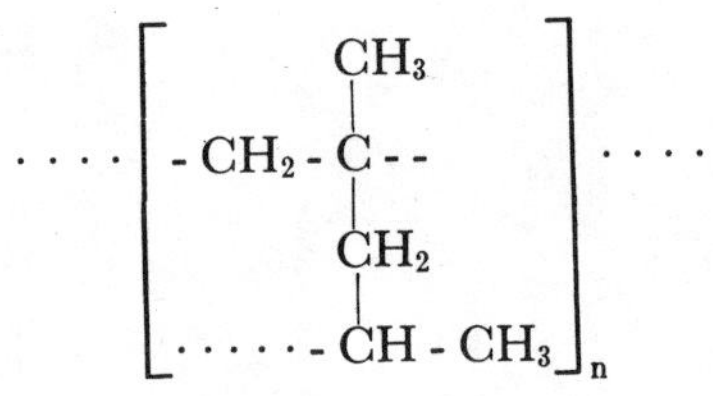

Branching may occur as a side reaction during polymerization (branched polypropylene) or as an intended reaction as in graft polymers.

Crosslinked polymers are those in which two or more polymer chains are linked together at one or more points, other than at their ends. This forms a network and these polymers are sometimes called *network polymers*.

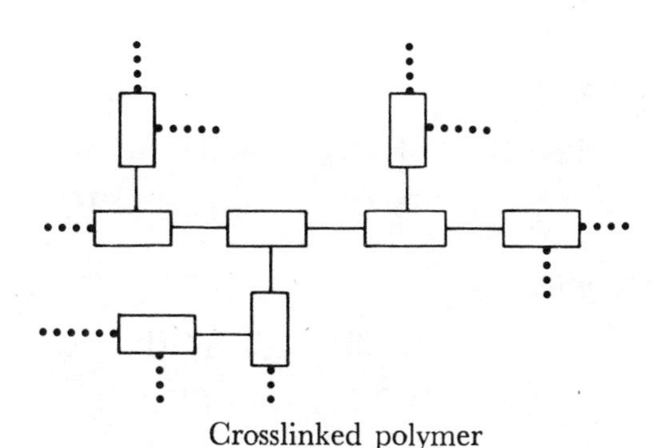

Crosslinked polymer

Crosslinking may be formed during the polymerization reaction when multifunctional groups are present as with phenol-formaldehyde resins or by the use of outside linking agents as in the vulcanization of rubber with sulfur. Because of the crosslinking, the polymer chains lose their mobility and do not melt or flow and cannot be molded. Thermosetting plastics are typical examples.

CLASSIFICATION

Synthetic polymers are classified in several ways. All the following classifications are used.

Chemical type of monomer:	Polyolefins, polyesters, polyamides, etc.
Type of formation reaction:	Condensation or addition polymerization
Type of process used:	Bulk, solution, suspension, emulsion
Type of utilization:	Thermosetting, thermoplastics, fibers, engineering plastics, sheeting, etc.

Polymer data are usually given under the name of the polymer, such as polyethylene and polystyrene, or by the chemical type monomer as acrylic or polyester. Polymers are generally distinguished by their end uses into *elastomers, plastics,* and *fibers.* It might appear that the classification of polymers emulates the general who jumped on his horse and rode off in all directions. This leads to confusion.

Elastomers. Elastomers are high molecular weight unsaturated hydrocarbons. The reversible elasticity is imparted to the polymer molecules by their long flexible chains with weak intermolecular forces. An isolated linear molecule has no shape because it is possible to have at least partial rotation of any two consecutive monomer units with respect to each other. Polymers of this type are predominantly amorphous with a certain degree of cross-linking to prevent them from slipping over one another.

Plastics. These are relatively tough substances of high molecular weight. A plastic usually becomes soft enough when heated to mold under pressure. Two types are commercial. *Thermoplastics* are those plastics which soften on heating. Examples of thermoplastic polymers are polyethylene and polypropylene. Thermoplastics usually possess moderate crystallinity and can undergo large elongation, but this elongation is not as reversible as with elastomers. *Thermosetting plastics* are usually rigid because of the large amount of crosslinking between polymer chains. Examples are unsaturated polyesters and urea- and phenol-formaldehyde resins.

Fibers. Fibers are polymers formed by building units joined either by addition polymerization or by condensation polymerization. The elasticity of the fiber is much less than for plastics and elastomers. Fibers undergo low elongation and are noted for their light weight, high tensile strength, high resistance to deformation, and low moisture absorption. Fiber polymers are highly crystalline due to secondary forces.

POLYMERIZATION REACTIONS

Synthetic polymers are formed by two general reactions: *Chain-reactions (addition polymerization)* and *step-reactions (condensation polymerization).*

Addition Polymerization. This is the most widely used type of polymerization reaction. Notable examples are in the preparation of polyethylene, polystyrene, and various elastomers. Table 11-1 contains the common addition polymers and their most important uses.

TABLE 11-1—Some Important Addition Polymers and their General Uses

Name of polymer	Monomer	Polymer	General major use
Polyethylene Low density, LDPE High density, DHPE	$CH_2{=}CH_2$	$\ldots CH_2{-}CH_2 \ldots$	Thermoplastic
Polypropylene, PP	$CH_3{-}CH{=}CH_2$	$\ldots CH(CH_3){-}CH_2 \ldots$	Thermoplastic
Polyisobutylene	$CH_3{-}C(CH_3){=}CH_2$	$\ldots C(CH_3)_2{-}CH_2 \ldots$	Liquid polymer, adhesive
Polyvinylchloride, PVC	$CH_2{=}CH{-}Cl$	$\ldots CH(Cl){-}CH_2 \ldots$	Thermoplastic
Polytetrafluoroethylene Teflon	$CF_2{=}CF_2$	$\ldots CF_2{-}CF_2 \ldots$	Thermosetting plastic
Polyacrylonitrile	$CH_2{=}CH{-}CN$	$\ldots CH(CN){-}CH_2 \ldots$	Fiber
Polymethylmethacrylate	$CH_2{=}C(CH_3){-}C({=}O){-}OCH_3$	$\ldots C(CH_3)(C({=}O){-}OCH_3){-}CH_2 \ldots$	Plastic
Polystyrene	$CH_2{=}C(C_6H_5)H$	$\ldots C(C_6H_5)(H){-}CH_2 \ldots$	Plastic
Polyisoprene	$CH_2{=}C(CH_3){-}CH{=}CH_2$	$\ldots CH_2{-}C(CH_3){=}CH{-}CH_2 \ldots$	Elastomer
Polybutadiene	$CH_2{=}CH{-}CH{=}CH_2$	$\ldots CH_2{-}CH{=}CH{-}CH_2 \ldots$	Elastomer
Polyformaldehyde	$H{-}C({=}O){-}H$	$\ldots C({=}O){-}CH_2 \ldots$	Plastic

Addition polymerization is usually initiated by free radicals, but cationic, anionic, and coordination catalysts also are used. The polymerization proceeds by the self-addition of unsaturated molecules one to another without loss of a small molecule. This type of polymerization proceeds by a chain reaction, although some addition polymers can be formed by a step reaction mechanism. A distinctive feature of chain addition polymerization is that high molecular weight polymers are formed immediately even at low conversions and the monomer is always present in appreciable amounts during polymerization.

Free Radical Polymerization. *Free radical initiators* are chemical compounds which possess a weak covalent bond that easily breaks into two free radicals when subjected to heat. Peroxides, hydroperoxides and azo compounds are examples of this type.

$$C_6H_5\text{-}\overset{O}{\overset{\|}{C}}\text{-}O\text{-}O\text{-}\overset{O}{\overset{\|}{C}}\text{-}C_6H_5 \rightarrow 2\ C_6H_5\text{-}\overset{O}{\overset{\|}{C}}\text{-}O\cdot$$

Benzoyl peroxide

$$C_6H_5\text{-}C(CH_3)_2OOH \rightarrow C_6H_5\text{-}C(CH_3)_2O\cdot + \cdot OH$$

Cumene hydroperoxide

$$CH_3\text{-}\underset{CH_3}{\overset{CN}{C}}\text{-}N=N\text{-}\underset{CH_3}{\overset{CN}{C}}\text{-}CH_3 \rightarrow 2\ CH_3\text{-}\underset{CH_3}{\overset{CN}{C}}\cdot + N_2$$

2,2'-Azobisisobutyronitrile

A simplified three-stage representation for the mechanism of free radical polymerization of vinyl monomers is: *initiation, propagation,* and *termination.*

Initiation. This stage involves the formation of a free radical from the initiator (peroxide or oxygen) and its addition to the monomer in one of two ways to produce a new radical.

$$\text{Peroxide} \rightarrow \text{Rad}^\bullet \quad \text{(free radical)}$$

$$\text{Rad}^\bullet + CH_2=CHY \xrightarrow{a} \text{Rad-}CH_2\text{-}\dot{C}HY \quad \text{(free radical)}$$

$$\text{Rad}^\bullet + CH_2=CHY \xrightarrow{b} \text{Rad-}CHY\text{-}\dot{C}H_2 \quad \text{(free radical)}$$

Y = H, Cl, methyl group, phenyl group, etc.

Oxygen and ultraviolet light are also used to produce free radicals. Path "a" is favored over path "b." This is the result of resonance and inductive effects of the substituent which increases the stability of the free radical.

Propagation. The free radical produced in the initiation reaction is very reactive and reacts with the unsaturated monomers very rapidly to form a growing chain—a *free radical chain reaction.*

$$\text{Rad-}CH_2\text{-}\dot{C}HY + nCH_2=CHY$$
$$\rightarrow \text{Rad-}CH_2\text{-}CH_2\text{-}(CH_2\text{-}CHY)_{n-1}\text{-}CH_2\text{-}\dot{C}HY$$

In each stage, the reaction of a free radical is accompanied by the formation of a new free radical to continue the chain reaction.

Termination. The reaction chain is terminated by reactions that do not form free radicals. These are the combination and disproportionation of two free radicals.

Combination

$$2\ \text{Rad-}(CH_2\text{-}CHY)_n\text{-}CH_2\text{-}\dot{C}H \rightarrow \text{Rad-}$$
$$(CH_2\text{-}CHY)_n\text{-}CH_2CHY\text{-}CHY\text{-}CH_2\text{-}(CHY\text{-}CH_2)_n\text{-Rad}$$

Disproportionation

$$2\ \text{Rad-}(CH_2\text{-}CHY)_n\text{-}CH_2\text{-}\dot{C}HY$$
$$\rightarrow \text{Rad-}(CH_2CHY)_n\text{-}CH_2\text{-}CH_2Y$$
$$+ \text{Rad-}(CH_2\text{-}CHY)_n\text{-}CH=CHY$$

The chains also may be terminated by contact with the walls of the container or in other similar ways.

Some compounds can act as *chain-transfer agents.* These are compounds which can transfer an atom to the growing polymer chain and themselves become free radicals with the ability to start a new polymerization chain. This is illustrated with styrene and carbon tetrachloride.

$$\ldots.\ CH_2\text{-}\underset{C_6H_5}{\dot{C}H} + CCl_4 \rightarrow \ldots.\ CH_2\text{-}\underset{C_6H_5}{CHCl} + \dot{C}Cl_3$$

Chain-transfer

$$\dot{C}Cl_3 + CH_2=\underset{C_6H_5}{CH} \rightarrow CCl_3\text{-}CH_2\text{-}\underset{C_6H_5}{\dot{C}H} \xrightarrow{styrene} \text{polymer}$$

The over-all result of chain-transfer reactions is a polymer with a lower average molecular weight.

When free radical chain reactions are to be stopped, an inhibitor is added. *Inhibitors* are compounds which react with the growing free radicals to produce new free radicals that are not reactive enough to add a new monomer and the chain is terminated. These inhibitors are frequently amines, phenols and quinones.

Because even traces of certain impurities can act as chain-transfer agents or inhibitors, the monomers used are among the purest petrochemicals produced.

Ionic Polymerization. These reactions are catalyzed by an ionic species or an ion pair which is formed with the monomer in an initial step and can add more monomers. Ionic initiators are either cationic, such as protonic and Lewis acids, or anionic, such as metal alkyls. In a simplified way cationic polymerization reactions are pictured as an electrophilic attack on the monomer molecules while anionic polymerization reactions are pictured as a nucleophilic attack.

$$I^+ + CH_2 = CHY \rightarrow I - CH_2 - C^+HY \quad \text{Cationic}$$
$$I^- + CH_2 = CHY \rightarrow I - CH_2 - C^-HY \quad \text{Anionic}$$

Ionic chain addition reactions are different from free radical chain addition reactions. In the former, a counter-ion is involved and is in close proximity to the propagating polymer chain, while in the latter, the free radicals are free. This situation affects the polymer termination and the reaction conditions, such as the concentration of the reactants. For example, in free radical polymerization termination could take place by combining two propagating free radicals. This is impossible with ionic polymerization. The effect of solvent polarity is pronounced in ionic polymerization, though very polar solvents cannot be used. This effect is negligible in free radical polymerization.

Cationic polymerizations are induced by chemical initiators or by irradiation. The active species are carbonium ions. A carbonium ion is a group of atoms that contains a carbon atom bearing only six electrons and a positive charge.

$$CH_3 - \overset{\overset{CH_3}{|}}{\underset{\underset{CH_3}{|}}{C^+}}$$

tert-Butyl carbonium ion

Like the free radical, the carbonium ion is an exceedingly reactive particle.

Cationic polymerization is used to produce polyisobutylenes, butyl rubbers, polyvinyl ethers, polytrioxanes, polyterpenes, petroleum resins, polyfuranes, V.I. improvers, additives, and lubrication oils. The relative importance of cationic polymerization in synthetic polymers in 1976 is illustrated in Fig. 11-1.[1]

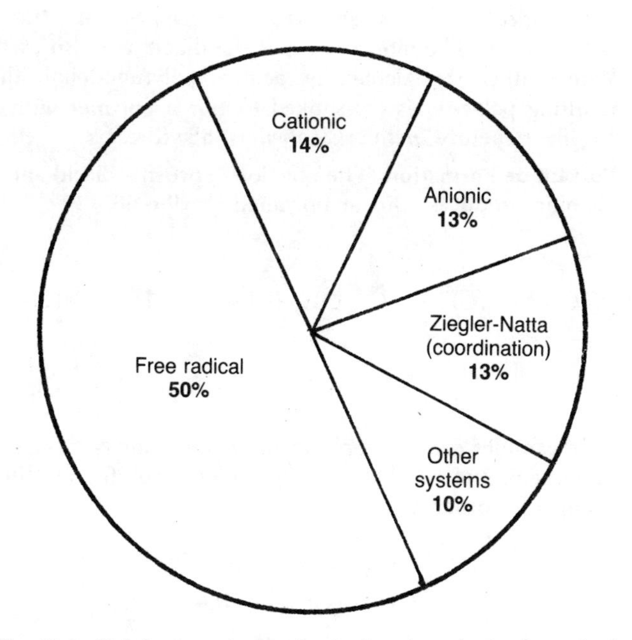

Fig. 11-1—Relative importance of cationic polymerization in synthetic polymer chemistry.[1]

In principle, vinyl monomers and dienes can be polymerized using cationic initiators. Cationic polymerization is not as useful as anionic polymerization in controlling the polymer structure. It is used more often when control is not necessary, as in the cases of phenol-formaldehyde resins and isobutylene polymers. The polymerization of dienes with cationic initiators gives low molecular weight polymers and cyclic compounds are formed.[2] Free radical polymerization, on the other hand, provides a simple way for polymer preparation since it can be used in all polymerization techniques. Only bulk and solution techniques are used with cationic polymerizations. Strong protonic and Lewis acids are initiators for cationic polymerizations.

Anionic polymerizations are initiated by bases such as $Li^+NH_2^-$:

$$Li^+NH_2^- + CH_2 = \underset{\underset{C_6H_5}{|}}{CH} \rightarrow$$

$$NH_2 - CH_2 - \underset{\underset{C_6H_5}{|}}{CH^-}Li^+ \xrightarrow{nCH_2 = CH - C_6H_5} \text{Polymer}$$

Organometallic compounds such as *n*-butyllithium are also effective as anionic polymerization catalysts.

Anionic polymerization can be used to polymerize organic oxides. Ethylene oxide, for example, is polymerized to a high molecular weight polyether in the presence of a small amount of methoxide ion.

$$CH_3O^- + \overset{O}{CH_2 - CH_2} \rightarrow$$

$$CH_3O - CH_2 - CH_2O^- \xrightarrow{n\overset{O}{CH_2 - CH_2}} \text{polymer}$$

The relative importance of anionic polymerization is indicated in Fig. 11-1.

Coordination Polymerization. In ionic polymerization the chain-carrier ions are always balanced by an ion of opposite charge. These ions are usually not closely associated with one another. If, however, the reactive center and the metal are appreciably covalent, the process is called *coordination polymerization.* This is still an ion type polymerization even if a separate anion is not present. Its reactivity is due to its anion-like character.

Ziegler-Natta catalysts are among the most important industrially useful coordination type catalysts and are complexes of transition metal halides with organometallic compounds. The most important combination is aluminum-titanium chloride.

The reaction involves nucleophilic addition to the carbon-carbon double bond of the monomer with the anion-like organic group of the growing organometallic compound as the nucleophile. The polymerization reaction consists of the insertion of an olefin molecule into the bond between the metal and the growing alkyl group.

$$M - CH_2 - CH_3 + CH_2 = CH_2 \longrightarrow$$

$$M - (CH_2)_3 - CH_3 \xrightarrow{CH_2 = CH_2}$$

$$M - (CH_2)_5 - CH_3 \xrightarrow{n\,CH_2 = CH_2}$$

$$M\ (CH_2)_m - CH_3$$

Polymerizations of this type have two distinct advantages over free radical polymerization:

1. It gives *linear* polymers.
2. It permits *stereochemical* control.

Free radical polymerization of ethylene, for example, produces a polymer with branched structure—low density polyethylene, *LDPE*. Polyethylene made by the coordination catalyst is essentially unbranched. The molecules fit closely together to produce high density polyethylene, *HDPE*, which has a high degree of crystallinity and greater strength than low density polyethylene.

When the monomer has one substitute such as propylene, CH_3-$CH=CH_2$, the use of coordination catalysts will allow the polymer to have any one of three different arrangements:

Isotactic, all methyl groups on one side of the extended polymer chain;

Syndiotactic, the methyl groups alternate regularly from side to side;

Atactic, methyl groups distributed at random.

Fig. 11-2 illustrates these three types of polypropylene.[3] By use of Ziegler-Natta catalysts, a higher percentage of the ordered isotactic structure is obtained.

Atactic polypropylene is a soft, elastic, rubbery polymer. Both isotactic and syndiotactic polypropylene are highly crystalline. Ziegler-Natta polymerization catalysts are also used to produce *cis*-polybutadiene and *cis*-1,4-polyisoprene. The *cis*-1,4-polyisoprene is essentially identical with natural rubber.

The mechanism of polymerization by use of these catalysts is not well understood. Most of them are complex, heterogeneous type systems of ill-defined structure containing more than one type of active center.[4, 5] Nevertheless, these catalysts have been of great value in many polymerization reactions which need control of the polymer structure.

Condensation Polymerization. *Condensation* or *step-reaction condensation,* in general, proceeds by reaction between two molecules with the elimination of a small molecule as the resulting molecule grows to a macromolecule. The small molecule is usually water. This type of polymerization is not a chain reaction. The steps are essentially independent and the reactants must be di- or poly-functional. In step polymerization, the monomer disappears much faster than in chain polymerization. A dimer is formed first, then a trimer, a tetramer, etc. High molecular weight polymers are not obtained until the end of the polymerization. Long reaction times are needed for high conversion.

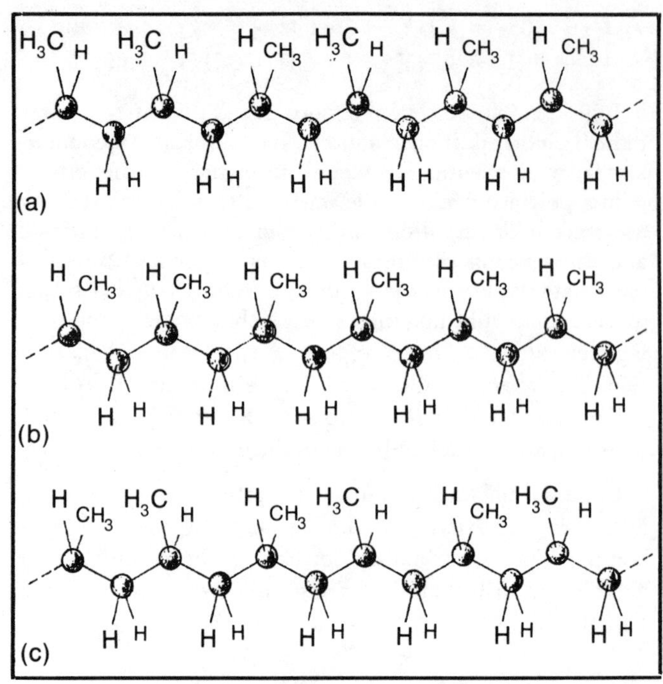

Fig. 11-2—Propylene can undergo polymerization in three different ways to form atactic (a), isotactic (b), or syndiotactic propylene(c).[3]

The most important polymerization reactions of this type are esterification, amide formation and various condensation reactions of formaldehyde.

$$CH_3O - \overset{O}{\overset{\|}{C}} - C_6H_4 - \overset{O}{\overset{\|}{C}} - OCH_3 + HOCH_2 - CH_2OH \rightarrow$$

$$\ldots . (- \overset{O}{\overset{\|}{C}} - C_6H_4 - \overset{O}{\overset{\|}{C}} - OCH_2 - CH_2O -) \ldots . + CH_3OH$$

Esterification. This reaction is classified as *trans-esterification.* The monoester and the diacid are also used. When either the alcohol or acid is polyfunctional, the resulting polyester is crosslinked to give a polymer with a net-like structure, *network polymers,* alkyd resins.

Polyamide Formation. The reaction between a diacid and a diamine produces a linear polyamide, nylon 66.

$$HO - \overset{O}{\overset{\|}{C}} - (CH_2)_4 - \overset{O}{\overset{\|}{C}} - OH + H_2N - (CH_2)_6 - NH_2 \rightarrow$$

$$\ldots . (- \overset{O}{\overset{\|}{C}} - (CH_2)_4 - \overset{O}{\overset{\|}{C}} - NH - (CH_2)_6 - NH -) \ldots . + H_2O$$

Polyamides are also produced by the anionic catalyzed ring opening of caprolactam molecules, nylon 6 and other lactams for other nylons.

$$\underbrace{CH_2 - (CH_2)_4 - \overset{O}{\overset{\|}{C}}}_{NH} \rightarrow$$

$$\ldots\ (-\overset{\overset{\displaystyle O}{\|}}{C}-(CH_2)_5-NH-\overset{\overset{\displaystyle O}{\|}}{C}-(CH_2)_5-NH-)\ \ldots$$

This reaction is similar to the ring opening of ethylene oxide and propylene oxide to form polyethers. The polymerization mechanism, either chain or step, is decided by studying the kinetics of the polymerization.

Formaldehyde Condensations. Formaldehyde condenses with itself and with other types of compounds to form thermosetting resins. The product of self-condensation is a polyether (an acetal, polyoxymethylene).

$$n\,H-\overset{\overset{\displaystyle O}{\|}}{C}-H \rightarrow \ldots (-CH_2-O-CH_2-O-CH_2-O-)\ldots$$

This polymer is unusually light and tough.

The first synthetic resin to be manufactured commercially, Bakelite, is produced by the acid or base catalyzed reaction between formaldehyde and phenol with the elimination of water. The reaction goes in several stages with the formation of *o*- or *p*-hydroxymethylphenol. This compound then reacts with another molecule of phenol with loss of water. The compound which is formed has two benzene rings joined by a $-CH_2-$ link. This process continues and yields a product of high molecular weight. The final product contains many crosslinks and has a rigid structure.

$$2\ C_6H_5OH + 2H-\overset{\overset{\displaystyle O}{\|}}{C}-H \rightarrow 2\ HOC_6H_4-CH_2OH \xrightarrow{-H_2O}$$

$$HOC_6H_4-CH_2-C_6H_3(OH)-CH_2OH \xrightarrow{-nH_2O}$$

$$\ldots CH_2-C_6H_2(OH)(CH_2\ldots)-CH_2-C_6H_2(OH)(CH_2\ldots)-CH_2-C_6H_2(OH)(CH_2\ldots)-CH_2\ldots$$

This type of polymer is a permanently hard, rigid, and infusible material that is resistant to chemicals and heat.

Formaldehyde also condenses with urea and melamine to form resins that are similar in properties and uses to the phenol-formaldehyde resins.

$$H_2N-\overset{\overset{\displaystyle O}{\|}}{C}-NH_2 \qquad C_3N_3(NH_2)_3$$

Urea — Melamine

Polyurethanes are addition polymers and not, in the strict sense, condensation polymers. Their structure, however, is similar to both polyesters and polyamides. Linear polyurethanes are formed by the reaction between a diisocyanate and a glycol.

$$CH_3-C_6H_3(N=C=O)_2 + HO-CH_2-CH_2-OH \rightarrow$$

$$\ldots[-OCH_2-CH_2O-\overset{\overset{\displaystyle O}{\|}}{C}-NH-C_6H_3(CH_3)-NH-\overset{\overset{\displaystyle O}{\|}}{C}-]\ldots$$

Polyurethanes are used to make foam rubber and similar types of flexible materials.

POLYMERIZATION TECHNIQUES

Addition polymerization reactions are generally exothermic. Therefore, it is necessary to control the heat of the reaction. This can be controlled to some extent by the use of an appropriate solvent, solution polymerization, or by limiting the reaction to a low conversion, as in bulk polymerization, or by the use of external cooling.

Another feature of addition polymerization which requires control is the progressive increase in viscosity. This is solved, as with the heat effect, by the use of a solvent or by carrying the reaction only to a low conversion.

The important processes used for polymerization reactions are bulk, solution, suspension and emulsion.

Bulk Polymerization. The absence of a solvent is the characteristic of this type of polymerization. The only species found after the start of the reaction are the monomers and polymer. Impurities in the polymer are low and depend on the purity of the original monomer. Heat and viscosity control is important and somewhat difficult to achieve. One way to control the temperature is by vigorous agitation and low conversions.

Although free radical addition reactions are highly exothermic, polymerization of ethylene, propylene, acrylonitrile, and vinylidene dichloride, for example, are customarily polymerized in bulk.

Solution Polymerization. This consists of polymerizing the monomer dissolved in an appropriate organic solvent. The control of heat and viscosity is easier than with bulk polymerization, especially if the polymer is also soluble in the solvent. Polar solvents are the most suitable for ionic polymerization.[6]

Many catalysts of the organometallic type are used in solution polymerization. These include alkyllithium and trialkylaluminum or a mixture of an organometallic type with a transition metal salt. The solvents used may show some activity with the polymer. This will affect the chain length of the polymer and its purity by chain transfer reactions.

In applications such as adhesives and coating, removal of the solvent from the polymer may not be necessary. Solution polymerization offers control of the polymer structure and is used extensively for the production of steriospecific elastomers such as *cis*-polybutadiene and *cis*-polyisoprene.

Suspension Polymerization. With suspension polymerization the monomer is dispersed in a liquid, usually water, by the use of mechanical agitation. Initiators are chosen that are soluble in the monomer—oil soluble. Stabilizers, such as talc, polyvinyl alcohol, and bentonite, are used to prevent polymer chains from adhering to each other and to keep the monomer dispersed in the liquid. The final polymer is obtained in a granular form, either bead or pearl.

Polymers produced by suspension are more pure than those produced by solution polymerization due to the absence of chain transfer reactions. Like solution polymerization, the dispersing liquid helps to control the heat evolved during the reaction. Suspension polymerization is used extensively for free radical olefin polymerization.

Emulsion Polymerization. In emulsion polymerization, the monomers are emulsified by use of an emulsifying agent with a free radical initiator such as cumene hydroperoxide. The emulsifying agent should have a finite solubility and is either non-ionic, usually polyvinyl alcohol, or an anionic type such as alkylsulfates, aryl sulfonates, and various types of soap.

The emulsion medium has a concentration capable of forming aggregates of micelles with large surface areas. These micelles can absorb the monomer droplets. X-ray and light scattering techniques have shown that the micelles increase in size after the addition of the monomer to the emulsion medium. This is an indication of the presence of polymer molecules which have been formed from monomers inside the micelles. For example, in the free radical polymerization of styrene the micelles increased 250 times their original volume after one minute from start of the reaction.[7]

Polymers in the form of emulsions such as floor polishes and paints can be made directly by emulsion polymerization and the emulsifying medium, water, is not separated. Emulsion polymerization is used extensively for producing polymers for the rubber industry.

Interfacial Polymerization. When acid chlorides, $R\text{-}\overset{\displaystyle O}{\overset{\|}{C}}\text{-}Cl$, are used for producing polymers such as polyesters and polyamides, the rates of reaction are much faster than when carboxylic acids or anhydrides are used. Low temperatures are used to control the rate of reaction and polymerization is effected by the addition of a solution of the second monomer to the solution of the acid chloride.

The reaction takes place at the interface between the two immiscible liquids. The polymer is continuously removed from the interface. This type of polymerization is used to produce heat sensitive polymers.

PHYSICAL PROPERTIES

Whether the polymer is going to be used as a plastic, elastomer, or as a synthetic fiber depends to a great extent on the nature of the polymer. Whether it is a homo- or copolymer, linear or branched, has a low or high molecular weight, etc., and how it is processed determines a polymer's physical, mechanical, and chemical properties. Weight average, molecular weight, and number average molecular weight are two parameters which are used extensively to indicate the physical and mechanical properties of polymers.

Average Molecular Weight. The molecular weight of a polymer affects the mechanical and physical properties of the polymer. Unlike low molecular weight substances, commercial polymers usually have molecular weights over 5,000 and will contain macromolecules with a range of molecular weights. The methods used to determine the average molecular weights of polymers include measurement of some colligative property such as viscosity or sedimentation. Different methods do not correlate well with each other, and thus usual practice is to determine the average molecular weight by more than one method. The following two methods are used.

Number Average Molecular Weight. This is defined as the average molecular weight according to the number of molecules present of each species. In other words, when one multiplies the total number of molecules of the polymer by M_n, the product is equal to the sum of the product of the number of each species multiplied by its molecular weight.

$$\overline{M}_n = \frac{W}{\Sigma N_i} = \frac{\Sigma N_i M_i}{\Sigma N_i}$$

i = the degree of polymerization (dimer, trimer, etc.)
N_i = the number of each polymer species
M_i = molecular weight of each polymer species
W = total weight of all polymer species.

The number average molecular weight can be determined by measuring such colligative properties as boiling point elevation, freezing point depression and vapor pressure lowering.

Weight Average Molecular Weight. This is the sum of the product of the weight of each of the species present and its molecular weight divided by the sum of the weights of all the species.

$$\overline{M}_w = \frac{\Sigma W_i M_i}{W} = \frac{\Sigma W_i M_i}{\Sigma N_i M_i}$$

where W = the total weight of the different polymeric species in the sample
W_i = the weight of each polymeric species
M_i = the molecular weight of each polymer species

Substituting $M_iN_i = W_i$, the weight average molecular weight, M_w, can be defined as

$$\overline{M}_w = \frac{\Sigma N_i M_i^2}{\Sigma N_i M_i}$$

Light scattering techniques and ultracentrifuging are used to determine weight average molecular weight.

The following simple example will illustrate the difference between number average molecular weight and weight average molecular weight. Suppose we have six macromolecules:

Two of them have a molecular weight of 600;
Three of them have a molecular weight of 1000;
One of them has a molecular weight of 2000.

The *number average molecular weight* is calculated as

$$M_n = \frac{2 \times 600 + 3 \times 1000 + 1 \times 2000}{6} = \frac{6200}{6} = 1033$$

The weight average molecular weight is calculated as

$$\overline{M}_w = \frac{2 \times (6 \times 10^2)^2 + 3 \times (10^3)^2 + 1 \times (2 \times 10^3)^2}{2 \times 6 \times 10^2 + 3 \times 10^3 + 1 \times 2 \times 10^3}$$

$$= \frac{2 \times 36 \times 10^4 + 3 \times 10^6 + 4 \times 10^6}{1200 + 3000 + 2000}$$

$$= \frac{7,720,000}{6200} = 1245$$

In a monodispersed systems $\overline{M}_n = \overline{M}_w$.

The difference in the value between $\overline{M}_n$ and $\overline{M}_w$ indicates the polydispersity of the polymer system. The closer $\overline{M}_n$ and $\overline{M}_w$, the narrower the molecular weight spread. Molecular weight distribution curves for polydispersed systems could be obtained by plotting the degree of polymerization, i, versus either the number fraction, n_i or the weight fraction, w_i.

Crystallinity. The freezing point of a pure liquid is the temperature at which molecules of the liquid lose translational freedom and molecules of the solid become more ordered within a definite crystalline structure. Polymers are considered nonhomogeneous and will not show a definite temperature at which the polymer crystallizes.

When a melted polymer is cooled, some polymer molecules will line up and form crystalline regions within the melt and are termed *crystallites*. The rest of the polymer is amorphous. The temperature at which these crystallites disappear when the crystalline polymer is gradually heated is described as the *crystalline melting temperature*, T_m. Another transition temperature is reached when the polymer is further cooled and the amorphous regions change to a glass-like material—this is termed the *glass transition temperature*, T_g.

The degree of a polymer's crystallinity is an important parameter in determining the mechanical and thermal behavior of the polymer and thus its end uses. For example, the highly disordered polymers (dominantly amorphous) show properties of an elastomer, while highly crystalline polymers give the ridigity needed for fibers.

The tendency of a polymer to be ordered and able to form crystallites is a function of the regularity of the chains, the presence or absence of bulky groups, their spatial arrangement, and the presence of secondary forces such as hydrogen bonding. For example, the random arrangement of phenyl groups in polystyrene (atactic polystyrene) shows low crystallinity and is highly amorphous, while the isotactic polystyrene with its phenyl groups arranged on one side of the backbone is crystalline. Linear polyethylenes with their simple structures can pack easily, and the polymer is highly crystalline. Hydrogen bonding appears in polymers such as polyamides (presence of amide groups) and affects their ability to crystallize. This is required in polymers for fiber production.

LITERATURE CITED

1. Kennedy, J.P. *CHEMTECH*, Vol. 6, No. 12, 1976, pp. 750-753.
2. Cooper, W. *The Chemistry of Cationic Polymerization*, Oxford: Pergamon Press Ltd., 1963.
3. Watt, G.W., L.F. Hatch and J.J. Lagowski. *Chemistry*, New York: W.W. Norton & Co., 1964, p. 449.
4. *Modern Plastics*, May 1975, p. 146.
5. Billingham, A.V. and R.J. Bondreau. *Journal of Polymer Science*, Vol. 51, 1961, p. 853.
6. Ravve, A. *Organic Chemistry of Macromolecules*, New York: Marcel Dekker, Inc., 1967.

12

Thermoplastics

Thermoplastics are organic long chain polymers that usually become soft when heated and can be molded under pressure. They are linear or branched chain polymers with little or no crosslinking. Table 12-1 lists the major thermoplastics, their chemical formulas, melting temperatures, and densities. Table 12-2 shows the projected consumption of the major thermoplastic polymers in the United States, Western Europe, and Japan. The past and future relationship between plastics and other petrochemicals is shown in Fig. 12-1.[1]

The growth of thermoplastics is attributed to certain attractive properties of the product such as lightness in weight, chemical corrosion resistance, toughness, and ease of handling. Appreciable technological development in plastics processing has also enchanced their growth.

Two major economic factors have had a favorable effect on the growth rate: **1.** Prices of many articles made of plastics are at least competitive or even less than prices of articles made from materials from natural sources, **2.** There is a limited supply of materials from natural sources.

As a result, the diversified market for materials made from paper, wool, cotton, leather, steel, wood, and concrete is now being replaced or the market is being penetrated by plastics. Fig. 12-2 shows the U.S. current use pattern for all types of plastics, of which thermoplastics represent 78-80 percent.[2] This replacement of older types of materials is also indicated by the increase in world per capita consumption. The projected per capita consumption per year is predicted to be 34.6 kg by 1990, compared to 11 kg in 1974. The demand for thermoplastics is predicted to grow at an average annual rate of 6 percent during the early 1980s, then slow to about 4.7 percent a year. Growth rate will be about 3 percent by 2000.[3] The past, present and future capacity and demand data for the European Community is given in Table 12-3.[4]

The packaging field is the largest plastics market and it is a steady market for commodity thermoplastics. Packaging consumed 3.23 million tons during 1977, and the figure is predicted to reach 5.75 million tons in 1985. The role played in packaging by the various thermoplastics is shown in Fig. 12-3.[5] The total amount of thermoplastics used for packaging in 1978 was 7.6 billion pounds.

One of the growing fields that uses commodity plastics is pipe production. *Engineering thermoplastics* include polymers with special properties such as high thermal stability, good chemical and weather resistance, and other useful characteristics. This type of plastic includes polycarbonates, polyethersulfones, polyacetals, nylons, and thermoplastic polyesters. Engineering plastics are discussed in Chapter 13.

The rise in oil prices and the tendency to find all possible ways to save energy have caused plastics to be considered as one of the alternatives to achieve this goal. By 1985 fleet average gasoline consumption should be 27.5 mpg. The average plastic share per car weight is about 4.5 percent, about 160 pounds/car. By 1980 the car plastic consumption will rise to 250-300 pounds or 10 percent of the car weight. Among the plastics used to reduce automobile weight are polyesters, polypropylene, polyvinyl chloride, polyurethane, polyethylene, nylon, and **ABS.** These will, in general, be used as reinforced plastic composites. Graphite reinforced plastics currently are the material of choice.[6]

The highest growth rate among those plastics is reinforced polyesters, with an estimated demand of 300,000 tons in 1982.[7] Following reinforced polyesters are polypropylene and polyurethanes, with an estimated demand of 250,000 metric tons/year for each.

With the expansion of the plastic industry and the larger rate of per capita consumption of plastics, problems associated with plastic wastes and pollution become more apparent. Plastic wastes comprise about 4 percent of the solid waste collected in the United States. Plastic in the environment is not biodegradable and can produce pollution problems. About 80 percent of municipal waste plastics is of the thermoplastic type. Remolding thermoplastics is possible. Western Electric has been reclaiming about 5 billion pounds

TABLE 12-1—Major Thermoplastic Polymers

Name	Abbreviation	Family	Formula	Melting temp.	Density
Low density polyethylene	LDPE	Polyolefin	$-CH_2-$	110 (Tm)	0.910
High density polyethylene	HDPE	Polyolefin	$-CH_2-$	120 (Tm)	0.950
Polypropylene	PP	Polyolefin	$-CH_2-CH(CH_3)-$	175 (Tm)	0.902
Polyvinyl chloride	PVC	Vinyl	$-CH_2-CH(Cl)-$	100 (Tg)	1.35
Polyvinyl acetate	PVA	Vinyl	$-CH_2-CH(O-\overset{O}{\overset{\parallel}{C}}CH_3)-$	....	....
Polystyrene	PS	Styrenic	$-CH_2-CH(C_6H_5)-$	100 (Tg)	1.05
Acrylonitrilebutadiene styrene	ABS	Styrenic	....	....	....
Acrylonitrile styrene	SAN	Styrenic	....	....	....
Polymethylmethacrylate	......	Acrylic	$-CH_2-C(CH_3)(\overset{O}{\overset{\parallel}{C}}OCH_3)-$	....	....
Polyhexamethylenediamide	Nylon 66	Polyamide	$-NH-(CH_2)_6-NH-\overset{O}{\overset{\parallel}{C}}-(CH_2)_4-\overset{O}{\overset{\parallel}{C}}-$	265 (Tm)	1.14
Polycaprolactam	Nylon 6	Polyamide	$-NH-(CH_2)_5-\overset{O}{\overset{\parallel}{C}}-$	225 (Tm)	1.14
Polyethyleneterephthalate	PET	Polyester	$-\overset{O}{\overset{\parallel}{C}}-C_6H_4-\overset{O}{\overset{\parallel}{C}}-OCH_2\,CH_2O-$	270 (Tm)	....
Polybutyleneterephthalate	PBT	Polyester	$-\overset{O}{\overset{\parallel}{C}}-C_6H_4-\overset{O}{\overset{\parallel}{C}}-O-(CH_2)_3-CH_2-O-$	250 (Tm)	1.3
Polycarbonates	PC	Polyester	$-O-R'-O-\overset{O}{\overset{\parallel}{C}}-O-R-O-\overset{O}{\overset{\parallel}{C}}-$	190 (Tg)	1.2
Polyacetals		Polyethers	$-O-CH_2-O-R-$	181 (Tm)	....

TABLE 12-2—Consumption of Major Thermoplastic Polymers in the United States, Europe and Japan*

Material	United States				Western Europe			Japan	
	1976	1977	1978	1980**	1976	1977	1983	1976	1982***
Low density polyethylene	2.636	2.955	3.227	3.63	3.20	3.05	4.36	1.39	1.3
High density polyethylene	1.410	1.682	1.818	1.96	1.18	1.14	1.72		
Polypropylene	1.136	1.227	1.364	1.73	0.765	0.73	1.45	0.669	1.10
Polystryene	1.450	1.564	1.705	2.14	1.435	1.42	....	0.875	1.30
Polyvinyl chloride	2.144	2.386	2.545	3.05	3.255	3.14	4.27	1.044	1.30

* In million metric tons
** Projections
*** Projections

and remolding it by an injection process. By 1980, plastic waste is expected to reach 8.4 million tons or 6 percent of the municipal load. The U.S. Bureau of Mines has developed a flotation process for separating plastics from other waste materials. This flotation process could possibly be used by automobile yards when the economics become favorable.

Another alternative to solving plastic waste problems is processing the waste and converting it to valuable chemicals by cracking. Cracking of plastics has been tried with a pilot plant study using a fluidized bed reactor of sand.[8] The recovered product depends to a great extent on the nature of the plastic used and the applied temperature. In an experiment where polyethylene was cracked at 740°C, the main products were methane, 16.2%; ethylene, 25%, **BTX** and other aromatics, 29%; and wax, 7%. When polystyrene was cracked at the same temperature, 71.6% of styrene was recovered. A third route being considered is to produce photodegradable plastics. One question arises, however: Is it worthwhile when only a small percentage of the total plastics appears as litter?

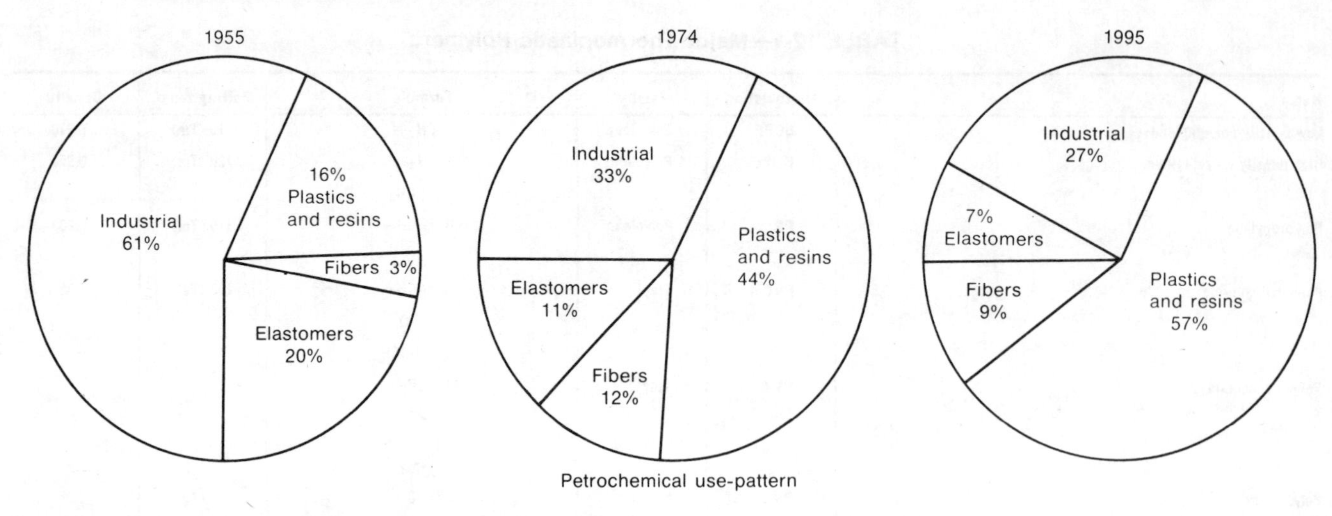

Fig. 12-1—The increasing importance of plastics in petrochemistry.[1]

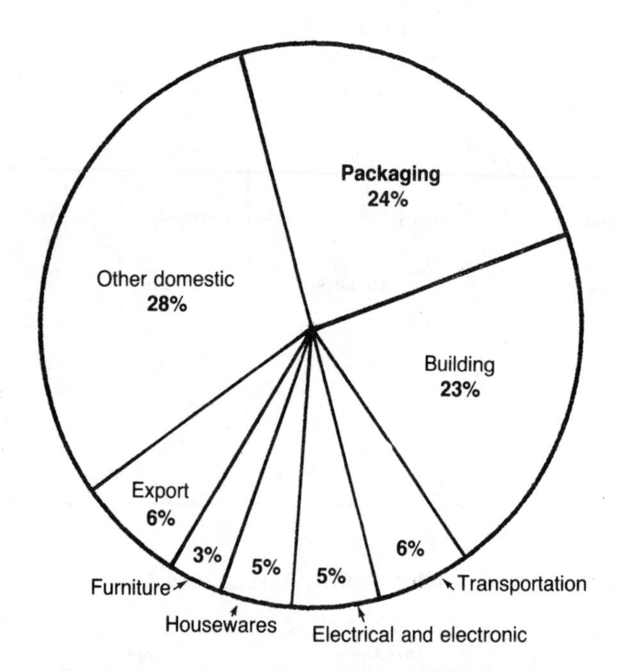

Fig. 12-3—The dominant role played by the polyethylenes in packaging.[5]

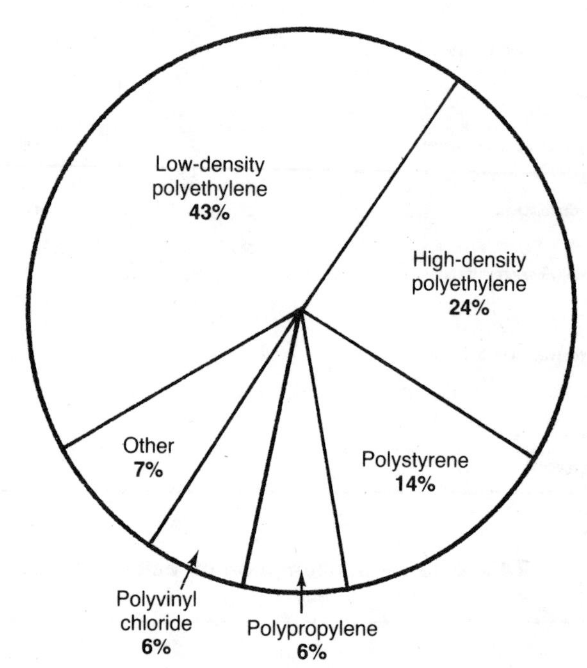

Fig. 12-2—The current use pattern for all types of plastics.[2]

TABLE 12-3—The European Community's Present, Past and Future Thermoplastic Capacity and Demand*[4]

	1976		1980		1982	
	Capacity	Demand	Capacity	Demand	Capacity	Demand
LDPE	4.14	2.44	5.23	3.08	5.35	3.30
HDPE	1.74	0.88	2.13	1.14	2.20	1.25
Polypropylene	1.07	0.66	1.75	1.16	2.17	1.26
PVC	4.38	2.77	4.87	3.35	5.92	3.66
Polystyrene	2.41	1.21	2.67	1.48	2.67	1.60

*Million metric tons/year.

Abbas, Knutsson, and Berglund have developed a model for reprocessing thermoplastics that helps to design in desired properties and have reported on its application in two cases: direct recycling at the injection molding machine and central mixing of recycled material with a constant fraction of virgin material.[9]

POLYETHYLENE

Polyethylene, $n(CH_2 - CH_2)n$ is the most extensively used thermoplastic. The ever-increasing demand for polyethylene is partly due to the availability of the monomer, ethylene, from the abundant raw material, associated gas, and naphtha. The ease of processing the polymer, its relative low cost, its resistance to chemicals, and its flexibility are also strong influences. All these and other factors lead to the strong market demand.

High pressure polymerization of ethylene was introduced in the 1930s. The discovery of new titanium catalysts by Karl Ziegler in 1953 opened a new era for the polymerization of ethylene at lower pressures.

The two most widely used grades of polyethylene are *low density polyethylene,* **LDPE**, which has branched chains, and *high density polyethylene,* **HDPE**, which is predominantly linear. Low density polyethylene is produced by a free radical initiated polymerization at high pressures while high density polyethylene is produced by a low pressure process with a metallic oxide catalyst of the Ziegler type.

The main difference between the two grades of polyethylene is that **LDPE** is more flexible because of its lower crystallinity. This lower crystallinity is caused by the presence of branches of two or four carbons along the backbone of the polymer. **HDPE** is more closely packed because of the absence of branches and thus the molecules become closer and less permeable to gases.

Several processes can produce polymers with a wide range of densities that cover both the low and high density ranges as well as medium density polymers.

Physical and Chemical Properties. Probably the most important property of the polyethylenes is the molecular weight and its distribution within a sample. Methods used to determine molecular weights are numerous. A widely used one is by viscosity determination. The melt viscosity and the melt flow index, **MFI,** measure the extent of polymerization. A polymer with a high melt flow index has a low melt viscosity, a lower molecular weight, and a lower impact tensile strength.

Polyethylene is to some extent permeable to most gases. The higher density polymers are less permeable than the lower density ones. Polyethylene, in general, has a low degree of water absorption and is not attacked by dilute acids and alkalis. However, it is attacked by concentrated acids. Its overall chemical resistance is excellent. Polyethylenes are affected by hydrocarbons and chlorinated hydrocarbons and swell slowly in these solvents. Tensile strength of polyethylenes is relatively low, but impact resistance is high. The use of polyethylenes in insulation is due to the excellent electrical resistance properties.

Table 12-4 gives some of the properties of the low and high density polyethylenes.

Production. The polymerization of ethylene is an exothermic reaction from which 850 calories are released for each gram of ethylene.[10] For this reason, an adequate method of removing the heat of reaction is needed. Polyethylene decomposes at high temperatures, even in the absence of air. Explosions might take place if the temperature is not controlled. The decomposition products are methane, carbon, and hydrogen. Reactions that take place in the reactor are:

Polymerization

$$nC_2H_4 \rightarrow 1/n\ (C_2H_4)n \qquad \Delta H = -22 \text{ kcal/mole}$$

Decomposition

$$C_2H_4 \rightarrow 2C + 2H_2 \qquad \Delta H = -11 \text{ kcal/mole}$$

$$C_2H_4 \rightarrow C + CH_4 \qquad \Delta H = -30 \text{ kcal/mole}$$

Low Density Polyethylene (LDPE). Most low density polyethylenes are made by high pressure processes. Either tubular or stirred autoclave reactors are used. In the *stirred autoclave reactor,* the heat of reaction is absorbed by the cold ethylene feed which is mixed with the reacting polymer. The stirring action keeps a uniform temperature throughout the reaction vessel. The *tubular reactor* consists of three zones—a heating zone, a reaction zone and a cooling zone. A large amount of the reaction energy is removed through the tube walls.

TABLE 12-4—Typical Properties of Polyethylenes

Polymer	Melting point range °C	Density g/cm³	Degree of crystal-linity %	Stiffness modules psi × 10³
Branched, Low density	107—121	0.92	60—65	25—30
Medium density	....	0.935	75	60—65
Linear, High density				
Ziegler type	125—132	0.95	85	90—110
Phillips type	....	0.96	91	130—150

The Arco process uses a tubular reactor.[13] Fig. 12-4 is a flow diagram for the Gulf process using a stirred autoclave reactor.[14] Highly purified ethylene (over 99 percent purity) is mixed with the purified ethylene from the low pressure recycle and is compressed to about 300 bars. In the intermediate pressure recycle the ethylene is introduced and the pressure raised to about 2000 bars using a secondary compressor. The free radical initiator, which is normally oxygen or a peroxide, and the chain transfer agent are fed at this stage into the reactor. The product leaves the reactor, expands through a let-down valve and enters the primary separator where the unreacted ethylene is cooled, passed through filters to remove waxes and recycled to the secondary compressor. Molten polymer goes to a secondary separator at near atmospheric pressure. The product is then fed into an extruder to be pelletized. Typical reaction conditions are:

Temperature	100-200°C
Pressure	1500-2000 bars
Ethylene, conversion/pass	10-25%
Over-all conversion	over 95%

The rate of polymerization is greatly accelerated by increasing the temperature, the initiator concentration and the pressure. Degree of branching and molecular weight distribution depend on pressure and temperature. Increasing the pressure and decreasing the temperature produce a polymer with a higher density and a narrower molecular weight distribution. However, the mean molecular weight can be decreased using a moderator. The crystallinity of **LDPE** can be adjusted by changing the reaction conditions and by adding comonomers such as vinyl acetate or ethyl acrylate. Copolymers have low crystallinity but better flexibility, and the resulting polymer has a better impact strength.[15] **LDPE** polymers produced using tube reactors are better suited for film manufacture, and products from the stirred autoclave reactor are more suitable for coatings. The autoclave will, in general, make all grades, but for film grades lower conversions are used to minimize long chain branching.

The economics of a process are determined by factors such as availability of raw materials and transportation of the raw materials and the finished products. One very important factor is unit size. A study by Imhausen, et al. on tubular reactors found that the relative investment cost could be

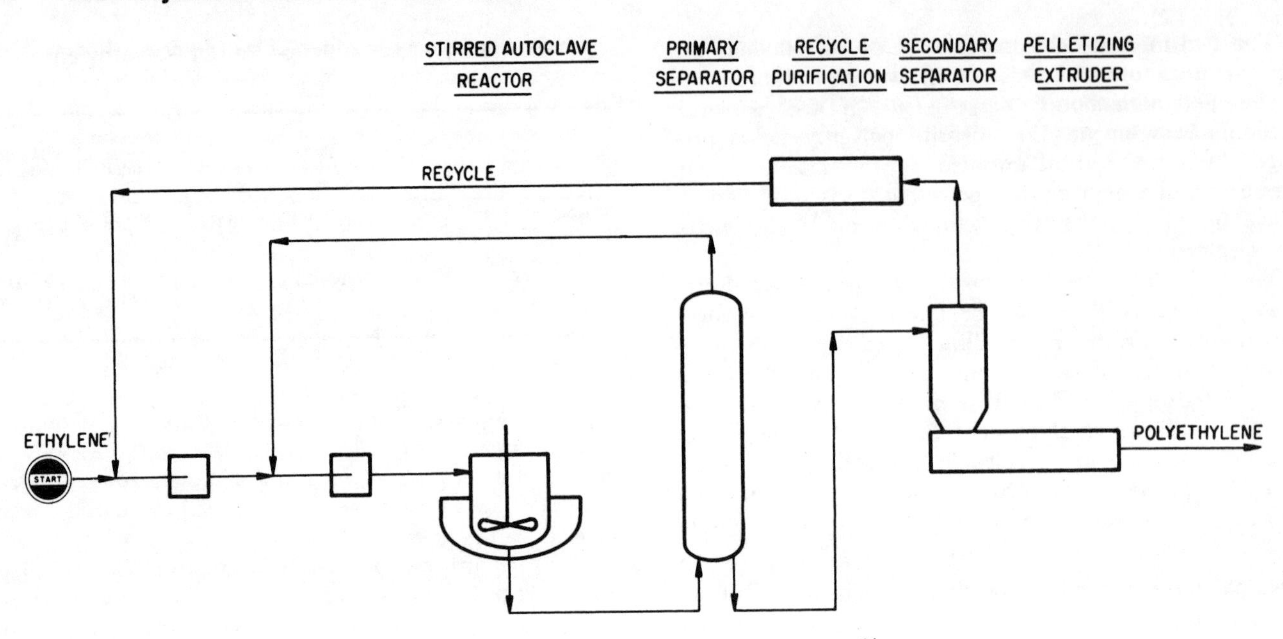

Fig. 12-4—The Gulf process for producing LD-polyethylene using the stirred autoclave reactor.[14]

reduced 50 percent by increasing the unit size from the conventional 60,000 tons/year to 180,000 tons/year.[16] In these larger units, the surface to volume ratio becomes smaller than in the smaller units and the heat removal becomes important. This effect could be compensated for by increasing the speed of flow of the reaction mixture. In addition to investment cost saving, operational costs are also decreased by such items as the energy needed for the high pressure pumps, etc.

A recent low pressure process produces low-density polyethylene.[17, 18] The process, developed by Union Carbide, reportedly reduces the capital investment by 50 percent and the energy requirement by 75 percent. In this process, ethylene and the catalyst are continuously fed to a fluidized bed reactor at a temperature below 100°C and a pressure range of 100-300 psi. Energy saving comes from the energy needed to maintain the high pressure required in conventional processes. Capital saving comes from elimination of the high pressure system needed for the high pressure processes. Fig. 12-5 is a flow diagram of the new low pressure process for the production of **LDPE.**[17] The effect of the new Union Carbide processes on high pressure processes has yet to be fully evaluated on the basis of the product quality and use.

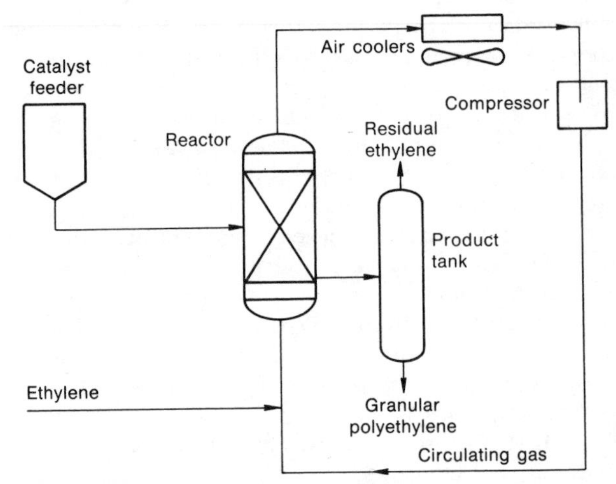

Fig. 12-5—The new low-pressure, gas-phase process for low-density polyethylene granular resins.

The Sumitomo Co. has reported a new high conversion autoclave reactor process and a tubular reactor process, **HC,** which produce a variety of **LDPE** economically.[19] Table 12-5 gives pertinent product data for both processes.

High Density Polyethylene. All high density polyethylenes are made by a low pressure process in a fluidized bed reactor. The catalyst is either a Ziegler type catalyst which is a complex between an aluminum alkyl and a transition metal halide, such as titanium tetrachloride, or a catalyst of silica or silica alumina impregnated with a small amount of a metal oxide, usually either chromium oxide or molybdenum oxide. Catalyst preparation and activation are very important. They not only determine the activity or efficiency of the catalyst but also the properties of the polymer.[20] Molecular weight may be controlled by introducing hydrogen to the circulating gas.

The heat of reaction is removed by circulating the gas through heat exchangers, then compressing it for recycling to the reactor. In the Hoechst process, ethylene recycle and monomer recovery systems are not required because of the almost complete conversion of the ethylene. Yield is more than 99 percent. Because of the small amount of catalyst used, its removal from the polymer is not necessary. The polymer is usually in the form of a powder or granules, depending on reaction temperature. Either solution or suspension reaction conditions may be used. Above 130°C, the hydrocarbon is a solvent for the polymer; below this temperature, the polymer forms a suspension. Typical reaction conditions for three of the major processes are given in

TABLE 12-5—Comparison of the Molecular Structure of Sumitomo LDPE of Different Polymerization Processes (Comparison for Samples with Same Melt Index [MI] and Density)[19]

Process	Autoclave reactor	Tubular reactor
Molecular weight distribution	• Broad • Spread toward high molecular weight region • Negligible amount of microgel	• Narrow • Having relatively large amount of low molecular weight portion • Relatively large amount microgel
Branching distribution	• Narrow • Having tailing-up	• Broad • No tailing-up
Branching dispersion	• Agree with theoretical distribution curve for low and medium molecular weight fraction • Reversion for high molecular weight fraction	• A little broader distribution compared with theoretical • No reversion
Degree of long-chain branching	• High	• Relatively low
Fine structure of long-chain branching	• Contain complex grafted chains having many short-chain branching	• Relatively simple

TABLE 12-6—Typical Reaction Conditions for Some Processes for the Production of HDPE

	Union Carbide	Hoechst AG	Naphtha Chimie
Temperature	85—105°C	80—90°C	60—100°C
Pressure	below 300 psig	below 147 psi	220—440 psi
Catalyst	supported chromium	....	....

Table 12-6. Fig. 12-6 shows the Union Carbide gas phase process[21] and Fig. 12-7 is the flow diagram of the Hoechst process[22] for the production of **HDPE** using a diluent.

The polymerization may be in the gas phase as in the Union Carbide process or in the liquid phase where a hydrocarbon diluent is added. This requires a hydrocarbon recovery system. Branching in **HDPE,** and accordingly its density, cannot be adjusted in the same way as with **LDPE.** Other molecules are put into basic polyethylene to modify its properties to specific applications. They are doing this in both the liquid phase and the vapor phase and at high and low pressures. Branching in **HDPE** is incorporated in the backbone of the polymer by adding varying amounts of comonomer. Hexene is frequently used as the comonomer.

"To anyone but an expert, the line that differentiates low density from high density polyethylene resins is becoming more blurred."[23]

Market and Uses of Polyethylene. Low density polyethylene, **LDPE,** is the largest volume thermoplastic. Its use is expected to grow from 7.1 billion pounds in 1978 to 7.8 billion pounds in 1980.[2] Consumption of **HDPE** was 4 billion pounds in 1978, and the demand is expected to reach 4.3 billion pounds in 1980.[2] Capacity will be 5.5 billion pounds.

Products made from polyethylene are numerous and range from building materials and electrical insulation to packing materials and sheets. It is an inexpensive plastic which can be molded into almost any shape, extruded into fiber or filament, and blown or precipitated into film or foil. Because it is more flexible and more transparent, the low density polymer is used in sheets, films, and injection molding. High density polyethylenes are extensively used in blow-molded containers. About 85 percent of the blow-molded bottles is produced from **HDPE.** Irrigation pipes made from polyvinyl chloride and high density polyethylene, **HDPE,** are widely used. Pipes made from **HDPE** are flexible, tough, and corrosion resistant. These pipes are used for carrying corrosive materials and/or abrasives, such as gypsum, slurry, and various chemicals. Corrosion resistant pipes are also used in well drilling and crude oil transfer. Spun-bonded polyethylenes are extremely fine fibers interconnected in a continuous network. Their uses include notebook and reference book

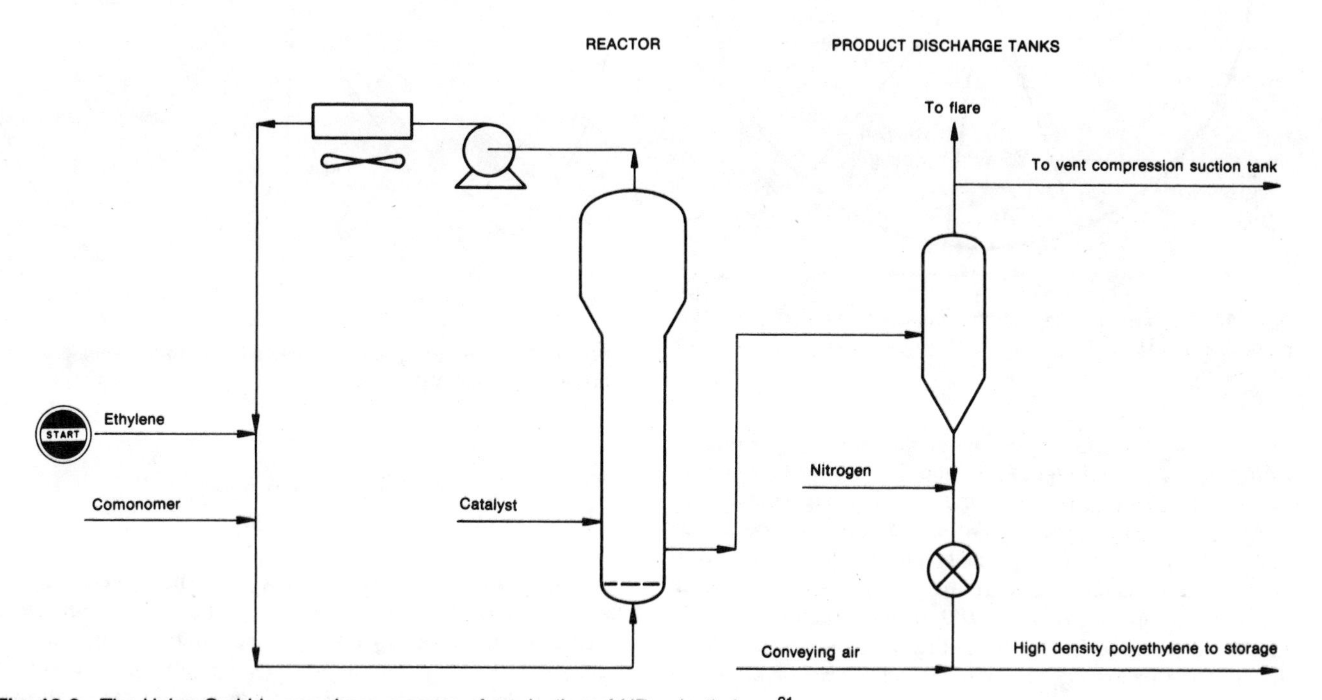

Fig. 12-6—The Union Carbide gas-phase process of production of HD-polyethylene.[21]

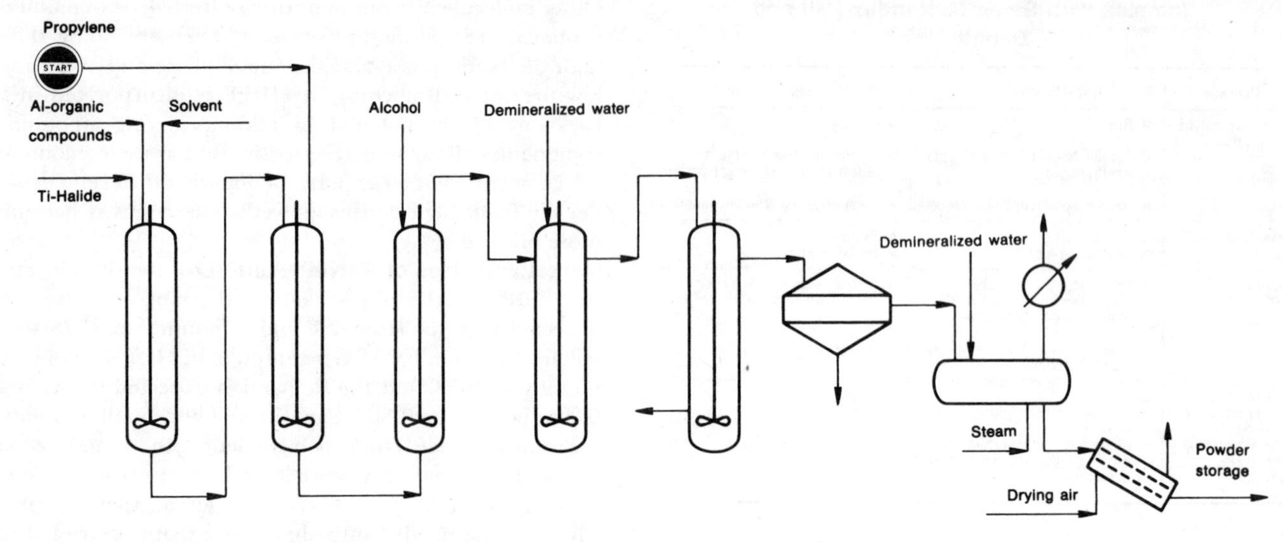

Fig. 12-7—The Hoechst AG process for the production of HD-polyethylene using a diluent.[22]

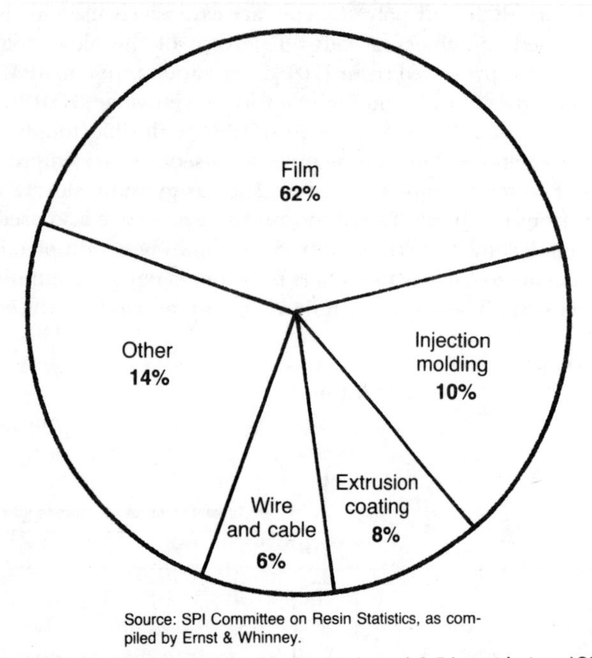

Fig. 12-8—Utilization pattern for the projected 3.54 metric ton 1980 production of LDPE.[2,24]

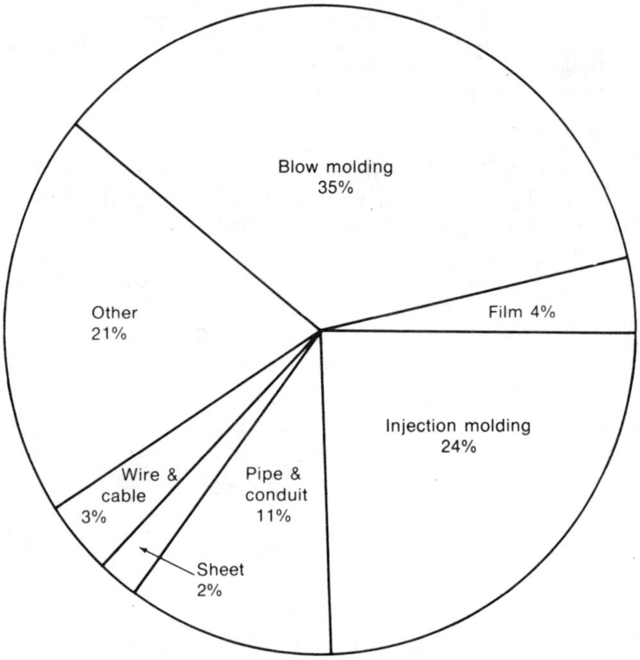

Blow molding 35%
Film 4%
Other 21%
Injection molding 24%
Wire & cable 3%
Pipe & conduit 11%
Sheet 2%

Fig. 12-9—Distribution of the projected 1.95 million metric ton HD-polyethylene demand in 1980.[25]

covers, wall coverings, tags, labels, etc. The point-bonded, spun-bonded polyethylenes are used, for example, as laboratory coats, aprons, garments, and sleeping bag liners.

Fig. 12-8[24] shows the breakdown of the projected 3.54 million metric tons of low density polyethylene demand in 1980.[2] There is little change in the pattern of current consumption except in films and sheets which show more **LDPE** will be used in this field. Fig. 12-9 shows the breakdown of the projected 1.95 million metric tons of **HDPE** in 1980.[25]

POLYPROPYLENE

Polypropylene, PP, n(CH_2-CH(CH_3))n, has been available commercially for more than 15 years, but only in recent years has it started to take its place among major thermoplastics. This delay in polypropylene production is due to the lack of direct production of the monomer. It is always a byproduct of either refinery operations or of the steam cracking of ethane and higher feedstocks for the production of ethylene.

Another factor which may have delayed polypropylene's rapid development was the lack of catalysts to produce a stereoregular polymer. Unlike ethylene, propylene has a methyl group in the molecule and, when polymerized, can have one of three structures—*isotactic,* where the methyl groups attached to the carbon chain are in the same plane; *syndiotactic,* where the methyl groups are alternately distributed in the same plane; and *atactic,* where there is a random distribution of methyl groups.[26]

The turning point in polypropylene production was the development of a Ziegler type catalyst by Natta to produce a polymer which is predominantly isotactic. Because of the stereoregularity of the polymer, it has a high crystallinity that enables the chains to pack closer, producing a resin with much better qualities than polymers produced by free radical initiators. High pressure, free radical chain polymerization of polypropylene has been tried, but the product was branched and lacked the stereoregularity and crystallinity needed for plastics and fiber processing.

Fig. 12-10 shows the growth of polypropylenes since 1972.[2, 27, 28] Polypropylene is considered the most dynamic large-tonnage polymer. The current capacity for polypropylene is 4.8 billion pounds/year, and the production is 70 percent of capacity.[2] Demand for polypropylene in 1980 will be about 3.85 billion pounds. In a more optimistic view, polypropylene may surpass high density polyethylene by 1980 with a projected demand of 4.95 billion pounds and by the turn of the century could pass low density polyethylene to become the "number 1" plastic.

Physical and Chemical Properties. The properties of commercial polypropylene vary according to the percentage of crystalline isotactic polymer and the degree of polymerization. Typically, polypropylene contains 50-60 percent crystalline isotactic, 20-30 percent amorphous, 20-30 percent amorphous isotactic, and 10-20 percent amorphous atactic. It has a high crystalline melting point of 170°C. The density of polypropylene is the lowest of all thermoplastics. (0.90-0.915 g/cc).

Articles made from polypropylene have good electrical and chemical resistance and low water absorption. It has a good heat resistance and can be sterilized at 100°C. Because of the resistance of polypropylene to flex fatigue, self-hinged articles made from polypropylene have long life expectancy. Other important properties of polypropylene are its toughness, high abrasion resistance, good dimensional stability, lack of toxicity, high impact strength, and transparency. Table 12-7 contains some of the important physical properties of polypropylene.

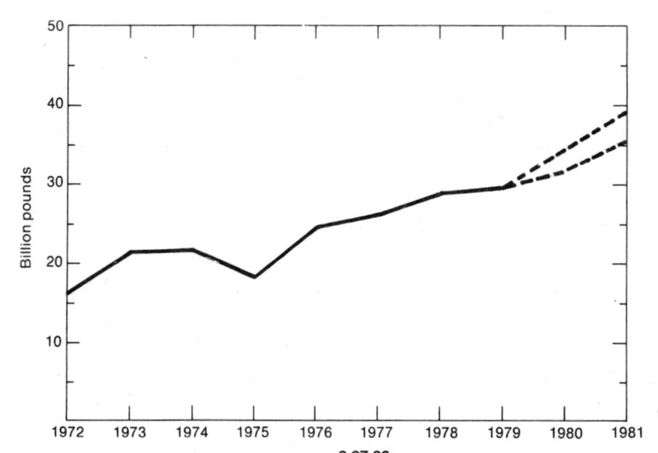

Fig. 12-10—Polypropylene growth.[2,27,28]

Production. Most polypropylene processes are at low pressure with a Ziegler-type catalyst—$TiCl_3$ or $TiCl_4$ and AlR_3—in solution. Although high pressure polymerization favors the yield, low pressure is used to increase the stereoregular isotactic configuration in the polymer. The product has qualities similar to high density polyethylene, but its brittle point (0° or higher) is higher than that for polyethylene.[29]

In the low pressure process, the practice is to feed highly purified propylene—with or without copolymer ethylene, the catalyst, and the hydrocarbon diluent—to the polymerization reactor at a temperature range of 50-100°C and a pressure range of 5-30 bars. The polymer is in the form of finely divided granules entraining some catalyst particles. Polymer is washed with either alcohol or water to decompose the catalyst. The solvent is then recovered from the polymer by either centrifugation, filtration or by steam treatment of the resin followed by hot air. The polymer thus obtained is in the form of powder which can be sold as molding powder after being homogenized, or can be processes to natural or colored pellets. Fig. 12-11 shows the Hoechst AG polypropylene process.[30]

Research in the catalyst field by Montedison and Mitsui has developed a new group of catalysts which are highly active and stereospecific.[31, 32, 33] These catalysts may move polypropylene to the position of being the "number 1" thermoplastic. The new catalysts are in the titanium aluminum alkyl family. Currently used catalysts require about one gram of catalysts for each 3,000 grams of polymer produced, while one gram of the new catalysts produces 300,000 grams of polymer. An important consequence of the higher yield is that the amount of catalysts entrained in the polymer is significantly less than it is for polypropylene produced by the older catalysts in which the titanium can occur at levels up to 300 ppm in the material coming from the reactor.[34] Reactor product from the new process contains less than 4 ppm of titanium and thus eliminates the need for a catalysts removal step.[35] The new catalyst is stereoselective and produces a polymer with an isotactic index of 93. *Isotactic index* is a measure of the percentage of the stereoregular isotactic form of the polymer that is insoluble in boiling heptane.

The high isotactic content eliminates the need to remove the atactic form of the polymer. Elimination of two stream units results in a 25 percent reduction in capital and operat-

TABLE 12-7—Properties of Polypropylene Plastic

Density, g/cm^3	0.90—0.91
Fill temperature, max. °C	130
Tensile strength, psi	3200—5000
Water absorption, 24 hr., %	0.01
Elongation, %	3—700
Melting point, Tm °C	176
Thermal expansion, 10^{-5} in. in. °C	5.8—10
Specific volume, cm^3/lb.	30.4—30.8

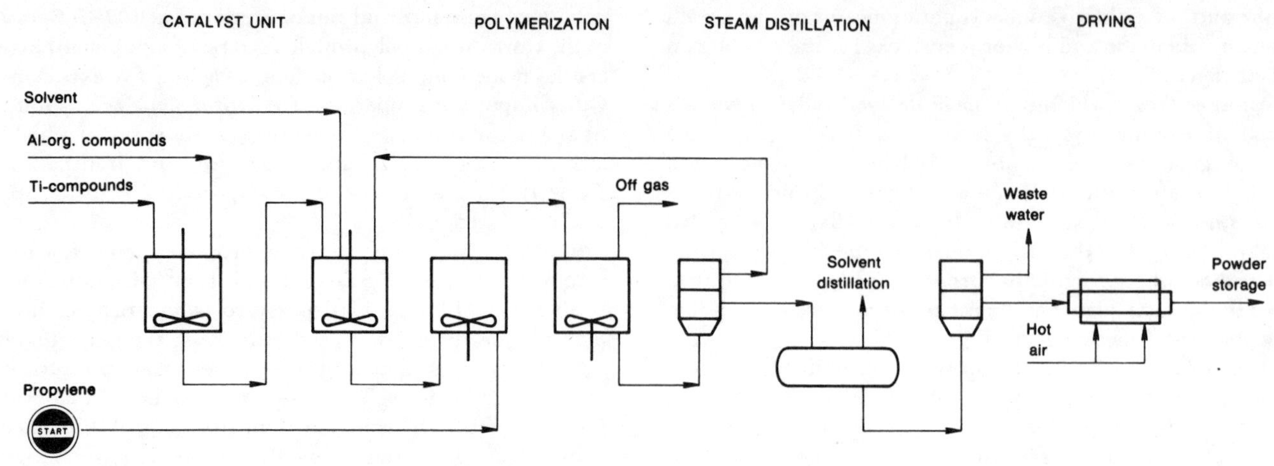

Fig. 12-11—The Hoechst AG process to produce polypropylene or copolymers from propylene.[30]

ing costs. The operational cost saving can be obtained from such items as monomer consumption costs, catalysts removal and effluent treatment, costs for the solvent recovery and atactic removal section. Fig. 12-12 shows a polypropylene unit which uses the new catalyst.[36]

Markets and Uses of Polypropylene. The 1979 consumption pattern for polypropylene is shown in Fig. 12-13.[37] Injection molding consumes about 39 percent of the production but blow-molded objects are made from molded container packaging field due to inherent physical properties which make thermoforming difficult by conventional techniques. This problem has been solved by Royal/Dutch/Shell Laboratories.[38] The "solid phase pressure forming" process can produce thin-walled containers which will enable **PP** to compete with **PVC, PS** and **ABS** in this field. In this process, special grades of polypropylene are extruded into sheets ranging in thickness from 0.2 to 2 mm. The sheet is fed to a continuous forming machine. About 22 million pounds of polymer were used in thermoform items in 1976.

Due to its light weight and toughness, polypropylene is being used extensively in automotive parts. This both reduces weight and cuts gasoline consumption. During 1976, 286 million pounds of polypropylene and its copolymers were consumed in passenger cars.[39] This figure is expected to reach 551 million pounds in 1982. They are used in automobile interior trim parts, finder liners, heat/air conditioner housings and ducts and battery cases.

Between 30 and 32 percent of polypropylene is used by the fiber industry. Low-cost fibers made from polypropylene are competing and replacing those made from sisal and jute. This has been attributed to improved techniques used in melt spinning and film filament processes. Properties such as low density, high stretch and high abrasion resistance, resistance to chemicals, no moisture absorption, and low cost are favorable for filament processing. However, its low softening point and dyeing problems are unfavorable.[40] Graft polymerization and crosslinking with monomers having polar groups, such as methyl methacrylate, are used to increase the softening point of polypropylene.

Improvements in the polypropylene field are continuing. A new material made from ethylene-propylene copolymers which is said to bridge the gap between true plastic and true elastomers—a thermoplastic elastomer—has been announced by DuPont. The material possesses rubber-like

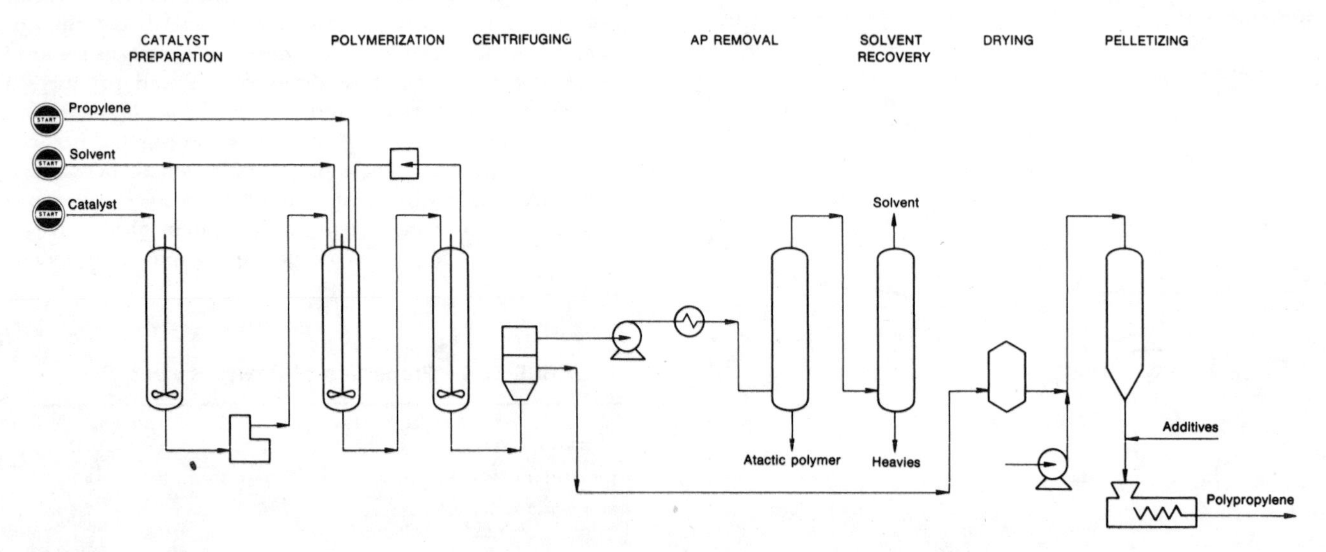

Fig. 12-12—A continuous process to produce polypropylene mono- and copolymers using HY-HS catalyst.[36]

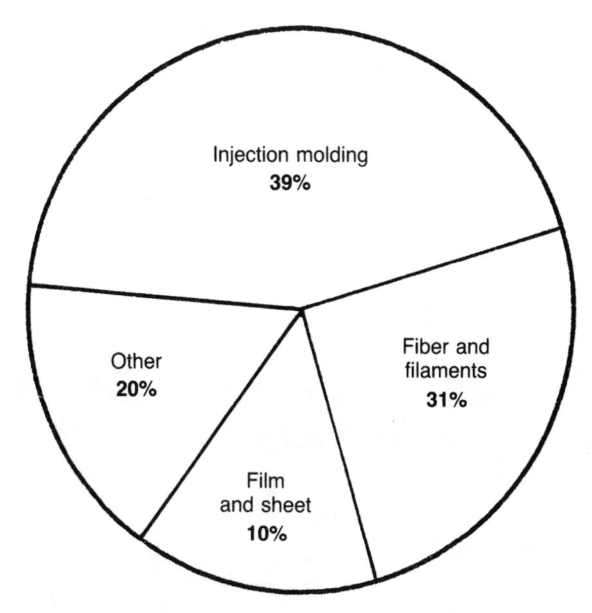

Fig. 12-13—The 1979 consumption pattern for polypropylene.[37]

properties and can be processed by conventional plastics processes. Products can be used for seals, tubing, shoe soles, and automotive parts.

POLYVINYL CHLORIDE

Polyvinyl chloride, PVC, $n(CH(Cl) - CH_2)n$, is one of the most widely used thermoplastics with several diverse applications. The U.S. consumption was 5.7 billion pounds in 1978, and the demand in 1980 is 7.1 billion pounds. Capacity was 7.3 billion pounds in 1979. However, the realistic capacity figure is believed to be much lower.[2] Worldwide capacity is over 26 billion pounds. The average annual growth rate for the next several years is predicted to be 7 percent.[41]

Physical and Chemical Properties. Two types of the homopolymer are available, the *flexible type* and the *rigid type.* Both types have a wide range of properties including self-extinguishing properties. This property is provided by the presence of chlorine atoms in the backbone of the polymer. Tensile strength of the rigid type ranges between 6,000 and 9,000 psi and that of the flexible type ranges between 1,000 and 4,000 psi.

Both types of **PVC** have excellent chemical and abrasion resistance. Materials made from the flexible type are soft and have a specific gravity ranging between 1.15 and 1.8. These materials can be stretched up to 4.5 times their original length. The flexible type is produced with high porosity to enable plasticizer adsorption. The rigid type has a specific gravity ranging between 1.3 and 1.6. Articles made from the rigid type polyvinyl chloride are hard and cannot be stretched over 40 percent of their original length. Polyvinyl chloride polymers, in general, have low crystallinity because of the random orientation of the chlorine atoms—atactic.

Vinyl chloride is copolymerized with other monomers to improve the quality of the resin. The copolymer with vinyl acetate is more flexible, possesses higher tensile strength, has a lower softening point, and is more stable to light and heat than the homopolymer of vinyl chloride. The copolymer of vinylidine chloride has a higher melting point than the homopolymer and may be spun into a fiber.

Production. Vinyl chloride monomer, **VCM,** is polymerized by using a free radical initiator with any of four general methods of polymerization—suspension, emulsion, bulk, and solution. However, about 80 percent of U.S. **PVC** production is by suspension polymerization. Water is used as the suspension and heat transfer medium along with a suspending agent such as polyvinyl alcohol. Polymerization temperature is generally between 40° and 70°C which corresponds to a vinyl chloride saturation vapor pressure of 6-12 bars. A design pressure of 18 bars provided an operating pressure safety margin.[42]

In a typical suspension process, **VCM** is distributed in the aqueous phase with the aid of protective colloids. The polymerization is batchwise in a water-jacketed autoclave where the initiator is added. After the reaction is completed, the suspension is transferred to a degassing unit in which the unreacted monomer is removed. Unreacted vinyl chloride is purified, condensed, and recycled. Polymer is separated from the suspension by centrifuging. Finally, the polymer is dried by hot air.

Although the usual reactor volume for **PVC** units is in the range of 6 to 40 cubic meters, reactors are now being built with a volume of 200 cubic meters. This size requires a better cooling and stirring system because of the smaller surface-to-volume ratio. This problem has been successfully solved by installing a reflux condenser on the top of the reactor.

These large reactor units can produce about 160,000 tons per year and have the advantage that they use only 65 percent of the space needed for the same capacity using several small units. Other savings are in capital cost reduction of 25 percent and in the reduction in manpower and maintenance. Fig. 12-14 gives a flow diagram of the new Huls **PVC** process using the 200-cubic-meter reactor.[43]

Polymerization of vinyl chloride in bulk using a water-cooled vertical autoclave stirred by a turbulent agitator gives a clear product. In a typical bulk process, **VCM** is pumped to the vertical autoclave prepolymerizer in which 8-12 percent of the monomer is converted into **PVC** seed. The reaction product is then transferred to a horizontal post-polymerizator where additional catalyst and monomer are added. Extent of conversion at the end of the polymerization cycle is between 80 and 85 percent. The polymer is finally degassed and classified to **PVC** grades. Fig. 12-15 as a flow diagram for a two-step bulk process by Rhone-Poulenc.[44]

Markets and Uses of Polyvinyl Chloride. Polyvinyl chloride resins are rarely used alone but are usually compounded with different types of additives. An important property of polyvinyl chloride is its acceptance of many types of compounding materials. Thus, its mechanical, physical, and chemical properties can be varied appreciably.

Flexible grades of **PVC** account for about 50 percent of the **PVC** production. It goes into such items as calendered sheet, tablecloths, shower curtains, furniture, automobile upholstery, and wire and cable. Rigid resins are used in many products such as pipes, irrigation pipes, fittings, roofing, blow-molded bottles, automobile parts, and siding. The plas-

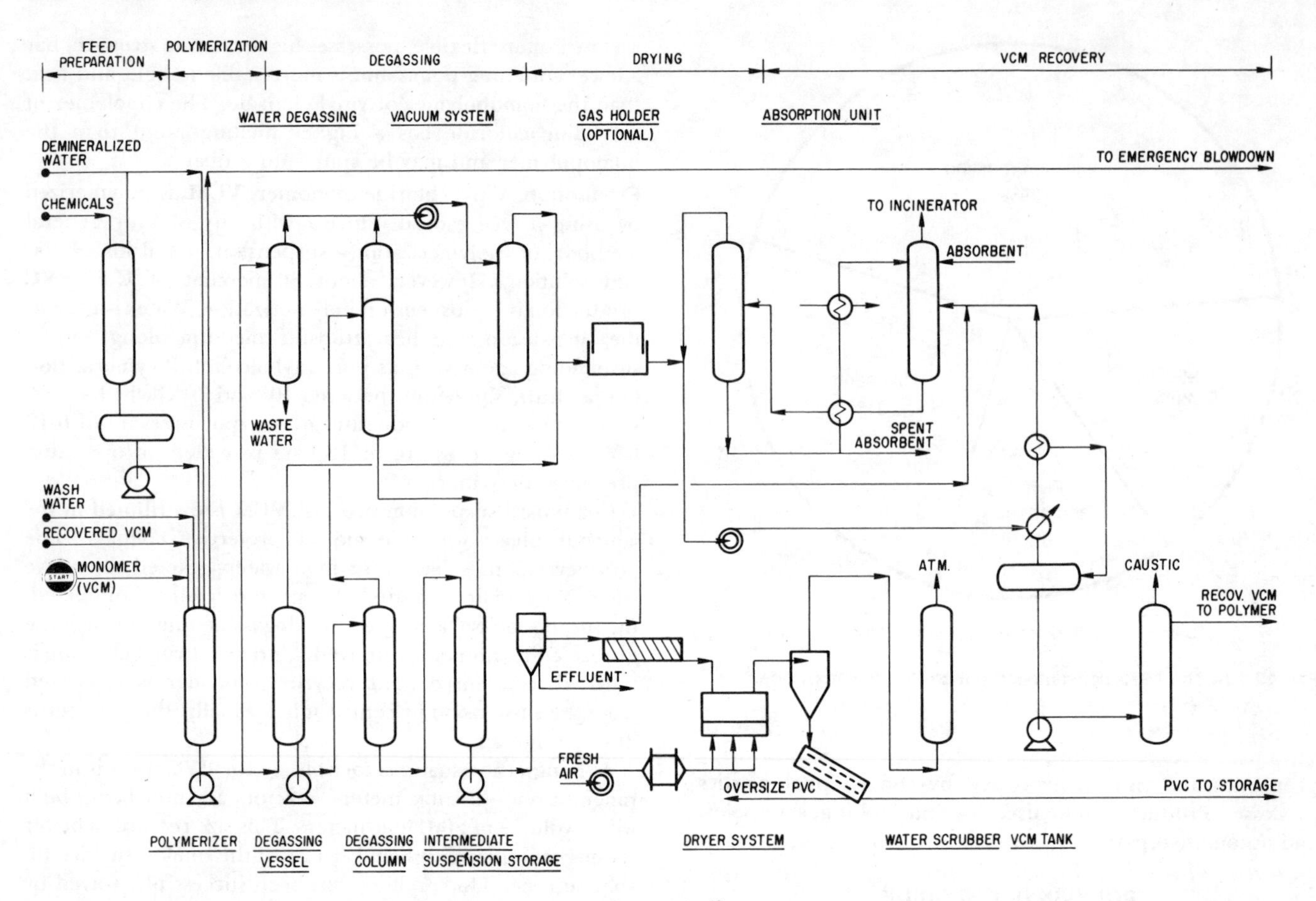

Fig. 12-14—Flow diagram of the Huels computer-controlled S-PVC process.[43]

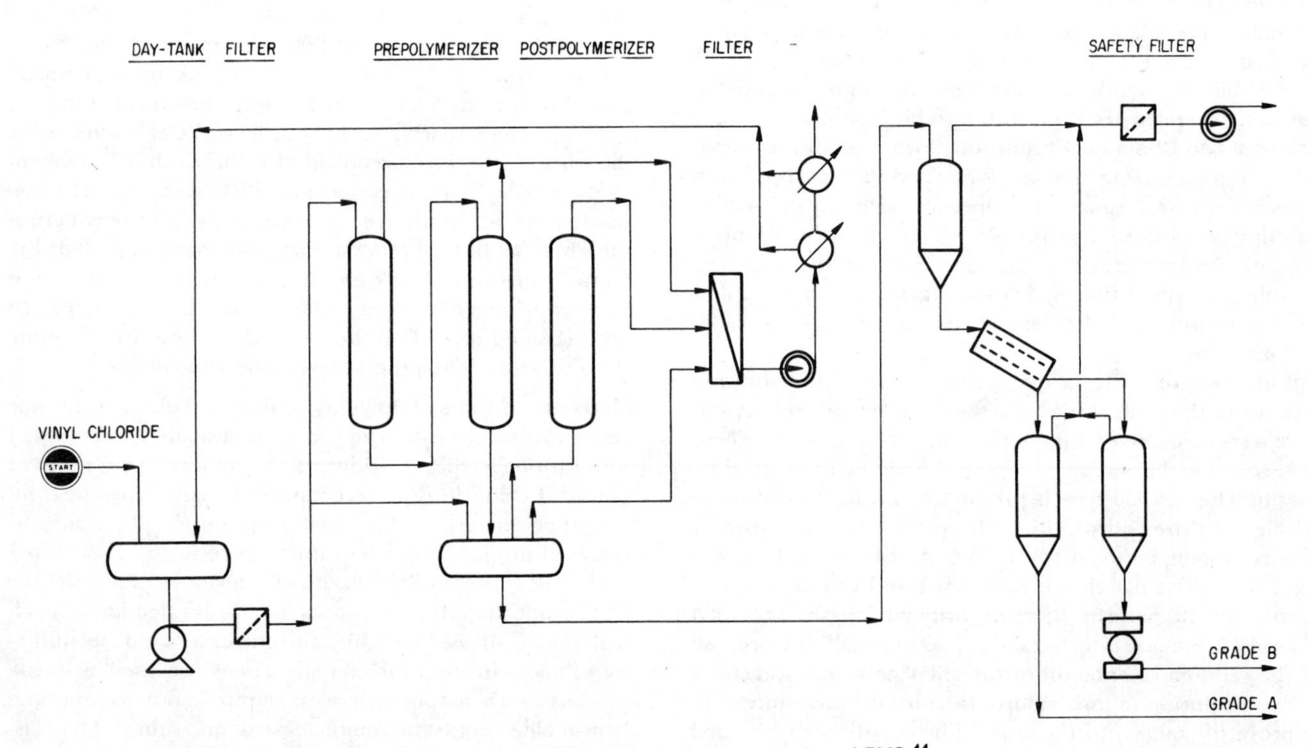

Fig. 12-15—Flow diagram for the two-step bulk process by Rhone-Poulenc for production of PVC.[44]

tic pipe market, the single largest market for **PVC,** is over 40 percent of the total **PVC** consumption, Fig. 12-16.[45] This represents 84 percent of the plastic pipe production.

Polyvinyl chloride self-skinning foam pipe is a new product trying to penetrate an intensely competitive nonplastic industry. Markets for foam pipes are those that do not require high pressure. Foam pipes are said to offer density reductions up to 40 percent and also a substantial cost reduction.[46] Siding is another field in which **PVC's** share is predicted to more than double by 1980. During 1975, 64 million pounds of **PVC** were used in siding. That demand is predicted to reach 160 million pounds in 1980 and 318 million pounds by 1985. Fig. 12-17 shows the major fabricated forms of **PVC.**[47]

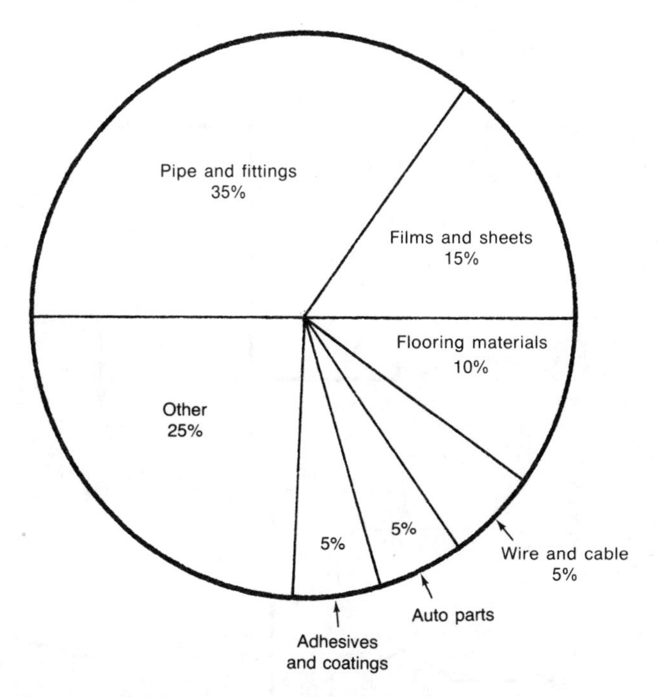

Fig. 12-17—The major fabricated forms of PVC.[47]

POLYSTYRENE AND STYRENE COPOLYMERS

Polystyrene, PS, $n(CH_2\text{-}\underset{|}{\overset{C_6H_5}{|}}CH)n$ is the third largest thermoplastic with a 1979 production at 4.1 billion pounds.[48] This large use of polystyrene results from its favorable properties. These include ease of fabrication, low specific gravity, thermal stability, and low cost. Polystyrene homopolymers, however, possess certain unfavorable physical properties such as brittleness and rigidity. These properties can be improved by copolymerication with other monomers and polymers.

When styrene is copolymerized with acrylonitrile, **SAN,** the polymer has a higher tensile strength than polystyrene alone. Styrene is also copolymerized with acrylonitrile and butadiene to produce **ABS,** which has special mechanical properties useful for engineering plastics. Copolymers of styrene and butadiene in a ratio of approximately 1 to 25 are the most important synthetic rubber, **SBR,** which is discussed in Chapter 15.

Physical and Chemical Properties. Polystyrene is a clear, transparent resin with a wide range of melting points and good flow properties which make it suitable for injection molding. It has excellent electrical properties (low electric losses) and is, therefore, useful in electronic applications. Polystyrene has a low water absorption which makes it a good electric insulator. It does, however, have some undesirable properties. Because of the inflexibility of the molecular chain in polystyrene, the homopolymer is brittle. Two other unfavorable characteristics are its thermal resistance and weather resistance. Crosslinking agents are used to increase heat stability. The products of crosslinking are considered thermosets and cannot be molded easily.

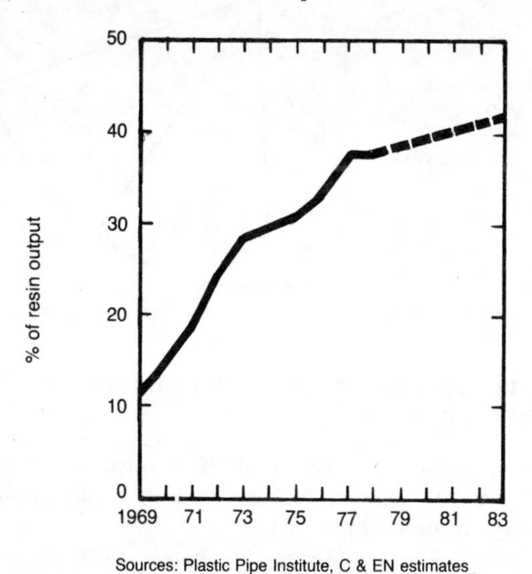

Fig. 12-16—The PVC demand for plastic pipe and fittings as a percentage of PVC total demand.[45]

Polystyrene crosslinked with divinylbenzene gives a resin which is used to prepare cation and anion exchange resins. Polystyrene has been crosslinked with 1,4-dimethyl-2,5-dichloromethylbenzene in acetic acid. The product has a higher melting point and a higher glass temperature than polystyrene.

Polystyrene has moderate resistance to chemicals but is attacked by aromatic and chlorinated hydrocarbons.

Production. Styrene is polymerized by either free radical initiators or by use of coordination catalysts. Bulk, suspension and emulsion techniques are used with free radical initiators and the polymer produced is atactic. Isotactic polystyrene is produced using Ziegler type catalysts.

In a typical suspension batch process, the monomer (styrene) is suspended in water using a suspension stabilizer and agitation. When polybutadiene is used as a copolymer, it is dissolved in styrene monomer prior to the suspension step. Catalysts and chain transfer agents are added.

After the polymerization reaction is completed, the polymer in the form of hard beads is transferred to an agitation tank. This tank is a feed tank to a continuous centrifugation process where the beads are separated from the water. Beads are dried by use of a rotary dryer, blended with different additives, and finally fed to an extruder and then to a pelletizing machine.

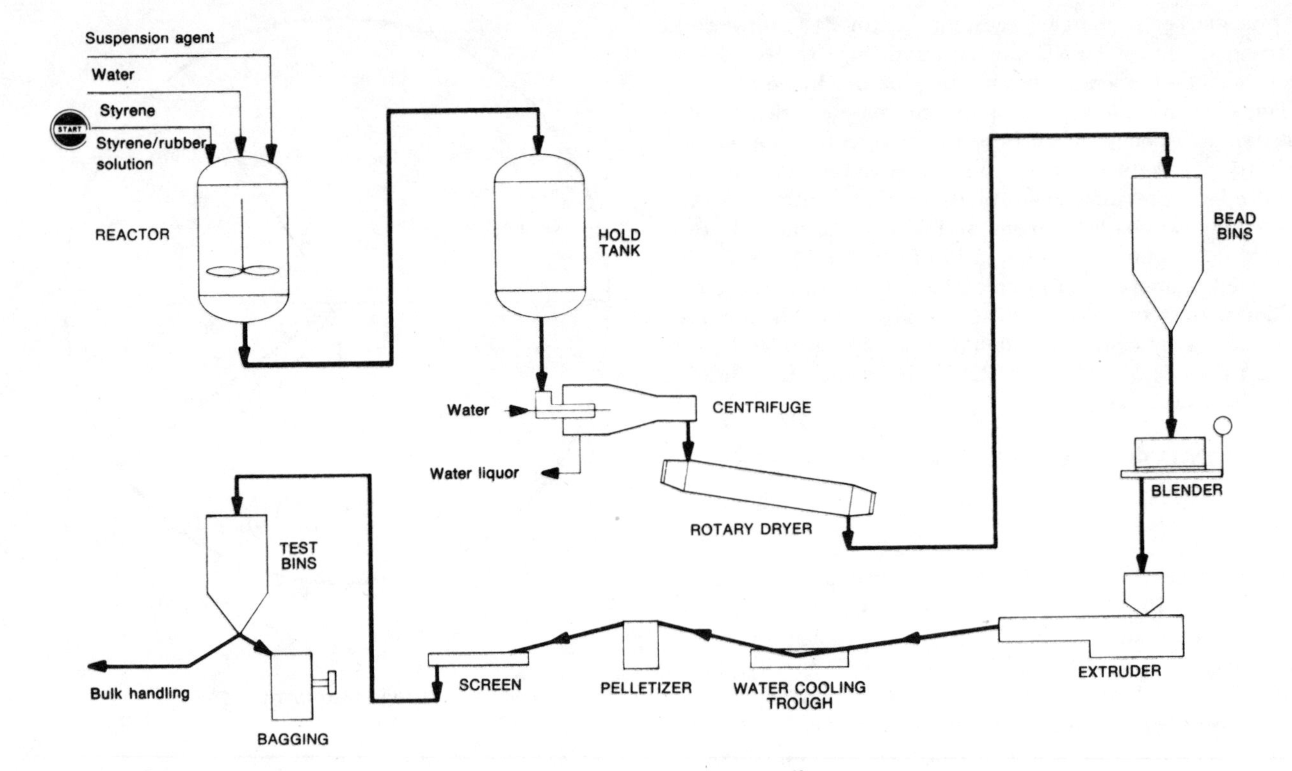

Fig. 12-18—The Cosden suspension process for polymerizing styrene to polystyrene.[49]

The polymerization reaction is controlled by the reactor's temperature and by chain transfer agents. Water is used as a cooling medium for the exothermic reaction and as the suspension medium. Fig. 12-18 is a flow diagram of a suspension process.[49]

Active homogeneous catalysts are difficult to separate from the reaction medium. Crosslinked polystyrenes are sometimes used as a catalyst substrate on which the catalyst is loosely held. Webster and Michalska have extensively reviewed this subject.[50]

Market and Uses of Polystyrene. The general uses of polystyrene are shown in Fig. 12-19.[51] Molded polystyrene is used in items such as automobile interior parts, furniture, home appliances, refrigerator parts, etc. Packaging uses plus specialized food uses, such as containers for carry-out food, are a growth area. This is result of changing eating habits which use increasing amounts of disposable containers. Most of the other uses of polystyrene are also for disposable items, but their life cycle is much longer.

Expanded polystyrene foams, which are produced by polymerizing styrene with a volatile solvent such as pentane, are used in packing. Because of the excellent thermal properties of expanded foam and its low density, it is also used as an excellent insulator and in flotation (life jackets). Foamed materials have a relatively higher growth rate than other polystyrenes. In general, polystyrene has a slower growth rate than other major thermoplastics.

Styrene-Acrylonitrile, **SAN.** The copolymer of styrene and acrylonitrile has better chemical resistance, higher heat resistance, and is stiffer than polystyrene. However, **SAN** is not as clear as polystyrene and is used in articles which do not

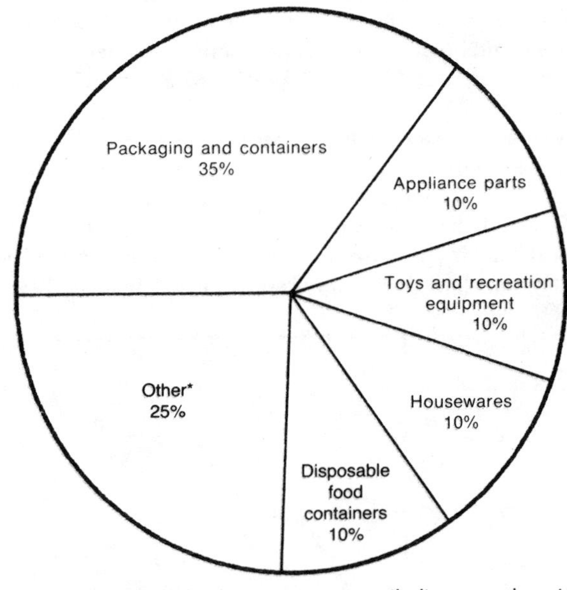

*Such as knobs, other furniture parts, cosmetic items, and custom-molded items.

Fig. 12-19—General use pattern for the 1978 3.75 billion pound production of polystyrene.[51]

require optional clarity. Compounding consumed 32 million pounds in 1977, while houseware and small appliances consumed 22 million pounds.

Acrylonitrile-Butadiene-Styrene Copolymers, **ABS.** These polymers are tough plastics with outstanding mechanical properties. They are one of the few thermoplastics which combine

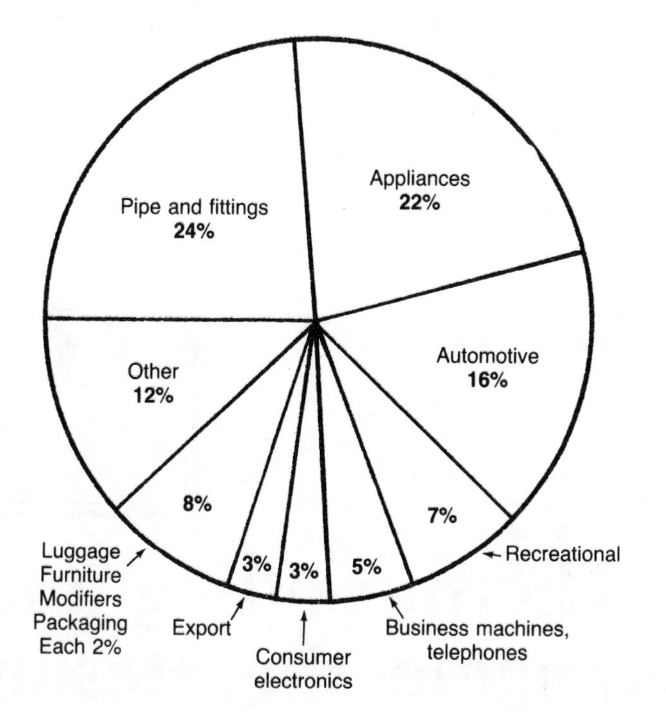

Fig. 12-20—The many and varied uses of the 1979 ABS production of 1.12 billion pounds.[52]

both toughness and hardness. It has been said that "standard grades meet physical properties required by a simple ballpoint pen while its specialty grades meet critical specification for space vehicle parts."

A wide variety of **ABS** modification is available with heat resistance comparable to or better than polysulfones and polycarbonates. The polymers are resistant to most chemicals and have good electrical properties. Another outstanding property of **ABS** is its ability to be blended, alloyed, with other thermoplastics to give improved properties. For example, when **ABS** is alloyed with rigid **PVC,** the product has better flame resistance. Flame-retardant **ABS** may gain new markets in TV sets.

Among the major applications of **ABS** are extruded pipes and pipe fittings—uses which will increase 20 percent per year; appliances such as refrigerator door liners, and in molded automobile bodies. Autos will use 25 pounds of **ABS** per car in 1985.[50] In 1978 only 17.5 pounds per car were used. Fig. 12-20 indicates the wide distribution of the 1.12 billion pound production of **ABS** in 1979.[52] The estimated 1982 capacity will be 2.02 billion pounds.[12]

LITERATURE CITED

1. *Chemical and Engineering News,* Sept. 18, 1978, pp. 10-11.
2. Rivoire, John. *Chemical Week,* Feb. 28, 1979, pp. 15-17.
3. *Chemical Age,* Sept. 23, 1977, p. 14.
4. *Chemical Week,* May 30, 1979, pp. 25-26.
5. *Chemical and Engineering News,* Sept. 17, 1979, p. 8.
6. *Chemical Week,* March 7, 1979, p. 22.
7. *Chemical and Engineering News,* Sept. 12, 1977, p. 15.
8. *Chemical and Engineering News,* July 28, 1975.
9. Abbas, K.B., A.B. Kuntsson and S.H. Berglund. *CHEMTECH.* Vol. 8, No. 8, 1978, pp. 502-508.
10. El Khadi, Mazahim and O.F. David. Second Arab Conference on Petrochemicals, United Arab Emirates, Abu Dhabi: March 15-22, 1976.
11. *Hydrocarbon Processing,* Vol. 55, No. 11, 1976, p. 159
12. Storck, W.J. *Chemical and Engineering News,* July 30, 1979, pp. 8-9.
13. *Hydrocarbon Processing, 1979 Petrochemical Handbook,* Vol. 58, No. 11, p. 219.
14. *Hydrocarbon Processing, 1979 Petrochemical Handbook,* Vol. 58, No. 11, p. 222.
15. Sittig, Marshall. "Polyolefin Production Processes." *Chemical Technology Review No. 79,* New Jersey: Noyes Data Corp., 1976, p. 9.
16. Imhausen, K.H., *et al. Hydrocarbon Processing,* Vol. 55, No. 11, 1976, p. 155.
17. *Chemical Engineering,* Dec. 5, 1977, p. 21.
18. *The Oil and Gas Journal,* Nov. 21, 1977, p. 73.
19. Tomura, Y., *et al. Hydrocarbon Processing,* Vol. 57, No. 11, 1978, pp. 151-154.
20. Sittig, Marshall. "Polyolefin Production Processes." *Chemical Technology Review No. 79,* New Jersey: Noyes Data Corp., 1976, p. 11.
21. *Hydrocarbon Processing, 1977 Petrochemical Handbook,* Vol. 56, No. 11, p. 212.
22. *Hydrocarbon Processing, 1977 Petrochemical Handbook,* Vol. 56, No. 11, p. 208.
23. *Chemical Week,* Nov. 7, 1979, p. 50.
24. *Chemical Week,* Nov. 7, 1979, p. 33.
25. Ponder, Thomas C. *Hydrocarbon Processing,* Vol. 56, No. 5, 1977, p. 178.
26. Stevens, P.M. *Polymer Chemistry,* Boston: Addison Wesley Publishing Co., Inc., 1975, pp. 52-53.
27. Greek, B.F., *Chemical and Engineering News,* June 5, 1978, pp. 8-10.
28. *Modern Plastics International,* Vol. 8, No. 11, 1978, p. 30.
29. Braunsteir, E.E., International Seminar on Petrochemical Industries, Bahgdad, Iraq: October 1975.
30. *Hydrocarbon Processing, 1977 Petrochemical Handbook,* Vol. 56, No. 11, p. 213.
31. *Modern Plastics,* Vol. 52, No. 6, 1975, p. 6.
32. *Chemical and Engineering News,* May 5, 1975, p. 17.
33. *Chemical Week,* May 5, 1976, p. 25.
34. *Chemical and Engineering News,* May 2, 1977, p. 28.
35. *CHEMTECH,* Vol. 6, No. 9, 1976, p. 582.
36. *Hydrocarbon Processing, 1979 Petrochemical Handbook,* Vol. 58, No. 11, p. 229.
37. *Chemical Week,* Oct. 24, 1979, pp. 29-30.
38. *Chemical and Engineering News,* Oct. 13, 1975, p. 25.
39. *Modern Plastics,* Vol. 54, No. 1, 1977, p. 43.
40. Thomas, O.G. International Seminar on Petrochemical Industries, Baghdad, Iraq: October, 1975.
41. *Chemical Week,* April 2, 1980, p. 13.
42. Terwiesch, B. *Hydrocarbon Processing,* Vol. 55, No. 11, 1976, p. 117.
43. *Hydrocarbon Processing,* Vol. 55, No. 11, 1976.
44. *Hydrocarbon Processing, 1979 Petrochemical Handbook,* Vol. 58, No. 11, p. 238.
45. *Chemical and Engineering News,* Mar. 19, 1979, pp. 15-18.
46. *Modern Plastics,* Vol. 54, No. 8, 1977, p. 50.
47. *Chemical and Engineering News,* Sept. 4, 1978, p. 13.
48. *Chemical and Engineering News,* Sept. 3, 1979, p. 15.
49. *Hydrocarbon Processing, 1977 Petrochemical Handbook,* Vol. 56, No. 11, p. 217.
50. *Chemical Week,* Sept. 20, 1978, p. 34.
51. *Chemical and Engineering News,* Sept. 4, 1978, p. 15.
52. *Modern Plastics International,* Vol. 9, No. 8, 1979, pp. 62-63.

13

Thermosetting and Engineering Resins

Thermosetting Resins. Thermosetting resins, *thermosets,* are a network of long chain molecules that are crosslinked which gives the polymer a three-dimensional, infusible structure. They polymerize irreversibly under heat or pressure to form hard, rigid masses. This is the basic difference between thermoplastic polymers and thermosetting polymers. An economic difference is that scrap from molding and from reject parts made of thermosetting polymers cannot be remolded. Table 13-1 lists major thermosets.

The crosslinking usually occurs during the curing reaction. The crosslinking may occur, however, during the polymerization reaction or after polymerization between the linear polymers. Such crosslinking concurrent with polymer formation is the case with phenol-formaldehyde resins. Crosslinking also may be obtained by the use of crosslinking agents as are used in the production of epoxy resins.

The consumption of thermosetting resins is much less than of thermoplastics and represents about 19 percent of the total plastic market.[1] Thermosets, however, hold specific markets. For example, urea and melamine resins are used in bonding and as adhesives for granulated wood.

Reinforcement of thermosets improves their physical properties. About 90 percent of the reinforced plastics, ***RP,*** is of the thermoset type with most of them made from polyester resins. Other thermosets that are reinforced are urethanes, phenolics, melamines, and epoxies. One of the largest uses of reinforced plastics is for chemical process equipment.

Although plastic reinforcement has been practiced for more than thirty years, development is continuing in the reinforcement technique to improve the quality, toughness, and corrosion resistance. The most widely used material for reinforcement is fiber glass in varying proportions ranging between 20 to 30 percent. Development in the reinforcement processing technique is also continuing. For example, sheet molding compounds, *SMC,* are made that allow the resin, glass reinforcement, pigment fillers, and thickeners to be sandwiched between two sheets of the plastic.

Engineering Thermoplastics. These resins include polymers with special properties such as high thermal stability, good chemical and weather resistance, transparency, good electrical properties, and other useful characteristics (Table 13-2).[3] Among this group are nylon, polycarbonates, polyvinyl acetates, polyethersulfones, polyacetals, and thermoplastic polyesters.

The U.S. consumption of engineering resins was 724 million pounds in 1978 and the projection for 1980 is 913 million pounds. This represents a growth rate of 12 percent per year. Fig. 13-1 shows the projected demand for the various engineering resins in 1980.[1] Figure 13-2 shows the distribution of the projected 274 million pound automotive market in 1982.[2] A recent prediction places general use of all types of plastics in automobiles in 1985 at 1.63 billion pounds, up 9 percent from the 1979 use of 1.49 billion pounds.[4] The 1990 models will use 2.1 billion pounds.

A major portion of engineering plastics is processed by injection molding and is used for replacement of die cast metals (zinc, magnesium and aluminum). Nylon is considered the number one engineering resin and accounted for 39 percent of the engineering resins in 1978.

THERMOSETS

Polyurethanes. These are produced by condensation reactions, but they are not considered condensation polymers, since a small molecule is not eliminated during their formation. Urethanes are the products of the reaction between alcohols and isocyanates.

$$R-OH + R'-N=C=O \rightarrow RO-\overset{\overset{\displaystyle O}{\|}}{C}-NH-R'$$

TABLE 13-1—Major Thermosetting Polymers

Name	Family	Formula
Polyurethane	Ester-amide	$-O-R'-O-\overset{O}{\overset{\parallel}{C}}-NH-R-NH-\overset{O}{\overset{\parallel}{C}}-$
Alkyd resins	Polyester	$C_6H_4(-\overset{O}{\overset{\parallel}{C}}-OCH_2-\underset{O}{\underset{\mid}{CH}}-\overset{O-}{\overset{\mid}{CH_2}})(-\underset{O}{\underset{\parallel}{C}}-)$; $O=C-C_6H_4-C=O$
Unsaturated polyesters	Polyester	$-O-\overset{O}{\overset{\parallel}{C}}-CH=CH-\overset{O}{\overset{\parallel}{C}}-OCH_2-CH_2-OCH_2-CH_2-$ $-O-\overset{O}{\overset{\parallel}{C}}-C_6H_4-\overset{O}{\overset{\parallel}{C}}-$
Epoxy resins	Polyether	$-O-R-O-CH_2-\overset{OH}{\overset{\mid}{CH}}-CH_2-$
Phenol formaldehyde	Phenolic	$-C_6H_3(OH)-CH_2-C_6H_2(OH)-CH_2-C_6H_3(HO)-$
Urea formaldehyde	Urea	$-HN-\overset{O}{\overset{\parallel}{C}}-\overset{H}{\overset{\mid}{N}}-CH_2-\overset{H}{\overset{\mid}{N}}-\overset{O}{\overset{\parallel}{C}}-\overset{H}{\overset{\mid}{N}}-CH_2-$
Melamine formaldehyde	Melamine	$-HN-C_3N_3(-NH-)-NH-(CH_2)_2-NH-C_3N_3(-)-NH-$

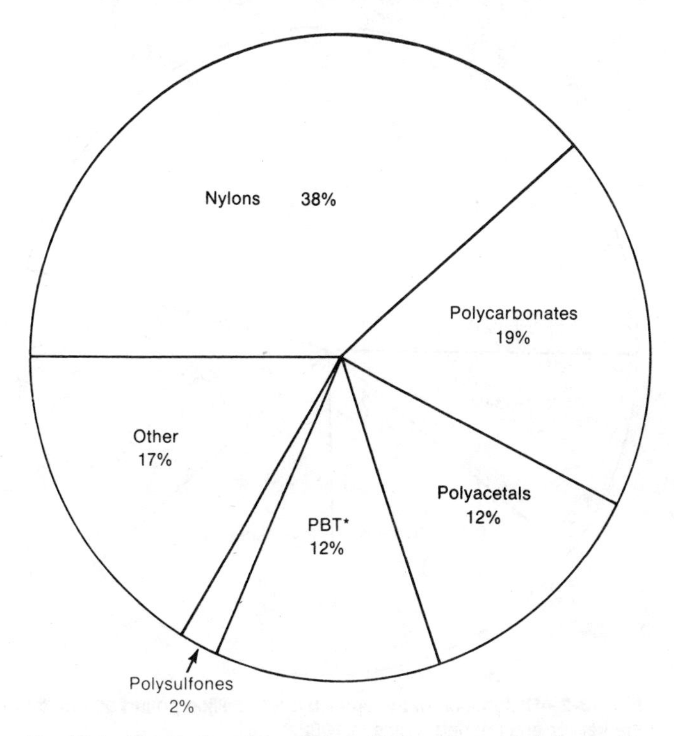

Fig. 13-1—Distribution of the projected 913 million pound demand for engineering resins in 1980.[1]

Polyurethanes are formed by the reaction between *diisocyanates,* R — C = N = O and a compound containing

R′ — C = N = O

more than two active hydrogen such as gylcerol and the polyglycols. Compounds generally used to provide the active hydrogen are polyglycols and polyether polyglycols. When a glycol and an isocyanate with more than two functional groups is used, a thermoset polymer is formed. Polyurethane polymers are either rigid or flexible, depending upon the type of polyol (polyhydroxy compound) used and the number of active hydrogens present. Flexible foams use diols, triols, polyols or a mixture of these polyols.

Triols derived from glycerol and propylene oxide are used for the production of block slabfoams. These polyols are characterized by moderate reactivity, since the terminal groups are predominately secondary. More reactive polyols which are used for the production of molding polyurethane foams are formed by the reaction between polyoxypropylene polyglycols with ethylene oxide to give the more reactive –OH groups.[5]

TABLE 13-2—Important Properties of Major Engineering Thermoplastics [3]

	Nylons		Acetals		Thermo-plastic polyesters (polybutylene)	Poly-carbonates	Polyether sulfones	Phenylene oxide
	6	66	Homopolymer	Copolymer				
Melting point								
T_m (crystalline)	225	276	181	175	232—267	...	...	105—120
T_g (amorphous)	...	...	...	...	...	150	230	...
Density (g/ml)	1.14	1.15	1.42	1.41	1.35	1.2	1.37	1.10
Tensile strength (psi)	11000—10000	12000—11000	10000	8800	8200	8000—9500	12200 at yield	7800—11500
Elongation (%)	100—300	60—300	25—75	40—75	50—300	100—130	30—80 at break	50—60
Flexal strength (psi)	14000—5000	17000—6100	18000	16000	12000—16700	12500	18650 at yield	12800—13000
Tensile modulus (10^5 psi)	3.8—1	...	5.2	4.1	2.8	3.0—3.5	3.5	3.5
Thermal expansion (10^{-5} in/in/°C)	8.3	8.0	10	8.5	6—9.5	6.6	5.5	5.2
Water absorption 24 hr (⅛ in thickness)	1.3—1.9	1.5	1.48	...	0.08—0.09	0.15—0.18	0.43	0.066
Dielectric constant (10Hz)	4.8	4.0	...	...	...	...	...	...
Dielectric strength (Short time V/mil)	440	385	...	...	...	...	...	...

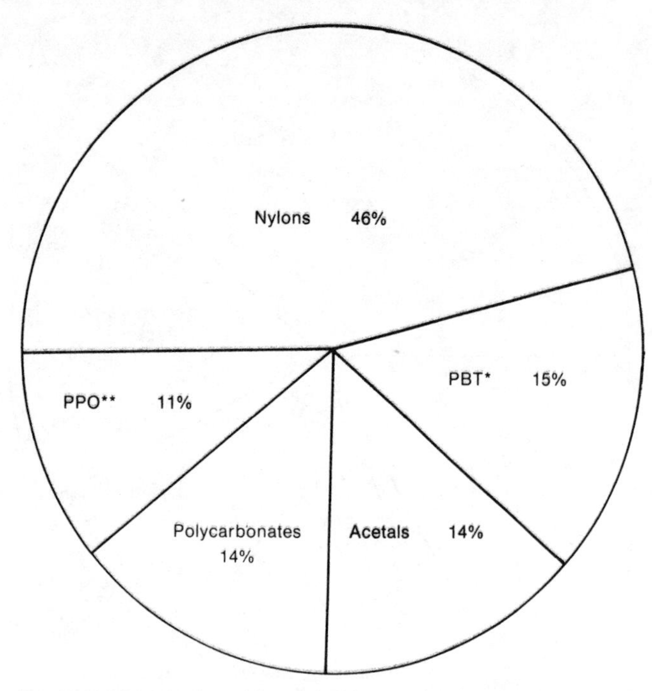

Fig. 13-2—Distribution of the projected 274 million pound automotive market for engineering resins in 1982.[2]

$$\begin{array}{l} CH_2-O-(CH_2-\underset{|}{\overset{CH_3}{CH}}-O)_n-CH_2-\overset{OH}{CH}-CH_3 \\ | \\ CH-O-(CH_2-\overset{CH_3}{CH}-O)_n-CH_2-\overset{OH}{CH}-CH_3 \\ | \\ CH_2-O-(CH_2-\overset{CH_3}{CH}-O)_n-CH_2-\overset{OH}{CH}-CH_3 \end{array}$$

$$+\ \overset{O}{CH_2-CH_2} \rightarrow$$

$$\begin{array}{l} CH_2-O-(CH_2-\overset{CH_3}{CH}-O)_{n+1}\ CH_2\text{-}CH_2\text{-}OH \\ | \\ CH-O-(CH_2-\overset{CH_3}{CH}-O)_{n+1}\ CH_2\text{-}CH_2\text{-}OH \\ | \\ CH_2-O-(CH_2-\overset{CH_3}{CH}-O)_{n+1}\ CH_2\text{-}CH_2\text{-}OH \end{array}$$

Other polyols may have polyethylene blocks in the backbone of the polyol.

The active hydrogen functionally of polyols used for polyurethane foams varies from three to eight. Polyols of propylene oxide are preferred because the use of ethylene oxide adversely affects the moisture resistance of the foams.

Equivalent weights of commonly used polyols range from 70 to 200. Triethanolamine, sorbitol and sucrose are polyhydric compounds with a functionality of three, six, and eight, respectively, and are used for producing polyols for rigid foams.[6] *4,4'-Methylene bis (phenylisocyanate), **MDI,*** is also used for the production of rigid foams. ***MDI*** is produced by the reaction between aniline and formaldehyde followed by the reaction of the diamine with phosgene to produce the diisocyanate.

$$2\ C_6H_5NH_2 + H-\overset{O}{\overset{\|}{C}}-H \rightarrow$$

$$H_2N-C_6H_4-CH_2-C_6H_4-NH_2 + H_2O$$

$$H_2N-C_6H_4-CH_2-C_6H_4-NH_2 + 2\ Cl-\overset{O}{\overset{\|}{C}}-Cl \rightarrow$$

$$OCN-C_6H_4-CH_2-C_6H_4-NCO + 4HCl$$

By varying the ratio of polyol to isocyanate, the physical properties of the polyurethane can be changed. For example, the tensile strength can be varied within a range of 1200 to 6000 psi and elongation between 150 and 800 percent.

A common mixture of diisocyanates used for flexible foams is one of *2,4-* and *2-6-toluene diisocyanate, **TDI.*** The polymerization of the diisocyanate and the polyol such as the reaction between polyethylene glycol and toluene diisocyanate to produce polyurethane takes place in solution. Impurities with very active hydrogens such as water, phenol, and acids are avoided to eliminate side reaction.

Properties and Uses. The physical and mechanical properties of polyurethanes can be varied widely. A wide range of densities can be produced from 1.0 to 6.0 lb./ft.^{3} for the flexible types to 1.0 to 50.0 lb./ft.^{3} for the rigid types. In general, polyurethane foams have good load-bearing characteristics, good abrasion resistance and low thermal conductance. They have moderate chemical resistance to organic solvents, but are attacked by strong acids.

The fire and smoke ratings of polyurethanes must meet furniture and insulation flammability regulations. Flame retardancy of polyurethanes is improved by using special additives, spraying a coating material such as magnesium oxychloride, or by grafting a halogen phosphorus moiety to the polyol used for urethane production. Polyols containing chlorine are now produced. Trichlorobutylene oxide is used in place of propylene and ethylene oxides to produce the polyol. It has been claimed that due to the absence of phosphorus in this formulation, the performance is better with respect to smoke generation.[7]

Fig. 13-3 shows the 1978 distribution of the 2.027 billion pound U.S. urethane consumption by types.[8] The major markets for polyurethane are furniture, transportation, and building and construction. (Fig. 13-4).[8] A wide range of other uses include domestic appliances, foundry core

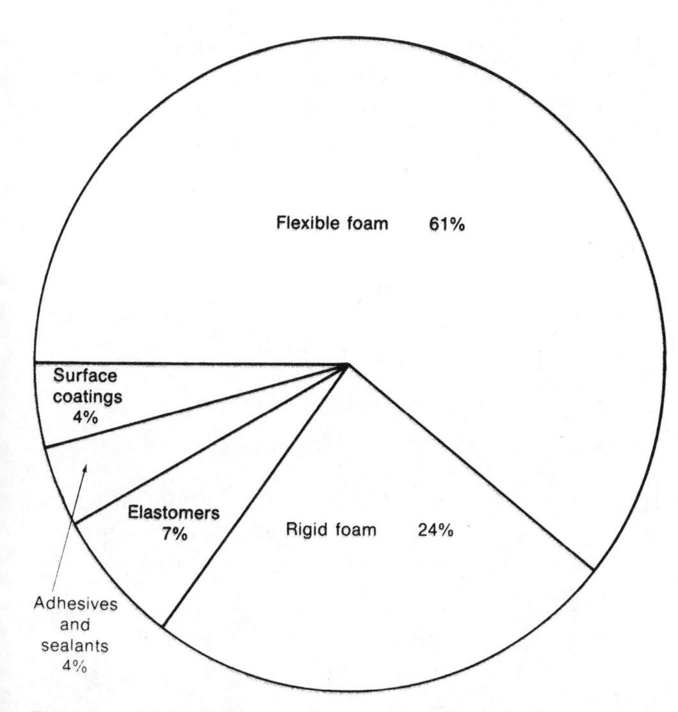

Fig. 13-3—Distribution of the 2.027 billion pound U.S. urethane demand in 1978 by type.[8]

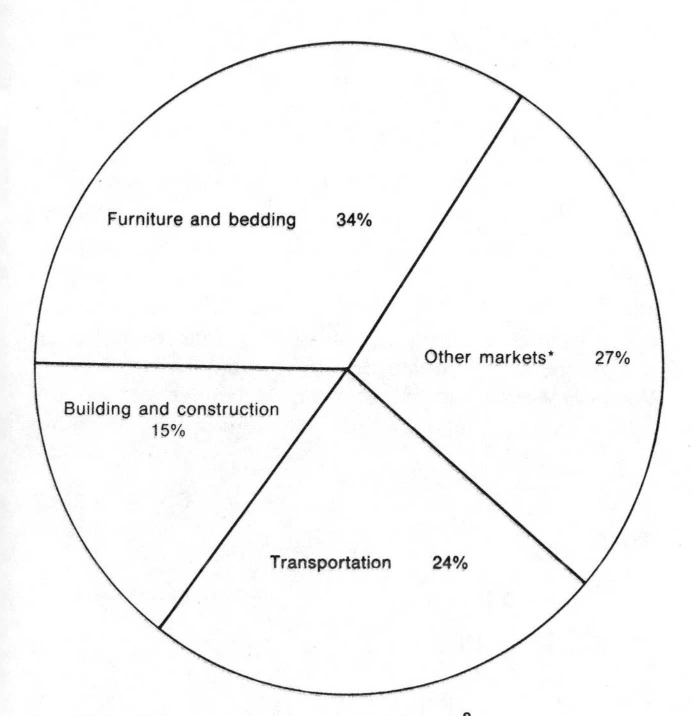

Fig. 13-4—The general markets for urethanes.[8]

binders, carpet underlay, textile laminates and coatings, footware, packaging, toys, and fibers.

The large use of rigid polyurethane for construction and industrial insulation is due to its high insulating property (Fig. 13-5), its good strength to weight ratio, and improved flame resistancy and smoke generation. Urethane foam is used for insulation of a variety of tanks including those used

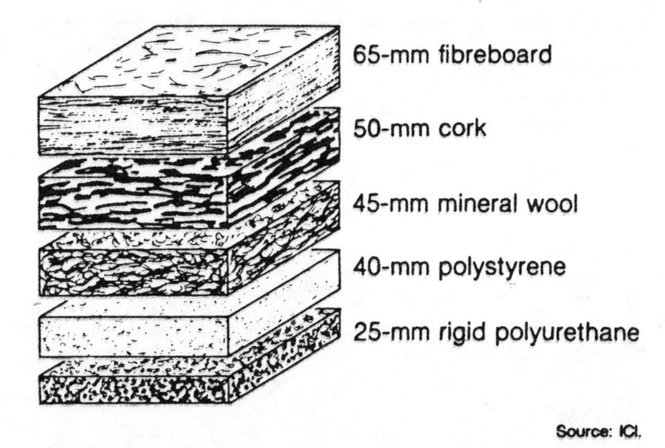

Fig. 13-5—The comparative thickness for the same degree of insulation (dry conditions).[9]

to maintain cryogenic temperatures, and vessels, pipes and domestic appliances such as refrigerators and freezers.

Flexible foam markets are furniture and transportation. More than 90 percent of U.S. market uses of flexible urethane foam is in cushioning. High resiliency, **HR,** flexible urethane foams offer better fire resistance than conventional foams.[10]

The transportation market is the second most important for flexible foams. The average per auto is 33 pounds, mainly in cushioning and padding. Front and rear metal bumpers are being replaced by molded urethane elastomeric bumpers. The weight saved by use of these light weight bumpers ranges between 20 and 40 pounds. The process for the manufacture of molded urethanes is "reaction injection molding," **RIM,** and is useful for the production of items such as bumpers, steering wheels, instrument panels, and body panels. The 1990 automobiles are projected to utilize 130 million pounds of urethanes.[4]

Polyurethane elastomers are attracting attention for their toughness, and resistance to abrasion, oils and oxidation. They are produced by the reaction of either **MDI** or **TDI,** with short chain polyols such as polytetramethylene glycol which is produced from 1,4-butanediol.[11] Peroxides, diamines and diols are used as crosslinking agents for elastomer production.

Spandex, a generic name for a manufactured fiber from polyurethane, is produced from *4,4'-methylene bis (phenylisocyanate),* **MDI.** These fibers have the property of being oriented when stretched but become disoriented when the force is released.

Polyurethanes are also good coating materials and come in various types. For example, urethane oils for marine finishes and wood coatings for chemical plants and other sites where protection against abrasion is required.

A summary of the uses of the different types of polyurethane is given in Fig. 13-6.[12] The United States demand in 1979 was 2.4 billion pounds, world-wide demand was 6.0 billion pounds.[13]

Urea and Melamine Resins. Urea, $H_2N-\overset{\overset{\large O}{\|}}{C}-NH_2$, is produced by the reaction between ammonia and carbon dioxide. It reacts with formaldehyde to form polymers called *urea-formaldehyde resins. Melamine,*

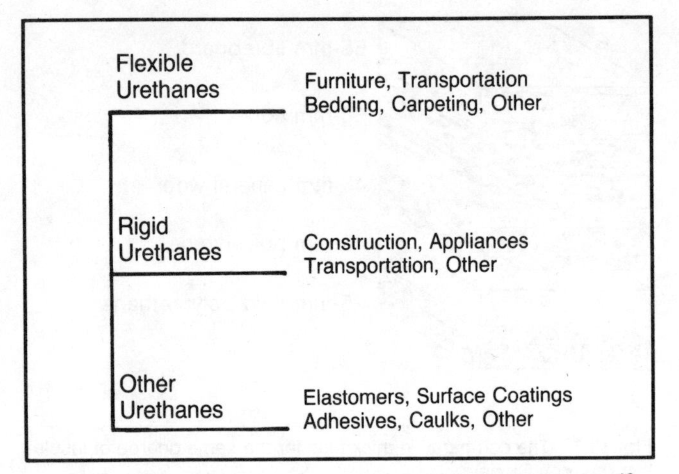

Fig. 13-6—The various uses of the different types of urethanes.[12]

```
          NH2
          |
          C - N
        //      \\
       N          C - NH2
        \        /
          C = N
          |
          NH2
        Melamine
```

is a heteocylic aromatic condensation product of urea. It reacts with formaldehyde to form resins, *melamine-formaldehyde resins.* Both melamine- and urea-formaldehyde resins have similar general properties although there are differences in their final applications. Both resins are called *amino resins (aminoplasts)*.

Urea-Formaldehyde Resins. The condensation reaction between urea and formaldehyde is as old as the condensation reaction between phenol and formaldehyde, but the commercial use of these resins did not start until the 1930's when fillers and additives were successfully mixed with the polymer to give a material that can be molded. The condensation of urea and formaldehyde takes place in several steps which are not fully understood. The ratio of urea to formaldehyde and the conditions under which the polymerization is carried out alter the course of the reaction.

When formaldehyde and urea are mixed in 1 : 1 mole ratio at room temperature, monomethylol urea and some dimethylol urea are formed by a nucleophilic addition of urea to formaldehyde.

```
         NH2                  H
        /                    /
2 O = C          + 3 O = C          →
        \                    \
         NH2                  H

       NH - CH2 OH              NH - CH2 OH
      /                        /
O = C                  + O = C
      \                        \
       NH2                      NH - CH2 OH
```

These compounds are crystalline and water soluble.

Condensation of methylol urea takes place if the mixture is heated under acidic conditions, giving polymeric products.

```
            NH - CH2 OH
           /              H+
2n O = C                  →
           \
            NH2

       O                     O
       ||                    ||
( n HN - C - NH - CH2 - NH - C - NH - CH2 - ) n +
                n - 1 H2O
```

It is believed that *imines*, $-C=NH$, are formed by the dehydration of monomethylol urea which then trimerizes, with subsequent condensation of methylurea and the amide groups of the trimer.

```
       NH - CH2 OH                 N = CH2
      /               H+          /
O = C                 →   O = C               + H2O
      \                           \
       NH2                         NH2

                             O = C - NH2
                                 |
         N = CH2                 N-CH2          O
        /                       /     \         ||
3 O = C              →      H2C        N - C - NH2
        \                       \     /
         NH2                     N-CH2
                                 |
                             O = C - NH2
```

In practice, the reaction is run under mild conditions with a formaldehyde-urea ratio greater than 3 to 2 to allow for crosslinking. A thick syrup is formed. Curing takes place by the addition of an acid. Fillers, plasticizers and pigments are added during or after curing.

Melamine-Formaldehyde Resins. Melamine is produced from urea under high pressure in an atmosphere of ammonia.

```
         NH2                     N = C - NH2
        /                       /         \
6 O = C         →    H2 N - C             N
        \                       \\        //
         NH2                     N - C - NH2
+ 6 NH3 + 3 CO2
```

Melamine is also prepared from cyanamide at high pressure and temperature.

```
                                 N = C - NH2
                                /         \
3 H2N - C≡N  →   H2 N - C              N
                                \\        //
                                 N - C - NH2
```

Melamine reacts with formaldehyde through a nucleophilic addition to give methylolamine. A variety of methylols are possible with methylamine because of the availability of six replaceable hydrogens. Tri- and hexamethylol melamines have been isolated, among others.

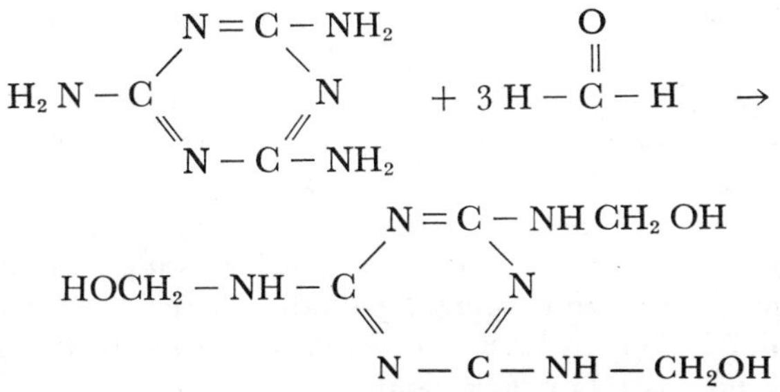

$$H_2N-C(=N-C(NH_2)=N-C(NH_2)=N) + 3\,H-\overset{O}{\overset{\|}{C}}-H \rightarrow$$

$$HOCH_2-NH-C(=N-C(NHCH_2OH)=N-C(NH-CH_2OH)=N)$$

Trimethylol Melamine

Other amino resins include formaldehyde condensates with toluene-*p*-sulfonamide, thiourea and aniline.

Properties and Uses. Articles made from amino resins are water-clear, hard, strong, but can be broken. They have good electrical properties and have better colors than articles made from phenolics. Amino resins are used for interior plywood applications only because they have a low weather resistance. The most important use of amino resins is in the production of adhesives for particle board and hardwood plywood.

Amino molding compounds are rigid and hard and are used for such products as radio cabinets, buttons, and cover plates. Melamine and urea resins are also used for textile treatment to increase shrink resistance. Because melamine resins have low water absorption and better chemical and heat resistance than urea resins, they are used for the production of dinnerware, and laminates such as formicas used to cover furniture. Alpha cellulose is a common filler. Other fillers include cotton fabric, asbestos, minerals, wood flour, glass fibers, and paper.

The current consumption of urea and melamine resins is 1.13 billion pounds of which about 67 percent is used as bonding and adhesive components for the production of fibrous and granulated wood, and as adhesive for plywood. Molding compounds consume about 10 percent and the rest is used for textile and paper treating and coating.[14]

Phenolic Resins. The condensation reaction between phenols and formaldehyde to produce *phenolic, phenoplast,* resins is the oldest condensation reaction and the product is the most important of the thermosets. Although many attempts were made to make use of the products and to control the reaction conditions for the acid catalyzed reaction described by Bayer in 1872, there was no commercial production of the resin until the exhaustive work by Baekeland. In his paper, published in 1909, he describes the product "as to Bakelite itself, you will readily understand that it makes a substance far superior to amber for pipe stems and similar articles. It is not so flexible as celluloid, but it is more durable, stands heat, does not smell, does not catch fire and at the same time is less expensive."[15] Convincing arguments.

Two types of products are produced by the reaction between phenol and formaldehyde—the base catalyzed polymers, *resols,* and the acid catalyzed *novalacs.*

When formaldehyde reacts with phenol in the presence of a base, the phenoxide ion is formed. The first step in the polymerization is the addition of the phenoxide ion to carbonyl carbon of the formaldehyde, giving a mixture of ortho- and para-substituted monomethylol plus di- and trimethylolphenol.

$$C_6H_5OH + OH^- \rightarrow C_6H_5O^- + H_2O$$

(resonance structures of the phenoxide ion: O^- ⟷ O ⟷ O ⟷ O)

$$C_6H_5O^- + O=CH_2 \rightarrow (CH_2O^-,\ H) \rightarrow C_6H_4(OH)CH_2OH$$

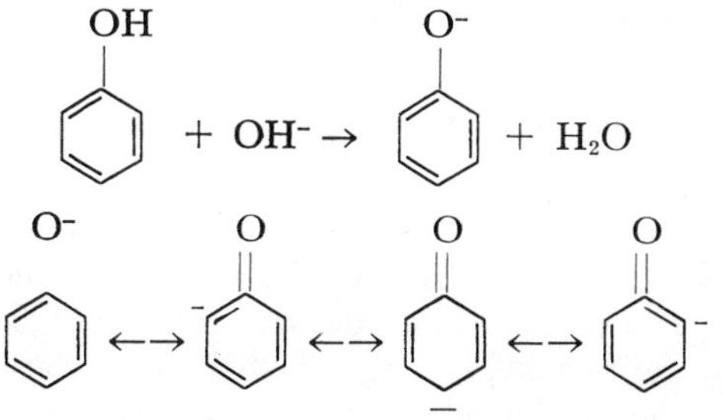

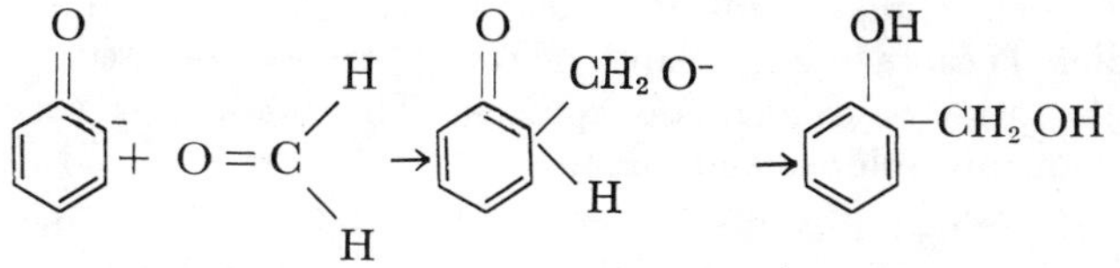

o—Methylolphenol

The next step is the condensation reaction between the methylolphenols with the elimination of water and the formation of fusible polymers.

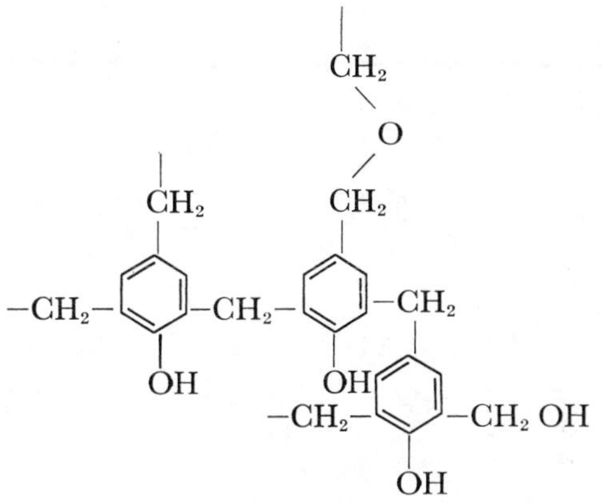

(structure: phenolic rings bearing OH linked by $-CH_2-$ bridges and a $-CH_2-O-CH_2-$ bridge, with terminal $-CH_2OH$)

Crosslinking occurs with the formation of bridges between the mono-, di-, and trimethylolphenols with the formation of a three dimensional network thermoset.

Two types of bridges have been identified in *resols*—methylene, $-CH_2-$, and ether, $-O-$, bridges. The methylene bridge is formed by the reaction between a methylol group and the phenoxide ion in the ortho and para positions. The ether bridge is formed by the condensation of two methylol groups. The curing of resols takes place by heating the mixture in neutral, basic or acidic medium where crosslinking takes place. Excess formaldehyde is used to allow for the needed crosslinking.

With *novalacs,* the acid catalyzed reaction leads to electrophilic substitution on the benzene ring in the ortho and para positions. The protonated formaldehyde adds to the benzene ring.

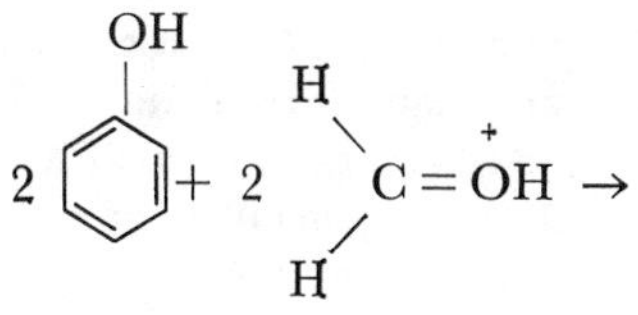

$$2\,C_6H_5OH + 2\,H_2C=\overset{+}{O}H \rightarrow$$

$$HO{-}C_6H_4{-}CH_2OH + HO{-}C_6H_4{-}CH_2OH + H^+$$

Polymerization which involves another electrophilic substitution takes place with the formation of methylene bridges. The ratio of formaldehyde to phenol is less than unity in the first stage of the reaction. This considerably reduces the likelihood of multifunctional units being formed. Unlike resols, these polymers are unable to crosslink because the available methylol groups are used in the primary condensation process. The product is permanently soluble and fusible.

Crosslinking is effected by molding the initial polymer with a small amount of *hexamine, hexamethylenetetraamine,* $(CH_2)_6N_4$. The hexamine decomposes in the presence of traces of moisture to give formaldehyde and ammonia. This results in the addition of methylol groups and subsequent crosslinking in the alkaline conditions provided by the ammonia and the formation of a thermoset resin.

$$-C_6H_3(OH){-}CH_2{-}C_6H_3(OH){-}CH_2{-}C_6H_3(OH){-} + (CH_2)_6N_4 \rightarrow$$

$$-C_6H_2(OH)(CH_2{-}NH{-}CH_2{-}){-}CH_2{-}C_6H_2(OH)(CH_2{-}NH{-}CH_2{-}){-}CH_2{-}C_6H_3(OH){-}$$
$$\quad -C_6H_2(OH){-}CH_2{-}C_6H_3(OH){-}$$

Phenolic resins with special properties are made by using other phenols such as cresols and resorcinol. Furfural also can be used.

Properties and Uses. The most important properties of phenolic thermosets are their hardness and rigidity and their acid and water hydrolysis resistance. They have excellent insulating properties and can be used continuously up to 150°C.

Phenolic resins are among the least expensive and easiest to mold. There are many formulations with various fillers and additives. The fillers include wood flour, oils and glass fibers. Fiber glass piping made with phenolic resins can operate up to 150°C and at pressures up to 150 psi.[17]

Molding compounds are used for producing switch gears and controls, handles, appliances, etc. Recently compression-molded, glass-filled phenolic disk brake pistons are replacing the steel ones in some automobiles because of their light weight and corrosion resistance. They are more economical than steel disks. Phenolics are also used as bonding, adhesive, and insulating materials, and as laminates for building, furniture, panels and automobile parts.

Although phenolic resins have seen little progress in new applications, they have never stopped growing in volume demand. The consumption pattern for the 1.25 billion pound demand in 1978 is shown in Fig. 13-7.[16] In 1979 the demand was 1.33 million pounds.

Epoxy Resins. These are general purpose polymers made by the reaction between a diphenol and *epichlorohydrin,* $CH_2{-}CH{-}CH_2Cl$ (with O bridging the first two carbons). Most commercial epoxy resins are made using *bisphenol* A, *2, 2′-bis (4-hydroxyphenyl) propane,* as the diphenol. Bisphenol A is the product of the reaction of two moles of phenol with one mole of acetone.

$$2\,C_6H_5OH + CH_3{-}\overset{O}{\overset{\|}{C}}{-}CH_3 \xrightarrow{HCl} HO{-}C_6H_4{-}C(CH_3)_2{-}C_6H_4{-}OH + H_2O$$

Bisphenol A is an important chemical intermediate which is used in the production of various plastics such as polycarbonates, polysulfones and epoxides. Epichlorohydrin is produced from allyl chloride.

The reaction between epichlorohydrin and bisphenol A takes place in a strong basic medium such as sodium

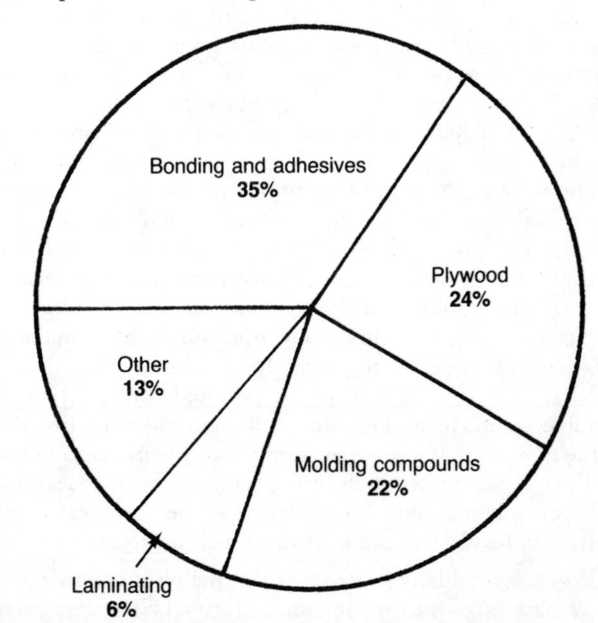

Fig. 13-7—The use pattern for the 1.25 billion 1978 demand for phenolic plastics.[16]

hydroxide. A nucleophilic addition between the phenoxide and the epichlorohydrin takes place which opens the epoxide ring. This is followed by chloride ion displacement by the internal alkoxy group and closure of the ring. The epoxy ring is again attacked by the phenoxide and the opening of the epoxy ring. The alkoxide is protonated, reacts again with another epichlorohydrin and the reaction continues until the desired polymer is obtained. The overall reaction produces a long condensation epoxide resin.

$$HO-C_6H_4-C(CH_3)_2-C_6H_4-OH + CH_2-CH-CH_2Cl\ (\text{epoxide ring, O bridging } CH_2-CH) \rightarrow$$

$$\overset{O}{CH_2-CH}-CH_2\left[O-C_6H_4-C(CH_3)_2-C_6H_4-OCH_2-CH(OH)-CH_2O\right]C_6H_4-C(CH_3)_2-C_6H_4-OCH_2-\overset{O}{CH-CH_2}$$

This linear polymer is cured by crosslinking either through the hydroxyl group using an acid anhydride or polyanhydrides, or through the terminal epoxide linkage by use of an amine. Lewis acids are sometimes used for crosslinking.

Cresols and other bisphenols also are used to prepare epoxy resins.

Properties and Uses. Epoxy novalacs are made by the reaction between epichlorohydrin and low molecular weight novalacs. They have superior thermal properties and better chemical resistance than ordinary epoxies.[18] Epoxy resins cured with anhydrides are more suitable for high temperature applications than those epoxy resins cured with amines, especially aliphatic amines. The curing temperature is above 20°C but ambient temperatures are frequently used.

The molecular weight of epoxy resins vary in a wide range (*ca* 300-10,000). The most important properties are the ability to withstand temperatures up to 500°C and their strong adherence to metal surfaces and their good resistance to chemicals and their dimensional stability.

Cured epoxy has low shrinkage properties and good strength to make them more suitable than other thermosetting polymers with respect to situations where cracking may be a problem. Improvement of the stress cracking properties and internal fracturing at vital locations in potted electrical units is made by the use of toughening agents. Carboxyl-terminated butadiene acrylonitrile liquid polymer, **CTBN,** is added to epoxy resins for this purpose. The carboxyl group reacts with the terminal epoxy ring to form an ester. The ester with its pendant hydroxyl group reacts with the remaining epoxide rings, then more crosslinking takes place by forming ether linkages. This material has electrical and thermal properties similar to epoxy resins but is more tough and is suitable for encapsulating electrical units.[19]

Epoxy resins are used in surface coatings for appliance finishes, can and drum coatings, auto primers and as an adhesive. Epoxies also find use in molding and laminating in the reinforced plastics field, for construction, for filament wound tanks and electrical laminates. Fig. 13-8 gives the use distribution for the 285 million pound demand for epoxy resins in 1978.[20] In 1979 the demand was 314 million pounds.

ENGINEERING RESINS

Nylon Resins. Nylon is the general name for a number of condensation polymers containing an amide, $-\overset{O}{\overset{\|}{C}}-NH-$, linkage. Nylons are more important as fibers than as plastics, at least in respect to tonnage use. Nylons for fiber production and their properties are noted in Chapter 14. The important thermoplastic nylons are nylon 6 and nylon 66. *Nylon 6* is produced by polymerization of *caprolactam*, $CH_2-(CH_2)_4-C=O$ (ring closed through NH), and *nylon 66* by condensation polymerization of *hexamethylenediamine*, $H_2N-(CH_2)_6-NH_2$, and *adipic acid*, $HO-\overset{O}{\overset{\|}{C}}-(CH_2)_4-\overset{O}{\overset{\|}{C}}-OH$. *Nylon 11* is produced by the condensation of *11 aminoecanoic acid*, $H_2N-(CH_2)_{10}-\overset{O}{\overset{\|}{C}}-OH$.

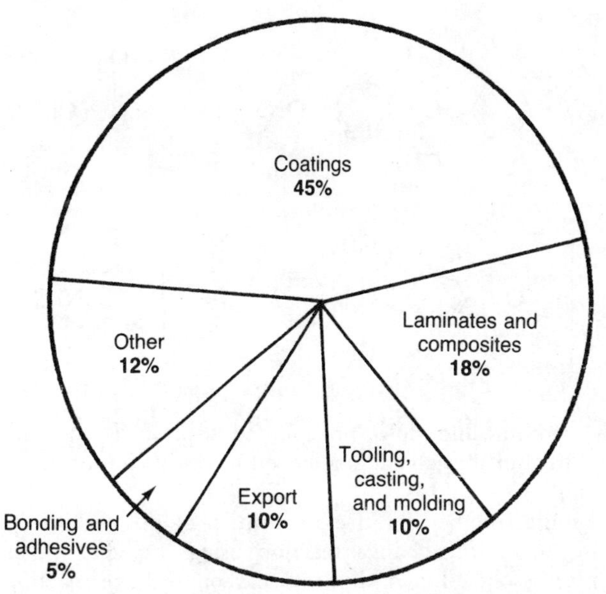

Fig. 13-8—The end-use pattern for the 285 million pound demand for epoxide resins in 1978.[20]

Properties and Uses. The important properties of nylons that make them excellent engineering thermoplastics are their high tensile and impact strength, toughness, good abrasion and wear resistance, and they are easily processed. Nylons can be reinforced with either glass or minerals to increase creep and dimensional stability. Because reinforced nylons have higher tensile and impact strength, and lower expansion coefficient than metals, they are used to replace metals in a wide variety of applications.

Because of the excellent resistance to chemicals and its high impact resistance and tensile strength, glass filled nylon has been used recently for making four-part housing and six-inch diameter motor assemblies for commercial air conditioning. This application allows a 50 percent price reduction.

Minerals used for reinforcement include calcium silicate, aluminum silicate, calcium carbonate, and mica. The mineral reinforced nylons with their low warpage are used extensively in automobiles for exterior appearance parts such as spoiler and spoiler extensions, hood scoop, etc. Minlon C (DuPont) is 40 percent mineral reinforced nylon 66 and is said to offer useful combinations of high impact, toughness, stiffness and temperature resistance.[21]

Nylon can be blow and injection molded. Molded gears made from nylon resins are self-lubricating and are used in places that are difficult to lubricate. Other uses are in bearings, bushings, and molded automobile parts and machinery parts. Extruded films are used for pharmaceutical packaging. Extruded nylons are also used for tubing and wire and cable insulation.

Nylon is the top thermoplastic engineering resin.

Polycarbonates. *PC* are thermoplastics which were developed separately in 1957 by Bayer in Germany and General Electric in the United States. Chemically, polycarbonates are polyesters of carbonic acid, $HO-\overset{\overset{\displaystyle O}{\|}}{C}-OH$, and are produced by condensation polymerization of the sodium salt of bisphenol A and phosgene in the presence of an organic solvent.

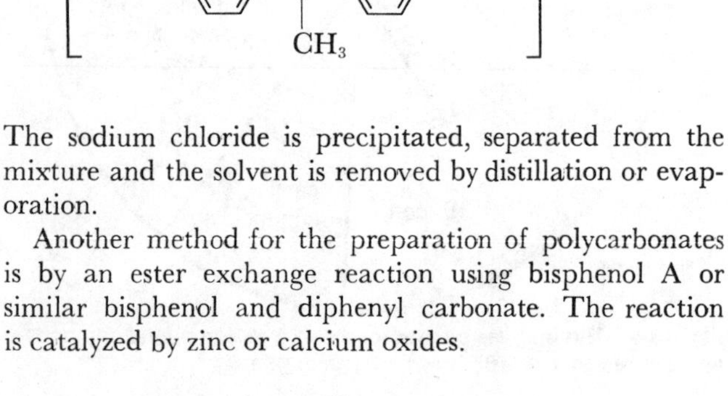

$$NaO-C_6H_4-C(CH_3)_2-C_6H_4-ONa + Cl-\overset{\overset{\displaystyle O}{\|}}{C}-Cl \rightarrow$$

$$\left[-O-C_6H_4-C(CH_3)_2-C_6H_4-\overset{\overset{\displaystyle O}{\|}}{C}-\right] + NaCl$$

The sodium chloride is precipitated, separated from the mixture and the solvent is removed by distillation or evaporation.

Another method for the preparation of polycarbonates is by an ester exchange reaction using bisphenol A or similar bisphenol and diphenyl carbonate. The reaction is catalyzed by zinc or calcium oxides.

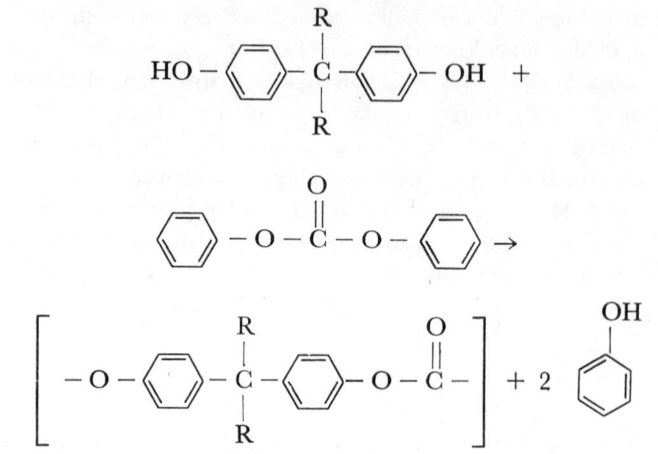

$$HO-C_6H_4-CR_2-C_6H_4-OH +$$

$$C_6H_5-O-\overset{\overset{\displaystyle O}{\|}}{C}-O-C_6H_5 \rightarrow$$

$$\left[-O-C_6H_4-CR_2-C_6H_4-O-\overset{\overset{\displaystyle O}{\|}}{C}-\right] + 2\ C_6H_5OH$$

The byproduct phenol is not readily removed from the reaction mixture and side reactions may take place. In order to produce a water-white polycarbonate, a highly purified, substantially colorless bisphenol-A is required.

Properties and Uses. Materials made from polycarbonates are transparent and are break- and heat-resistant. They are resistant to light, water, oxidation, fats, and salts. They are attacked by weak alkalies and acids and are soluble in many solvents. Polycarbonates are very tough over a wide range of temperatures. They are thermally stable and can be molded at temperatures as high as 550°C. They are rated as self-extinguishing and have good electrical properties.

The polymers can be injected, blow molded, and extruded. They are especially suitable for extrusion of intricate and exacting profile shapes.[22] Polycarbonates are used in a variety of articles such as laboratory safety shields, street lighting globes, safety helmets, sunglasses, solar heat collectors, and school windows. They can replace elements in some applications—gears and bushings, for example. The maximum continuous usage temperature is 125°C. A wide range of melting points is available, depending upon the alkyl groups in the bisphenol.

Growth rate for polycarbonates is around 10-5 percent and the forecast for 1980 is 200 million pounds of polycarbonates. However, the demand in 1978 was already 190 million pounds.[16]

Polyvinyl Acetate. Vinyl acetate, $CH_2=CH-O-\overset{\overset{\displaystyle O}{\|}}{C}-CH_3$, is produced from ethylene and acetic acid by a catalytic oxidation process. Essentially all of this production (95 percent) is polymerized. Both suspension and emulsion processes are used, and the reaction is initiated by a free radical initiator, usually an organic peroxide. The polymer is highly branched, amorphous and atactic.

Polyvinyl acetate is colorless, odorless and nontoxic. It is not a big volume plastic with a 1979 production of 1.86 billion pounds.[23] Polyvinyl acetate cannot be used for molded plastics because of its low molding temperature. It is used in a group of specialty products including adhesives (25 percent) and coatings (25 percent) and in textile treating (10 percent). In the area of textile and non-woven emulsions, vinyl acetate is predicted to have 55 percent of the market by 1987.[24] Fig. 13-9 shows polyvinyl acetate's share of various

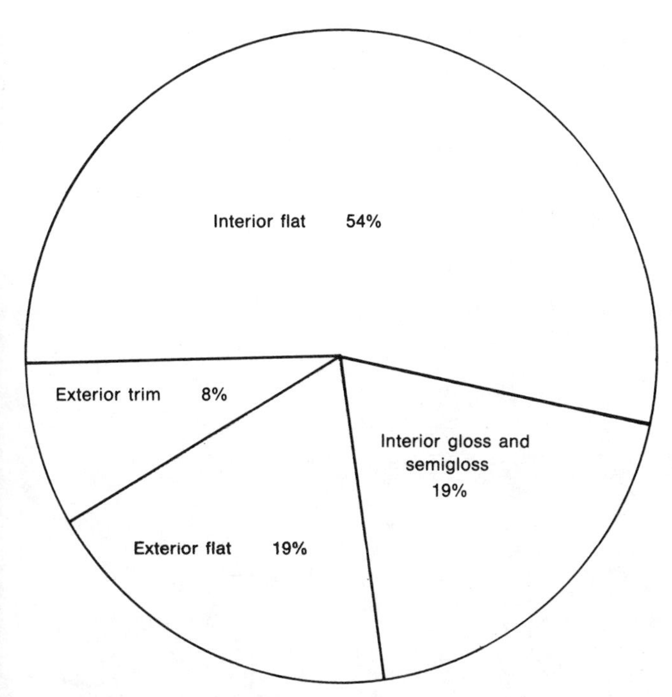

Fig. 13-9—The predicted polyvinyl acetate share of the 1987 paint market.[24]

types on paint.[24] Vinyl acetate's use as a co-polymer accounts for about 15 percent of current production.

An important use of polyvinyl acetate (20 percent) is the production of polyvinyl alcohol which is made by the hydration of the acetate. Uses of polyvinyl alcohol are similar to those of polyvinyl acetate.

Aromatic Polyethersulfones. PES are a class of engineering thermoplastics generally used for objects which have continuous use at temperatures around 200°C and which also can be used at low temperatures with no change in their physical properties. The desired flexiblity and mobility of the polymer chain are obtained by the introduction of ether groups.

Polyethersulfones can be prepared by using the sodium or potassium salt of bisphenol A and 4,4′-*dichlorodiphenylsulfone,* $Cl - C_6H_4 - SO_2 - C_6H_4 - Cl$. Bisphenol A acts as a nucleophile in the presence of the deactivated aromatic ring of the dichlorophenylsulfone.

$$KO-C_6H_4-C(CH_3)_2-C_6H_4-OK + Cl-C_6H_4-SO_2-C_6H_4-Cl \rightarrow$$

$$\left[-C_6H_4-C(CH_3)_2-C_6H_4-O-C_6H_4-SO_2-C_6H_4-O-\right] + 2\ KCl$$

Friedel Crafts catalysts may be used and the reaction is an electrophilic attack by the dichlorophenylsulfone on the bisphenol A. Only catalytic amounts of catalyst are required.

One monomer may be used for preparing polyethersulfones.

$$nCL-SO_2-C_6H_4-O-C_6H_4 \rightarrow$$

$$\left[-C_6H_4-O-C_6H_4-SO_2-\right]_n + HCl$$

Copolymers of polyethersulfones with various ratios of monomers have different properties and thermal resistance. In general the properties of polyethersulfones are similar to those of polycarbonates, but they can be used at higher temperatures. Fig. 13-10 shows the maximum use temperature for several engineering thermoplastics in comparison with aromatic polyethersulfones.[25]

Aromatic polyethersulfones are resistant to alkalies, acid, and inorganic salts, but are not resistant to highly polar and chlorinated hydrocarbons. These polysulfones are sometimes used to replace metal because of their high tensile strength which is more than the tensile strength of polycarbonates and nylons. Aromatic polyethersulfones can be extruded into thin films and foils and injection molded into various objects which need high temperature stability.

Polyacetals. Polyacetal, —CH_2—, is produced by the polymerization of formaldehyde under carefully controlled conditions. The polymer is actually a polyether but by convention the designation "polyether" is reserved for polymers with two or more —CH_2— groups between the other group. For example, *polyethylene oxide,* —O—CH_2—CH_2—O—, is a polyether, which is also called polyethylene glycol because of the terminal —OH groups.

A typical process for the production of polyformaldehyde consists of passing anhydrous formaldehyde into an inert liquid hydrocarbon at a low temperature. A continuous process is shown in Fig. 13-11.[26]

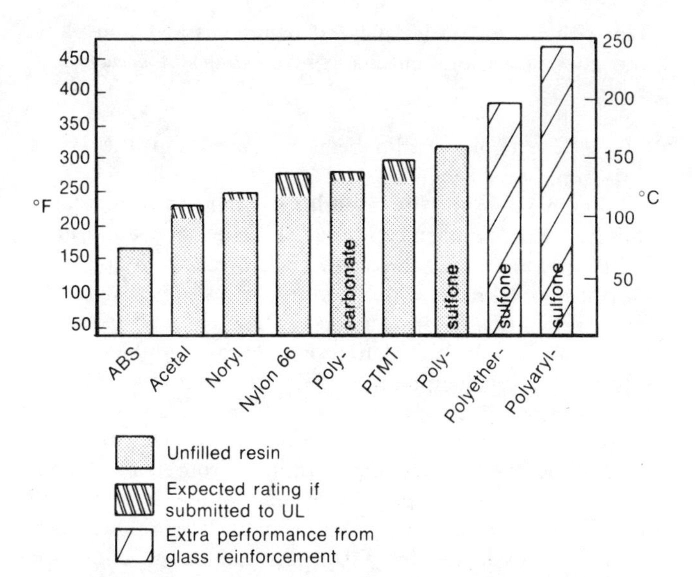

Fig. 13-10—Maximum continuous use temperatures of some engineering thermoplastics based on Underwriters Laboratories ratings.[25]

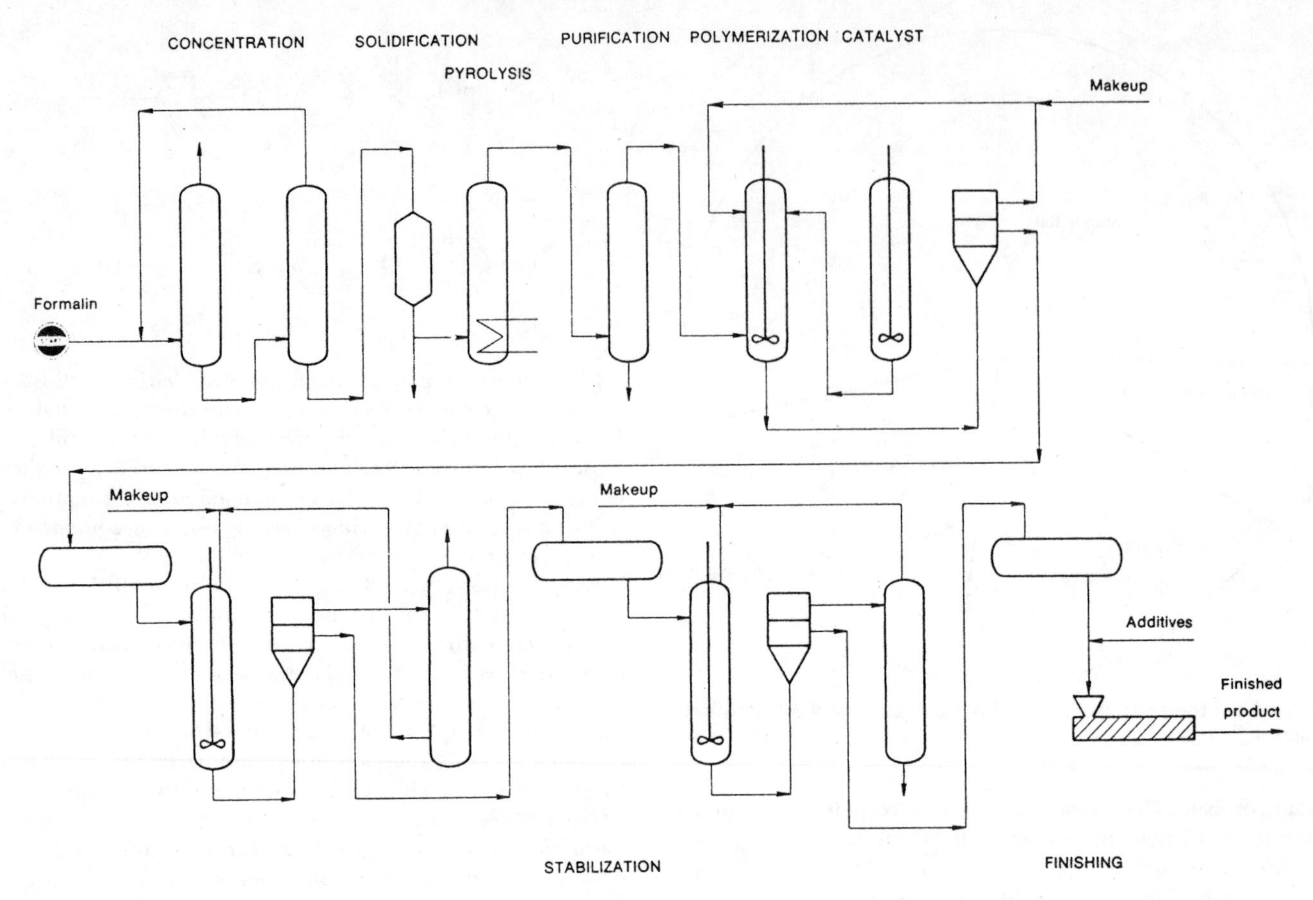

Fig. 13-11—A flow diagram for the continuous production of polyformaldehyde acetal.[26]

$$H-\overset{\overset{\displaystyle O}{\|}}{C}-H \rightarrow \left[-O-CH_2-O-CH_2-\right]$$

This polymerization is catalyzed by amines and phosphine and goes by an ionic mechanism. Polymers of *acetaldehyde*, $CH_3-\overset{\overset{\displaystyle O}{\|}}{C}-H$, have also been produced but they have little commercial value.

Polyacetals are highly crystalline with good dimensional stability, high impact resistance, high fatigue endurance limits, great strength and a low coefficient of friction. The polymers are linear with good wear resistance and low stress relaxation characteristics. They are also very resistant to chemicals and hydrolysis and insoluble in any solvent at room temperature. They are attacked by strong acids and oxidizing agents.

Articles made from acetals are numerous and vary from door handles to gears and bushings, from aerosol containers to carburetor parts. Outsert molding is used with acetals when metals are the main materials in the outsert with the acetal used for hubs, pins and bosses. This type of molding is used for electronic equipment, business machines, appliances, etc. Molded grades account for about 80 percent of acetal production with extrusion grades accounting for 8 percent and special grades 12 percent.

Structural foam acetals are primarily used in the automotive field for such things as foam-molded seat backs, and for lawn and garden and agricultural equipment.

The consumption of acetal resins was 1.32 million pounds in 1978 and the forecast is for a demand of 2.42 million pounds by 1983.[27]

Thermoplastic Polyesters. Polyethylene terephthalate, **PET,** is an important synthetic fiber, but it is also a thermoplastic material used as film for magnetic tape. Films made from **PET** have good abrasion and chemical resistance, low water absorption and low permeability for gases. Barrier bottle grade **PET** is used for carbonated beverage bottles which are becoming popular because of their light weight. The production of fiber grade **PET** is noted in Chapter 14.

Polybutylene Terephthalate. PBT is a new thermoplastic polymer that became commercially available around 1971. It is becoming an important engineering thermoplastic.[28] The polymer, mp 224°C, is produced by either bulk or solution condensation polymerization of *1,4-butanediol*, $HO-CH_2-(CH_2)_2-CH_2-OH$, and *terephthalic acid*, $HO-\overset{\overset{\displaystyle O}{\|}}{C}-C_6H_4-\overset{\overset{\displaystyle O}{\|}}{C}-OH$, or its methyl ester.

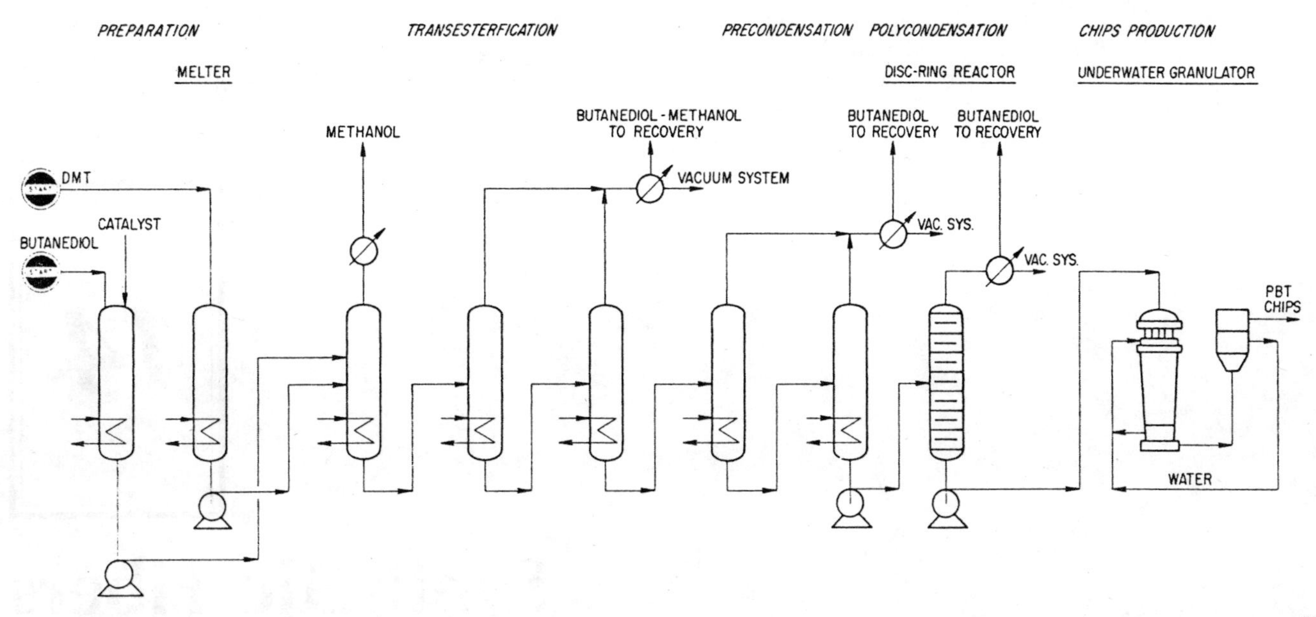

Fig. 13-12—A continuous process for production of polybutylene terephthalate, PBT, from 1,4-butanediol, BD, and dimethyl terephthalate, DMT.[29]

$$HO-CH_2-(CH_2)_2-CH_2-OH+HO-\overset{\overset{\displaystyle O}{\|}}{C}-C_6H_4-\overset{\overset{\displaystyle O}{\|}}{C}-OH \rightarrow$$

$$\left[-O-CH_2-(CH_2)_2-CH_2-O-\overset{\overset{\displaystyle O}{\|}}{C}-C_6H_4-\overset{\overset{\displaystyle O}{\|}}{C}-\right] + H_2O$$

A flow diagram for a continuous process is shown in Fig. 13-12.[29]

Thermoplastic polyesters are among the fastest growing engineering plastics with **PBT** being the one with the greatest growth rate. It is predicted that **PBT** will grow by 25 to 30 percent per year over the next five years.[28] Consumption in 1978 was *ca* 55 million pounds. Applications are now developing in electronics, sporting goods and other fields.

LITERATURE CITED

1. *Modern Plastics International,* Vol. 8, No. 7, 1978, pp. 58-59.
2. *Chemical Age,* Feb. 14, 1978, pp. 10-11, 19.
3. *Guide to Plastics,* New York: McGraw Hill Inc., 1976.
4. *Chemical Week,* Feb. 20, 1980, p. 54.
5. Patton, J.T., Jr. *CHEMTECH,* Vol 6, No. 11, 1976, p. 783; *Chemical and Engineering News,* Feb. 27, 1978, p. 11.
6. Hill, B.G. *CHEMTECH,* Vol. 3, No. 10, 1973, p. 613.
7. Davis, J.C. *Chemical Engineering,* June 9, 1975, p. 44.
8. "Urethane '78," Upjohn Polymer Chemicals.
9. *Modern Plastics International,* Vol. 9, No. 4, 1979, pp. 8-10.
10. *Modern Plastics International,* Vol. 8, No. 2, 1978, pp. 46-47.
11. *Chemical and Engineering News,* Jan. 3, 1975, p. 13.
12. Landau, Ralph, G.A. Sullivan and D. Brown. *CHEMTECH,* Vol. 9, No. 10, 1979, pp. 602-607.
13. *Chemical and Engineering News,* Oct. 15, 1979, pp. 12-13.
14. *Modern Plastics International,* Vol. 8, No. 1, 1978, p. 55.
15. Baekeland, L.H. *The Journal of Industrial and Engineering Chemistry,* March 1909; *CHEMTECH,* Vol. 6, No. 11, 1976, pp. 40-53.
16. *Modern Plastics International,* Vol. 9, No. 1, 1979, p. 51.
17. *Chemical Engineering,* Sept. 15, 1975, p. 106.
18. Doresy, J.S. *Chemical Engineering,* Vol. 82, No. 9, 1975, pp. 104-114.
19. Walker, J.M., W.E. Richardson and C.H. Smith. *Modern Plastics,* May 1976, p. 62.
20. Stinson, S.C. *Chemical and Engineering News,* Aug. 6, 1979, pp. 10-11.
21. *Modern Plastics International,* Vol. 8, No. 3, 1978, p. 28.
22. *Modern Plastics International,* Vol. 8, No. 11, 1978, pp. 52-54.
23. *Chemical and Engineering News,* Oct. 30, 1978, p. 13.
24. *Chemical and Engineering News,* Mar. 20, 1978, p. 11.
25. Leslie, V.J., J. Rose, G.O. Rudkin and J. Fitzin. *CHEMTECH,* Vol. 5, No. 5, 1975, pp. 426-432.
26. *Hydrocarbon Processing, 1979 Petrochemical Handbook,* Vol. 58, No. 11, p. 211.
27. *Modern Plastics International,* Vol. 8, No. 5, 1978, pp. 56-57.
28. *Modern Plastics International,* Vol. 8, No. 4, 1978, pp. 22-24.
29. *Hydrocarbon Processing, 1975 Petrochemical Handbook,* Vol. 54, No. 11, p. 179.

14

Synthetic Fibers

A polymer that is at least 100 times as long as it is wide is considered a *fiber.* Historically, the fibers for fabric production were linen, cotton and wool. The first two are from plants and are composed of cellulose, and the third is of nonplant origin and is protein in nature. Man-made fibers are derived either from plants, the *cellulosic* fibers, or from petroleum which provides the monomers from which the fiber polymers are produced. The fibers from petroleum-based monomers are called *synthetic fibers.* Man-made fibers account for 76 percent of the United States textile mill consumption, cotton 23 percent and all the others 1 percent.[1]

Rayon and cellophane are "regenerated cellulose." A better name would be "modified cellulose" because they have shorter chains than the original cellulose. This is caused by the degradation of the original cellulose by the alkali treatment used in rayon production. Other modified celluloses include cellulose nitrate, cellulose ethers, and cellulose acetate. The first two are used as plastics. Cellulose acetate is used as a fiber, *acetate rayon.* Rayon and acetate rayon have less than ten percent of the fiber market.

This chapter is limited to those fibers that are entirely synthetic. The most important synthetic fibers are polyesters, polyamides (nylons), polyacrylates and, to a lesser extent, polyolefins. Fig. 14-1 shows the estimated distribution of the 8.1 billion pound 1979 U.S. production of these synthetic fibers.[2] While the growth has been rapid in the past with a 9 percent growth in 1978 and an estimated 7 percent growth in 1979[3], it is predicted that this growth rate will slow to about 2 percent per year by 1990.[4] Current worldwide capacity utilization is predicted to be 29.9 billion pounds with the following distribution: polyesters 52 percent, polyamides 27 percent, acrylics 21 percent.[5] The overall distribution of the predicted 70.2 billion pound production of fibers is synthetics 42.6 percent and other 57.4 percent as shown in Fig. 14-2.[5] The Western Europe 1978 production of 5.0 billion pounds of synthetic fibers had the following distribution: polyester 35 percent, polyamides 32 percent, and acrylic 33 percent.[6]

The manufacture of synthetic fibers basically involves the physical conversion of a linear polymer with the polymer chains in a relatively disordered state into a low denier continuous filament with molecular chains parallel to the fiber's longitudinal axis.[7] The most important properties of polymers to be used for fiber production are:[8]

1. High melting point, preferable above 100°C
2. Linear and symmetrical structure and high molecular weight.

General physical and mechanical properties of some of the important synthetic fibers are given in Table 14-1.[9]

The two major processes used for fiber production are *melt spinning* and *solution spinning.* Melt spinning is

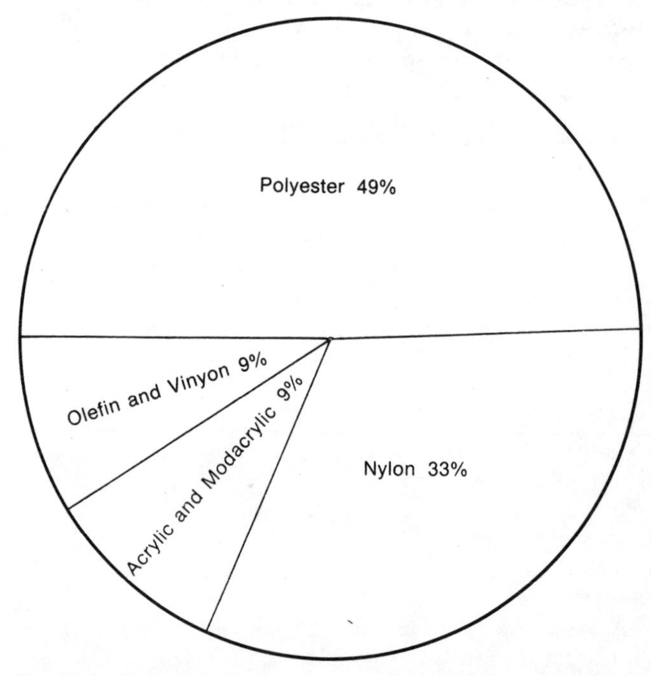

Fig. 14-1—Estimated distribution of the 8.1-billion-pound 1979 U.S. production of synthetic fibers.[2]

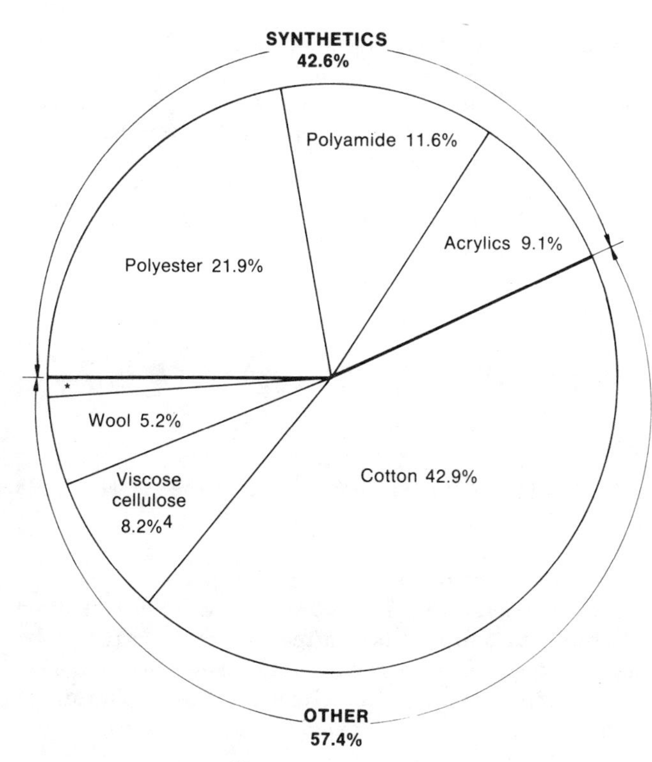

Fig. 14-2—The distribution of the predicted 1983 world-wide production of 70.2 billion pounds of fibers.[5]

TABLE 14-1—Properties of Various Synthetic Staple Fibers[9]

Property	Nylon 6	Polyester	PAN	Modacryl
Tensile strength, g/den				
dry	4.7—6.7	4.7—6.0	2.5—4.5	2.5—4.5
wet	3.9—5.7	4.7—6.0	2.0—4.5	2.0—4.5
Knot strength, g/den	3.7—5.5	4.0—5.0	2.0—3.5	2.0—3.5
Elongation, %				
dry	38—50	35—50	27—48	27—48
wet	40—53	35—50	27—48	27—48
Elastic recovery at 3% elongation in %	95—100	90—95	90—95	90—95
Apparent Young's modulus, kg/mm	80—250	310—620	260—650	260—650
Specific gravity	1.14	1.38	1.14—1.18	1.26—1.37
Water content, 20°C, 65% RH	3.5—5.0	0.480—0.500	1.2—2.0	0.5—0.8
Softening point °C	180	238—240	190—240	150
Melting point °C	215—220	255—260	not clear	210—240
Weatherability	Strength lowers, turns yellow	Almost no lowering of strength	No lowering of strength	No lowering of strength
Dispersibility	Dispersible, acid	Dispersible, high temperature or carrier of naphthol vat	Dispersible, basic acidic	Dispersible, basic acidic

used for polymers which can be melted into a highly viscous liquid and which do not decompose during processing. The melt is extruded through spinnaret holes, either individually to form monofilaments or in bundles to form multifilaments. The filaments are then mechanically stretched in order to provide them with the required tensile strength.

A solvent is used in solution spinning. After filtering and degassing the polymer solution, it is fed by use of a screw or gear pump through spinnarets while exposed to hot air to evaporate the solvent. The fibers are washed and subjected to mechanical stretching by passing them through rotating rollers.

Technical developments in the processing of fiber polymers have created a new generation of fibers. Bulk and texture are improved by thermal or mechanical texturing using processing such as false twist or jet texturing. Combining polymers with different shrinkage characteristics give fabrics with attractive luster and feel. New developments also include the processing and dyeing of polypropylene fibers.

POLYESTER FIBERS

Carothers (1930) was the first to try to synthesize a fiber-forming aliphatic polyester.[10] These polymers were not suitable for synthetic fiber production because of their low melting points. However, he was successful in preparing the first synthetic fiber, a polyamide (Nylon 66). His procedure was followed by Whinfield and Dickson in the field of aromatic polyesters and their first patent for preparing polyethylene-terephthalate appeared in 1946.[11] This polyester is still the dominant linear homopolymer polyester for fiber production.

Commercial polyesters are condensation polymers and are produced by several methods. The most important are

1. The reaction between a diol, $HOCH_2-(CH_2)_n-CH_2OH$, and an ester of a dicarboxylic acid—*trans esterification.*

$$R'O-\overset{\overset{\displaystyle O}{\|}}{C}-R_n-\overset{\overset{\displaystyle O}{\|}}{C}-OR' + HOCH_2-(CH_2)_n-CH_2OH \rightarrow \left[-\overset{\overset{\displaystyle O}{\|}}{C}-R_n-\overset{\overset{\displaystyle O}{\|}}{C}-OCH_2-(CH_2)_n-CH_2O-\right] + 2R'OH$$

2. The reaction between a diol and a dicarboxylic acid —*esterification.*

$$HO-\overset{\overset{\displaystyle O}{\|}}{C}-R_n-\overset{\overset{\displaystyle O}{\|}}{C}-OH + HOCH_2-(CH_2)_n-CH_2OH \rightarrow \left[-\overset{\overset{\displaystyle O}{\|}}{C}-R_n-\overset{\overset{\displaystyle O}{\|}}{C}-OCH_2-(CH_2)_n-CH_2O-\right] + 2H_2O$$

3. The reaction between a diol and an acid dichloride.

$$Cl-\overset{\overset{\displaystyle O}{\|}}{C}-R_n-\overset{\overset{\displaystyle O}{\|}}{C}-Cl + HOCH_2-(CH_2)_n-CH_2OH \rightarrow \left[-\overset{\overset{\displaystyle O}{\|}}{C}-R_n-\overset{\overset{\displaystyle O}{\|}}{C}-OCH_2-(CH_2)_n-CH_2O-\right] + 2HCl$$

Less important methods are self-condensation of omega-(ω)-hydroxy acids, and the ring opening of lactones and cyclic esters. In the self-condensation of ω-hydroxy acids, cyclization might compete seriously with linear polymerization, especially when the hydroxyl group is in a position to give five or six membered ring lactones.

Production

Polyethyleneterephthalate, (PET), (Dacron), (Terylene) is prepared either by the reaction between dimethyl terephthalate (**DMT**) and ethylene glycol (**EG**) (transesterification) or between terephthalic acid and ethylene glycol (esterification). To stabilize and terminate the polymer, excess glycol is used so that the two end groups in the polymer are —OH. The presence of excess diol also helps in driving the reaction to completion. An alternative is to use monofunctional terminators. Both transesterification and esterification take place in two steps which occur at different temperatures. In the first step, the temperature is moderate (150-200°C) and either water or an alcohol is released, depending upon whether the monomer is the diacid or the dimethyl ester.

Transesterification

$$CH_3O-\overset{O}{\overset{\|}{C}}-C_6H_4-\overset{O}{\overset{\|}{C}}-OCH_3+HOCH_2CH_2OH\rightarrow$$

$$HO(CH_2)_2-O-\overset{O}{\overset{\|}{C}}-C_6H_4-\overset{O}{\overset{\|}{C}}-O(CH_2)_2OH+2\ CH_3OH$$

Esterification

$$HO-\overset{O}{\overset{\|}{C}}-C_6H_4-\overset{O}{\overset{\|}{C}}-OH+HOCH_2CH_2OH\rightarrow$$

$$HO(CH_2)_2O-\overset{O}{\overset{\|}{C}}-C_6H_4-\overset{O}{\overset{\|}{C}}-O(CH_2)_2OH+2H_2O$$

With diacids the reaction is self-catalyzed, but acid catalysts are used to compensate for decrease in diacid concentration as the esterification nears completion. The yield of the diglycol terephthalate [bis(2-hydroxyethyl) terephthalate] is increased by removal of released water or alcohol to shift the reaction to the right. Fig. 14-3 is a flow diagram of a two-step batch process for the production of PET from ethylene glycol and either terephthalic acid or dimethylterephthalate.[12]

The extent of polymerization is monitored by the melt viscosity which indicates the average molecular weight of the polymer. In self-catalyzed polycondensations, the reaction is third order corresponding to rate = k $[COOH]^2$ [OH], especially over the later stages, but when the reaction is catalyzed with strong acids, the reaction is second order and corresponds to rate = k[COOH][OH].

In addition to the catalysts and the terminator, certain other additives are used, such as color improvers and dulling agents. For example, fiber grade **PET** is delustered by the addition of titanium dioxide.

Batch polymerization is still widely used, but the use of continuous polymerization and direct spinning is increasing. The production of high tenacity industrial yarns such as tire cords with high molecular weight cannot be manufactured in batch processes due to competitive reactions of thermal degradation. It has recently been noted that there is a worldwide trend to use purified terephthalic acid instead of dimethyl terephthalate intermediate.[13]

The reaction between fiber grade terephthalic acid, **TPA**, and ethylene oxide, **EO**, is also used to prepare bis(2-hydroxyethyl) terephthalate.

$$HO-\overset{O}{\overset{\|}{C}}-C_6H_4-\overset{O}{\overset{\|}{C}}-OH+2\overset{\quad O}{CH_2-CH_2}\rightarrow$$

$$HO-(CH_2)_2-O-\overset{O}{\overset{\|}{C}}-C_6H_4-\overset{O}{\overset{\|}{C}}-O-(CH_2)_2-OH$$

The second step or polymerization step takes place at a higher temperature (270-300° C) and under reduced pressure (0.1 mm). At first, a reaction between two molecules of the diglycol terephthalate takes place and produces a dimer and ethylene glycol is released. This dimer further reacts with a monomer to produce a trimer and on to form a polymer.

$$HO-(CH_2)_2-O-\overset{O}{\overset{\|}{C}}-C_6H_4-\overset{O}{\overset{\|}{C}}-O-(CH_2)_2OH$$

$$+HO-(CH_2)_2-O-\overset{O}{\overset{\|}{C}}-C_6H_4-\overset{O}{\overset{\|}{C}}-O-(CH_2)_2-OH$$

$$HO-(CH_2)_2-O-\overset{O}{\overset{\|}{C}}-C_6H_4-\overset{O}{\overset{\|}{C}}-O-(CH_2)_2-O-\overset{O}{\overset{\|}{C}}-C_6H_4-\overset{O}{\overset{\|}{C}}-O-(CH_2)_2-OH+HO-CH_2-CH_2-OH$$

$$\left[-O-CH_2-CH_2-O-\overset{O}{\overset{\|}{C}}-C_6H_4-\overset{O}{\overset{\|}{C}}-\right]$$

Polymer

Properties and Uses

Polyester fibers contain crystalline and non-crystalline regions and X-ray diffraction studies show that the unit cell is made of one repeating unit.[14] The degree of crystallinity and molecular orientation is important in determining the tensile strength of the fiber (between 18-22 denier) and its shrinkage ability. The degree of crystallinity and molecular orientation can be obtained by use of X-ray diffraction and optical methods.[15] Another important property of polyester fibers is the ability to be blended with natural fibers such as cotton and wool. **PET** staple fiber is blended with cotton in 50:50 or 65:35 ratios of **PET** to cotton. Polyester fibers resist environ-

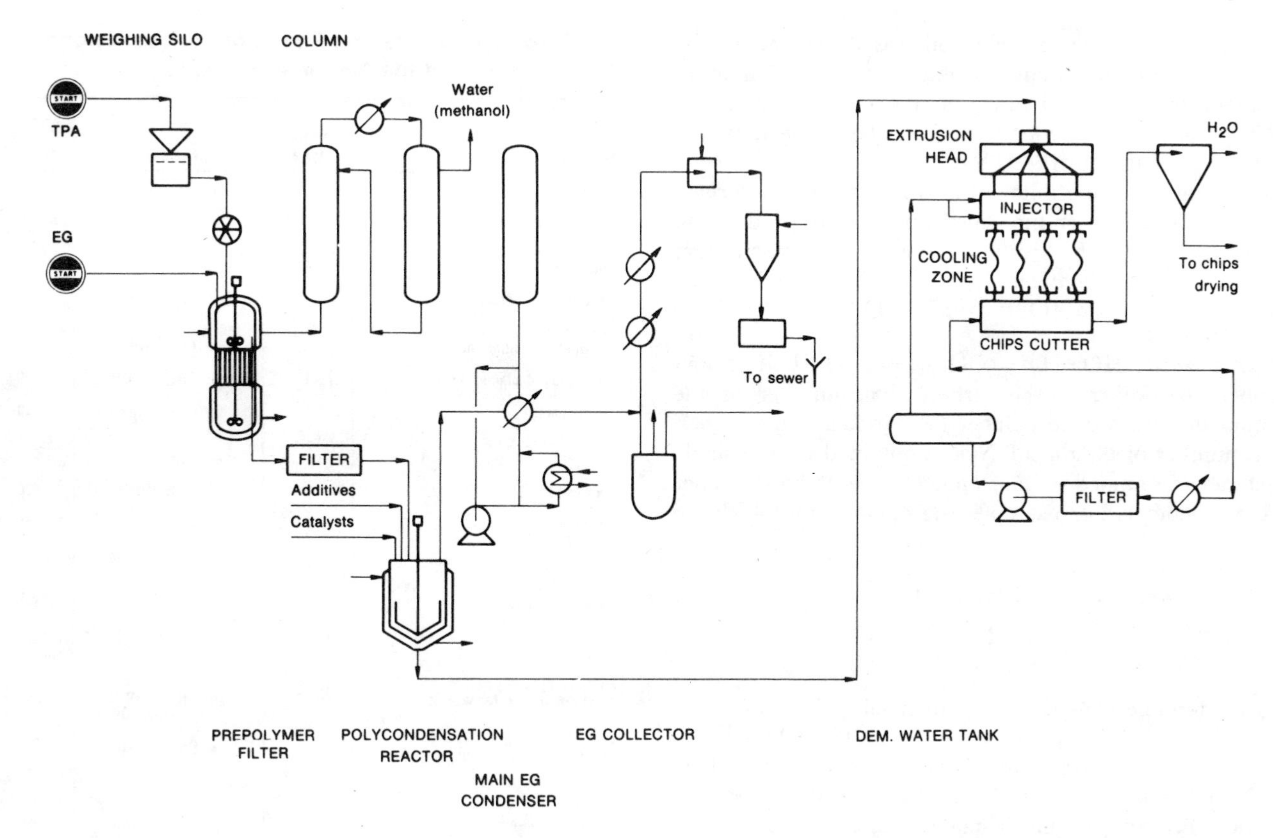

Fig. 14-3—Process for the production of polyethyleneterephthalate, *PET*, starting with ethylene glycol, *EG*, and terephthalic acid, *TPA*, or dimethylterephthalate, *DMT*.[12]

mental conditions well. Resistance to sunlight is exceeded only by acrylics.

Polyesters have a relatively high melting point (Table 14-2) and for this reason melt spinning is preferred since it is less expensive than solution spinning processes. Polyester fiber is strong, durable and easy to process, but ionic dyes are hard to penetrate because of the hydrophobic property of the fiber's surface. Linear chain diacids such as adipic acid are sometimes added in small quantities. These diacids copolymerize and disturb the fiber crystallinity and dye penetration is thus improved. An alternative approach is the use of sulfonated terephthalic acid residue which provides anionic dye sites for cationic dyes.

Important uses of polyester fibers are durable press varieties blended with cotton and rayon for men's and women's wear, pillow cases, bed spreads, and broadloom carpeting, etc. High tenacity filament for tire cord reinforcement is increasingly used. Cords are equivalent in strength to nylon tire cords and are superior because they do not flat spot. V-belts and fire hose are also made from industrial filaments.

Double knits of spun yarn have a tendency to bag, but this is overcome by texturing the yarn and then using it in apparel manufacture. Fiber fill made from polyester is used in mattresses, pillows, sleeping bags, etc; deep pile with a fur-like appearance is used increasingly for coats and jackets, bath mats, soft toys, etc. Polyester films are used for electrical insulation, audio tapes, vacuum packing of foodstuffs, etc.

The major uses of polyester fiber are apparel (61 percent), home furnishings (17 percent), and tire cord (8 percent).[16]

TABLE 14-2—Physical Properties of Polyester Fibers[14]
(Mechanical properties at 21° C, 65% relative humidity, using 60% minute strain rate)

	Polyethyleneterephthalate PET				Poly (cyclohexanedimethylene terephthalate) staple and tow
	Filament yarns		Staple and tow		
Property	Regular tensile strength	High tensile strength	Regular tensile strength	High tensile strength	
Breaking strength (g/denier)*	2.8—5.6	6.0—9.5	2.2—6.0	5.8—6.0	2.5—3.0
Breaking elongation (%)	19—34	10—34	25—65	25—40	24—34
Initial modulus (g/denier)*	75—100	115—120	25—40	45—55	24—35
Elastic recovery (%)	88—93 (at 5% elongation)	90 (at 5% elongation)	75—85 (at 5% elongation)	....	85—95 (at 2% elongation)
Moisture regain (%)**	0.4	0.4	0.4	0.4	0.34
Specific gravity	1.38	1.38	1.38	1.38	1.22
Melting temperature (°C)	265	265	265	265	290

* Grams per denier-grams of force per denier. Denier is linear density, the mass for 9000 meters of fiber.
** The amount of moisture in the fiber at 21°C, 65 percent relative humidity.

POLYAMIDES

Polyamides are condensation polymers formed by the reaction between a dicarboxylic acid and a diamine (Nylons 66 and 610), a ring opening of lactams (Nylons 6 and 12), or by the polymerization of *omega* amino acids (Nylons 7 and 11).[17] Nylon 66 was the first synthetic fiber which appeared on the market (1940) with the culmination of research under the direction of Carother. Nylon 6 was developed in Germany by I. G. Farben at about the same time as Nylon 66 was developed in the U.S.A. Because of patent restrictions and raw material considerations, Nylon 6 has been the nylon most extensively produced in Europe and Nylon 66 is the most extensively produced in the U.S.A.

The number which follows the word "nylon" is derived from the monomer or monomers used in its production. For example, nylon 610 is the product of the condensation of *hexamethylenediamine,* $H_2N-(CH_2)_6-NH_2$, and *sebacic acid,* $HO-\overset{O}{\overset{\|}{C}}-(CH_2)_8-\overset{O}{\overset{\|}{C}}-OH$. The first number (6) indicates the carbon atom number of the diamine and the second number (10) indicates the carbon atom number of the diacid. Nylons obtained from a single monomer are named so as to indicate the carbon number of the monomer. For example, the monomer for Nylon 6 is *caprolactam,* $CH_2-(CH_2)_4-\overset{O}{\overset{\|}{C}}$ (ring closed through NH), with six carbon atoms. Monomers used for commercial polyamides, the formulas and melting points are given in Table 14-3.[17]

Nylons produced on a large commercial scale are nylons 66, 6, and 610. Nylons 4, 11 and 12 are also produced commercially but on a much smaller scale.

TABLE 14-3—Melting Points of Various Nylons and the Monomer Formula[17]

Monomer	Nylon type	Formula	Approximate M.P. °C
Caprolactam	6	$CH_2-(CH_2)_4C=O$ (ring through NH)	223
Adipic Acid	66	$HOOC-(CH_2)_4COOH$	265
Hexamethylenediamine		$H_2N-(CH_2)_6-NH_2$	
Sebacic Acid Hexamethylenediamine	610	$HOOC-(CH_2)_8COOH$ $H_2N-(CH_2)_6-NH_2$	215
11-Aminoundeclanoic Acid	11	$H_2N-(CH_2)_{10}-COOH$	190
Laurolactam	12	$CH_2-(CH_2)_{10}-C=O$ (ring through NH)	119
Acrylamide	3	$CH_2=CHCONH_2$	320
2-pyrrolidone	4	$CH_2-(CH_2)_2-C-O$ (ring through NH)	265
Valerolactam	5	$CH_2-(CH_2)_3-C=O$ (ring through NH)	260
7-aminoheptanoic Acid or Ethyl 7-aminoheptanoate	7	$N_2H-(CH_2)_6-COOH$ $H_2H-(CH_2)_6-COOC_2H_5$	233
Caprylactam	8	$CH_2-(CH_2)_6-C=O$ (ring through NH)	200
9-Aminopelargonic acid	9	$H_2N-(CH_2)_8-COOH$	209
10-Aminodecanoic acid	10	$H_2N-(CH_2)_9-COOH$	188

Production

Nylon 66 (polyhexamethyleneadipamide) is produced from monomers *adipic acid,* $HO-\overset{O}{\overset{\|}{C}}-(CH_2)_4-\overset{O}{\overset{\|}{C}}-OH$, and *hexamethylenediamine* (**HMDA**), $H_2N-(CH_2)_6-NH_2$, Adipic acid is produced from benzene by at least four different routes utilizing either cyclohexane or cyclohexanol oxidation. In the cyclohexane route either nitric acid or air oxidation is used whereas in the cyclohexanol route nitric acid is used as the oxidizing agent. Hexamethylenediamine is produced either by hydrogenating adiponitrile, produced from adipic acid, or from butadiene. More than 50 percent of **HMDA** produced is based on butadiene. Fig. 14-4 shows the various routes to adipic acid and hexamethylenediamine.[17]

Hexamethylenediammonium adipate salt is produced by the stoichiometric mixing of adipic acid and hexamethylenediamine. It is in the form of a dilute salt solution and is concentrated to 60 percent and charged with 0.5 to 1.0 mole percent acetic acid. The acetic acid is a monofunctional stabilizer. The reaction mixture is heated and water is continuously taken off which drives the polymerization to completion.[18] The temperature is then increased to approximately 270 to 300°C after which the pressure is raised to 250 psig. The pressure is finally reduced to atmospheric pressure to permit further water removal. After a total of 3 hours, nylon 66 is extruded under nitrogen pressure through a bottom valve of the autoclave.

Nylon 6 (polycaproamide) needs only the monomer caprolactam for production. Benzene, the primary petrochemical source of caprolactam, is first reduced to cyclohexane which is oxidized in turn to cyclohexanone. The cyclohexanone is either treated with *hydroxylamine sulfate,* $(H_2N-O)_2SO_4$, at *ca* 95°C to form the oxime which converts to caprolactam by a Beckman rearrangement catalyzed by an acid such as sulfuric acid (oleum)[19] or boric acid,[20] or changed to caprolactone by treatment with peracetic acid. The caprolactone is then treated with ammonia to produce caprolactam, Fig. 14-5.[17] Another route to cyclohexanone oxime is the treatment of cyclohexanone with nitrosylchloride, NOCI, under the influence of ultraviolet light. The oxime is then treated as previously with sulfuric or boric acids to give caprolactam.

Toluene is also a primary source for producing caprolactam. Toluene is oxidized to benzoic acid and then hydrogenated to cyclohexane carboxylic acid. The carboxylic acid is treated with nitrosylsulfuric acid, $NOHSO_4$, to give caprolactam directly without forming the oxime.[21] Fig. 14-5 shows the various routes for the production of caprolactam.[17] Ninety percent or more of the caprolactam production is used to produce nylon 6 fibers. Small amounts are used to produce resins for molded products, films, polish formulations and adhesives.

The polymerization of caprolactam starts by the addition of water which opens the ring to give omega amino acid.

$$\underbrace{CH_2-(CH_2)_4-C=O}_{\text{ring through NH}} + H_2O \rightarrow HO-\overset{O}{\overset{\|}{C}}-(CH_2)_5-NH_2$$

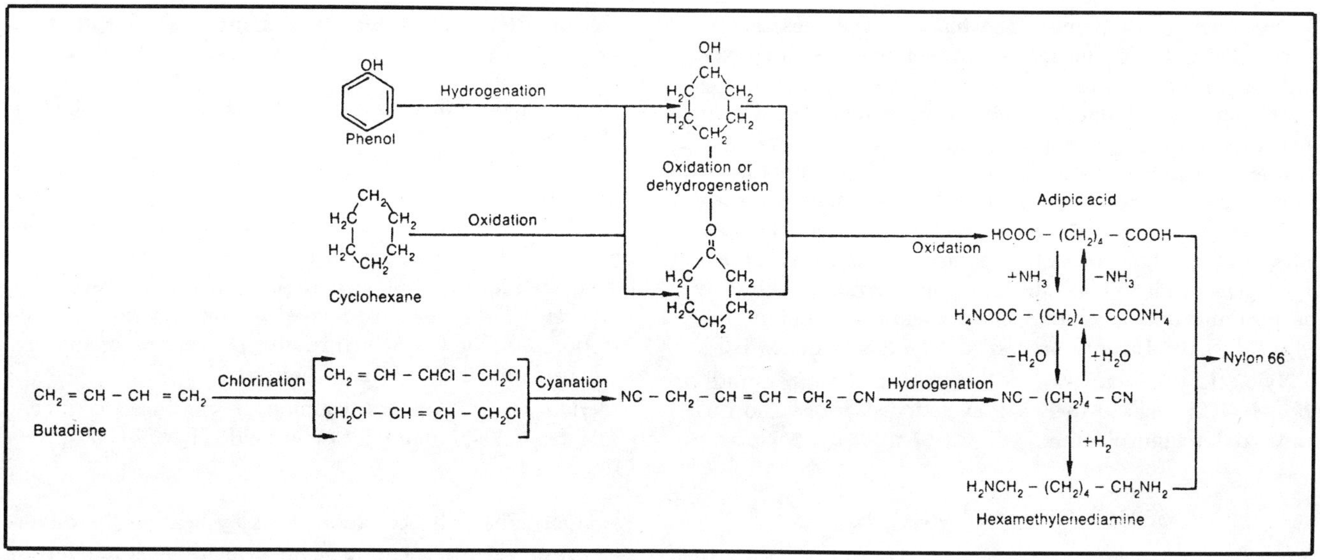

Phenol
Hydrogenation
Oxidation or dehydrogenation
Cyclohexane
Oxidation
Adipic acid
Oxidation
$HOOC - (CH_2)_4 - COOH$
$+NH_3$ $-NH_3$
$H_4NOOC - (CH_2)_4 - COONH_4$
$-H_2O$ $+H_2O$
Nylon 66
$CH_2 = CH - CH = CH_2$
Butadiene
Chlorination
$CH_2 = CH - CHCl - CH_2Cl$
$CH_2Cl - CH = CH - CH_2Cl$
Cyanation
$NC - CH_2 - CH = CH - CH_2 - CN$
Hydrogenation
$NC - (CH_2)_4 - CN$
$+H_2$
$H_2NCH_2 - (CH_2)_4 - CH_2NH_2$
Hexamethylenediamine

Fig. 14-4—Various routes to adipic acid and hexamethylene-diamine, monomers for nylon 66.[17]

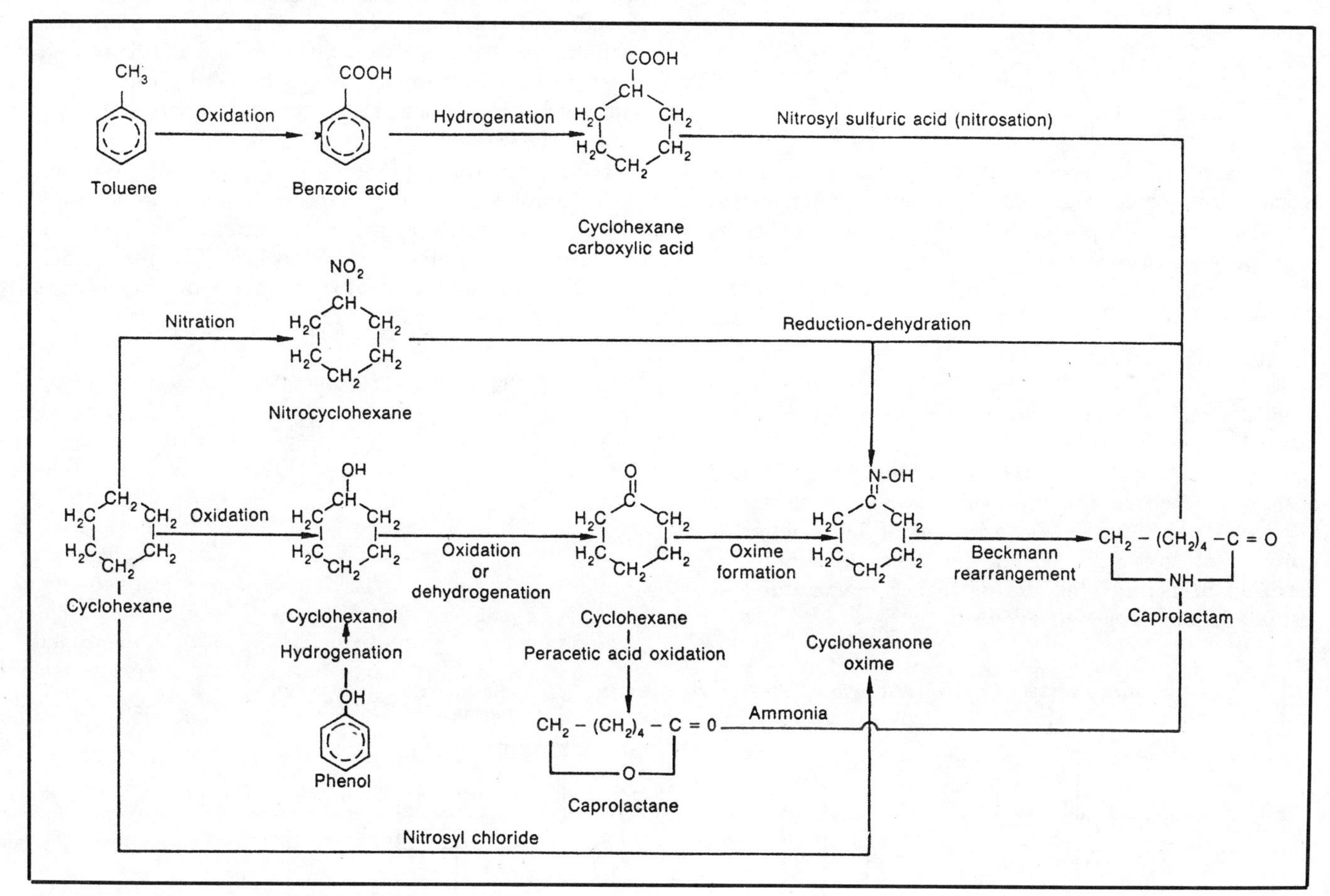

Fig. 14-5—Various routes to caprolactam.[17]

The amino acid is then polymerized at elevated temperatures (250-280° C). The polymer, which is in equilibrium with about 10 percent of the monomer, is continuously removed by washing with hot water. An extraction column is used where the polymer chips are in counter flow to the hot water extract. The free amino acid reacts with caprolactam and ring opening takes place.

$$HO-\overset{O}{\overset{\|}{C}}-(CH_2)_5-NH_2+\underbrace{CH_2-(CH_2)_4-C=O}_{NH}$$

$$\rightarrow HO-\overset{O}{\overset{\|}{C}}-(CH_2)_5-NH-\overset{O}{\overset{\|}{C}}-(CH_2)_5-NH_2$$

Some polymerization may take place by condensation of omega amino acid molecules between the free carboxyl and the amino groups.

The control of monomer and low molecular weight polymer content in the polycaprolactam resin is one of the most serious problems in nylon 6 production. Low temperatures favor high polymer homogeneity but reaction times are excessive. Activators (cocatalysts) such as water, acids, amine salts, and alcohols tend to stabilize the polymer at low molecular weights. The rate of depolymerization is directly proportional to reaction temperature and water content. Fig. 14-6 shows the Inventa process for producing nylon 6.

Nylon 12 (polylaurylamide) is produced from butadiene which is cyclized to 1, 5, 9-cyclododecatriene and this converted to laurolactam.

$$3CH_2 = CH - CH = CH_2 \rightarrow$$

```
     CH              CH2
   /    \\          /    \
H2C       CH - CH2        CH
 |                        ||
H2C       CH - CH2        CH
   \    //          \    /
     CH              CH2
```

This triene can be epoxidized with peracetic acid or acetaldehyde peracetate. The epoxide is hydrogenated and the saturated epoxide is rearranged to the ketone with magnesium iodide at 100° C.[24] The oxime of this ketone is then rearranged to *laurolactam*, $CH_2 - (CH_2)_{10} - C = O$ (with CH_2 and C joined through NH). Nylon 12 possesses a lower water absorption capacity than other nylons for the same reasons mentioned earlier.

The polycondensation of laurolactam can be carried out in the melt in the presence of water or salts that liberate water.[25] The polymerization of laurolactam is slower than for caprolactam but thermal stability of laurolactam permits the use of higher temperatures to assure adequate polymerization rate.

Nylon 610 is produced in a manner similar to that used for nylon 66 from sebacic acid, $HO - \overset{O}{\overset{||}{C}} - (CH_2)_8 - \overset{O}{\overset{||}{C}} - OH$, (source; castor oil) and hexamethylenediamine. Nylon 610 is not used extensively as a textile fiber but it has several properties which make it suitable for mono filaments to be used in brushes, sports equipment, etc. These uses are related to the low moisture absorption of nylon 610 which results in retention of stiffness and mechanical properties when wet.

Nylon 11 (polyundecanylamide) is produced commercially from 11-aminoundecanoic acid, $H_2N - (CH_2)_{10} - \overset{O}{\overset{||}{C}} - OH$. The polymerization starts by heating the monomer in a water suspension. The suspension is fed into a three-compartmented reaction column. The monomer melts in the first compartment, much of the water is removed, and polycondensation begins. The condensation continues in the second compartment and the final stage of polymerization is carried out in the third compartment. The third compartment also serves as a reservoir for the spinning process.

Nylon 4 (polybutyramide) is produced from the monomer 2-pyrrolidone by several routes. Three of these routes start with acrylonitrile, two with butrolactone, one with butadiene, and another with acetylene. The preparation from 2-pyrrolidone and acrylonitrile by one route is

$$H_2C = CH - CN + HCN \rightarrow$$

$$NC - CH_2 - CH_2 - CN \xrightarrow{H_2,\ H_2O}$$

```
CH2 - CH2
 |     |
CH2   C = O
   \ /
    N
    |
    H
```

2-pyrrolidone

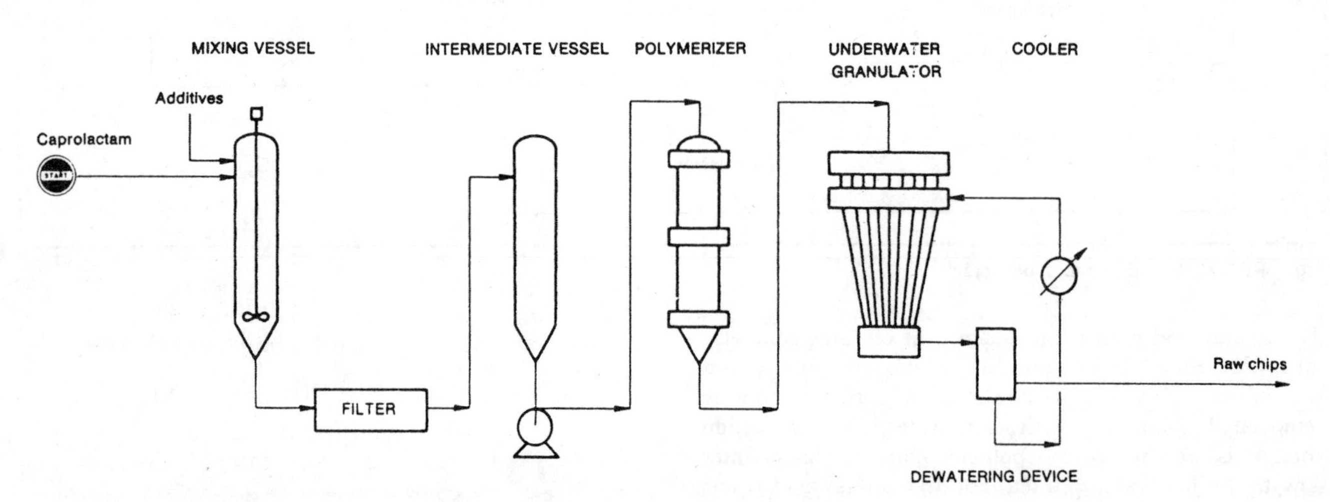

Fig. 14-6—The Inventa AG process for the production of nylon 6 from caprolactam monomer.[22]

The polymerization of 2-pyrrolidone is not easy with difficulties such as low yield, low molecular weight, and low stability. Anionic polymerization is used and catalysts are needed to activate the reaction. Activators, such as the N-acyl type, are used. Carbon dioxide has been reported to be an excellent activator for the polymerization. The following scheme shows the polymerization, activated by CO_2 and using potassium hydroxide.

$$\begin{array}{c} CH_2 - CH_2 \\ | \qquad\quad | \\ CH_2 \quad C{=}O \\ \diagdown \;\; \diagup \\ N \\ | \\ C{=}O \\ | \\ O^- K^+ \end{array} + \begin{array}{c} CH_2 - CH_2 \\ | \qquad\quad | \\ CH_2 \quad C{=}O \\ \diagdown \;\; \diagup \\ N \\ | \\ H \end{array} \rightarrow K^+ \, {}^-O - \overset{\displaystyle O}{\overset{\|}{C}} - \overset{\displaystyle H}{\overset{|}{N}} -$$

$$(CH_2)_3 - \overset{\displaystyle O}{\overset{\|}{C}} - N \begin{array}{l} \diagup CH_2 - CH_2 \\ \qquad\qquad\;\; | \\ \diagdown \underset{\displaystyle O}{\underset{\|}{C}} \text{——} CH_2 \end{array}$$

Properties and Uses

Linear polyamides are highly crystalline. The degree of crystallinity in polyamides depends upon factors such as the polymer structure and the distance between the amide linkages. The tensile strength, abrasion resistance and modulus of elasticity increase with an increase in crystallinity.

The melting points and the amount of water absorption of polyamides are especially important for fiber applications. The melting points and moisture absorption of nylons are related to the extent of hydrocarbon character in the macromolecule, i.e. the distance between the amide linkages. For example, the melting point of nylon 11 (11 carbons) is 190° C while the melting point of nylon 6 (6 carbons) is 223° C. Nylon 11 is less water sensitive because of its greater hydrocarbon character. All synthetic fibers lack the hydrophilic characteristic of natural fibers. The most important property of nylon 4 is its similarity to cotton as its moisture absorption is nearly that of cotton.

The degree of hydrogen bonding in polyamides affects the melting point and the degree of crystallinity. Polyamides with odd numbers of carbon atoms are arranged so that carbonyl groups face the amide group in another macromolecule giving a perfect arrangement for hydrogen bonding. Nylon 7, for example, has a melting point of 233° C while nylon 8 and nylon 10 have melting points of 200° C and 188° C, respectively.

Nylons, however, are to some extent subject to deterioration by light. This has been explained on the basis of chain breaking and crosslinking. For high pressure applications, nylon 11 may be ideal. In addition to its resistance to deterioration, it is an ideal lubricant. It has been reported that at contact pressure of 120,000 psi and room temperatures, its friction coefficient was only half that of polyfluoroethylene.

No single nylon excels all others in desirable properties; they are remarkably similar in properties. The choice of the appropriate nylon is usually dictated by economic considerations except for specialized uses.

Tire cord is the most important application for nylon fibers followed by apparel manufacture from nylon stretch to woven fibers. Elastic textured yarns are used for ladies' undergarments. By use of a heating process, nylon can be given a permanent set or crease useful for preshaped garments such as hosiery, permanently pleated pants, etc. Nylon staple and filaments are extensively used for the carpet industry. About 75 percent of the total carpet fiber market is nylon. Nylon fiber is also used for a variety of other articles such as seat belts, monofilament finishes, and knitwear. Because of its high tenacity and elasticity, it is a valuable fiber for ropes, parachutes, glider tow ropes, and underwear. The remarkable resistance to abrasion accounts for its use for driving belts and conveyor belts.

ACRYLIC FIBERS

The acrylics are the third major class of synthetic fibers and they were developed about the same time as polyesters. Commercial acrylic fibers (Orlon) were developed in the United States by DuPont in 1949 while modacrylic fibers (Dynel) were developed by Union Carbide in 1951. The two types of fibers use acrylonitrile as one of the monomers in the polymerization but in different proportions. Acrylic fibers should contain at least 85 percent acrylonitrile while modacrylics should have between 35 and 85 percent acrylonitrile. Acrylonitrile is produced by the ammoxidation of propylene with ammonia and oxygen.[27]

Production

Acrylonitrile can be polymerized using free radical or anionic type initiators at low temperatures.[28] The commerical polymerization of acrylonitrile takes place by addition reaction and it is usually copolymerized with other monomers such as *methacrylate, vinyl acetate, vinyl chloride,* and *acrylamide.* The polymerization takes place continuously in an aqueous medium with a redox catalyst such as ammonium persulfate. The resulting polymer is filtered, washed, and dried.

Melt spinning is not used with polyacrylics since they are sensitive to high temperatures. Solution spinning (wet or dry) is the process of choice. A highly polar solvent is necessary to dissolve polyacrylics, and dimethylformamide is most often used. Modacrylics are spun from an acetone solution. Various additives are used, such as dyes, lusterants, and brighteners, which may be added to the solution before spinning.

A typical modacrylic fiber, Dynel, is produced by copolymerizing acrylonitrile and vinyl chloride. The spinning techniques for modacrylics are similar to acrylics, but additional steps, such as annealing, are required. In the dry spinning process, the solution is degassed, filtered, sent to spinnerettes at temperatures of around 100°C, then to a column with circulating hot air to evaporate the solvent. The fibers are finally washed and subjected to stretching which extends

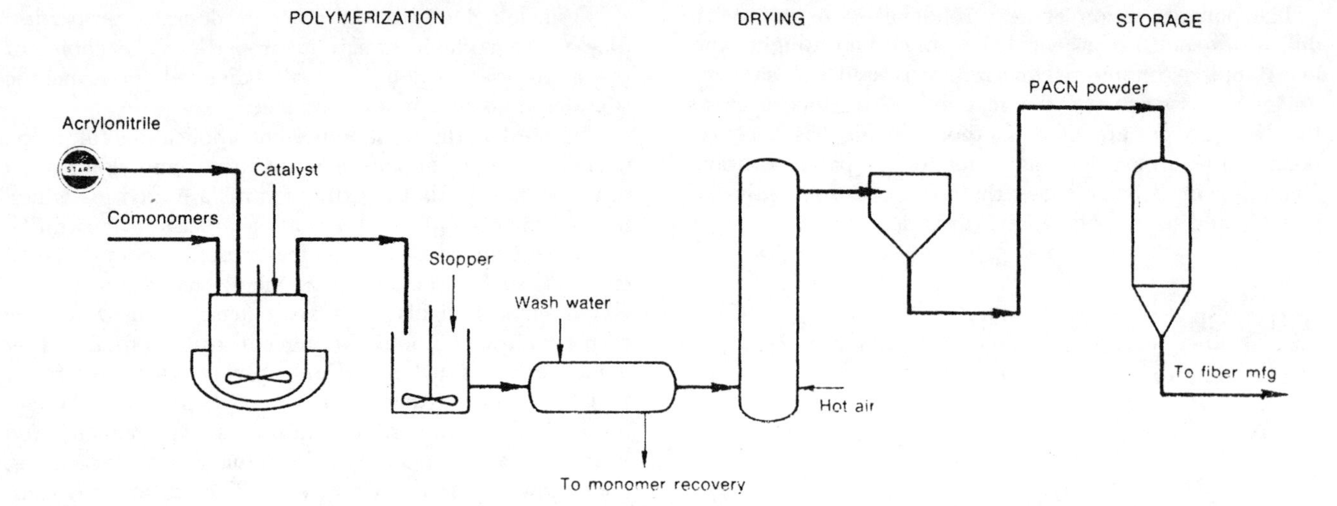

Fig. 14-7—A process for the production of copolymer polyacrylonitrile, PACN.[29]

the fiber up to 4-10 times the original length. Fig. 14-7 shows an acrylonitrile polymerization process for the production of a copolymer, PACN.[29]

Properties and Uses

The commercial names of acrylic fibers are Orlon, Acrilan, and Courlelle. These fibers possess the popular properties of softhand, resistance to creasing, and quick drying. They have properties similar to wool and have replaced wool in many markets such as for blankets, sweaters, etc.

These fibers are highly resistant to degradation to sunlight and chemicals. They were originally intended for industrial usage and for fabrics such as used for awnings which are exposed to direct sunlight.

A new development still in the laboratory stage is the production of new types of acrylics which are lighter in weight and have a high water absorption power. The new fiber is being developed by Bayer AG and is said to have properties that closely match and in some cases surpass those of wool. These new fibers can absorb up to 30 percent by weight of water in comparison to 5 percent for conventional fibers and 40 percent for cotton and wool. The absorbed water by the new acrylic-fibers does not cause the fiber to swell as in the case of cotton and wool.[4]

Commercially available fibers are Dynel and Verel. Dynel is copolymerized with vinyl chloride. Because of the polymer's high chlorine content, it is less flammable than most synthetic fibers. It is also resistant to inorganic chemicals and microorganisms. Dynel is used for protective clothing, hospital blankets, and other uses.

The two major uses of acrylic fibers are apparel (66 percent) and home furnishings (32 percent).

POLYPROPYLENE FIBERS

Polypropylene fibers utilize only 30 percent of the total utilization of polypropylene and they represent only a small percent of the total synthetic fiber production (Fig. 14-1).

Production

The production of isotactic polypropylene, the type of polymer used for polypropylene fibers, has been noted in Chapter 9.[31]

Properties and Uses

The physical properties of fiber grade polypropylene are given in Table 14-4. This polypropylene has a melting point *ca* 171°C and a degree of crystallinity ranging between 60 and 85 percent. Melt spinning technology is used for polypropylene fibers.

The favorable general properties of polypropylene for fiber use are its high abrasion resistance, strength, low static build up and resistance to chemicals. Unfavorable properties are low softening point, low resilience and dyeing problems. Table 14-5 shows the range of properties of commercial polypropylene fibers.[9]

TABLE 14-4—Properties of Fiber Grade Polypropylene[9]

Property	Fiber-grade homopolymer	Fiber-grade copolymer
Specific gravity at 23°C	0.905—0.910	0.895—0.905
Flow rate at 230°C, 2160 g load g/10 min.	6	3
Tensile yield at 2 in./min psi	5000	4000
Stiffness in flexure 10^3 psi	190	150
Unnotched Izod, impact at 0°F-ft-lb/in.	<10	>20
Melting point, dilatometer °C	172	170
Water adsorption, 24 hr. %	<0.01	<0.01
Environmental stress cracking % failure	none	none

TABLE 14-5—Range of Properties of Commercial Polypropylene Fibers[9]

	Multifilaments	Staple
Breaking strength, psi	60,000—95,000	45,000—70,000
Tenacity, dry and wet, g/den.	5—7	4—6
Elongation to break, %	15—35	20—35
Elastic recovery on 5%, elongation, %	88—98	88—95
Modulus of elasticity on 10% extension, g/den.	70—90	20—40
% shrinkage in boiling water after 20 min.	0—3	0—3

Polypropylene fibers are used for face pile of needle felt and tufted carpets, upholstery fabrics and underwear blends. These fibers are the lowest cost synthetic fibers and are used as replacements for sisal and jute.

LITERATURE CITED

1. *Chemical Week*, March 28, 1979, pp. 19-20.
2. Kiefer, D.M. *Chemical and Engineering News*, Dec. 18, 1978, pp. 24-26.
3. Greek, Bruce. *Chemical and Engineering News*, March 26, 1979, p. 10.
4. *Chemical and Engineering News*, Sept. 18, 1978, pp. 10-11.
5. *Chemical and Engineering News*, Feb. 26, 1979, p. 20.
6. O'Sullivan, D.A. *Chemical and Engineering News*, Dec. 18, 1978, pp. 30-33.
7. Langanke, H. International Seminar on Petrochemical Industries, October 25-30, 1975, Baghdad, Iraq.
8. Goodman, I. *Encyclopedia of Chemical Technology*, Vol. 16, 1969, p. 181.
9. Brownstein, E.E. International Seminar on Petrochemical Industries, October 25-30, 1975, Baghdad, Iraq.
10. Carothers, W.H. and J.W. Hill, *J. Am. Chem. Soc.*, Vol. 54, 1932, p. 1577.
11. Whinfield, J.R. and J.T. Dickson. British Patent 578,079, 1946.
12. *Hydrocarbon Processing, 1977 Petrochemical Handbook*, Vol. 56, No. 11, p. 203.
13. *The Oil and Gas Journal*, May 17, 1976, p. 94.
14. Brown, A.E. and K.A. Reinhart. *Science*, Vol. 173, No. 3994, 1971, p. 290.
15. Farrow, G. and J. Bagley. *Texas Research Journal*, Vol. 32, 1962, p. 587.
16. *Chemical and Engineering News*, Dec. 4, 1978, p. 10.
17. Hatch, Lewis F. *Studies on Petrochemicals*, New York: United Nations, 1966, pp. 511-522.
18. Odian, G. *Principles of Polymerization*, New York: John Wiley & Sons, 1970.
19. Inventa, A.-G. French Patent 1,337,717.
20. Badische Aniline and Soda-Fabrik A.-G. German Patent 1,155,132.
21. Snia Viscosa. U.S. Patent 3,022,291; Belgian Patent 582,793.
22. *Hydrocarbon Processing, 1977 Petrochemical Handbook*, Vol. 56, No. 11, p. 189.
23. Wilke, G. *Angewandte Chemie*, Vol. 69, 1957, p. 397.
24. Studiengesellschaft Kohle. German Patent 1,075,610.
25. *Chemical and Engineering News*, April 23, 1962, p. 60.
26. Peters, E.M. and J.A. Gervasi. *CHEMTECH*, Vol. 2, NO. 1, 1972, pp. 16-25.
27. Hatch, Lewis F. and Sami Matar. *Hydrocarbon Processing*, Vol. 57, No. 6, 1978, pp. 149-162.
28. Beaman, R.G. *J. Am. Chem. Soc.*, Vol. 70, 1948, p. 3115.
29. *Hydrocarbon Processing, 1975 Petrochemical Handbook*, Vol. 54, p. 175.
30. *Chemical and Engineering News*, Dec. 4, 1978, p. 12.
31. Hatch, Lewis F. and Sami Matar. *Hydrocarbon Processing*, Vol. 58, No. 9, 1979, pp. 175-187.

15

Synthetic Rubber

Natural rubber is composed mainly of high molecular weight hydrocarbon made up of *isoprene*, $CH_2=C(CH_3)-CH=CH_2$, units. The isoprene units are linked in a *cis*-1,4-configuration

$$\begin{matrix} CH_3 & & H \\ & C=C & \\ -H_2C & & CH_2- \end{matrix}$$

which gives natural rubber the outstanding properties of high resilience and strength. The use of natural rubber is currently about 30 percent of total rubber utilization. It will have increased to about 40 percent of rubber utilization by 1980. This increase will be the result of a large planting of rubber trees reaching maturity.[1]

Elastomers, synthetic rubbers, are polymers with physical and mechanical properties similar to those of natural rubber. Elastomers are unsaturated and can be vulcanized or similarly processed into materials that can be stretched at 20°C to at least twice their original length and after being so stretched and stress removed will return to approximately their original length.

The 1978 United States consumption of synthetic rubber was 2.37 million metric tons[2] while for Western Europe the consumption was 1.58 million metric tons.[3] Fig. 15-1 shows the distribution of the 1979 consumption of synthetic rubber for the United States.[2] Fig. 15-2 shows the 1979 distribution of the non-communist production of 5.83 million metric tons of synthetic rubber.[4] The total rubber production—synthetic plus natural—is 65 percent utilized, on a dollar basis, for tires with 34 percent in non-tire consumption. The non-tire consumption was 7 billion dollars in 1978.[5] Fig. 15-3 shows the dollar distribution of the non-tire use of rubber.[5]

Dr. Robert A. Krueger, B.F. Goodrich, Chemical Division, has predicted there will be no new general-purpose elastomers within the foreseeable future. The present general types will be modified to upgrade green strength, building tack, crack resistance, and hysteresis. The specialty elastomers market will have available improved processibility and curability; heat, age, oil, chemical, solvent, and corrosion resistance; low temperature flexibility; and vibration flexibility. More elastomers will be available in new forms including powder, crum, pellet, stick and chip, friable bale, and liquid.[5]

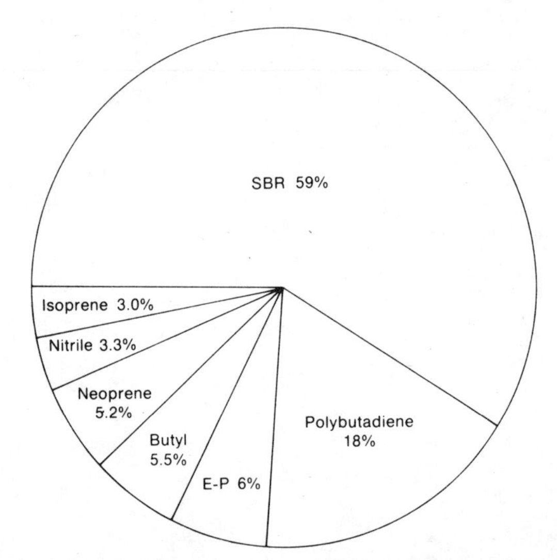

Fig. 15-1—Distribution of the estimated 2.29 million metric ton U.S. production of synthetic rubber.[2]

PRODUCTION

Synthetic rubber is produced commercially by the polymerization of mono-olefins such as isobutylene, and conjugated diolefins such as butadiene and isoprene, and by the copolymerization of mono-olefins and conjugated diolefins as with styrene-butadiene, SBR, and two mono-

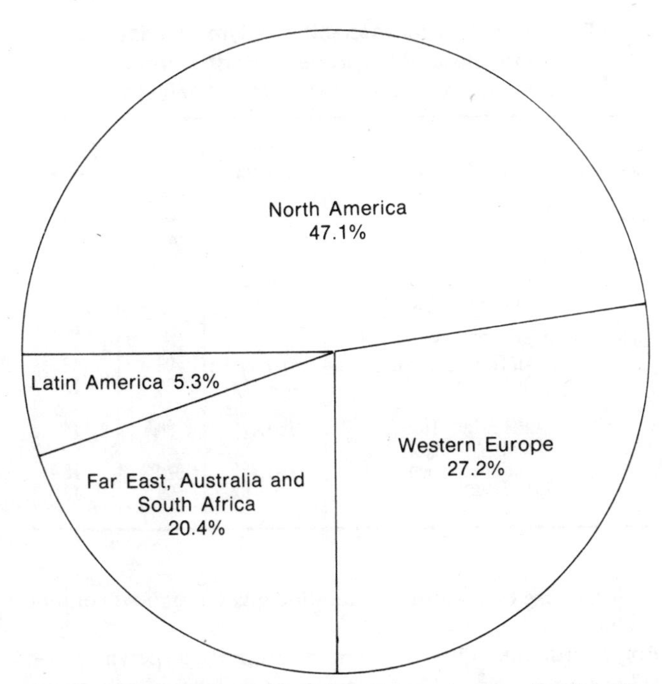

Fig. 15-2—Breakdown of noncommunist production of 5.83 million metric tons of synthetic rubber in 1978.[4]

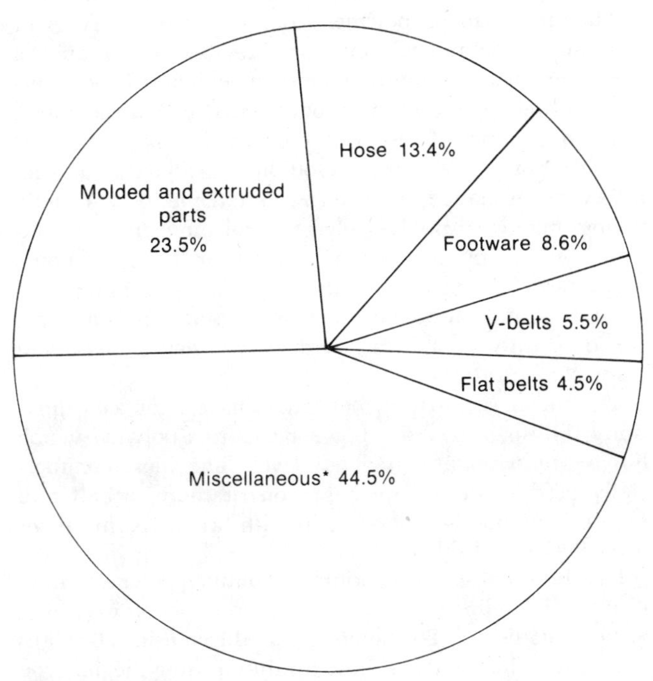

*Numerous low-volume miscellaneous products such as plastics impact modifiers and plasticizers, motor oil viscosity additives, adhesives, and coolants.

Fig. 15-3—The 1978 distribution of the non-tire use of rubber.[5]

olefins as with ethylene-propylene or by the ring opening and then polymerization as with cyclopentene.

The most important monomer used for synthetic rubber is *butadiene,* $CH_2 = CH - CH = CH_2$. It is produced commercially by cracking naphtha and other hydrocarbons, dehydrogenation of C_4 hydrocarbon mixtures, and as a by-product of ethane cracking.[6]

Before 1960 most synthetic rubber was produced by an emulsion process using a free radical initiator. These polymers consisted of a mixture of isomers in different proportions. For example, the polymerization of butadiene using free radicals produced predominately *trans*-1,4-, with lesser amounts of *cis*-1,4- and 1,2—addition products. The following is a representation of the free radical pathway:

$$I^{\bullet} + CH_2 = CH{-}CH = CH_2 \rightarrow ICH_2 - CH = CH - CH_2{\cdot}$$

$$ICH_2 - CH \cdot CH - CH_2^{\bullet} + n\,CH_2 = CH \cdot CH = CH_2 \rightarrow$$

$$\left[\begin{array}{ccc} H & & CH_2 - \\ & C = C & \\ -CH_2 & & H \end{array}\right]_x + \left[\begin{array}{ccc} -CH_2 & & CH_2 - \\ & C = C & \\ H & & H \end{array}\right]_y +$$

trans-1,4   *cis*-1,4

$$\left[\begin{array}{c} -CH_2 - CH - \\ | \\ CH \\ \| \\ CH_2 \end{array}\right]_z$$

1.2-

The product is still used in the formulation of styrene-butadiene, **SBR,** rubber, the synthetic rubber with the largest consumption (about 60 percent).

The progress that following the discovery of lithium and Ziegler-Natta catalysts for the polymerization of isoprene to a predominantly *cis*-1,4-addition product has led to the production of stereoregular rubber that has similar and sometimes better physical properties than natural rubber.

Catalysts. For the synthesis of high *cis* stereoregular rubber, two types of catalysts are generally used:

- Lithium metal and lithium alkyls (anionic catalysts)
- Coordination catalysts (transition metal compounds and complexes)

Anionic catalysts. All alkali metals have been tried for the polymerization of conjugated dienes. It has been observed that lithium gives a higher 1,4-addition product than the other alkali metals, which gave high levels of 1,2 addition products when butadiene was polymerized in pentane.[7] When the polymerization was conducted in a solvating medium such as tetrahydrofuran, all alkali metals gave high levels of 1,2 addition products.

Polybutadiene, **BUNA,** which was synthesized using sodium or potassium metal by the Germans and Russians during World War II, has a high 1,2 content. In general, rubbers produced by other alkali metal catalysts are of inferior quality than those produced using lithium. Lithium differs markedly from other alkali metals in its small ionic size and the ionization power of the lithium-carbon bond which is weaker than with other alkali metals. Alkyllithium compounds are also used for polymerization of dienes of which *n*-butyllithium is more extensively used. In addition to the alkali metal, the solvent has an effect on the stereospecificity of the produced polymer. The structure of the diene monomer is also important. For example, when isoprene is polymerized in pentane using *n*-butyllithium as the initiator, the *cis*-1,4-polymer is obtained almost quantitatively. This has been attributed to the ability of lithium alkyl to hold isoprene molecules in a cisoid conformation by pi-complex formation.

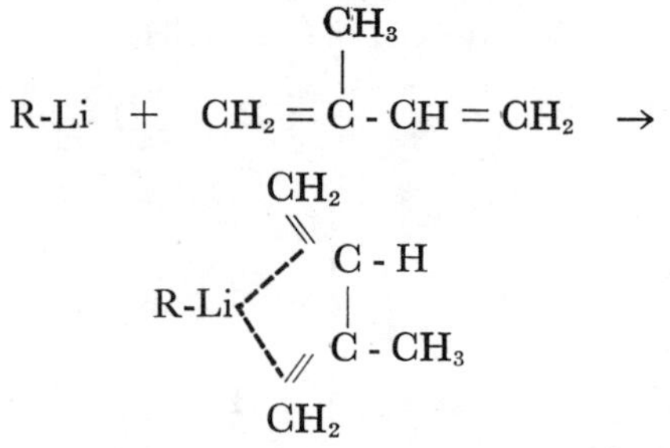

$$R\text{-}Li + CH_2{=}C(CH_3)\text{-}CH{=}CH_2 \rightarrow$$

$$\text{R-Li} \cdots \begin{matrix} CH_2 \\ \| \\ C\text{-}H \\ | \\ C\text{-}CH_3 \\ \| \\ CH_2 \end{matrix}$$

On the other hand, when butadiene is polymerized using lithium metal in tetrahydrofuran, very high 1,2-addition results.[8] The polymerization of 1,3-butadiene using lithium compounds in a nonpolar solvent gives a high *cis*-1,4 but is not as extensive as with isoprene. This is due to the fact that butadiene exists mainly in a transoid conformation at room temperature.[9]

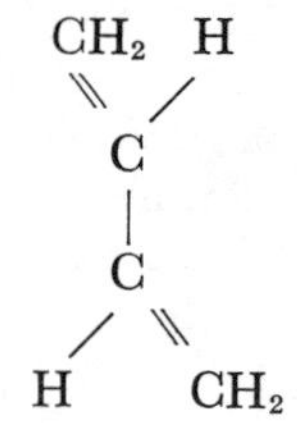

$$\begin{matrix} CH_2 & & H \\ & \backslash\backslash \; / & \\ & C & \\ & | & \\ & C & \\ & / \; \backslash\backslash & \\ H & & CH_2 \end{matrix}$$

Alkyl lithium compounds and their complexes are among the most favored anionic initiators because of their high activity and excellent rate of polymerization.

Coordination Catalysts. The first coordination catalysts used for the polymerization of conjugated diolefins was titanium tetrachloride coupled with tri-isobutylaluminum, the same Ziegler-Natta catalyst that was used for ethylene polymerization.[10] With this catalyst over 96% *cis*-1,4-polyisoprene could be obtained.

Many other transition metal compounds have been investigated for polymerization of butadiene and isoprene. Among the most useful ones for the production of predominantly *cis*-polybutadiene are titanium, cobalt, nickel and uranium compounds. Also certain other catalyst systems can give stereoregular polymers with high *trans*-1,4-levels or high 1,2-levels. Table 15-1 shows the catalyst systems used for the stereospecific polymerization of isoprene and butadiene. Not only the nature and position of the transition metal in the periodic table are significant in the preparation of a stereoregular polymer but also the halogen moiety and the polarity of the solvent. Polybutadiene prepared using titanium tetrachloride and aluminum alkyls gives a moderate *cis* content while the use of titanium tetraiodide gives a high *cis* content (Table 15-1).

TABLE 15-1—Stereospecific Polymerization of Butadiene and Isoprene in Hydrocarbon Solvents Using Coordination Catalysts

Monomer	Catalyst system	Isomer structure	% yield	Reference
Butadiene	$R_3Al + TiI_4$	*cis*-1,4	93-94	11
	$R_3Al + TiCl_4$	*cis*-1,4	60-70	12
	$R_3Al + CoCl_2$	*cis*-1,4	96-97	11
	$R_3Al + VCl_4$	*trans*-1,4	97-98	11
	$R_3Al + Cr(C_6H_5CN)_6$	1,2–	~100	13
Isoprene	$R_3Al + TiCl_4$			
	$Al/Ti > 1$	*cis*-1,4	96	13
	$Al/Ti < 1$	*trans*-1,4	95	13
	$(AlHN\ n\text{-}Bu)_nTiCl_4$	*cis*-1,4	96.5	14
	$Ti(OR)_4/Al_2Et_3Cl_3$ + silicon compound	*cis*-1,4	97-98	15
	$R_3Al + VCl_3$	*trans*-1,4	99	13

Polybutadiene. This is the most important polymer for rubber processing. This position of polybutadiene stems from the availability of butadiene, the ease by which butadiene is polymerized and copolymerized with other monomers and polymers, polybutadiene's ability to be blended with **SBR** and natural rubber, and its widespread utilization.

Butadiene can be polymerized using different types of initiators. Free radical initiators are used to produce a predominantly 1,4-addition product which is copolymerized with either styrene to produce **SBR** or with acrylonitrile to produce **ABR**. Heterogeneous *alfin* catalysts which consist of an alkenyl sodium, an alkoxide and an alkali metal halide, polymerize butadiene very rapidly giving mainly *trans*-1,4-polymer. Polymerization is initiated either by a radical anion or by an anionic mechanism.[16] Anionic catalyst, especially lithium and alkyl lithium compounds, are very useful in preparing predominantly 1,4-polymer with varying levels of *cis* and *trans* configurations.

Butadiene is also polymerized using certain coordination catalysts in hydrocarbon solvents to a polymer which has a predominantly high *cis* level. The most commercially used processes are based on titanium, cobalt and nickel compounds cocatalyzed with aluminum alkyls (over 90% *cis*), Table 15-1.

Fig. 15-4 shows the continuous solution polymerization process of Phillips Petroleum for the production of a high *cis*-polybutadiene.[17] Polybutadiene produced using alkyl lithium differs slightly from those produced using Ziegler type catalysts in their narrow molecular weight distribution. Medium *cis*-1,4-polybutadiene is produced by solution polymerization in the presence of butyl lithium catalyst.

Properties and Uses. The most important use of polybutadiene, *ca* 95%, is in the tire industry. This is the result of polybutadiene's abrasion resistance, high resilience and its ability to be blended with natural rubber and **ABR.** Table 15-2 shows the important physical properties of rubber polymers.[18]

These polymers also have good physical properties such as low glass transition temperatures. The *cis*-1,4-con-

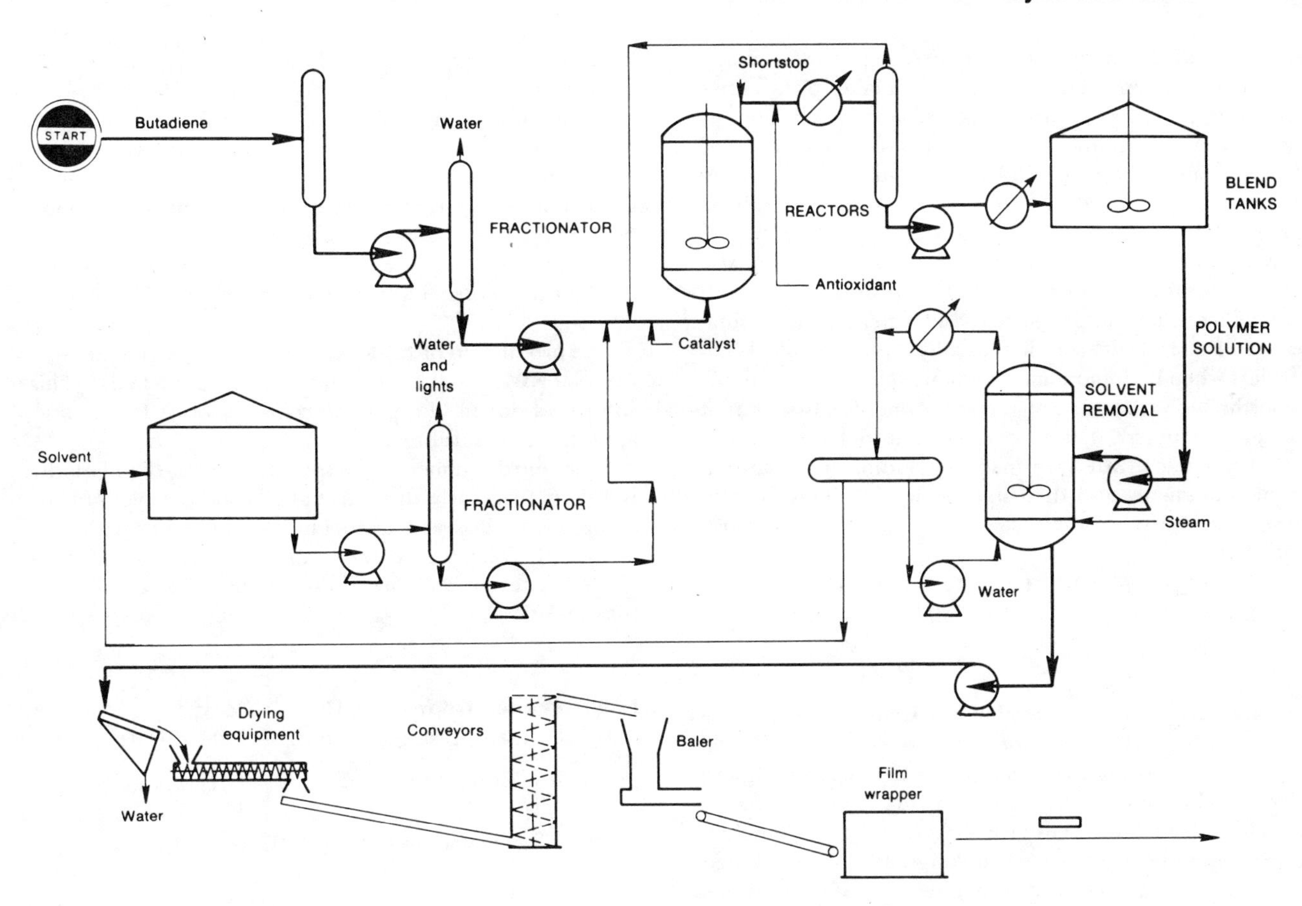

Fig. 15-4—Phillips' solution polymerization for the production of high *cis*-polybutadiene rubber.[17]

TABLE 15-2—Some Properties of Selected Elastomeric Compounds[18]

	Durometer hardness range	Tensile strength at room temp, psi	Elongation at room temp, %	Temp range of service °C	Weather resistance
Natural rubber	20—100	1000—4000	100—700	−55—80	Fair
Styrene-butadiene rubber (SBR)	40—100	1000—3500	100—700	−55—110	Fair
Polybutadiene	30—100	1000—3000	100—700	−60—100	Fair
Polyisoprene	20—100	1000—4000	100—750	−55—80	Fair
Polychloroprene	20—90	1000—4000	100—700	−55—100	Very good
Polyurethane	62—95 A 40—80 D	1000—8000	100—700	−70—120	Excellent
Polyisobutylene	30—100	1000—3000	100—700	−55—100	Very good

figuration is characterized by a lower glass-transition temperature ($t_g = -108°$ C) than the *trans* configuration ($t_g = -14°$ C). This results in a requirement of less rubber to give equal impact strength of a graft copolymer, especially at low temperatures. A medium *cis*-1,4-52% *trans*-1,4- and 12% 1,2- (vinyl) configuration is the elastomer most widely used in impact polystyrene. Its glass-transition temperature is between −80° C and −96° C.[19]

Polybutadiene was the first thermoplastic elastomer to be polymerized in blocks with polystyrene, block polymerization. The poor mechanical strength of *cis*-1,4-polybutadiene is improved by incorporating *cis, trans*-block copolymer or *cis*, 1,2- (vinyl) blockcopolymer in the polybutadiene matrix.[19]

In the United States, polybutadiene demand was 409,760 metric tons in 1978, and in 1979 it was 2.5 percent higher. The major market for polybutadiene is tires and tire products (91%). The remaining nine percent is mostly used for high-impact resin modifiers.

Styrene-Butadiene Rubber, SBR. Styrene-butadiene rubber is the most important synthetic rubber used by the tire industry. This is the result of its good mechanical and physical properties coupled with its favorable cost. **SBR** represents over 60% of the 8.4 metric tons total world production. **SBR** is made by copolymerizing *ca* 75% butadiene and 25% styrene. The polymerization takes place either in emulsion using free radical initiators such as peroxides or in solution using anionic initiators or coordination catalysts. When polymerization takes place in emulsion, a random copolymer is produced. The microstructure of butadiene units is 14-19% *cis*-1,4, 60-68% *trans* 1,4, and 17-21% 1,2-configuration. Random **SBR** with ordered sequence can also be made in solution using butyllithium provided that both monomers can be charged slowly.[20] Polymerization of butadiene and styrene in solution usually produces block copolymers when both monomers are charged with the anionic or coordination catalyst. Butadiene polymerizes first until it is almost consumed, then styrene starts to polymerize. Triblock copolymers, **SBS,** are made by adding in succession styrene, butadiene, and then styrene again.[15] This method allows strict control of the molecular weight. These triblock copolymers:

$$[-(CH_2-CH{=}CH-CH_2-)_3(-CH_2-\underset{\underset{C_6H_5}{|}}{CH}-)]$$

are now referred to as thermoplastic elastomers which can be processed without vulcanization and possess unique physical properties such as high resilience, high tensile strength and reversible elongation. The rigid polystyrene end segments associate with each other to give large domains. At ambient temperatures these act as crosslinking sites. When heated above 100°C, the glass transition temperature of polystyrene, the domains soften allowing the block polymer to flow and regain its tensile strength, when cooled.[21] The use distribution of the 1.39 metric million ton production of **SBR** is shown in Fig. 15-5.[2]

Acrylonitrile-Butadiene Rubber, NBR. This is produced by copolymerizing varying proportions of butadiene and acrylonitrile in an aqueous emulsion. Emulsifier and catalyst addition rates vary according to reaction temperature. The molecular weight of the polymer is controlled by the addition of modifiers and inhibitors when the reaction is 65%, complete. On completion of the polymerization process, vacuum distillation with saturated steam recovers unreacted monomer.[22] Fig. 15-6 shows an **NBR** process.[22]

Oil resistance of this rubber increases with increase of the acrylonitrile moiety but with a loss of it plasticizers compatibility.

The production of **NBR** latex, which utilizes about 30% of the total **NBR** production, requires large particle size. This is controlled during the polymerization process by the use of appropriate emulsifiers.

NBR, nitrile rubber, is resistant to hydrocarbon liquids. It is used in places where contact of oils and petroleum hydrocarbons is needed such as fuel hoses, fuel cell liners, gaskets, etc. Nitrile latexes find applications in textiles and paper. United States production of nitrile rubber was 77,000 metric tons in 1979.

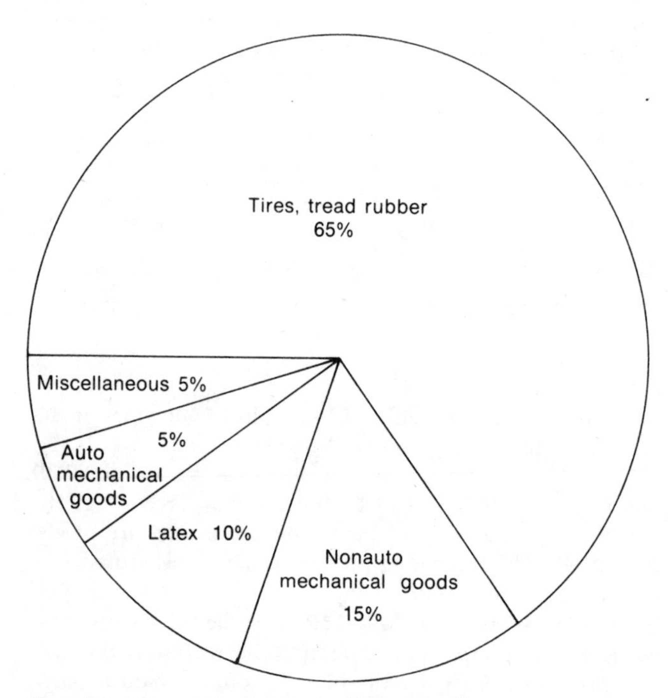

Fig. 15-5—The major markets for SBR rubber.[2]

Polyisoprene. Isoprene, $CH_2{=}\overset{\overset{CH_3}{|}}{C}{-}CH{=}CH_2$, can be produced by many processes which include the dehydrogenation of *2-methylbutene-1,* $CH_3{-}CH_2{-}\overset{\overset{CH_3}{|}}{C}{=}CH_2$, and *2-methylbutene-2,* $CH_3{-}CH = \overset{\overset{CH_3}{|}}{C}{-}CH_3$. These amylenes are recovered from a C_5 feed from catalytic cracking units. Other methods include the dimerization of propene followed by cracking, and the IFP process of cracking dimethyl dioxane at atmospheric pressure. The dimethyl dioxane is produced by the reaction between isobutene and formaldehyde.

As with butadiene the polymerization of isoprene gives a mixture of isomers if the conditions are not controlled for stereospecific addition. Since the isoprene molecule is not symmetrical, the addition can take place in 1,2-, 1,4- and 3,4- positions. Six tactic forms are possible from both 1,2- and 3,4 addition and two geometrical forms from 1,4-addition, *cis* and *trans* 1, 4.

$$\left[-CH_2-\overset{\overset{CH_3}{|}}{\underset{\underset{CH=CH_2}{|}}{C}}-\right] \qquad \left[-CH_2-\overset{\overset{H}{|}}{\underset{\underset{CH_3-C=CH_2}{|}}{C}}-\right]$$

1,2-addition 3,4-addition

$$\left[\begin{array}{ccc} -CH_2 & & CH_2- \\ & C=C & \\ CH_3 & & H \end{array}\right] \qquad \left[\begin{array}{ccc} CH_3 & & CH_2- \\ & C=C & \\ -CH_2 & & H \end{array}\right]$$

cis-1,4-addition *trans*-1,4-addition

Polyisoprene is produced commercially by the polymerization of isoprene using either lithium and lithium compounds as catalysts, or coordination catalysts. At present

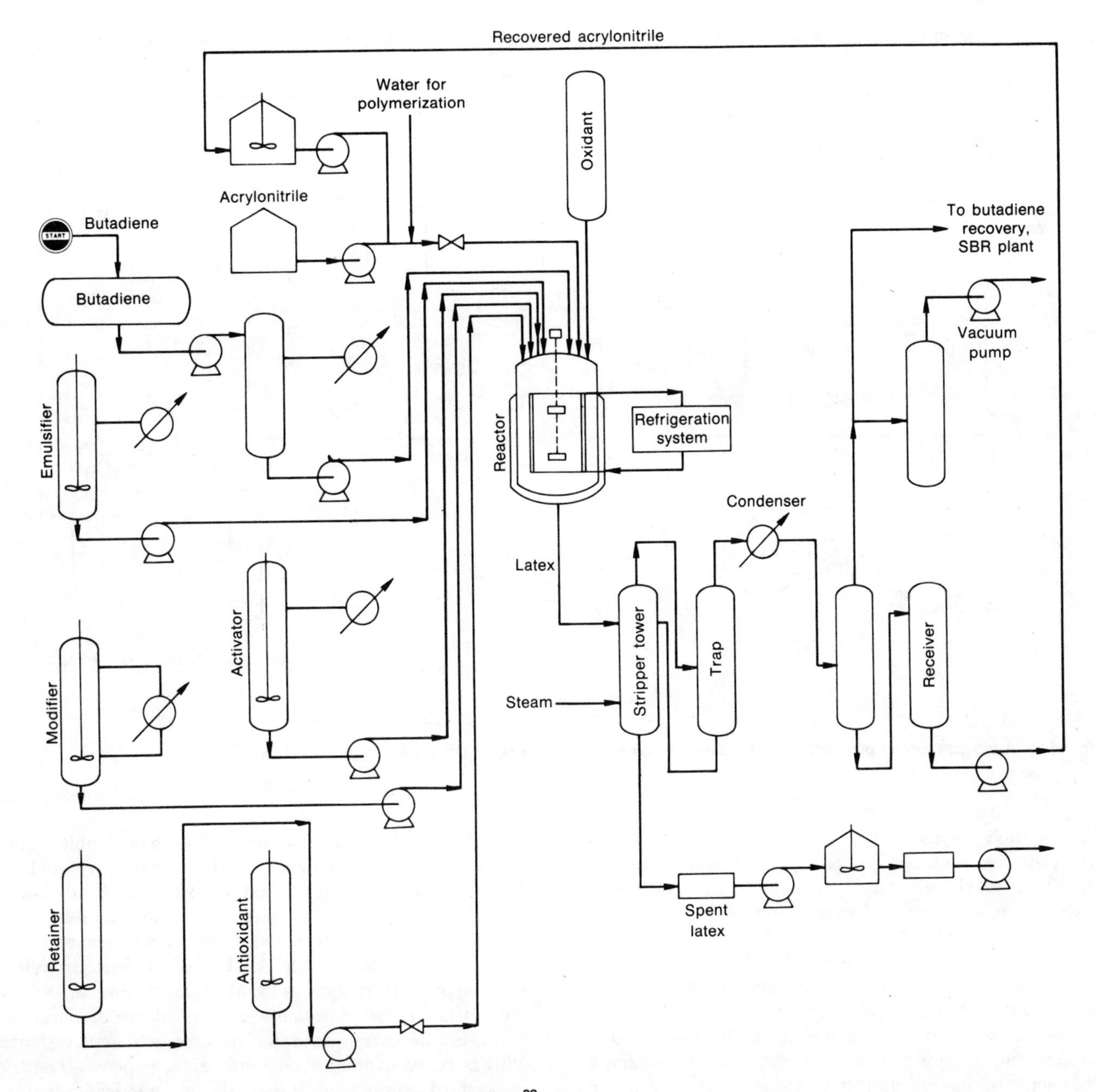

Fig. 15-6—A flow diagram of a process for the production of NBR.[22]

only the Shell group in Europe uses lithium alkyl catalysts. In this process no gel is formed even at moderate intrinsic viscosities and the product is a high molecular weight polymer with a narrow molecular weight distribution.

Ziegler type catalysts are used to produce a very high level of *cis* structure. The most important catalyst is α-$TiCl_3$ cocatalyzed with aluminum alkyls. The polymerization rate and *cis* content depends upon the Al/Ti ratio which should be greater than one. Lower ratios produce predominantly the *trans* structure. Fig. 15-7 shows Goodrich continuous solution process for the production of a high *cis*-1,4-polyisoprene (over 99% *cis*).[23]

Transpolypentamer Rubber (TPR). A new type of rubber (TPR) is produced by the ring cleavage of cyclopentene. The monomer cyclopentene is prepared from naphtha or gas oil which contains a low percent of cyclopentene, cyclopentadiene, and dicyclopentadiene. Cyclopentene is polymerized using organo-metallic catalysts and the double bond is retained in the macromolecule.[18]

$$n \; \text{(cyclopentene)} \xrightarrow{\text{catalyst}} \left[-CH_2-CH_2-CH_2-\overset{H}{\overset{|}{C}}=\underset{\underset{H}{|}}{C}- \right]_n$$

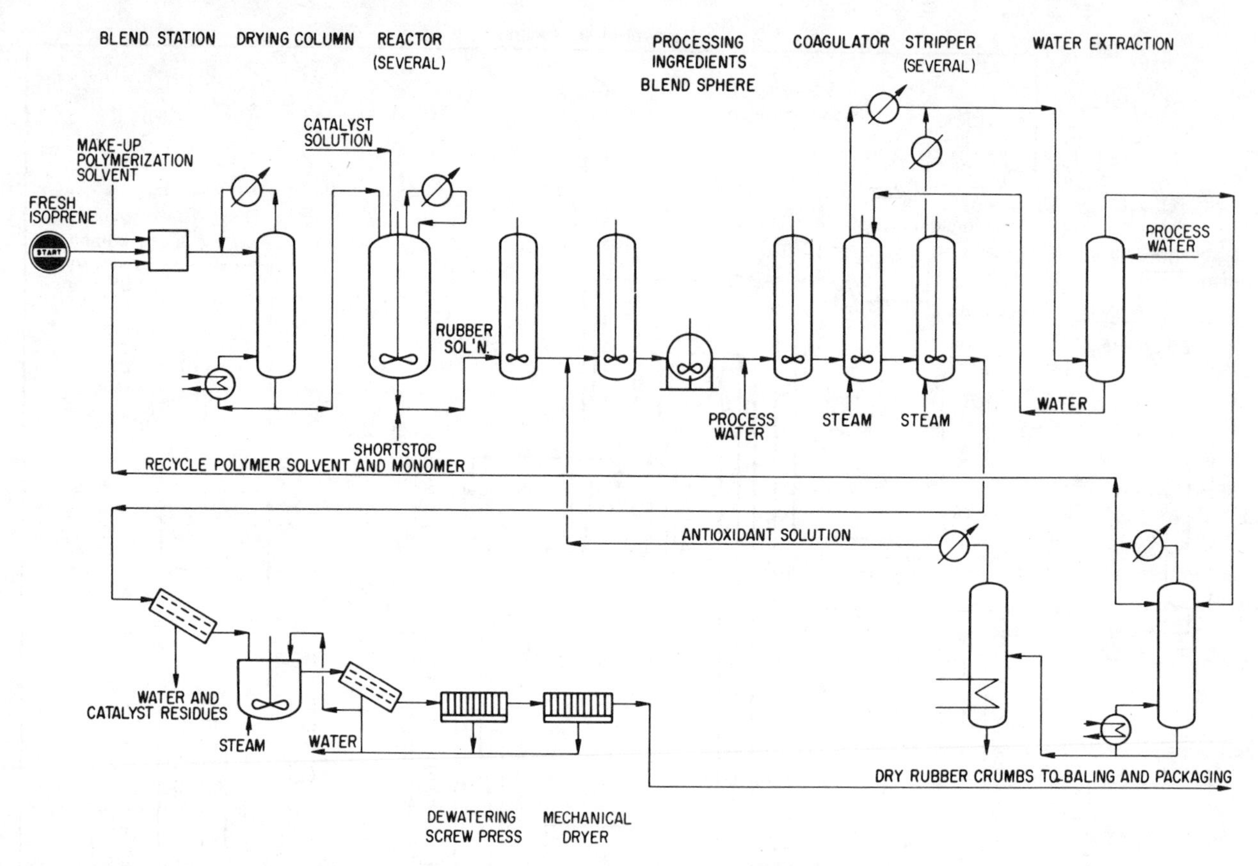

Fig. 15-7—A process for the production of 99% plus 1,4-polyisoprene by a continuous solution polymerization.[23]

The solution polymerization is effected by use of a catalyst complex such as tungsten hexachloride and an aluminum alkyl. A cocatalyst is used to increase the activity of the complex and to give higher degrees of conversion. The product is gel free and has a *trans* content of 90%. The proportion of *trans* double bonds is affected by the polymerization temperature and the catalyst component ratio. The rate of polymerization is affected by the monomer concentration and a 10-20 percent monomer concentration is practical for preparing the elastomer. Polymer can be crosslinked using ordinary crosslinking agents since it possesses double bonds. The most important property of **TPR** is that it is essentially made of *trans* structure, amorphous at normal conditions, and has a glass transition temperature (T_g) about 90°C.[19]

Among the important properties of *trans*-polypentamer are its excellent processibility, high abrasion resistance, and high mechanical properties. It has, however, some disadvantages as a general purpose rubber such as low skid resistance and hardening at low temperatures (below 20°C). The polymer is almost linear with no branching and has a low density, 0.85 g/cm,[3] which makes it a favorable competitor to other rubbers on a volume basis.

Ethylene-Propylene Rubber (EPR). Solution polymerization is used for producing the polymer with aluminum alkyls and certain vanadium compounds as the catalyst complex. Polymerization takes place at low to moderate temperatures.

Elastomers of this type do not possess double bonds necessary for crosslinking (as in the case of butyl rubber) and third monomer such as 1,4-hexadiene or ethylidene norbornene is used to provide the residual double bond. The product elastomer, which is sometimes termed ethylene-propylene terpolymer (**EPT**) or ethylene propylene diene monomer rubber (**EPDM**), can be crosslinked using sulfur. These elastomers have good abrasion, oxidation and heat resistance but possess a slow cure rate. This is being overcome but with extra expense. The use of a third component is not always practiced. For example, the Enjay Co. uses a peroxide to impart the needed crosslinking.

The use of **EPT** rubber for general purpose (as **SBR**) tires is being evaluated taking into consideration the extra cost needed for fast curing. Up until now the polymer is more expensive than **SBR** but blends of both rubbers (**SBR** and **EPR**) are customarily used. The main uses of **EPT** are in mechanical goods, tire production and cable coatings. **EPDM** main uses are in automotive parts, tires, appliance parts, and wire and cable coatings. The consumption of **EPR** is estimated to be 148 thousand metric tons, an increase of 5 thousand pounds from the 1977 consumption.[2]

Butyl Rubber. Butyl rubber is produced by the cationic copolymerization of isobutylene (97.5%) and isoprene (2.5%). The polymerization is carried out at a low temperature using a Lewis acid catalyst ($AlCl_3$) and a very small amount of water as a cocatalyst.

$$AlCl_3 + H_2O \rightarrow H^+(AlCl_3)OH^-$$

$$CH_3-\overset{\overset{CH_3}{|}}{C}=CH_2 + H^+(AlCl_3)OH^- \rightarrow$$

$$(CH_3)_3C^+ - AlCl_3)OH^-$$

The polymerization reaction is highly exothermic and the temperature has to be controlled. Antioxidants are used during the process. A small percentage of isoprene is copolymerized with the isobutylene to provide the double bonds needed for vulcanization.

Butyl rubber's important properties are its low permeability to air and its resistivity to many chemicals, solvents and to oxidation. These properties qualify butyl rubber to be used effectively for the production of tire inner tubes. The market for inner tubes is stagnant because of the extensive use of tubeless tires in passanger cars. More than 50% of butyl rubber is used in the tire industry. Other markets include cable covers, mechanical goods, adhesives, etc. Chlorinated butyl is a low molecular weight polymer used as an adhesive and sealant. The consumption of butyl rubber was 127 thousand metric tons in 1979, down from 140 thousand metric tons in 1977.[2]

Polychloroprene (Neoprene). Polychloroprene is the oldest synthetic rubber. The monomer, *2-chlorobutadiene*,

$$CH_2=\overset{\overset{Cl}{|}}{C}-CH=CH_2,$$

is polymerized in water emulsion with potassium persulfate as a catalyst.

$$CH_2=\overset{\overset{Cl}{|}}{C}-CH=CH_2 \rightarrow \left[-CH_2-\overset{\overset{Cl}{|}}{C}=CH-CH_2-\right]$$

Anionic polymerization of chloroprene has also been tried and the polymer obtained is completely insoluble. The structure of this polymer is mainly *trans* 1,4-.[20]

Neoprene rubber can be used under extreme service conditions because of its good resistance to oil and heat. The major uses of polychloroprene are in adhesives, gaskets, conveyor belts and cables. The consumption of neoprene in 1979 was 120 thousand metric tons, the same as for 1978 and an increase of 2 thousand metric tons from 1977.[2]

LITERATURE CITED

1. Mullins, L. *Chemical and Engineering News*, June 16, 1975, p. 23.
2. Anderson, E.V. *Chemical and Engineering News*, March 5, 1979, pp. 8-11.
3. Morris, N. *Chemical Age*, April 6, 1979, p. 3.
4. *The Oil and Gas Journal*, May 7, 1979, p. 60.
5. *The Oil and Gas Journal*, Oct. 16, 1978, p. 7.
6. Hatch, L.F. and S. Matar. *Hydrocarbon Processing*, Vol. 57, No. 1, 1978, pp. 135-139; Vol. 57, No. 3, 1978, pp. 129-139.
7. Tobolskvand, A.V. and C.E. Rogers. *J. Polymer Science*, Vol. 40, 1960, p. 73.
8. Kutz, I. and A. Berber. *J. Polymer Science*, Vol. 42, 1960, p. 299.
9. Stevens, M.P. *Polymer Chemistry*, London: Addison Wesley Publishing Co., 1975, p. 156.
10. Horne, S.E. Jr., *et al. Ind. Eng. Chem.*, Vol. 48, 1956, p. 785.
11. Natta, G. *J. Polymer Science*, Vol. 48, 1960, p. 219.
12. British Patent 848,065 to Phillips Petroleum Co., April 16, 1956.
13. Natta, G.L., A.C. Porri and G. Stoppa. *J. Makromal Chem.*, Vol. 77, 1964, p. 144.
14. Mazzer A., *et al. Makromal Chem.*, Vol. 12, 1969, p. 168.
15. British Patent 168,397, Sept. 19, 1967.
16. Stevens, M.P. *Polymer Chemistry*, London: Addison Wesley Publishing Co., 1975, p. 173.
17. *Hydrocarbon Processing, 1977 Petrochemical Handbook*, Vol. 56, No. 11, p. 199.
18. Hall, Dana and E. Allen. *Chemistry*, Vol. 45, No. 6, 1972, pp. 6-12.
19. Platzer, N. *CHEMTECH*, Vol. 9, No. 1, 1979, pp. 16-20.
20. Platzer, N. *CHEMTECH*, Vol. 7, No. 8, 1977, p. 637.
21. Platzer, N. *CHEMTECH*, Vol. 6, No. 1, 1976, p. 58.
22. Congram, G.E. *Oil and Gas Journal*, Sept. 22, 1975, pp. 94-95.
23. *Hydrocarbon Processing, 1975 Petrochemical Handbook*, Vol. 54, No. 11, p. 194.

Index

Following the page numbers, "t" refers to Tables; "f" refers to Figures.

R

S

T

Alcohol –(OH)

Acid – (COOH)